Drinking Water

Vladyslav V. Goncharuk

Drinking Water

Physics, Chemistry and Biology

Springer

Vladyslav V. Goncharuk
Institute of Colloid and Water Chemistry
National Academy of Sciences of Ukraine
Kyiv
Ukraine

ISBN 978-3-319-35312-8 ISBN 978-3-319-04334-0 (eBook)
DOI 10.1007/978-3-319-04334-0
Springer Cham Heidelberg New York Dordrecht London

Printed on acid-free paper

Springer is part of Springer Science+Business Media (www.springer.com)

Preface

The intellect of any nation
is determined by the quality of its drinking water
and the progress of a civilization depends
on the level of water supply and sewage systems.

Recent attention to water and water supplies has generated tremendous interest in its properties and inspired many new myths connected with water. A great many modern publications in mass media and scientific journals have entertained readers with stories of fantastic properties and capabilities of water that are not proved by science. Nevertheless, newly funded research results on water properties has afforded an opportunity to explain some earlier incomprehensible and extraordinary water characteristics.

Among numerous riddles which are still not completely understood there are two global issues: *How did life emerge on the Earth?; and, Why is water, one of the tiniest and lightest molecules, which consists of two widely common elements: hydrogen and oxygen (occupying first and third place in abundance in the Universe respectively) a "matrix" of life?*

The most enigmatic riddle is the presence of a vast amount of water on the Earth—not just water molecules containing hydrogen and oxygen atoms but with a particular ratio of hydrogen isotopes. How could it be explained that millions years ago such a particular ratio of protium to deuterium appeared to be: ~ 150 ppm of deuterium in a protium water? This concentration of deuterium determines the maximum biological activity in water.

Life on Earth became possible only because of such a ratio of hydrogen isotopes. This ratio also determines physicochemical properties of water: from the temperatures of boiling and freezing of water to its light refraction factor, viscosity, density, velocity of sound and other properties such as reaction rate constant, biological processes which take place in living organisms.

Water is described as one of the most explored substances on Earth but it seems to receive little study in contemporary fundamental science. Almost 100 years ago water was known to be a complex substance. Excepting its light component—hydrogen atoms—there are some heavy isotopes; but nobody understood their function. We didn't just made a suggestion but also proved that all physicochemical parameters of water determined by the presence of heavy hydrogen—deuterium. It was clarified that altering the concentration of deuterium in water would annihilate all known notions about this substance. The boiling temperature (if a pure isotopes composition is considered), freezing point and other properties of water can be greatly modified depending on the hydrogen isotopes ratio. Water can't exist in a liquid phase if deuterium is removed from its molecules. Thus, the structure and physicochemical parameters of water are defined by the presence of deuterium. The results we received about the influence of deuterium on the properties of water form the basis for discovery of a new method of controlling water quality. I want to emphasize a special relevance of this discovery for validating the quality of potable water.

For thousands of years people consumed only natural water. The earliest water pipelines were made of wood and granite, which was satisfactory because these materials are natural. After replacement of such natural materials by anthropogenic ones, some victims of progress appeared. For example, ancient Romans used lead pipes and lead tableware. Some archaeologists, biologists and toxicologists point to this fact as a reason for a quick degradation of the Roman nation and attribute the high rate of death-to lead poisoning. This is one of many examples of the consequences of an anthropogenic approach to the water-supply system, and the utilization of toxic water. However, broad awareness of the severity of the problem came only in the twentieth century. More than 150 years ago, global industrialization began; megalopolises emerged and this caused ecological problems including environmental pollution. This in turn induced people to apply chemical reagents to surface water in order to produce drinking water. The first standards for drinking water were established in 1853. The pollution of surface water sources was such that the quality of drinking water was determined by nine components. Over recent decades chemical the composition of water has changed. A vast amount of anthropogenic components have emerged; these substances were made by people and never existed in nature. In reviewed publications more than 35 million of such anthropogenic components were found. In addition in 1 year nearly one and a half million new compounds are synthesized and all it could be found in the aqueous medium. The majority of surface water which forms our centralized water supply is characterized by chemical and bacteriological pollution. Antibiotics are the most dangerous, as they sterilize the water and change the surroundings. It is impossible for a higher biological form of life to exist in an antibiotic medium. It's not the way nature works.

The centralized water supply of Ukraine is drawn mostly (70 %) from surface water sources. According to data of 1994–1997 most sources of surface water had third level of quality and according to international classification, fourth and fifth quality level. What does the fourth level of water quality mean? This corresponds to waste water, and the third level of quality corresponds to diluted sewage. There are no clean rivers or other sources of water where drinking water could be taken from. The same situation is true in the USA, France and other countries.

Many years ago I expressed a negative attitude towards state drinking water standards in effect at that time. Nowadays the standards which define the quality of drinking water have been adjusted by widening the enumeration of parameters subjected to control and toughening reliability-and-quality baselines. I am convinced that under present-day conditions this approach cannot solve the problem of the quality of drinking water; it is a dead-end. Why? The World Health Organization today recommends monitoring 95 components in drinking water, while 150 years ago there were only 9 parameters! USA standards list 102 indicators. In Ukraine there are only 29 components! How could we speak about the quality of the water by controlling only a scant fraction of indicators? WHO data supports this point. Every year 5 million people die because of poor water quality; this is 10 times more than in war time. Around 40 countries have an ultimate deficiency of water (Middle Eastern countries, Africa, India, China, etc.). Nearly 20 % of European population have no ecologically clean water. Approximately 20 % of Americans (~50 million people) drink polluted water.

What we have found in a water supply system: (1) instead of natural microflora, micromycetes emerged. These are fungi that could cause numerous human diseases. Specialists in the area of micromycosis assert that more than 90 % of people have such an illness. Drugs currently produced do not heal mycosis. Fungi live peacefully in a niche created by people as a result of utilization of chemical reagents for water purification: chlorination kills microflora which are natural and safe for people. Even if we find a protection against parasitic fungi some other unconquerable hazards could appear. This will surely happen as our world is build on a balance of forces.

An elementary question arises: What is a solution is to this emerging situation? My answer: implementation of a completely new concept of water supply system. This concept is based on three fundamental elements: new state standards for sources of drinking water, completely new standards for drinking water, and a unique method of control of drinking water quality.

One of the basic steps to ensuring a high quality of drinking water is a new standard of qualification.

It will take a good deal of time to assess, adopt and implement a fundamentally new approach to drinking water quality which will have a positive physiological influence on people. Exploration of a solution of a problem is not easy as comes to drinking water. What basic principles apply to quality assessment of drinking water? The answer to this question was found in a classic definition of life: "Life" is a cellular albumen form of existing of matter. The work of the famous Russian physiologists I. I. Mechnikov, I. M. Sechenov, I. P. Pavlov, biochemists A. I. Oparin, geneticist N. V. Timofeev-Resovski, A. A. Zavarzin emphasizes that the health of human beings starts with the health of cells. If everything starts with the health of cells, the quality of water in cells should be estimated. More complex living organisms have very strong immune systems. The biological object without immune system would react drastically to the quality of drinking water: it either lives or dies. When I apprehended this, I realized I had found the method of estimation of water quality. A principally new method of water quality monitoring was developed by us: the behavior of a cell is controlled on a cytogenic level and its abnormal behavior is detected. If the water is good cell division follows classical biology laws. If the water is toxic, cells exhibit anomalous division—instead of one nucleus they have two or more. This is a characteristic feature of genetic changes. Moreover, medicine registers the beginning of oncological disease when a cell division is very fast, and we know which pesticides and antibiotics provoke what processes.

As a method of assessment of a drinking water quality, biotesting allows new standards for the quality of packaged drinking water.

Bad quality of drinking water caused an abrupt increase in demand for packaged drinking water and correspondingly the appearance of numerous producers of packaged drinking water. Every manufacturer has its own technology of production but the inevitable stage is water conservation. Water can't be stored; it dissolves some quantity of everything it comes into contact with. The maximum safe period for storing water in polyethylene containers is 1 month. In order to lengthen storage time, antibiotics are used as preservative agents and to prevent microflora reproduction. Our research of 31 different types of packaged drinking water showed that all of them, except "Morshynska" (not carbonated) are not safe. "Morshynska" is perfect from the point of view of physiological influence on humans.

To sum up I want to point out water prices are high today and will be even higher in future as water is essential for life processes. There are good reasons for making a prognosis that drinking water will be a problem of the twenty-first century.

All the aforementioned became a reason for writing this book. My profound belief that water is essential not just for human health but for forming a level of human intellectual development defines the main idea of the book: the quality and characteristics of drinking water from point of view of chemical physical and biological fundamental scientific states. I endeavored to present this information in a manner which is understandable for the general reader. Only in a few chapters are some details that could be a bit difficult for laymen to understand. This is dictated by the necessity to show scientific proofs of our new research results. How much I succeeded only the reader can judge.

The author expresses his thanks to reviewers for criticism offered while the book was in prepress.

Contents

List of Figures

List of Tables

Chapter 1
Role of Water in Human Life

Abstract The challenging analysis of the sustainable development of civilization in the twenty-first century has been delivered. The issues of the water influence on global ecological and climate processes have been discussed. It has been examined that water is the buffer of the planet and its immune system. It has been revealed a relationship between the quality of drinking water and the level of intellectual development of a person, and his/her health. The ways to maintain stable development of water ecosystems have been identified. Current problems related to the quality of drinking water worldwide have been highlighted. The problems of getting physiologically and genetically safe drinking water, and also peculiarities and disinfection problems of drinking and bottled water have been covered. In the context of particular magnitude of the drinking water quality a detailed analysis of the sources conditions of drinking water supplies has been affected.

Keywords Ecosystem · Sources of drinking water supply · Quality of drinking water · Physiologically safe drinking water · Bottled water · Toxicants · Micromycetes · Disinfection

1.1 Water Resources of the Planet and Their Quality

The bulk of the planet's water is concentrated in the seas and oceans and constitutes around 1,350 million km^3. This is salt water, with an average mineralization of 35.0 g/l, which makes it unsuitable not only for drinking, but also for agriculture needs. Fresh water is localized mainly on the North and South poles, and in the mountains in the form of glaciers. This resource constitutes 30–50 million km^3. However, this source of fresh water is almost unobtainable for human use.

The earth's rivers and lakes contain about 0.4 million km^3 of fresh water that is accessible for mankind. Subterranean regions of the planet have substantial stocks of water, mainly saline and salt ones. At a depth of 800 m and up to 1,600 m the water resource constitutes around 4 million km^3.

Thus, our planet has only 3 % of fresh water of its total amount, with its bulk in the pack ice of the Arctic and Antarctic. The resources of fresh water accessible to human constitute only 0.06 %, or 0.8 million km^3.

V. V. Goncharuk, *Drinking Water,* DOI 10.1007/978-3-319-04334-0_1,

At present, more than 40 countries of the world experience an absolute water deficiency (e.g. the Near East, Africa, Indochina, and Australia). A fifth of the population in Europe and the Americas drink contaminated water, which does not meet international standards criterion. According to World Health Organization (WHO) official data, around 80 % of human diseases around the globe are related to the consumption of low-quality drinking water [1].

The first attempts to establish priorities for conservation and functioning resilient natural resources were defined in the documents of the United Nations (UN) Conference in Rio-de-Janeiro (June 1992) and the World Summit on Sustainable Development in Johannesburg "Johannesburg Agreement on Sustainable Development" (August 2002) [2]. "Every inhabitant of the planet has the right to clean drinking water," is the formula declared by the UN [3, 4]. This is a humane statement is unfortunately not supported by any specific actions. An ill-advised economy, the pursuit of profits without accounting for ecological problems, and the harsh competitive struggle of monopolies has resulted in the majority of surface and underground drinking water supply sources being depleted and heavily polluted. As a result, polluted water penetrates rivers, seas, and oceans. A fourfold increase of the world population within twentieth century alone—from 1.5 to 6 billion people has dramatically exacerbated this deficiency of fresh water, which is used not only for drinking purposes, but also for industrial activity. All of this combined has resulted in the worsening of the global ecological situation.

Artesian groundwater is a source of drinking water that is reliably protected from human impacts. We currently know of more than 150 different types of freshwater and saline drinking water treatment.

The most abundant impurities in underground water are two-valence iron and manganese, fluorine ions, nitrates, ammonia, hydrogen sulfide, hardness, and an increased salt content. Normally, these impurities substantially exceed maximum allowable concentrations for drinking water. Their presence is caused by natural factors of a geological nature. At the same time, one has to remember that high-quality water for human health should contain a wide spectrum of micro-impurities, vitally important, biologically active elements, and natural organic compounds. It is these compounds dissolving in water that provides its taste, smell, transparency and physiological properties.

Surface level fresh water are the second most common source of the drinking water supply. However, the global nature of human activities in the twentieth century in terms of industrial development, progress in agriculture, transportation, urban utilities, the formation of megalopolises, and the growth of cities and settlements resulted in the wide-scale pollution and contamination of surface water. The composition of wastewater is constantly changing due to the synthesis of new chemical substances, often possessing toxic, carcinogenic and mutagenic properties. This created biologically difficult to remove, which effectively rules out the possibility of natural and inherent self-purification of water bodies.

One can identify the following most hazardous types of pollutants entering the environment which eventually affect the water supply:

- Chemical pollutants
  - Inorganic compounds
    - Heavy metal ions
    - Salts
    - Toxic, biologically active substances
  - Organic compounds
    - Oil products
    - Phenols
    - Pesticides
    - Surface active substances
    - Chlorine-containing compounds
    - Xenobiotics
- Bactericidal and viral pollutants;
- Radioactive substances of natural and anthropogenic origin
  - Isotopes of elements;
- Mutagenic compounds of organic and inorganic origin;
- Mycotic pollutants.

Seas and oceans may serve also as the source of drinking water supply. It is known that the average salt content in water constitutes about 35 g/l. It is natural that such water cannot be used for drinking. There are various approaches intended to desalinate water. In practice, desalinization technologies such as distillation, membrane, and electrochemical methods are used.

Advantages and drawbacks of water desalination by reverse osmosis for drinking water purposes Desalination of sea (ocean) water is one of the fastest growing sectors of the economy. The introduction of reverse osmosis technology has had a big impact on the rapid development of desalination. Reverse osmosis is used for water treatment and purification of drinking water from heavy metals, nitrates, nitrites, surfactants, phenols, hardness, organic and microbiological contaminants, and organochlorine compounds.

At present, this is method in one of the most popular methods of purifying drinking water, including artesian well water for the industrial production of bottled drinking water. The reverse osmosis method of water purification produces water which contains virtually no contaminants. This method has more economical operation and maintenance than any other technology. The effectiveness of using reverse osmosis technology for the desalination of sea water is three times more effective in terms of energy consumption and the degree of purification than the other methods.

At the same time, the obvious disadvantage of reverse osmosis is the lack of selectivity of the membranes to remove contaminants. This means that not only toxic impurities and microorganisms are removed, but also beneficial minerals and microelements. Will this water fulfill physiological requirements? Of course not! After all, purity of drinking water is not the only criteria of water quality. Also important is the extent to which drinking water is the source of elements necessary for normal

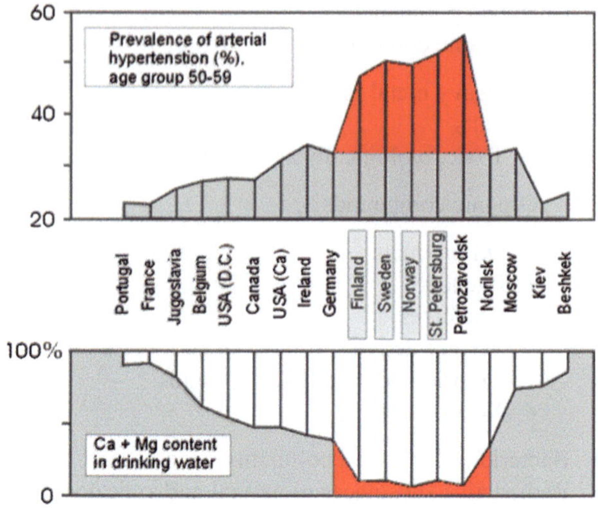

Fig. 1.1 Effect of calcium and magnesium in the drinking water on the incidence of hypertonic disease [5]

body functioning. Water obtained by the reverse osmosis method cannot be called drinking water; one gets somewhat distilled water.

The results of studies on biotesting demineralized (distilled) water show that it affects both the physiological and intracellular processes in the body (see Chap. 6, Sect. 6.5).

The required presence in the optimal drinking water content of calcium and magnesium is confirmed in Fig. 1.1. Regions supplied by water with a low concentration of calcium and magnesium exhibit a significant (30–40 %) increase of the incidence of hypertonic disease (see Fig. 1.1).

It is necessary to do adjustments to the salt composition in water obtained by reverse osmosis [6]. This is achieved by the introduction of salts required for normal human activity: calcium, magnesium, sodium, and potassium. Taking into account this fact, factories using the reverse osmosis method need to provide special mineralizers, in order to restore the optimal mineral content of water. Thus, an artificial formation of water for drinking purposes is created. This is not the best option, but is vitally necessary where there is no other source of fresh water.

The proof of the failure of world's approach to assessing the quality of drinking water and new scientifically-based approaches to assessing its quality Life on earth is only possible as a result of the presence of water. The emergence and development of biological diversity on our planet is due to the presence of water, which has unique physical and chemical properties. Although almost a hundred years it is known that water—not a simple compound where apart of protium there are also heavy isotope patterns however no one has ever thought about why they exist and what their role.

The unique properties of water are determined by its isotope composition and, first of all, by the ratio of protium and deuterium. The natural concentration of deuterium in the world's oceans, in salt and fresh water varies within the interval

90–180 ppm. The stable optima; concentration of deuterium in water constitutes 150 ppm. It is within this concentration interval that our ordinary water possesses maximum biological activity both in sea and in fresh water bodies. A decrease or an increase of the deuterium concentration in water results in a radical change of its physicochemical and biological properties [7–9].

Water is currently viewed from fundamentally new positions. The comprehensive study of water has yielded quite new results [10]. It has been found that if the deuterium concentration in water is adjusted all ideas about its fundamental properties accepted today are changed drastically. Both the melting point and the freezing point, and other water properties are substantially changed depending on the relation of the isotope composition of hydrogen. If deuterium is completely removed from water, then it will not exist on our planet in its liquid state. Consequently, water structure, just as all physicochemical parameters, is determined by the presence of deuterium. The results regarding the influence of deuterium on water properties became the basis for the discovery of a fundamentally new method for monitoring water quality.

Here questions arise. What is drinking water? How can the quality of drinking water be assessed? What are the optimal parameters that water should possess in order to be completely safe according to biological and physiological points of view?

The development and the setting of regulations and standards for drinking water quality in different countries since the late nineteenth and early twentieth centuries have changed from simply regulating the macrocomponents of natural water to more insightful knowledge about the impact of anthropogenic contaminants and toxic micro-components of water on human organisms [11].

In connection with the worsening of the ecological state of surface and underground sources, the issue of water quality control, used by people for drinking purposes, became more acute [12, 13]. As a result of the low quality of drinking water, real threats to the sanitary-epidemiological situation in various regions of the planet crop out.

An increase of the quality of the matter being controlled in national standards of various countries of the world does not solve the issue of obtaining safe drinking water at centralized water-treatment stations, either. All this calls for ever more expensive equipment and the complication of technological processes. Special attention should be paid to the fact that over the last 2–3 decades, an uncontrolled sharp increase of chemical compounds in the environment took place [11].

While the environment experiences these catastrophic changes, the quality control of dozens of substances in tap water does not guarantee that it really is drinkable and safe for human health. A high level of a technogenic load on water bodies, the use of imperfect technologies of water treatment and secondary contamination of the water in distribution networks results in the ingress to the drinking water of a substantial amount of inorganic and organic pollutants, whose joint effect on the human organism causes the effect of synergism known in chemistry and biology, threatening human health. In the situation there is no possibility to provide the population of any country and continent with quality drinking water that is safe for human health. In addition, drinking water obtained from surface sources is unsafe due to the presence in it of microtoxins. As a result of our research, we found the presence in surface water and distribution systems of micromycets of various classes including very toxic ones, which are not disinfected by current techniques used at centralized water treatment stations, even when chlorinated with high doses.

Hence, classic approaches to the assessment of drinking water are useless. The notion of "normalizing maximum allowable concentrations in drinking water of different toxicants" purportedly safe for human health, is, as a matter of fact, immoral.

1.2 Quality of Drinking Water: Problems and Solution

The harmonic development of *Homo sapiens* concurrently with the planet's biosphere calls for the immediate prevention of environmental pollution. Our generation has witnessed qualitative changes of natural and drinking water. For the first time in the history of mankind, the necessity of introducing quality standards for drinking water has appeared. These standards over time were improved, and new ones were introduced. Substances both of natural and anthropogenic origin were hazardous for human health and were strictly controlled. General human and worldwide approaches to drinking water quality evaluation were established. The first countries in the world that developed state standards for drinking water quality were the USA and the USSR. Owing to a high level of bacterial pollution of surface water, drinking water were disinfected with chlorine. This was one of the greatest mistakes of mankind, since water always contains organic compounds. Its chlorination inevitably results in the formation of very toxic, mutagenic and carcinogenic organochlorine compounds. For the first time in the history of mankind, people began to drink chlorinated water hazardous for human health—technogenic water!

At the beginning of the twentieth century, the level of surface water contamination with chemical and bacterial components did not achieve a critical state. Therefore, small doses of chlorine, used for disinfection of water, do not lead to the formation of substantial concentrations of hazardous organochlorine compounds. However, the turbulent development of industry, agriculture, the formation of megapolises, and a dramatic increase of the population size of the world (a fourfold rise over the twentieth century) resulted in a catastrophic level of bacterial and chemical contamination of water sources used for conditioning drinking water. It entailed the necessity of using large doses of chlorine for the disinfection of water. As a result, there was an increase in the concentration of organochlorine compounds in drinking water. Given such a dangerous factor, new, more sophisticated technologies of water treatment began to be developed. These technologies include preliminary filtration from suspended particles, primary chlorination, and chemical water treatment by coagulation concurrent with aluminum, iron, and flocculants of organic and inorganic origin, then filtration on sand and carbon filters. For suppressing the development of microorganisms in pipelines, the water again is treated with chlorine of such a concentration that at the outlet of the faucet of every consumer the content of residual activated chlorine was within the range 0.3–0.5 mg/dm^3.

Schematic chlorine dioxide, possessing a higher oxidizing potential, was used instead of chlorine in a number of countries, according to regulations. Potentials of some oxidants are given below:

Chemical substance and its formula		Oxidizing potential (eV)
Chlorine	Cl_2	1.36
Chlorine dioxide	ClO_2	1.57
Hydrogen peroxide	H_2O_2	1.78
Ozone	O_3	2.07
Atomic oxygen	O	2.42
Hydroxyl radical	HO•	2.80

The current technology has its advantages and disadvantages. Advantages include the fact that the decontamination process occurs more effectively. However, it also means that chlorine dioxide may produce a broader set of organochlorine compounds.

An especially negative side of the abovementioned technologies is the use of coagulants containing aluminum. On the one hand, this final introduction into the initial water of many types of contaminants is a necessary stage of treatment. On the other hand, it introduces a new, very dangerous contamination of drinking water with residual aluminum compounds. It has been common knowledge for a long time that aluminum ions contained in drinking water are exceptionally toxic, and affect human health. Several publications deal with this research. This is why the World Health Organization from year to year toughens its requirements regarding the concentration of residual aluminum in drinking water.

Based on these two important prerequisites, namely, the presence in drinking water of organochlorine compounds and aluminum, I allow myself to discuss drinking water not as tap water, but as technogenic water. Never before has mankind consumed water for drinking purposes which contained very toxic organochlorine compounds and also aluminum ions, which are byproducts of the modern water treatment technology.

Contaminations of the initial water reached such a level that water treatment facilities built in compliance with the effective world standards already are not capable of preventing the ingress of substances to drinking water. In this regard their joint effect became a real threat to human health [14].

All this results in the necessity of a search for new alternative water treatment technologies, whose use would make it possible to obtain safe water. Here, first and foremost, one should pay attention to natural, ecologically safe oxidants, which could be used in advanced technologies for obtaining high quality drinking water from the severely fouled sources.

Early in the last century, for the first time, Russian scientists proposed a fundamentally new approach in water treatment technology—the use of ozone instead of chlorine. The world first ozonizer was developed in Lomonosov Moscow State University and the first ozonation station for water treatment was built and commissioned in St. Petersburg before World War I. Since then, many years passed before the world community started to treat this technology with due attention.

Below are specified the potential oxidants used today in water treatment, which are the most promising for creating new technologies.

The analysis of the cited data is evidence of the fact that the maximum reactivity is possessed by a hydroxyl radical, which is capable of oxidizing any organic compound into carbon dioxide and water. It is formed in interaction of the ozone with water or hydrogen peroxide with water by the following mechanism:

$$O_3 + H_2O \longrightarrow H_2O_2 + O_2;$$

$$H_2O \xrightarrow{h\nu} HO^{\cdot} + HO^{\cdot};$$

$$O_3 + HO^{\cdot} \longrightarrow HO_2^{\cdot} + O_2;$$

$$O_3 + HO_2^{\cdot} \longrightarrow HO^{\cdot} + 2O_2;$$

$$O_3 + H_2O_2 \xrightarrow{\text{Very slowly}} HO_2^{\cdot} + O_2 + HO^{\cdot};$$

$$O_3 + HO_2^{-} \xrightarrow{\text{Quickly}} HO^{\cdot} + O_2^{\dot{-}} + O_2;$$

$$O_3 + H_2O_2 \longrightarrow HO_3^{\cdot} + O_2^{\dot{-}} + H^{+};$$

In aqueous alkaline solutions ozone, decomposes with the formation of $O_3^{\dot{-}}$ according to the reaction

$$O_3 + HO^{-} \rightarrow O_3^{\dot{-}} + OH^{\cdot}.$$

Then the process follows the schematic with separation of the anion radical of atomic oxygen $O_2^{\dot{-}}$ superperoxide anion radical $O_2^{\dot{-}}$, and then molecular oxygen:

$$HO^{\cdot} + OH^{-} \rightarrow O^{\dot{-}} + H_2O;$$

$$2O^{\dot{-}} + H_2O \rightarrow O + 2OH^{-};$$

$$O^{\dot{-}} + O \rightarrow O_2^{\dot{-}};$$

$$2O_2^{\dot{-}} + H_2O \rightarrow HO_2^{\dot{-}} + OH^{-} + O_2.$$

Acid conjugate to anion radical $O_2^{\dot{-}}$, i.e. radical $HO_2^{\dot{-}}$ is a weaker donor of electrons than $O_2^{\dot{-}}$, while in the system

$$O_3^{\dot{-}} = O_2 + O^{\dot{-}}$$

The most active particle is the anion radical of atomic oxygen. Therefore many reactions of $O_3^{\dot{-}}$ are actually reactions of $O^{\dot{-}}$. In an aqueous solution radicals $O_3^{\dot{-}}, O_2^{\dot{-}}, O^{\dot{-}}$ are also formed in photolysis or γ-radiolysis.

According to reactivity the active particles may be arranged in the following series

$$HO^{\bullet} > O_2^{\dot{-}} > O_3^{\dot{-}} > O^{-} > HO_2^{\bullet}.$$

It has been found that impurities of hydrogen peroxide produce a rather strong catalytic effect on ozone decomposition. Schematically it may be displayed as the following:

$$HO_3^{\bullet} > HO^{\bullet} + O_2$$
$$O_3 + HO_2^{\bullet} > HO^{\bullet} + 2O_2;$$
$$O_3 + O_2^{\dot{-}} > O_3^{\dot{-}} + O_2;$$
$$O_3^{\dot{-}} + H_2O > HO^{\bullet} + HO^{-} + O_2.$$

The results of our research confirmed that the identified processes are substantially accelerated in the presence of catalysts (both homogenous and heterogeneous) and under ultraviolet radiation.

It is exactly these results that were used for creating fundamentally new water treatment technologies for any types of pollutants in combination with a set of physical-chemical and biological methods. The research of the kinetics and the mechanism of all processes taking place in aqueous systems provided us with the possibility to develop not only new complex technologies, but new equipment for water treatment. These methods may help obtain high quality drinking water from effectively any source of drinking water supply as far as pollution is concerned. They have an exceptionally great significance for solving the issue of water disinfection from any microbiological and chemical source of pollution.

We offer three ways to solve these problems in the selection of the appropriate parameters—concentration, temperature, pH environment, and other factors.

The first is a simple water treatment with ozone or hydrogen peroxide.

The second is photooxidative disinfection:

$$O_3 + H_2O \xrightarrow{h\nu};$$
$$H_2O_2 + H_2O \xrightarrow{h\nu};$$
$$O_3 + H_2O_2 + H_2O \xrightarrow{h\nu};$$

The third is the combinated effect of three factors simultaneously:

- Photocatalytic disinfection with ozone

$$O_3 + H_2O \xrightarrow{h\nu + cat};$$

- Photocatalytic disinfection with hydrogen peroxide:

$$H_2O_2 + H_2O \xrightarrow{h\nu + cat};$$

- Photocatalytic disinfection using the combined effects of ozone and hydrogen peroxide:

$$O_3 + H_2O_2 + H_2O \xrightarrow{h\nu + cat},$$

- Photocatalytic disinfection with chlorine:

$$Cl_2 + H_2O \xrightarrow{h\nu + cat}.$$

To date, we are unaware of more efficient and environmentally friendly methods of water treatment.

1.3 Water and Human Health

The wide-scale use of chlorine in the technology for centralized drinking water supply stations is determined by the following three main reasons. The use of chlorine ensures:

- Primary decontamination of raw water;
- Substantial improvement of the coagulation process whereby phyto- and zooplankton and many organic and inorganic substances are removed from water;
- Prevention of the biological fouling of treatment facility tanks, filtering media, water system networks and equipment.

However, as was indicated above, water treatment with chlorine is accompanied by the formation of a whole range of highly toxic halogen-containing organic compounds such as chlorinated methanes, phenols, aldehydes, ketones, acids and especially hazardous polychlorinated biphenols up to dioxins—the most toxic matter known on earth. Many years of medical-biological research demonstrated that the long term use of such drinking water results in an emergence of neurotoxins, cardiovascular and oncological diseases of liver, kidneys, and the hematopoietic system of human, and produces mutagenic effects. It is these compounds that lead to the development of impotence. Therefore, the technology of water ozonation has been used increasingly in water treatment systems all over the world. Ozone as an alternative reagent simultaneously plays both the role of a disinfectant and an oxidant. In addition, as was noted above, it possesses a higher oxidizing potential, especially the products of its interaction with water. Therefore, the rate of its interaction with all classes of organic compounds is much higher as is the depth of their destruction. It decolorizes water very well, eliminates fetid smells and odors, removes iron and manganese, suppresses the growth of algae, etc. At the same time one should remember that the use of ozone in the technology of water treatment is possible only in conjunction with other physical-chemical methods.

1.3.1 Ecology—the Health of Human

Having addressed all the above problems, the world-wide issue of determining the degree of risk in the ecology–health system becomes a high priority one [4]. The hygienic standards and regulations for the levels of environmental pollutants effective today are mostly of a declarative nature. In each specific country, in every region of the world, one needs to assess the priorities of pollutants by the risk levels to human health.

As sources of the risk I would separate four factors playing the most important part in the life of human:

1. Drinking water;
2. Air;
3. Soil;
4. Foods.

It is necessary to carry out a complex monitoring of all environmental aspects. Why is monitoring needed? First of all, for finding priority pollutants in each specific region that produce maximum negative impact on human health. Certainly, one may follow the path of eliminating all types of pollutions, but this requires fantastically large financial costs. The most expedient formula of the ecological policy in every country should be the following: min costs → max results, starting from that of the highest priority, the most dangerous pollutants to human health.

One should isolate such main priorities of risk levels:

- Chemical pollutants;
- Bacterial pollutants;
- Radioactive and isotopic pollutants;
- Mutagenic pollutants.

One should not forget that complex pollution of the environment with different classes of toxic substances very often results in synergism of their effects, when each toxicant's individual effect is much weaker than a blend of different toxicants. The most glaring among the known examples of the synergetic effect is a joint presence of radioactive matter and asbestos, which strengthen the negative effect of each other sixfold.

As risk indicators let us points out the following five factors whose presence may help unambiguously assess the state of the environment:

- Genetic violations;
- Disease incidence;
- Birth rate;
- Death rate;
- Average longevity.

1.3.2 Bottled Water: Issue of Disinfection and Preservation

Over the last 2 decades the technology of water bottling and its sale through a wide network of shops has become very popular [14, 15]. There are just a few who would think about the enormous scientific and technical issues of preserving the quality of such water. First of all, the water being subjected to bottling should be immaculate in terms of chemical, biological and organoleptic indicators and in addition should preserve its qualities in a closed state.

We talked about the issue of water disinfection above, but the issue of its preservation requires a more detailed consideration. Carbon dioxide is the most common preservative, which diluting in the water, forms carbonic acid. The taste of such acid water is experienced by everybody who drank aerated water in their lifetime. This water is used for extreme situations of quenching one's thirst rather than for its systematic consumption.

The other effective reagent preserving water is silver or, to be more accurate, silver ions. Since ancient times the technology of water storage in silverware has been known. Such water can be drunk without worry since the concentration of silver ions is meaningless, but the disinfection process is rather lengthy. It can be amplified a thousand times using the electrochemical technology of silvering water. However, the excessive silver concentration is dangerous. Over time silver ions are adsorbed on the walls of the vessel and rather quickly lose their disinfecting properties. Under these conditions, secondary bacterial contamination is possible.

Unscrupulous companies may use as preservatives special chemical substances whose functional properties have effects similar to those of antibiotics, which suppress or destroy microorganisms. These substances present enormous hazard for human health. The consumption of such water by drinking leads to the sterilization of the digestive tract causing a known disease—disbacteriosis.

The chemical nature of the containers used is a very serious issue in the process of water bottling. There are a sufficiently large number of investigations, whose results are an evidence of the toxicity of polymer containers (see Chap. 6, Sect. 6.3). The safest containers are glassware, glass containers, enameled and ceramic ware.

1.4 "Green" Chemistry

The contemporary world early in the twentieth century found itself on the verge of the global ecological crisis, which threatens the very existence of civilization. Industrial production still remains high-waste. That is why a new science has appeared and continues to develop effectively. This is the science of ecological economy, which tries to unite social, economic, and ecological systems. Ecological economy should take into account ecological costs when assessing the economic performance of production. A modern economy should develop while accounting for the ecological factor.

It is quite clear: the correction of the unfavorable situation that has taken shape does not look to be an easy or quickly resolved issue. This results from the impossibility of allocating sufficient investments for the solution of the ecological issues facing mankind.

In the world science this area of science and technical investigations has been referred to as "green chemistry".

When we talk about drinking water quality we, first of all, should think and care about the quality of drinking water supply sources, and this is almost the entire water basin of the planet. The solution to the existing issue of protecting the whole ecosystem from the products of the mankind's life activity lies in the quest for fundamentally new technical solutions. These solutions aim not only to overcome the consequences, but should prevent the causes, leading to the unfavorable ecological aftermath.

The most widely known water contamination, referred to as water "color," is caused by blue-green algae—cyanobacteria. An important adverse influence on the water basin is related to their ability to synthesize biologically active substances possessing very high toxicity. The ingress to water of such biogenic elements as phosphorus, nitrogen, potassium, sulfur and also organic matter from industrial, domestic, rainfall and waste water of the agricultural production created the prerequisites for intensive development of algae of various systematic groups, which results in the formation of water "color" of various degrees.

The facts of the formation of toxins by the algae have been known for a rather long time, as early as since the 1960s, while the systematic research of the causes of this phenomenon and factors increasing the toxicity started about 20 years ago. The most serious attention at present has been drawn to the toxicant microcystine, whose producers are cosmopolite cyanobacteria, causing water "blooming" in water bodies of virtually of all countries of the world. In terms of biological activity, toxins of cyanobacteria surpass by many times such known substances as strychnine, curare, a range of fungi toxins, and potassium cyanide. They exceed the toxicity of potassium cyanide by as much as 50–1,000 times. Such a strong poison as curare exceeds them by least ten times. Hence, it is quite obvious that such water presents a serious threat to the life of fish, birds, and humans. Even swimming in "blooming" water pools is dangerous for human health, not to mention the use of this water as a source of drinking water supply.

All these facts indicate the necessity of forming the Global World Program for the Protection of the Planet's Water Basin against Pollution.

1.5 The Biosphere and Civilization—the Issue of Compatibility

Completing the analysis of the state of the art in the biosphere indirectly through one of its components—the hydrosphere—and its interaction with the development of society, the formation of a quite new artificial reconstruction of the biosphere into the noosphere and then into the technosphere, one may, with a rather high degree of confidence state the transfer of the civilization to a some special state. The biosphere in its classical state no longer exists. We are witnessing the apocalypse of the biosphere, which started in the middle of the twentieth century. Transforming the environment—the biosphere, whose part is human himself—society underestimated the fact that creating the technogenic environment by its existence violates the basic laws of the biosphere, the laws of the universe.

Nowadays we live in the transition period of the coexistence of two worlds—the biosphere and the noosphere. There is no return to the biosphere. Man as a creature of the biosphere origin should find new ways for his survival and development. The gene pool of the biosphere has started to intensively rearrange itself—the turbulent process of evolution has begun, which has given birth to ecological changes on a global scale. Mass extinction of the existing species and emergence of new ones is observed across the board. Especially intensive is the evolution of infections. This is evidenced by a widely known fact of the adaptation of pathogenic microflora to antibiotics. This is an illustrative manifestation of mutagenesis—the emergence of a new, earlier absent quality—resistance to antibiotics. Complex genomes are created by nature during millions of years, while in modern laboratories they are created within a year. Man has designed a new artificial habitat—the technosphere, kept safe from all biological processes of the biosphere, without connection with which he simply cannot exist.

The issues of nuclear energy and its impact on the environment bear a complex nature. Since the radioactivity phenomena have been discovered, the irreversible processes of dramatic increase of background radiation all over the planet have been seen [16].

The intensive development of the nuclear industry led to the emergence of a considerable number of objects presenting potential radiation danger: nuclear power plants with reactors of various types, chemical mining plants, and plants for processing nuclear fuel. The radionuclide composition of pollutants emitted to the environment during the operation, and also in emergency situations substantially differ, which makes the use of a single flow chart of purification impossible in all mentioned cases.

Among the numerous aspects of protecting the water basin from technogenic pollutants, one of the most important aspects is achieving the effective purification of various types of pollutants, including radioactive pollutants. Achievement of this task is hampered by the general worsening of the ecological situation, accompanied by the appearance in the water of an ever larger amount of various toxicants of organic and inorganic nature. Together with the necessity of development and the employment of more complex and sophisticated technologies to achieve purification levels to meet the existing standards, this begs the question of the applicable range of the standards themselves. Here the ever increasing restraint, which currently manifests itself, despite the sufficient theoretical and practical substantiation, at the moment of adoption, is determined by insufficient accounting of the possibility for aggravation of the adverse influence of toxicants on live organism in their synergetic effect. The synergetic effect of pollutants of various types and radionuclides seems to be one of the most dangerous for human health.

The adverse effect of radioactive elements of the environment is of a complex nature. Despite the fact that it is well studied, the issue of the processing and burial of radioactive wastes, which today like "a nuclear Gene" are either on the sea bottom or in the mines under earth and are waiting for their time, is still outstanding.

There is one more aspect of this issue which has not been sufficiently studied yet. I mean the issue of the impact of nonradioactive isotopes on the development of biological objects and, first of all, hydrogen isotopes, in particular, deuterium. The impact of the isotopic composition of substances on kinetic characteristics of numerous reactions has been known for a long time.

It turns out that deuterium strongly affects the chemical parameters of chemical and biological processes since it forms a stronger connection with oxygen in a water molecule, unlike light hydrogen—protium. It is also known that an increased concentration of deuterium in drinking water results in acceleration of the organism aging process. Deuterium manifests toxic properties with respect to living organisms [17].

Nowadays one of the most serious issues is increased levels of background radioactivity and an increase of the concentration of heavy isotopes with respect to their light prototypes. First of all, it concerns vitally important elements such as hydrogen, oxygen and carbon, which are a basis of protein life in our biosphere and affect the mechanism of biological evolution.

Thus, ecology as a science studied, strictly speaking, the state of the biosphere, where human is its indispensable part. However, today, when it is quite obvious that the biosphere is being transformed, in an increasing degree, to the noosphere transforming into the technosphere, one cannot help taking into consideration that the world has really changed. It is necessary to have a clear notion of a place of human in a quickly changing world. Ecology transforms into a new geological phenomenon, into a new branch of science investigating the interaction of three globally coexisting constituent components of the universe: biosphere, noosphere, and technosphere. The decisive role in their interaction is played by human. The future of our civilization is determined by the level of the awareness of the changes that has taken place in the surrounding world.

References

1. World Health Organization (2004) Guidelines on drinking water quality control: recommendation, vol 1. WHO, Geneva
2. United Nations (2002) Report of the world summit on sustainable development. United Nations, New York
3. World Water Day (2010) Clean water for a healthy world. http://www.unepcom.ru/?go=razdel&level=2&cid=66. Accessed 1 Jun 2010
4. Global Environment Outlook (2007) GEO4. Environment for development. United Nations environment programme. Progress, Malta
5. Koshcheev A (2011) How to turn water into the water. Mineral additive "SEVERYANKA". http://www.koshcheev.ru/2011/10/20/severyanka/. Accessed 20 Oct 2011
6. Goncharuk VV (2010) SOS: drinking water. J Water Chem Technol 32(5):255–283
7. Lobyshev VN, Kalinichenko LP (1978) Izotopnyye effekty D_2O v biologicheskikh sistemakh (Isotope effects of D_2O in biological systems). Nauka, Moscow

8. Somlyai G (1998) The biological effect of deuterium-depleted water, a possible new tool in cancer therapy. J Oncol 30(4):91–94
9. Goncharuk VV, Kavitskaya AA, Romanyukina IY et al (2013) Revealing water's secrets: deuterium depleted water. Chem Cent J 7(1):103
10. Goncharuk VV, Lapshin VB, Burdeinaya TN et al (2011) Physicochemical properties and biological activity of the water depleted of heavy isotopes. J Water Chem Technol 33(1):8–13
11. Figueras MJ, Borrego JJ (2010) New perspectives in monitoring drinking water microbial quality. J Environ Res Public Health 7:4179–4202. doi:10.3390/ijerph7124179
12. Shannon MA, Bohn PW, Elimelech M et al (2008) Science and technology for water purification in the coming decades. Nature 452:301–310
13. Stedmon CA, Seredynska-Sobecka B, Boe-Hansen R et al (2011) A potential approach for monitoring drinking water quality from groundwater systems using organic matter fluorescence as an early warning for contamination events. Water Res 45(18):6030–6038
14. Loo S-lL, Fane AG, Krantz WB et al (2012) Emergency water supply: a review of potential technologies and selection criteria. Water Res 46(10):3125–3151
15. Reimann C, Ayräs M, Chkushin A et al (1998) Environmental geochemical atlas of the central Barents region. Special publication of the Central Kola Expedition, Geological Survey of Finland and Geological Survey of Norway
16. Sobotovich EV, Dolin VV (eds) (2012) Tritij y biosferi (Tritium in the biosphere). Nauk Dumka, Kyiv
17. Goodall KB (2003) In search of the fountain of youth. Preliminary analysis of deuterium's role in DNA degradation. http://www.springboard4health.com/notebook/health_deuterium_analysis.html. Accessed 22 July 2003

Chapter 2
Life-Forming Role of Water in the Latest Discoveries and Hypotheses

Abstract The analytical review about life-forming role of water in conjunction with its unique properties which are predicated upon water structure peculiarities has been performed. The water role has been investigated from the perspective of different archebiosis scenarios on the Earth. Special attention has been paid to the isotopic water composition and deuterium influence on water structure, its properties and destruction of deoxyribonucleic acid. The isotopic composition and properties of heavy, light and ubiquitous water have been examined.

Keywords Water properties · Water structure · Isotopic water composition · Deuterium · Protium · Heavy water · Light water · Deoxyribonucleic acid

2.1 Unique Properties of Water

Throughout the entire history of our planet's existence, water has affected all aspects of the globe. It is this main building material, and the environment which contains it, that ensures the propagation and evolution of life on Earth.

For many centuries, scientists have studied this chemical substance, composed of the simple formula, H_2O, and known as one of the smallest and lightest molecules. Despite its simplicity and size, water somehow plays the main role in all biological processes and is considered "the matrix of life". The large heat capacity, high heat conduction and enormous amount of water in organisms contributes to heat regulation and prevents local temperature fluctuations, thereby enabling us to more easily control our body temperature. The high latent heat of evaporation ensures resistance to dehydration and high cooling in evaporation. Water, due to the polarity of its molecules, its high dielectric constant and its small sized molecules, is a good solvent primarily for polar ionic compounds and salts. It has unique properties of hydration with respect to biological macromolecules (especially to proteins and nucleic acids), which determine their three-dimensional structures and, consequently, functions in the solution. This hydration causes gels to form, which can be reversibly subjected to the sol–gel transition lying behind many cellular mechanisms.

Water is ionized and secures a light flow of proton exchange and is thereby conducive to the variety of ionic interactions in biology. "Water is incredibly multi-faceted, no other liquid has this property", says F. Gaiger, a leading water expert from Dortmund University. He believes that "… the world of liquids has been divided

V. V. Goncharuk, *Drinking Water*, DOI 10.1007/978-3-319-04334-0_2,

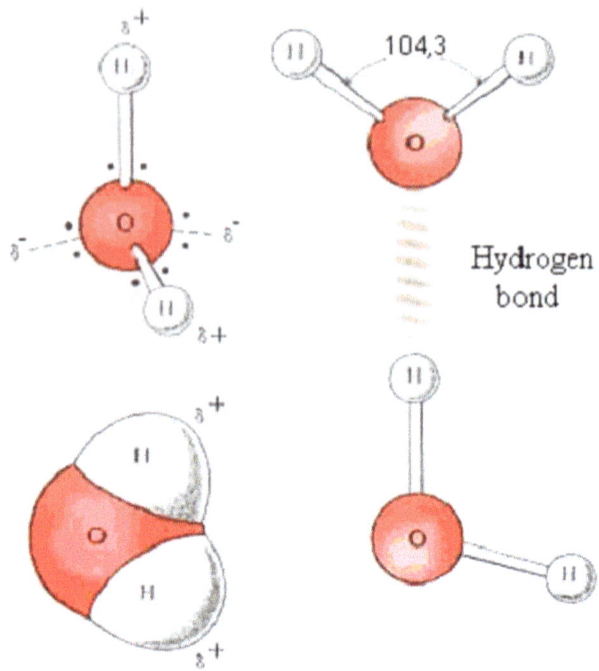

Fig. 2.1 Formation of hydrogen bonding

into two parts. On the one hand—water, on the other—all other substances". To this day there are a number of fundamental questions with no conclusive answers. Why does this substance behave, in many respects, atypically? Why does water in its crystalline state, i.e. in the form of ice, have so many "faces"—15 [1]?

The complicated behavior of this substance is determined by the structure of its molecules (Fig. 2.1). In the water molecule, two atoms of hydrogen and an atom of oxygen are arranged at an angle of 104.3° to each other. Owing to their electric polarity, both water molecules may create a special link between themselves; positive partial charges on hydrogen atoms and negative charges on oxygen atoms attract each other, forming hydrogen bridges. This link is 20 times weaker than the one which connects atoms of hydrogen and oxygen inside a molecule, but it exceeds the force of the normal attraction between the molecules, the Van der Waals force, by 60 times. The presence of the hydrogen bond in the water is necessary, but it is an insufficient condition for explaining its main properties. While the hydrogen bond is a necessary component of the water molecule, it alone does not represent the main properties of water. This can only be done when the structure of water is viewed as a single system [2].

Liquid Water In terms of statistics, particles of water in a liquid state form three and a half hydrogen bridges with their neighbors. These bridges collapse and line up again at a periodicity of a billionth of a fraction of a millisecond. In 1916, new notions about the structure of a liquid were proposed. The first X-ray structural

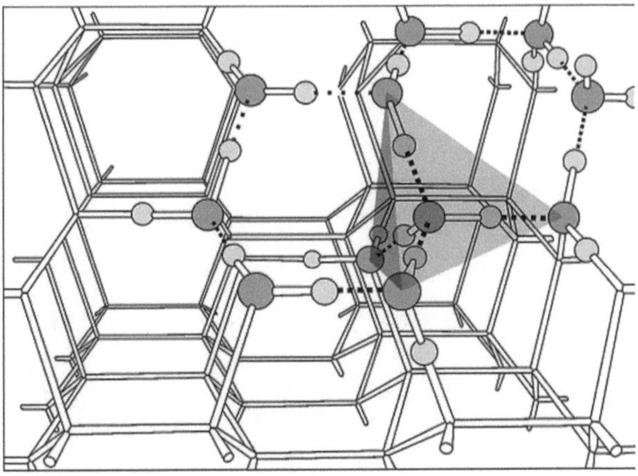

Fig. 2.2 Crystalline structure of ice

investigations of water were carried out in 1922 by Dutch scientists W. Keese and J. de Smedt. They proved that liquid water is characterized by the ordered arrangement of water molecules, i.e., water has a certain regular structure. The water structure in a living organism in many ways resembles the crystalline structure of ice (Fig. 2.2).

Special "adhesiveness" of a molecule is a reason why H_2O remains liquid at an extreme temperature of 100 °C. Chemically kindred materials in this sense go through heat at the maximum 25 °C [1]. Every water molecule in the crystalline structure of ice is involved in the formation of four hydrogen bonds directed to the apexes of the tetrahedron (Fig. 2.3). The tetrahedral center contains an oxygen atom, and two apexes each have a hydrogen atom whose electrons are involved in the formation of a covalent bond with the oxygen atom. Two remaining apexes are occupied by pairs of valence electrons from the oxygen atom, which do not take part in the formation of intramolecular links. In the interaction of the proton of one molecule with a lone pair of oxygen electrons from the other molecule, a hydrogen bond appears weaker than the intermolecular link but sufficiently powerful to hold together neighboring water molecules. Every molecule may simultaneously form four hydrogen bonds with other molecules at strictly definite angles, equaling 109°28′. These bonds are directed to the apexes of the tetrahedral, which does not allow them at freezing to form a dense structure.

The first theory on the structure of water was put forward by English researchers J. Bernal and R. Fowler. They created a concept about the tetrahedral structure of water and defined the role of hydrogen bonds in water. It was found that there are covalent and hydrogen bonds in water. The covalent bonds do not break in phase transitions of the water: water–steam–ice. Only electrolysis, a process of heating water on iron and similar processes, break water covalence bonds. Hydrogen bonds are 24 times weaker than covalence bonds. In melting ice and snow, hydrogen bonds partially remained in the water, and in a steam state they are all broken (Fig. 2.4) [2].

Fig. 2.3 Tetrahedral structure of water molecules

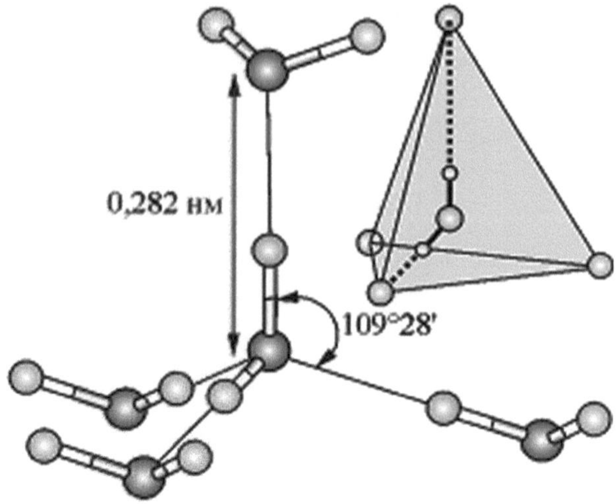

Later, an idea was developed, according to which liquid water was proposed as a pseudocrystal (Fig. 2.5); in it, individual tetrahedral molecules of H_2O are said to be linked to each other by directional hydrogen bonds, forming hexagonal structures such as in the ice structure.

Ice Ih In the abbreviation *Ih*, the letter *h* denotes hexagonal. This type of ice is drifting in seas and oceans; it is one of 15 types of the crystalline states of the water known today, and this shape is encountered on the Earth more often than others. The molecules are arranged in such a way that each element has 4 neighbors to which they are linked by hydrogen bridges. As a result, a lattice with a hexagonal base is formed by oxygen atoms [1].

X-ray crystallographic research carried out by J.P. Morgan and B.E. Warren showed that a structure like that of ice is inherent in water. In water, like in ice, every atom of oxygen is surrounded by oxygen atoms as in the tetrahedron. The distance between neighboring molecules is not identical. At 25 °C, every water molecule in the framework has one neighbor at a distance of 0.277 nm and three at a distance of 0.294, in average 0.290 nm. The mean distance between the nearest neighbors of the molecule is approximately 5.5 % greater than between the ice molecules. The other molecules are at distances which are intermediate between the first and second neighboring distances. The distance 0.41 nm is a distance between atoms O–H in an H_2O molecule.

The main differences between the structure of liquid water and the structure of ice are a more diffuse arrangement of the atoms in the lattice and a disturbance of the remote order. Heat fluctuations result in the bending and breakage of hydrogen bonds. Water molecules of the equilibrium positions get into neighboring voids of the structure and remain there for some time since these voids have the corresponding relative minimums of potential energy. This leads to an increase in the coordination number and to the formation of the lattice defects, whose presence determine

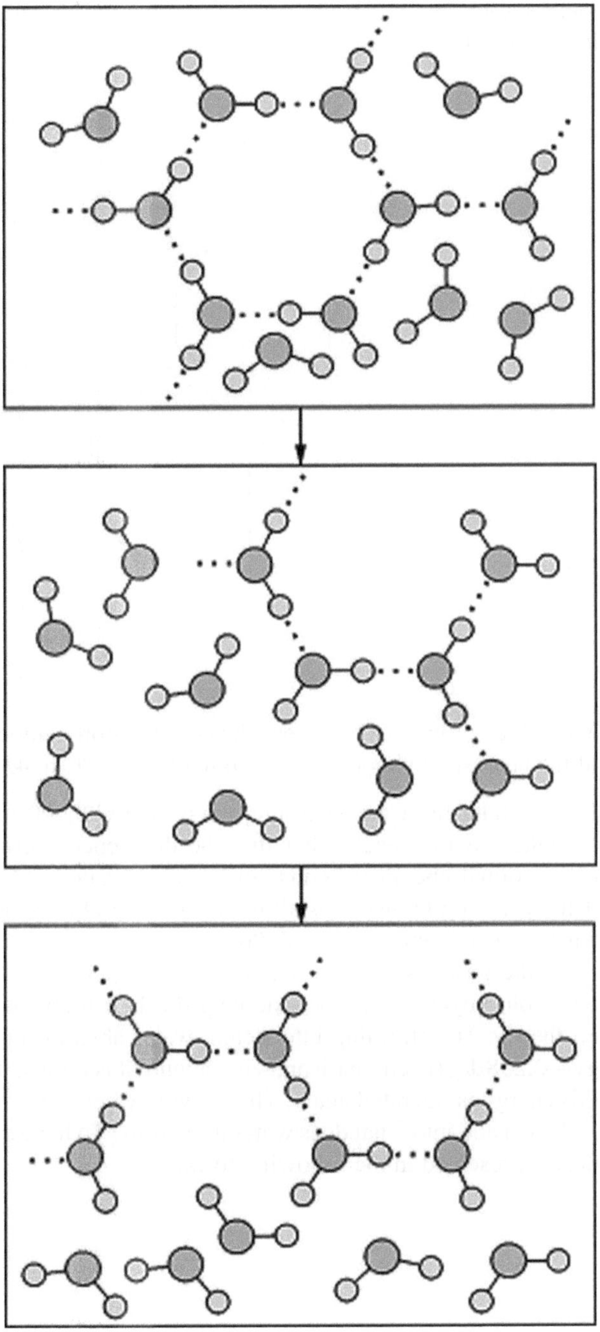

Fig. 2.4 Scheme of partial destruction and formation of hydrogen bonds at ice thawing (hopping time is 10–12 s.)

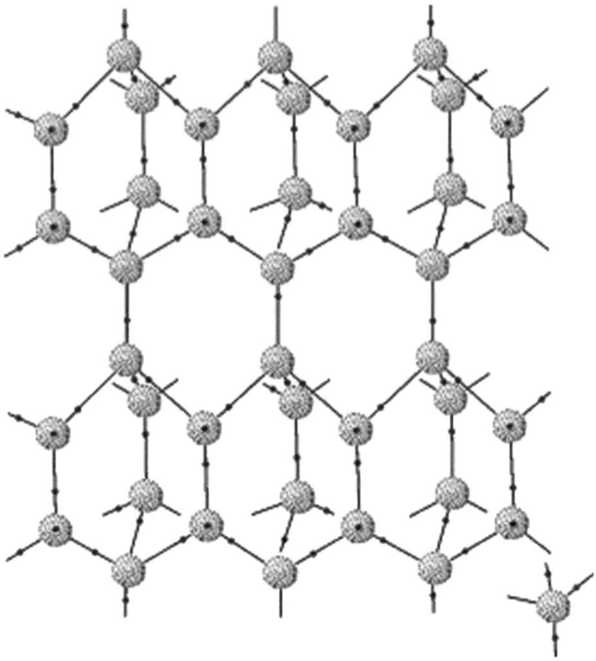

Fig. 2.5 Water as a pseudo-crystal

anomalous properties of water. The coordination number of molecules (the number of the nearest neighbors) varies from 4.4 at 1.5 °C to 4.9 at 83 °C.

Ice X Under the water pressure, magnified 600,000 times, water molecules snag so tightly against each other that the differences between hydrogen bridges and conventional chemical bonds completely disappear. Ice X possesses the greatest density among known crystalline forms of H_2O (2.51 g/cm^3). Its pressure remains hard even at a temperature of 500 °C.

In the gaseous state, water molecules cut loose from each other and embark on a solo voyage. Only occasionally do their individual elements join each other on the fly. Transforming into steam, water absorbs an enormous amount of energy—2,258 J/g (in comparison, with ethanol it is only 854 J/g). During condensation, this energy is liberated again. That is why water vapor may cause serious burns.

When and into what does water transform? To have a more graphic picture, these data is presented in the following form:

The state of the water under different conditions				
Temperature °C				
From −50 to +100	0	+4	+100	+374
In this temperature range at a pressure of 200 MPa, the running water splits into two structures of different density	This is the melting and freezing point. When freezing, the water "explosiveness" sharply increases all at once by 11 %	The water reaches it's the greatest density	This is water's boiling point under conditions of normal pressure. On Everest, for instance, due to low pressure, water starts to boil as early as 70 °C	Higher than this critical point and at a pressure of 22.1 MPa, the difference between the liquid and gaseous state of the water levels out. The matter then takes some intermediate form

Nearly all temperature scales have to do with water, including Kelvin's temperature scale which is based on the triple point of water, defined as exactly 273.163 K or 0.01 °C. It uses the same graduations as in Celsius's scale.

The above data give a graphic example of the anomalous behavior of water; for example, "abnormal" temperatures of water melting and boiling are shown. However, this is far from the only example of water abnormality. The reason for abnormal properties of water is the hydrogen bonding and peculiarities of water structure. The present cluster model of water (for details, see Chap. 3) explains many of its abnormal properties.

Thus, a hydrogen bridge (link) is a linking element, a normal, so-called, covalence link between one molecule and a normal link for the molecule of the weak Van der Waals attraction force. In a water drop the myriads of particles form an infinite network of tetrahedrons, while hydrogen bridges are organized into an ordered system. In doing so, they are flexible and compliant, which ensures variation of the water. The extreme lightness of ice is related to the fact that hydrogen bridges weave water molecules in a crystal into volumetric networks in such a way that much space remains between particles. During the melting phase, the regular network is partially broken and turns into liquid, which increases the density of the water.

The aforementioned phenomenon can be explained by discussing hydrogen bonds (bridges). The diversity of macroscopic solid forms of H_2O is reflected at the molecular level. The variants of the H_2O crystals demonstrate complexity and sophistication which cannot be compared with any other molecule. As has already been noted, 15 types of crystals are known. Christopher Zalzman from Oxford University announced the last two discoveries in March 2006. In paper [3], the summarized information about properties of all 15 types of water is presented. Thomas Lepting, from the Institute of General Inorganic and Theoretical Chemistry at the Innsbruck University, has been researching ice since 1980. "Even under normal pressure there are two forms of ice", he states. These are the so-called hexagonal and cubic ice, *Ih* and *Ic,* respectively. On Earth hexagonal ice is almost exclusively found. Snowflakes, ice covering lakes and ponds and ice cubes in a glass with a drink consist of this ice. Its particles form a hexagonal structure. The basic form of the other kind of ice is a cube. Such crystals which have a cubic shape can be found only at high altitudes of the earth's atmosphere, where deathly cold reigns. However, both variants of ice are united by one property—third-dimensionality.

It is not simple to account for possible types of crystalline varieties of ice. Water is also capable of forming glassy and amorphous forms. Scientists believe that 99.9 % of ice in space is in the amorphous form. It covers dust particles in interstellar space and is incorporated by comet composition.

The difference between crystalline and amorphous ice lies in their internal structures. In a crystal, water molecules are arranged in a regular way in all directions with identical interspaces. In the amorphous state they are chaotic, just as in the liquid state—as if the liquid had solidified instantaneously. The most well known representative of such form of matter can be found on an ordinary window pane.

According to one scientific interpretation, ice has become the basis of life on the Earth. Its flexible structure makes it possible to free such elements as nitrogen, oxygen and carbon. Upon interacting with each other, these substances form simple biomolecules. On Earth, universal vitrified ice may be obtained if over hundredth of a fraction of a second one is able to decrease the temperature by hundreds of degrees Celsius or in a high-pressure press. A piece of ice obtained in the Innsbruck Laboratory under the pressure 9.0 GPa at a sharp cool down is in fact amorphous ice—HAD (high-density amorphous ice). Externally, it does not differ at all from ordinary ice, but if it is thrown into a glass of water, it will sink because it is heavier than water. When heated, HAD is transformed into one more variant of vitrified ice, which, due to its low density, was referred to as LAD (low-density amorphous ice). However, when the scientists heated HAD in the press, having increased the temperature from 196 to + 105 °C, the sample shrank. This contradicts the generally accepted idea that substances subjected to an increased temperature should expand. The scientists therefore stated that they had discovered the third variant of vitrified ice—very high-density amorphous ice (VHAD).

The state of the water when there is no difference between liquid and gas is achieved at the point with the temperature 374 °C and pressure 22.1 MPa. Scientists believe that at low temperatures such a point should also exist. Under experimental conditions, one manages to lower the water temperature to − 38 °C, and in this case it does not freeze. Such phenomenon is seemed to be unreal however it does exist. Light clouds at an altitude of several kilometers above the Earth's surface form drops, and liquid remains at approximately the same temperature (− 38 °C). When the temperature decreases, the "overcooled" water behaves in a more bizarre manner. Scientists imply that at a temperature of − 50 to + 100 °C and a pressure of 200 MPa the second critical point should be found (M. Chaplin, see [3] sticks to another viewpoint). Below this mark, in compliance with the theory, there are two varieties of liquid water with different densities; at a higher density, these variants are not distinguished.

It is assumed that two forms of ice, HAD and LAD, correspond namely to these liquid forms after their fast freezing processes. However, the result is that within one temperature range in which, despite all tricks of experimentalists, water drops

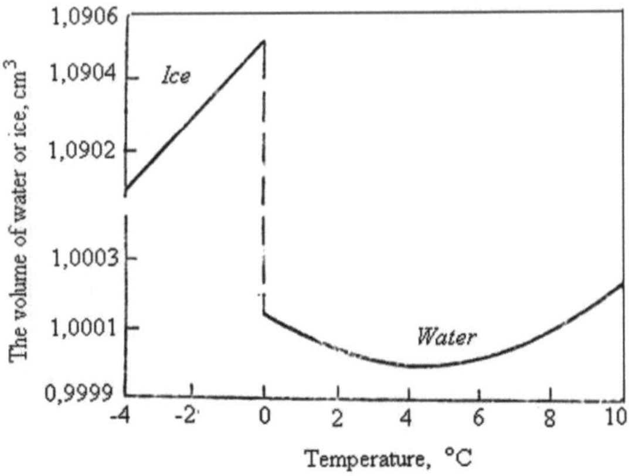

Fig. 2.6 Relationship between the specific volume of ice, water and temperature

always freeze. Should several forms of liquid water exist [1]? This is one of many questions for which there still no conclusive answer.

2.1.1 Anomalies of Water

It should be noted that anomalous properties of water, like anomalous properties of oxygen molecules, are important factors for the existence of the Earth's biosphere [4]. So, how many anomalies are present in water? Some scientists have counted 40 anomalies characteristic of water [1]. According to [3] there are 63 of them. These scientists have tried to explain such anomalous properties of water; to some they seem exhausting, to others they are controversial, and to the rest they are unsatisfactory.

How important the role of the anomalous properties of water in life is can be understood, for example, by considering its special properties at 4 °C (Fig. 2.6). For direct determination of the ice structure and liquid water, the data of the X-ray diffraction analysis and the scattering of slow neutrons are used [5–9]. In these methods, the main structural characteristic is the distribution function of the distance between the oxygen atoms. For water at 1.5 °C, the maximum of this function corresponds to the distance between the nearest water molecules in 0.290 nm, which slightly exceeds the distance between the molecules of the water in ice (0.276 nm). In this case, the average coordination number (n) of each water molecule equals 4.4. In other words, the destruction of hydrogen bonds in melting ice does not imply an increase in the density of liquid water, and the reason for this lies in a large coordination number of molecules. This is due to the fact that they partially occupy hollows, which are vacant in ice.

Close values of the distances between molecules of liquid water and ice as early as 1933 made it possible for a number of authors to express the opinion that water is a pseudocrystal [10] (Fig. 2.5). In turn, one author succeeded to qualitatively explain why water density at 4 °C reaches a maximum. Therefore, in accordance with

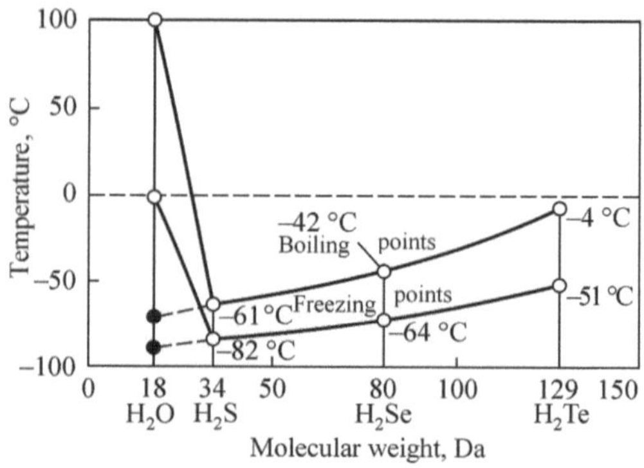

Fig. 2.7 Anomalies of boiling and freezing water points compared with other hydrogen compounds

this model, the water in the liquid state close to 0 °C represents a mixture of three components with various structures, which correspond to hexagonal ice, crystalline quartz and the densely packed structure of liquid water. As the temperature rises, ice-like structures are destroyed, and the contribution of the densely packed structure of the water increases. At $T > 4\,°C$, the effect of increasing the distance between the molecules starts to prevail. It is determined by the heat caused by movement which leads to the decrease of water density. In the following years, several types of cluster models were developed. These models made it possible to explain the bulk of anomalies both of pure water and the water containing ions or molecules [11–16].

The maximum density of water at 4 °C, together with a low density of ice, result in that prior to freezing, it is necessary that the entire volume of fresh water (not only its surface) is close to 4 °C. Since the freezing of rivers, lakes and oceans moves from top to bottom, there is a possibility for the bottom ecosystem to survive, insolating the water from further freezing, reflecting the sunlight into space and thawing the water's surface.

Thanks to heat convection, which is controlled by density, seasonal mixing in deeper water of the moderate climate occurs, carrying life-supporting oxygen to its depths. The high heat capacity of oceans and seas makes it possible for them to act as heat reservoirs in such a way that marine temperatures vary only by one third. This is caused by the change of the land's temperature, which determines our climate (for example, the Gulf Stream carries tropical heat to north-western Europe).

Water compressibility decreases the level of the sea by approximately 40 m, which gives us 5 % more land. A high surface tension of water, plus its expansion when freezing, stimulates the erosion of stones, thus forming soil for agriculture.

The opposite properties of hot and cold water (see Fig. 2.7) are noted among its anomalies, which have already been mentioned in Sect. 2.1. These properties are expressed more vividly at lower temperatures, when the properties of the overcooled water deviate from hexagonal ice. If the water—oxygen hydride H_2O were a normal monomolecular compound, such as its analogies of the sixth group of D. I. Mendeleyev Periodic System of Elements, sulfur hydride H_2S, selenium hydride H_2Se, tellurium hydride H_2Te, then water in the liquid state would range from −90 to −70 °C. With such properties of water, life on Earth would not exist.

As the pressure rises, the molecules of cold water move faster, whereas the molecules of hot water move more slowly. Hot water freezes faster than cold water, and ice thaws in compression, except under high pressures, when liquid water freezes under compression. No other material exists under ordinary conditions in the form of a solid, liquid and gas [3].

For the whole biosphere, an important feature of water is its ability, when freezing, not to decrease but to increase its volume, i.e., to reduce density. This anomaly of water is referred to as the *anomaly of density*. Initially, this water feature attracted the attention of G. Galileo. In the transition of any liquid (except gallium and bismuth) to a solid state, its molecules are arranged more densely, and the substance itself, while decreasing its volume, becomes denser. This is true for all liquids but water.

During the cooling process, water behaves at first as other liquids; it gradually becomes compact and reduces its volume. Such a phenomenon can be observed to +4 °C (or to be precise, to +3.984 °C) (see Fig. 2.6). It is at temperature +3.984 °C that water has the greatest density and the smallest volume. Further cooling of water gradually results in an increase of the volume rather than a decrease. The gradual trend of this process is suddenly terminated, and at 0 °C the volume sharply increases by nearly 10 %! At this instant the water turns into ice.

The unique features of water behavior in the processes of cooling and ice formation play an exceptionally important role in nature and life. These water features prevent the freezing of all Earth's bodies of water in winter, and thereby save aquatic life on Earth.

Unlike fresh water, sea water behaves differently during the cooling process. It freezes not at 0 °C but at −1.8 to −2.1 °C, depending on the concentration of salts dissolved in it. It has maximum density of −3.5 °C rather than +4 °C. Thus, it turns into ice without attaining the greatest density. If vertical stirring in fresh water bodies terminates in cooling the whole mass of water to +4 °C, then vertical circulation occurs in sea water at temperatures below 0 °C. The exchange process between the upper and lower layers of water proceeds as such, continuously creating favorable conditions for the development of animals and vegetable organisms.

Melt water, formed in the melting of glaciers and icebergs, are an especially favorable medium for the inhabitants of the seas and oceans.

According to its thermodynamic properties, water noticeably differs from other substances. The most significant among these substances is the anomaly of specific heat. The high temperature of water makes the seas and oceans a gigantic regulator of our planet, and as a result, a sharp climatic difference takes place from winter to summer and from day to night. The climate in coastal areas is mild; in these areas, temperature differences throughout the seasons are low. Powerful atmospheric currents containing enormous amounts of heat are absorbed in the course of vaporization, and gigantic oceanic currents play an essential role in the creation of the weather on our planet.

The anomaly of water heat is elucidated in the following description (see Fig. 2.8). In the heating of any substance, heat capacity continuously increases. However, this is not true for the water. As water temperature rises, its heat capacity

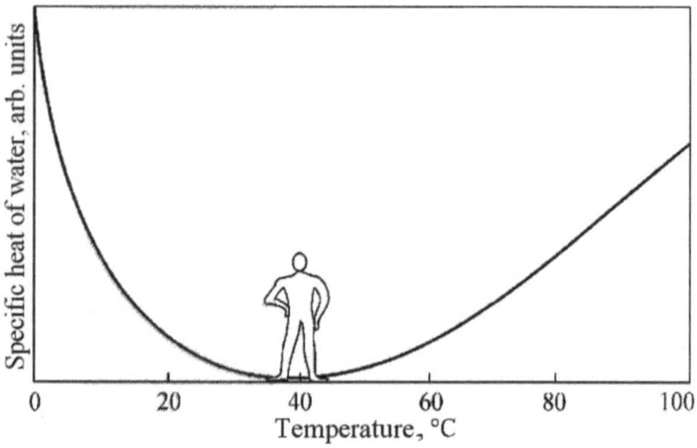

Fig. 2.8 Anomaly of water heat

varies abnormally: from 0 to 37 °C, heat capacity decreases, and from 37 to 100 °C it grows steadily. For temperature ranges close to 37 °C, water heat capacity is minimal. This temperature is close to the temperature of human body. The physics of water within the temperature interval 35–41 °C (the limits of normally proceeding physiological processes which are possible in human organisms) register the probability of achieving a unique state of water; when masses of quasicrystalline and volumetric water are equal to each other and the capacity of one structure to go into the other, variability is maximized. This extraordinary capacity of water predetermines the equiprobability of running reversible and irreversible biochemical reactions in human organisms and securing "easy control" of them.

The exceptional capacity of water to dissolve any matter is common knowledge. Here it demonstrates anomalies extraordinary for a liquid, namely those of the water dielectric constant. This is related to the fact that the dielectric constant (or permittivity) constitutes 81, while for other liquids it does not exceed 10. In accordance with the Coulomb law, the interactive force of two charged particles in water will be as 81 smaller as that, for instance, in the air, where this characteristic equals unity. In this case, the strength of intramolecular links is decreased in 81 times, and, under the effect of the heat, motion is dissociated with the formation of ions. It should be noted that, due to its exceptional capacity to dissolve other matter, water is never strictly pure.

A wonderful anomaly of water is its incredibly high surface tension. Out of all known liquids, only mercury has a higher surface tension than water. This property is manifested in fact that water always tries to reduce its own surface area. Uncompensated intermolecular links of the outer (surface) layer of water molecules, caused by quantum-mechanical forces, create the outer elastic film, which prevents the immersion of many heavier objects. For example, if a steel needle is carefully placed in water, it does not sink. Also, the density of steel is as much as eight times greater than the density of water. Owing to the high surface tension, a drop of water, for instance, takes a ball-like shape in free fall.

Surface tension and wetting form the basis of a special property of water and aqueous solutions, referred to as capillarity. Capillarity is of great importance in the life of the vegetable and animal world, the formation of the structure of natural minerals, and the fertility of the earth. In channels which are many times narrower than a human hair, water acquires wonderful properties: it becomes more viscous; it compacts 1.5 times; it freezes at −80 to −70 °C. The cause of supernormal capillary water is intermolecular interaction, whose secrets have not yet been revealed.

Scientists and experts are aware of so-called porous water. A very thin film lines the surface of the pores, microhollows and minerals of the Earth's crust, and other objects of live and inanimate nature. Bound to intermolecular forces and the surface of other bodies, it possesses a special structure [2], just like capillary water.

According to M. Chaplin [3], water can be characterized by 63 anomalies. He grouped anomalous properties of water into the following five groups: phase anomalies, anomalies of density, anomalies of water as a substance, thermodynamic anomalies, and physical anomalies.

Phase Anomalies Water has an unusually high melting point, boiling point and critical point. In solid form, it exists in a wider variety of stable (and metastable) crystalline and amorphous structures than other materials. Ice heat conductivity decreases as pressure increases, while the structure of liquid water at high pressure varies. The overcooled water has two phases, and the second critical point is approximately at −91 °C. Liquid water easily overcools but solidifies with difficulty; it exists at very low temperatures and freezes at heating. Hot water may freeze faster than cold water (the Mpemba effect), and warm water vibrates (fluctuates) longer than cold water.

Anomaly of Density Ice density increases when heated (up to 70 K). Water compresses when thawed (melting). Pressure decreases the melting point. Liquid water has a high density, which increases when heated (up to 3.984 °C). Pressure decreases the temperature of the maximum density. Overcooled water has a minimal density. Water is characterized with a low expansion coefficient (thermal coefficient of volumetric expansion—TCVE), which substantially decreases (becomes negative) at low temperatures. The water TCVE increases as the pressure increases. The number of the nearest neighbors in ice and thawing (melting) increases with rising temperatures. Water has unusually low compressibility, which decreases with an increase of the temperature to 46.5 °C. In the temperature–compressibility relationship there is a maximum. The speed of sound in water increases with an increase of the temperature to 74 °C. The "fast sound" was detected at high frequencies; it is nonuniform at a higher pressure. The NMR of the spin-lattice relaxation is very short at low temperatures. The refractive index of water has a maximum value at temperatures below 0 °C.

Anomaly of Water as a Substance No water solution is ideal. The liquids D_2O and T_2O differ greatly from H_2O in terms of their physical properties; H_2O and D_2O differ in their phase behaviors (detailed information on the D_2O properties can be seen in Sect. 2.3). Dissolved substances affect such properties as density and viscosity in a different way. The solubility of non-polar gases in water decreases with an increase in temperature to the minimum and then increases again. Water permittivity is high

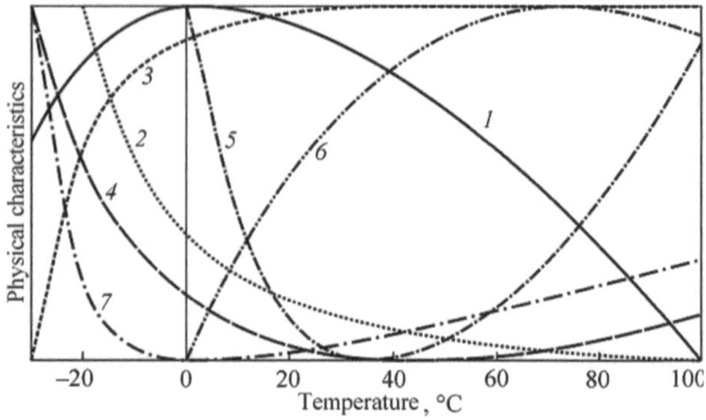

Fig. 2.9 Temperature dependences of physical characteristics of water [7]: *1* density, *2* viscosity, *3* change of viscosity with a pressure, *4* compressibility, *5* heat capacity C_p, *6* velocity of sound, *7* thermal expansion

and demonstrates the temperature maximum. The mobility of protons and ions of hydroxonium are anomalously high in the electric field. The water conductance increases to a maximum of approximately 230 °C. Acidity constants of weak acids have temperature minimums. The diffraction of X-rays demonstrates an unusually detailed structure. Under high pressures, water molecules move even further away from each other as the pressure mounts.

Thermodynamic Anomalies The maximum temperature variation of water melting heat is − 17 °C. The water specific heat is twice as much as that of ice or steam. The water specific heat (C_P, C_V) is unusually high [3]. The parameter C_P is minimal at 36°C and maximum at approximately − 45 °C; there is a minimum with respect to pressure. The specific heat C_V has a maximum. High heat of evaporation and sublimation also exist; this is the entropy of evaporation. Water heat capacity is high and increases to the maximum, which is approximately 130 °C (Fig. 2.9).

Physical Anomalies Water has an unusually high viscosity, which increases greatly when the temperature drops and decreases with an increase of the pressure at temperatures below 33 °C. At a decrease of the temperature, diffusion weakens appreciably; at low temperatures, water self-diffusion increases with an increase of the pressure and density. Heat diffusion increases to a maximum of approximately 0.8 MPa. Water has an unusually high surface tension–concentration (the Johnson–Ray effect) and prevents the coalescence of small bubbles.

This is how the author comments on the classification of anomalous properties of water he proposed: “Whether or not the properties of water are seen to be anomalous depends upon which materials water is to be compared and the interpretation of ‘anomalous’”. For example, it could well be argued that water possesses exactly those properties that one might deduce from its structure. Comparisons between water, liquid sodium, argon and benzene appear to Franks to indicate several of the abovementioned properties as not being anomalous. However, these materials are perhaps not the most typical of liquids. My list gives the unusual properties generally understood to make liquid water (and ice) stand out from “typical” liquids (or solids).

It is therefore very difficult to obtain pure water (e.g. $< 5\ ng\ g^{-1}$). Note that ice, in contrast, is a very poor solvent and may be used when purifying water (e.g. degassing) by implementing successive freeze-thaw cycles.

Some scientists attribute the low temperature anomalous nature of water to the presence of a second critical point; they pose an interesting if somewhat unproductive hypothesis (as the attribution mixes cause with effect). Water's anomalies do not require this as an explanation.

The temperature range from "hot" to "cold" water varies in these examples; see the individual entries for details.

The anomalies of water are divided into groups, but clearly some anomalies may be included under more than one topic, and there may not be universal agreement for the groupings.

2.2 The Role of Water in the Scenario of Development on Earth

One of the most difficult issues to which there is no unequivocal answer is as follows: how did the life appear on Earth? What could be an inception which made life possible on Earth? The current study is in search of this inception, referred to in paper [17] as "molecules of protolife", as the development of science changes the scenario of the emergence of life on Earth.

So what happed 4.6 billion years ago and how was life born? For the genuine scientist C. Darwin's works are not a dogma, but rather guidance for action. The history of evolution contains plenty of ambiguities, which says nothing about the global issue of the origin of life on the Earth. At one of the last scientific sessions of the General Conference SO RAN [18] Academician N. Dobretsov asserted the following: the Earth's age, like other bodies of the Solar System, constitutes approximately 4.6 billion years, which has been confirmed by the analysis of meteorites. Two important conclusions that can be made today, based on numerous data, is the fact that the evolution of life is determined by the irreversible processes of the Earth's cooling, which started about 4 billion years ago, and the oxidation of upper shells, namely the hydroatmosphere, which became the impetus for the beginning of evolution.

Upon discovering the structure of deoxyribonucleic acid (DNA), scientists started to search for the molecules responsible for initial life. The discovery of the DNA structure has become one of the major scientific achievements of the twentieth century. It was done by an international group of researchers: Fr. Crick, J. Watson and M. Wilkins received high-quality DNA X-ray photographs; M. Nirenberg from the Bethesda National Institute of Health, Maryland, succeeded in deciphering the genetic code; R. Holy from Cornell University cleared up the structure of transfer RNA; a research group from the Massachusetts Technological Institute, headed by Indian biochemist G. G. Koran, synthesized an artificial gene for the first time in the history of the world. All the researchers for these experiments were awarded

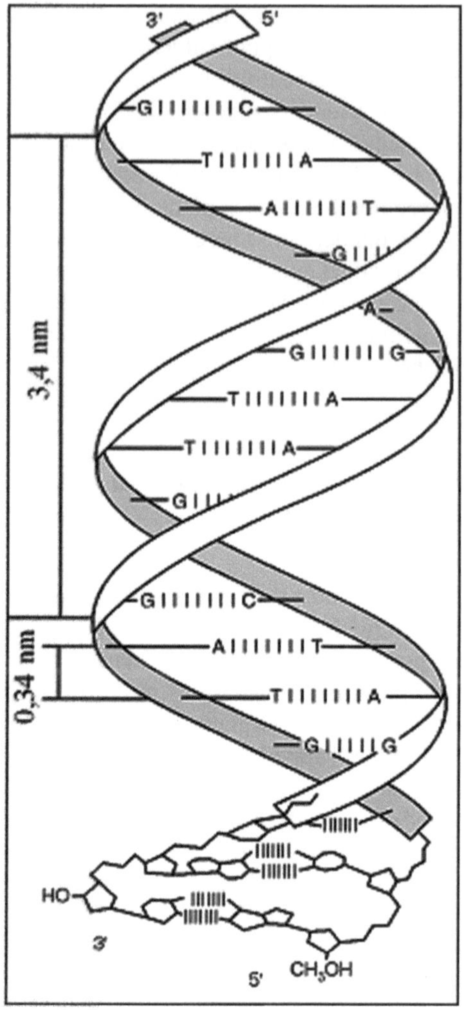

Fig. 2.10 The structure of DNA

Noble Prizes [19]. For the first time, the structure of the hereditary information carrier, DNA, was presented in the article of US geneticist J. Watson [20] and English physicist F. Creek [21]. In compliance with the definition of the American Chemical Society, DNA are large molecules representing a double helix of atoms held together and located inside cells of effectively all living beings capable of creating, storing and transferring biological properties to successive generations (the schematic of DNA's molecular structure is given in Fig. 2.10).

Deoxyribonucleic acid is a polymer whose monomers are nucleotides, one of the most well known polymer molecules today. In some organisms, the DNA polymer chain consists of a hundred million links. The length of such a molecule reaches several centimeters, and this is a very large value for molecular objects. Since the molecular cross section is only 2 nm (1 nm = 10^{-9} m), its proportions may be compared with a rail of tens of kilometers long (Fig. 2.10).

The succession of polynucleotides in an unramified polynucleotide chain, the DNA primary structure, is strictly individual and specific for every natural DNA. They represent a code form of recording biological information, known as the genetic code. In every molecule, polynucleotide chains are united into one double helix, in which the bases are arranged in pairs opposite to each other and are linked with hydrogen bonds. The pairs can be formed complementarily, i.e., with the bases matching each other. The chain is twisted into a helix around an imaginary axis with the bases perpendicular to it. Such a double helix is called the secondary DNA structure. The tertiary structure is formed, for example, in chromosomes, by double helixes linked with proteins by ionic bonds. Cell division causes DNA doubling or replication (repletion, self-doubling), and as a result, every new cell obtains DNA bearing the same genetic information, which was available in the initial cell. The succession of the main DNA determines the order of the arrangement of amino-acid residues in the molecules of proteins. For realization of genetic information based on the DNA coding chain in the course of transcription (the transfer of the genetic information from DNA to RNA), an RNA chain, a replica of the given section of DNA, is formed. The sequence of RNA bases is the matrix for the synthesis of protein encoded in the gene itself.

Proteins were the first candidates for the role of "the molecules of protolife". In 1924, A.I. Oparin asserted that in the atmosphere of a young Earth consisting of hydrogen, methane, ammonia, carbon dioxide, and water vapors, amino acids might be synthesized and then spontaneously joined into proteins and formed into "clots" resembling a primitive cell.

By the mid-1960s, scientists knew the details of two functioning molecules, which were more suitable than proteins for the role of "the molecules of protolife," DNA and RNA [22]. Fig. 2.11 gives the structures of DNA and RNA for comparison.

Differences between the macromolecules of RNA and DNA boil down to the fact that the RNA sugar-phosphate skeleton includes sugar ribose whereas DNA "loses" one oxygen atom and is transformed into deoxyribose. In addition, instead of thymine (T) the RNA composition includes uracil (U), which differs from thymine nearly as little as ribose from deoxyribose: it lacks only the lateral methyl group ($-CH_3$). However, such minimal differences in the structures of RNA and DNA result in a substantial difference in the structure and functions of these molecules. Late in the twentieth century, one more breakthrough occurred in the theory of the origin of life. In this breakthrough, RNA should be "blamed" which was thought to be thoroughly investigated and rather predictable substance.

This story started in the 1970s, when unusual enzymes were found in the cells of certain organisms. The enzymes included in their composition, apart from protein, an RNA molecule. Late in the 1970s, American biochemists Thomas Check and Sidney Altman, independent of each other, studied the structure and functions of such enzymes. One of their tasks was to clear up the role of RNA incorporated as part of the enzymes. At first, following the received opinion, the scientists believed

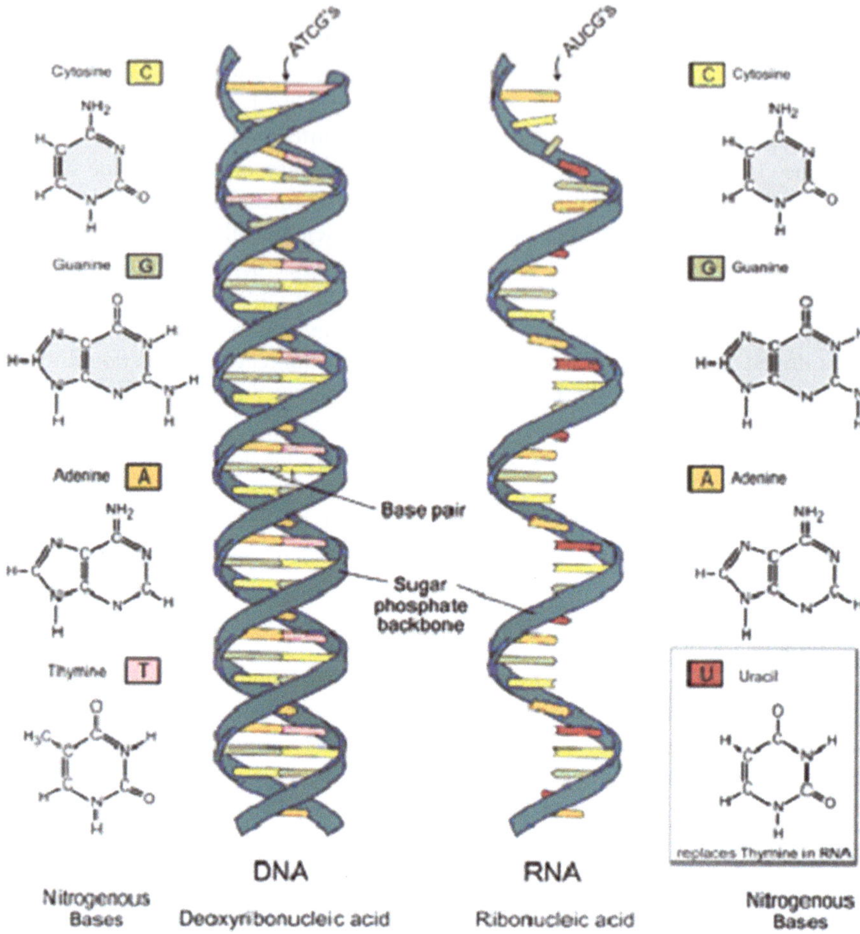

Fig. 2.11 The structure of DNA and RNA [23]

that the RNA molecule in such complexes was only an auxiliary element which may be responsible for the construction of the correct structure of the enzyme or for the correct orientation in the interaction of the enzyme and the substrate (i.e., that molecule, which is in fact subjected to the change) while the reaction being catalyzed is performed by protein. In order to clear up the situation, the researchers separated the protein and RNA components from each other and investigated their ability to perform catalysis. To their great amazement, they saw that even after the removal from the enzyme of the protein, the remaining RNA was capable of catalyzing its specific reaction. Such a discovery would mean a breakthrough in molecular biology. Earlier it had been believed that reactions could be catalyzed only by proteins rather than nucleic acids.

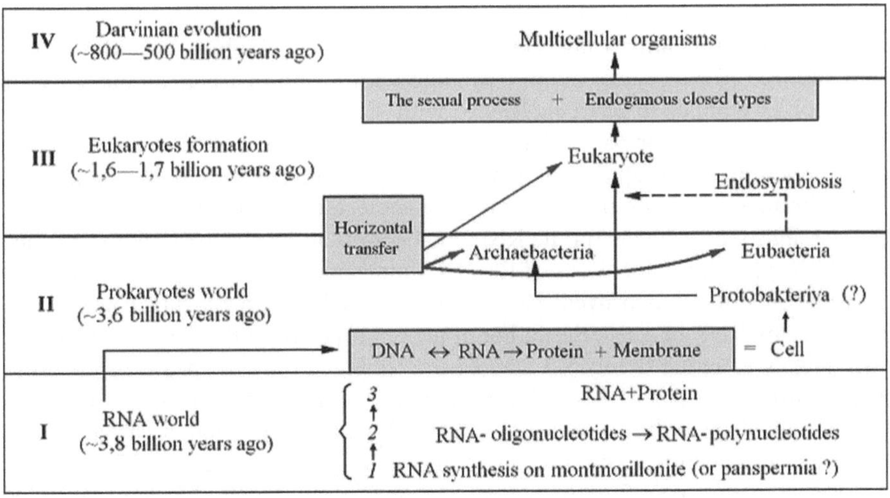

Fig. 2.12 The main stages of occurrence and life evolution on the Earth [6]

However, could it be that this was ineffective purification of RNA, leaving some intangible, minimal pieces of protein which are capable of implementing catalysis? The last and the most convincing proof of RNA capability for catalysis was a demonstration of the fact that even artificially synthesized RNA, as part of the enzymes being studied, may catalyze the reaction on its own. The RNA molecules, capable of catalysis, were called ribozymes (by analogy with enzymes, i.e. protein ferments). For their discovery in 1989, T. Check and S. Altman were awarded the Nobel Prize for chemistry. It did not take long for these results to shed light on the theory of the origin of life; the RNA molecule became "the favorite". Indeed, a molecule capable of carrying genetic information and other material to catalyze chemical reactions had been detected! It would be difficult to imagine a more suitable candidate for generation of protocellular life. Therefore, the scenario of the evolution of life has been transformed [19].

The scenario of the commencement of life on Earth consists of four stages, and at the beginning is RNA (Fig. 2.12) [22]. Paper [22] considers all four stages of the formation and evolution of life on the Earth in detail. Is there a place in this scenario for water as a participant in the origin of life on the Earth? With regard to this, the first stage as the very beginning of the origin of life on the Earth is most interesting.

In all modern living systems, from viruses to higher animals, DNA or RNA "enjoy the services" of protein enzymes in order to quickly and effectively, by means of catalysis, transfer its information to a number of generations. None of the nucleic acids in the modern living systems are able to replicate themselves.

The idea of a world composed of RNA, i.e., the existence of independent RNA at the first stage, which started ~3.8 billion years ago, was established less than 20 years ago. In 1993, the Institute of Protein of the Russian Academy of Sciences (RAS) empirically proved that RNA molecules are capable of forming molecular colonies similar to bacteria on gels or other solid media if they are provided with conditions for replication (repetition). Such molecular colonies on solid or semi-solid surfaces (for instance, montmorillonite with a film of water on the surface) consist of assemblies of RNA molecules with different ribosome activity and can be the first evolutionizing acellular ensembles. The evolution of RNA ensembles in acellular colonies sped up due to the fact that these colonies were not isolated from the environment and could easily exchange molecules and their genetic material. The possibility of exchanging the RNA molecules via the air due to the spontaneous nonenzyme recombinations of the RNA molecules in the case of collisions in the water medium was proved. Having summarized known functions, which are performed by RNA in a cell, the author of paper [22] came to the conclusion that all basic processes in the living cell may be implemented only by means of the presence of different types of RNA. Furthermore, the famous triad DNA $\leftrightarrow$ RNA $\rightarrow$ protein took shape only at substage I of stage II of the origin and evolution of life on the Earth, probably 3.6 billion years ago (Fig. 2.12). The classical definition that "life is a cellular protein form of matter existence" means that both macromolecules and the cell, i.e., the membrane with its complex and varied functions, are important for its existence. After the appearance of the cell, the bacterial world developed very fast.

Hence, at what phase of the development of the Earth, as a planet, do these transitions occur? According to the opinion of some scientists, the origin of life could be in space. This rather popular theory of panspermia asserts that life is carried in all directions in the Universe in the form of "spores". These may be organic compounds sometimes of a rather complex structure, whose traces were found in the meteorites that had fallen on the Earth. According to astrophysicists, organic matter exists in great amounts in gas-and-dust clouds and space dust, which is constantly engulfed by the Earth. According to the author of paper [22], life from space is carried only in the form of short oligonucleotides, which may be frozen into ice of any composition (methane, aqueous) and transform into a favorable medium. This cycle starts anew every time: the synthesis on montmorillonite, the emergence of the RNA and DNA molecules and protein, and finally, the emergence of a cell.

The scientists tend to come to a conclusion that the presence of water in space objects is most likely a rule, rather than an exception. American space probes detected water on two of Saturn's satellites and on Mars. From this evidence, the existence of life on these planets in the past or at present can be assumed. In 2001, the scientists of the IMS Scientific and Research Center under NASA and the division of California University in Santa Cruz (the USA) carried out an experiment under conditions similar to those for the formation of the Solar System. The blend of various substances (water, methanol, carbonic acid and carbon dioxide) was cooled down to 10 K (−263.16 °C). It was irradiated by ultraviolet light with the wavelength similar to the wavelength in a dense molecular cloud from which the Solar System allegedly formed. The formation of organic molecules took place. Self-organizing structures of 10 μm in size were identified. By shape these were bubbles, which resembled cells. In July 2005, the American scientists found at least three areas covered with ice on the surface of the Temple comet's core. According to the researchers, the water, which changed to ice, contains a lot of impurities. Scientists long ago came to the conclusion that comets are fragments of the initial Solar System at an instant of its inception 4.6 billion years ago.

An important conclusion, which one may make today based on numerous data, is the fact that the evolution of life is determined by the irreversible processes of the Earth cooling, which started about 4 billion years ago, and by the oxidation of the upper shells, in the first place, the hydrosphere, which served as a prime cause for the beginning of evolution. A sensational conclusion was made by the scientists of a laboratory under the US Department of Power in the course of research related to an attempt to explain the change of every 200 thousand years of the direction in the Earth's magnetic field [18]. In the center of the Earth there is a core of the melted mixture of uranium and plutonium, maintaining a constant nuclear reaction, rather than melted iron and nickel. This core, almost 8 km in diameter, is a "natural gigantic nuclear reactor". As a result of its "work" around the Earth, a powerful nuclear reactor, which protects the planet from dangerous cosmic rays capable of annihilating all biological life within several seconds, appears. The natural reactor also supplies energy to the movement of the continental platforms and manifests itself in the eruption of volcanoes. It is obvious that a number of additional conditions are necessary for the inception of the process. One of them is the presence of water. Water (12–15 %) is necessary as a moderator of neutrons, which form in the case of spontaneous division of uranium, since the reaction may occur only under the effect of slow neutrons.

One more interesting hypothesis lies in the fact that life originated near emissions of hot volcanic water, where, due to temperature and the presence of large concentrations of biogenic molecules, the reaction of the formation of biomolecules may occur at high speeds [22]. In addition, substantial temperature differences could facilitate the processes of the matrix synthesis of nucleic acids. High temperatures (near the source) were conducive to the decomposition double filar nucleic acids to unifilar acids, during which (as a result of a decrease of the temperature) the other cycle of synthesis may occur. Such a scenario resembles the technology of the repeated replication of nucleic acids, referred to as a polymerized chain reaction (PCR), and developed in the mid-1980s [17].

So far the theory of the RNA world is full of contradictions and ambiguities. Many of them, no doubt, will be solved within the framework of the "classic" hypothesis implying that life's source was RNA as we know it in the modern form. However, how can we explain within the framework of this theory the evidence of archeologists and paleontologists who have found the remnants of the first primitive cells in layers from the period of 3.5 to 3.8 billion years ago? At the same time, one may consider that life could not originate earlier than 4 billion years ago because before that the Earth had been intensively "shelled" by meteorites and comets. According to more radical data, it's "shelling" ended even later, about 3.8 billion years ago. Thus, effectively there was no time for the development of the pre-cellular world. Prominent adherents and founders of the RNA world hypothesis Thomas Chech and Lesley Orgel [19] agreed to this.

In paper [18], Academician V. Parmon put two rather daring hypotheses about the origin of life. Firstly, the planets, according to V. Parmon, were formed under conditions of the catalytic reaction process. They formed from a gas-and-dust cloud, which looked like a huge catalytic reactor since in the absence of chemical reactions dust particles cannot hold together. Secondly, natural selection started as early as the chemical stage of evolution, while primary organic compounds, from which life may originate, occurred prior to the formation of the planets. Neither of the existing theories about the origin of life clarifies how the first RNA and DNA appeared. The mathematical likelihood of the random "self-creation" of these molecules is too feeble. Hence, numerous speculations about the extraterrestrial origin of life appeared. However, according to Academician V. Parmon, the start of evolution could be triggered by a catalytic reaction. The simplest schematic of such a reaction is as follows: the molecule of "food" plus the molecule of an "assistant" (autocatalyst) is provided by two autocatalyst molecules.

In the course of experimenting with the concentration of "food" and assuming that the molecules of the autocatalyst are capable of mutating, one may find an analogue of natural selection; the molecules which can do with a minimal amount of "food" can survive. This does not correspond to the familiar theory of "thick bouillon", according to which life originated due to the excess of nutrients. Even so, the assumption that hunger is a "driving force of the progress" seems to be quite logical.

However, is there a reaction in chemical science which corresponds exactly with this example? The answer is yes, there is at least one. Butlerov's reaction was described by a prominent Russian chemist as early as in 1864. This is a synthesis of various sugars from formaldehyde which appear in the presence of ions of calcium or magnesium. It is well-known that the Butlerov's reaction is autocatalytic although the nature and action mechanism of autocatalysts is still unknown. The reaction is uncontrollable; each time very different sugars are obtained at very different proportions. The experiments dealing with the study of this reaction, when the amount of the "food" decreased, so far were not carried out. Similar research has started at the Institute of Catalysis.

It is considered as a fact that ribose (a type of sugar) in the Butlerov's reaction appears consistently. Additionally, ribose is the basis for the creation of both RNA and DNA. Hence, an important conclusion follows: the required "spare parts" for the assembly of the first RNA and DNA could appear in the course of the Butlerov's reaction. Moreover, as geologists believe, there was an explosive atmosphere composed of methane, water vapor, carbon dioxide and ammonia on ancient Earth. With such a cocktail for the formation of formaldehyde, which is necessary in the above reaction, only the presence of a hot surface was needed. Therefore, the basis of life could not be protein at all, as is a commonly accepted belief; instead it was probably sugar, which appeared in the course of the auto catalytic reaction.

The author of paper [23] constructed a scenario, based on his own polytetrameric model of the water structure, of the earliest stage in the development of life on Earth—the synthesis of prebiological organic matter. He considered the involvement of water not only as a universal and the only possible solvent but also as a matrix which forms when water participates in the coding of proteins through nitrogen bases of DNA. Behind such an approach, as the author writes, lies the fact that "I succeeded to graphically reveal the existence of left-hand and right-hand aqueous tetramers playing in the usual water the role of molecules …where the bilateral symmetry of the structural units of the water… should somehow affect the prebiological synthesis of organic matter". The scenario proposed in paper [23] of the early stage in the development of life on the Earth is based on the phenomenon of randomness and in the absence of experimental data on the existence of bilateral symmetry of the structural units of water in real conditions, when life originated on the planet Earth.

According to contemporary ideas about the origin of life on Earth [22], the choice elected by organic molecules of a certain type of bilateral symmetry served as a main prerequisite of their survival and subsequent reproduction. However, the question of how and why the evolutionary selection of this or another mirror antipode occurred remains one of the largest riddles of science. Soviet scientist L. L. Morozov proved that a transition to chiral orderliness could take place in the case of a certain sharp phase change rather than an evolutionary change. Academician V. I. Goldansky referred to the period during which life originated on the Earth as a chiral catastrophe. So, how did the conditions for the phase catastrophe, which caused the chiral transition, appear? The most important was the fact that organic compounds melted at 800–1,000 °C in the Earth's lower crust, while compounds in the upper crust cooled down to the temperature of space, i.e., the absolute zero. Under such conditions, the temperature difference reached 1,000 °C. Organic molecules on the lower crust melted under the effect of high temperatures and were completely destroyed while the top crust layer, where organic molecules were frozen, remained cold. Gases and water vapors which seeped from the Earth's crust varied the chemical composition of organic compounds. Gases carried heat which caused the melting boundary of the organic layer to shift up and down, creating a gradient. At very low atmospheric pressures, the water on the Earth's surface was only in the form of steam and ice. When the pressure reached the so-called triple point of water (60 Pa), water for the first time could be in the form of liquid. Certainly, it is only empirically that one can prove what caused the chiral transition: earth or space reasons. However, in any event, at some instant chiral ordered molecules (namely, left-hand rotating amino acids and right-hand rotating sugars) turned out to be the more resistant which lead to the continuous growth of their amount—known as chiral transition [18, 24].

The most wonderful thing in the structure of water lies in the fact that its molecules at low negative temperatures and high pressures inside nanotubes may crystallize in the form of a double helix resembling DNA. This was proved by computer experiments of American scientists, headed by Syao Chen Zen at the Nebraska State University. The scientists expected to see that water in all the cases forms a thin tubular structure. However, the model demonstrated that, given the tube diameter 1.35 nm and the pressure 4.0 GPa, hydrogen bonds warp, which results in the formation of a helix with a double wall. The inner wall of this structure is a helix twisted fourfold; the outer one consists of four double helixes like the structure of the DNA molecule [18]. The last fact affects not only the evolution of our concepts about water, but also the evolution of the early life and the DNA molecule itself. If one assumes that in the time of the origin of life cryolite clay rocks had the shape of nanopipes, then a question arises: could the water absorbed in them serve as a structural basis (matrix) for the synthesis of DNA and reading the information? Possibly, therefore, the helix structure of DNA repeats the helix structure of the water in nanopipes. To confirm or refute the existence of such macromolecules may only lead to similar experimental investigations under the ambient conditions.

2.3 Water Isotopic Composition: The Role of Deuterium in the Destruction of DNA

Only in 1931 did the existence of heavy water in nature become known. Its composition, although in very small amounts, incorporates the heavy hydrogen isotopes—deuterium and tritium. This year was marked by the discovery of heavy hydrogen—deuterium, by American physicochemist Harold Urey. H. Urey entrusted his assistant to evaporate 6 L of liquid hydrogen, and in the last fraction of 3 cm^3 by the spectral analysis, he detected a heavy isotope of hydrogen with an atomic mass twice the mass of known protium [24, 25].

The scientists arrived at the conclusion that perhaps there was a heavy isotope of hydrogen with an atomic mass of 2. In 1932, H. Urey and E.F. Osborn for the first time found in natural water a heavy one. Two year later Harold Urey was awarded the Nobel Prize. The discovery of the third superheavy hydrogen isotope, tritium, with the atomic mass 3, was kept a secret for 3 years due to strategic considerations. In 1951, tritium water was obtained and researched.

In natural water the atomic fraction of tritium is insignificant at only 10^{-18}. Nonetheless, it is contained in the water we use to drink, and over the years it substantially harms our genes, causing ageing and diseases.

Deuterium water is obtained with a miniscule amount of tritium water by its concentration in the leftovers of electrolyte after electrolytic decomposition of natural water and also in fraction distillation of liquid hydrogen. Industrial fabrication of heavy water increases with every year in nearly all countries, and especially in states possessing nuclear arms.

Heavy water is mainly used as a moderator of fast neutrons in splitting radioactive elements in nuclear reactors. The prospects of using heavy water or human needs are grandiose. It may become an inexhaustible source of energy: 1 g of deuterium gives tens of millions times greater energy than the burning of 1 g of coal.

Tritium water so far is of limited use and is used, mainly, in thermonuclear reactions and also in physicochemical and biological investigations as tracer radioactive molecules.

Six oxygen nuclides have been detected: ^{14}O, ^{15}O, ^{16}O, ^{17}O, ^{18}O, ^{19}O. Three of them—^{16}O, ^{17}O, ^{18}O are stable, while ^{14}O, ^{15}O and ^{19}O are radioactive. Stable oxygen isotopes are contained in all natural water. Their ratios are as follows: per 10 000 portions of ^{16}O fall 4 parts of ^{17}O and 20 parts of ^{18}O.

What is Earth's water from the viewpoint of the isotope composition of the elements forming it? For characterization of the water isotope composition, three standards are used: the generally accepted Viennese Standard SMOW, and also GISP and SLAP, which are given below.

International standards of the water isotopic composition [26]	
SMOW—Viennese standard of average ocean water	$D/H=(155.76\pm0.5)\times10^{-6}$ $^{18}O/^{16}O=(2005.2\pm4.5)\times10^{-6}$
GISP—the water standard from Greenland ice	$D/H=(124.6\pm4.5)\times10^{-6}$
SLAP—the water standard from Antarctic ice	$D/H=(90.5\pm1.0)\times10^{-6}$

SMOW standard (Standard Mean Ocean Water) corresponds to the depth water of the World Ocean with the content of 997.0325 g/kg (99.73 %) of light water $^{1}H_2{}^{16}O$. The SLAP standard (Standard Light Antarctic Precipitation) corresponds to natural water from the Antarctic, where the fraction of light water constitutes 997.3179 g/kg (99.76 %).

The isotope composition of the water on Earth with the account of only stable isotopes looks as follows:

Water isotope composition [26]		
Natural water		
Stable molecules		
$H_2{}^{16}O$	$H_2{}^{17}O$	$H_2{}^{18}O$
$HD^{16}O$	$HD^{17}O$	$HD^{18}O$
$D_2{}^{16}O$	$D_2{}^{17}O$	$D_2{}^{18}O$
Light water, volumetric fraction, %	Heavy oxygen water, volumetric fraction, %	Heavy water, volumetric fraction, %
$H_2{}^{16}O$—99.727	$H_2{}^{18}O$—0.20 $H_2{}^{17}O$—0.04	$HD^{16}O$—0.033 $H_2O+D_2O=2HDO$

As can be seen from the data shown, the combination of stable isotopes yields nine varieties of molecules. In this case the bulk water falls on the molecules of protium (light) water with oxygen 16–99.727 % and on heavier molecules only 0.273 %. In this case, heavy molecules are distributed in the following way: $H_2{}^{18}O$—73.3; $H_2{}^{17}O$—14.7; $HD^{16}O$ is 12.1 % of the total volume. In fresh water sources the content of heavy water usually constitutes about 330 mg/dm^3 (per the HDO molecule) and in a heavy-oxygen molecule ($H_2{}^{18}O$)—about 2 g/dm^3. This can be compared with or even exceed the allowable content of salts in drinking water. In addition, it was noted that natural isotope variations of D and ^{18}O are abnormally high. If we consider deuterium as a microelement, incorporated not only into water, but also into important organic compounds, then in terms of significance it can rank as one of the first if not the first. Among the other elements in human organism, deuterium appears right after sodium. The amount of deuterium in blood plasma is four times higher than potassium, 6 times higher than calcium, 10 times higher than magnesium and much higher than the content of such important trace elements as fluorine, iron, iodine, copper, manganese and cobalt.

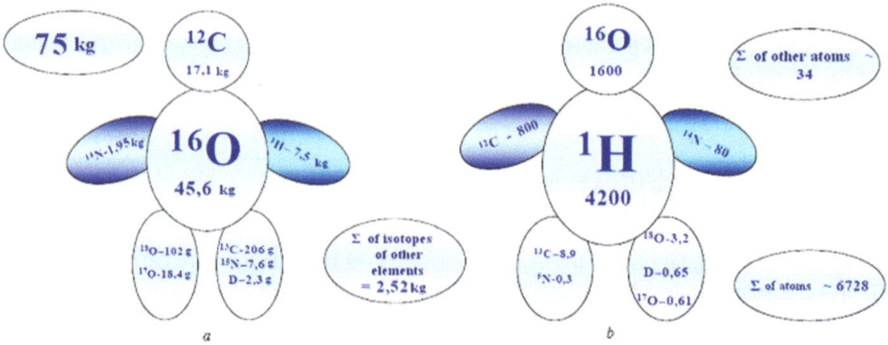

Fig. 2.13 The isotopic composition of the human body and number atoms of elements forming it (unit—number of atoms × 10^{24})

As shown in paper [26], based on natural variations of deuterium, one may infer that its content in human organisms varies greatly, from 9 to 16 mmol/dm^3, even if one considers that within a lifetime it does not build up in organisms, owing to reactions of isotopic exchange; however, perhaps, this is not so. In human blood plasma the deuterium concentration is higher than in consumed drinking water, which is confirmed by the following data:

Content of trace elements in human blood plasma [26]

Microelement	Content, mmol/dm^3	Microelement	Content, mmol/dm^3	Microelement	Content, mmol/dm^3
Na^+	130–156	Ca^{2+}	2.30–2.75	Fe^{2+}	0.012–0.032
Br^-	17	Mg^{2+}	0.7–1.2	I^-	0.000275–0.00063
D^+	16	F^-	0.37	Mn^{2+}	0.00007–0.0004
K^+	3.4–3.5	Cu^{2+}	0.071–0.074	Co^{2+}	0.00002–0.0006

In plasma the content of deuterium exceeds 16 mmol/dm^3. Since hydrogen is not only part of the water but also macromolecules, which make up proteins, fats and hydrocarbons. Therefore, it may be assumed that these oscillations will not remained unheeded by the organism, so that by the number of atoms in a human body hydrogen convincingly holds the first place, though by the mass it is only the third (Fig. 2.13) [26].

Heavy water at present has been sufficiently well studied. Numerous investigations involving heavy and radioactive isotopes of oxygen have proved that the water, enriched with deuterium and tritium, is harmful for the all forms of life.

Biological effects of heavy water manifest themselves in: a reduction in the rate of biochemical reactions, tissue breathing, an increase of the viscosity of the protoplasm of the cells and the rate of organism ageing; other reactions include the induction of mutations, lesion of the gene pool, diseases like cancer, inhibition of cell division, the reduction of growth rate, and the death of high vertebrates.

The fact that 97 % of the water mass is represented by light and stable isotopes H and O pose an advantage to life.

"The prime" moment of tritium water has not yet come because there is very little of it on the Earth, only about 25–30 kg. However, its amount in the Earth's water continuously grows since it is formed during the bombardment of the nuclei of nitrogen and oxygen with cosmic rays. Unlike protium and deuterium, tritium is a radioactive element with a half-life period of 9 years. The properties of superheavy tritium water differ from those of light protium water more than from deuterium water.

Light water is the deuterium depleted water whose molecules contain heavy isotopes of hydrogen and oxygen in minimal quantities. In the first approximation, such water may be considered as a substance with the chemical formula $H_2{}^{16}O$ [27].

The protium water $H_2{}^{16}O$ in a pure form, i.e. without any infinitesimal impurities of the other isotope varieties, should be considered as classical. Stable hydrogen isotopes with stable oxygen isotopes form 9 isotope varieties of water molecules, namely: $H_2{}^{16}O$, $H_2{}^{17}O$, $H_2{}^{18}O$, $HD^{16}O$, $HD^{17}O$, consisting of light nuclides $HD^{18}O$, $D_2{}^{16}O$, $D_2{}^{17}O$, $D_2{}^{18}O$. In a quantitative respect, the bulk of the water of natural sources is represented by the $H_2{}^{16}O$ molecules, consisting of light nuclides D, ^{17}O, and ^{18}O. The amount of water molecules containing heavy nuclides D, ^{17}O, and ^{18}O depends on the concentration of these nuclides, which in natural water varies within the limits recorded in the main standards of the isotope composition of the hydrosphere. Though the content of protium water substantially exceeds the contents of all types taken together, there is no pure $H_2{}^{16}O$ to be found under these ambient conditions.

The water with a substantially increased fraction of $H_2{}^{16}O$, has a smaller molecular weight, possesses a lower density and is the lightest in the category of isotope light water. Such water is defined as especially pure.

Features of Light Water The main feature of light water is the isotopic purity of the substance, which is unlike ordinary water. Light water does not contain impurities like heavy (D_2O) and semiheavy water, whose molecular weight exceeds 18 amu. The water containing $H_2{}^{16}O$ displays properties which are different from the water with a natural content of isotopologues.

As a universal medium, where all biological reactions are directed, the water with an increased content of $H_2{}^{16}O$ speeds up these reactions, compared to the water from natural isotopic composition. Such an effect is known in fundamental science as "the kinetic isotopic effect of a solvent". During this effect, heavy isotopes inhibit or delay the reactions more than light isotopes. When heavy isotopes are eliminated from the water, it triggers the reactions; therefore, light water generally activates biological reactions of live systems. This property of light water is useful under conditions when the rate of metabolic reactions is low; it can be applied, for example, in ageing, metabolic diseases like diabetes, insulin resistance and the metabolic syndrome.

Density and temperatures of phase transitions of light water is lower than that of heavy water (see Table 2.1).

Table 2.1 Change of physical properties of the water upon isotope substitution [26]

Indicator	Water isotope composition		
	$H_2{}^{16}O$	$D_2{}^{18}O$	$H_2{}^{18}O$
Density at 20 °C, g/cm³	0.9970	1.1051	1.1106
Temperature of maximum density, °C	3.98	11.24	4.30
Melting temperature at 1 atm, °C	0	3.81	0.28
Boiling temperature at 1 atm, °C	100	101.42	100.14
Steam pressure at 100 °C, mm Hg	760.00	721.60	758.10
Viscosity at 20 °C, mPa s	1.002	1.247	1.056

Water vapor's equilibrium pressure from different isotope composition differs and attracts acute attention. The lighter a water molecule, the higher the pressure of the vapor will be, and this means that the vapor, which is in equilibrium with the water, is always enriched with light isotopes of oxygen and hydrogen.

The reaction of biosystems when they are subjected to water may vary depending on quantitative and qualitative changes of the H_2O isotope.

The positive biological activity of water with a reduced degree of the deuterium concentration is described in papers [28–35].

Biological effects of light water:

- Optimization of the biological reactions rate;
- Stimulation of cell division and growth of organisms;
- Radioprotective effect;
- Antimutagenic effects;
- Medical-prophylactic and anti carcinomatous effect.

Production of Light Water For the production of isotope light water, methods and units proposed aimed to lower the amount of only heavy isotope hydrogen, mainly deuterium, in water, preserving the initial isotope composition of oxygen in H_2O. The method of obtaining "melt-water" and "relict" water with a reduced content of heavy isotopes of deuterium and tritium includes the operations of water cooling and subsequent thawing of the frozen water. However, the degree to which the amount of deuterium can be lowered in this case is rather small. Rather heavy water depletion by deuterium isotopes is achieved by the method of electrolysis, but the capacity of such an apparatus is very low.

Though the reduction of deuterium moiety in the water results in an increase of the $H_2{}^{16}O$ fraction, water deuterium depletion only limits the possibility of further increasing $H_2{}^{16}O$. A substantial increase of the $H_2{}^{16}O$ in the water is possible only upon removal of the maximum number of molecules containing heavy nuclides D, ^{17}O, and ^{18}O. Therefore, it is preferable that the content of $H_2{}^{16}O$ in commercially produced water increase due to the removal of molecules containing not only deuterium but also heavy oxygen nuclides (^{17}O, ^{18}O).

The Role of Deuterium in Destruction of DNA Isotope effects of heavy water in high concentrations, not existent in field conditions, are successfully studied at present. The extrapolation of the obtained data to strong dilution of heavy water does not allow tangible effects in lowering the concentration of heavy isotopes below the natural level. Nonetheless, in the past decade it has been shown that natural water, depleted by heavy isotopes of hydrogen and oxygen, possesses a stimulating effect for different biological objects and even contains medicinal properties [32]. Thereby, it was confirmed that a live cell is capable of responding to infinitesimal changes of the deuterium content in water.

As early as in 1974, an assumption was expressed that deuterium was a possible reason for ageing. According to available data, ageing is related to the gradual build-up of DNA errors arising in connection with the break-off of the strand, with the errors occurred during DNA replication or with the dysfunction of the mechanisms of its reparation. Paper [20] considered factors which unfavorably affect DNA and also represent the research results in removing these factors from the cellular medium. It is believed that the most abundant mutagen, which causes damage to DNA, is low-level solar radiation; DNA is also negatively affected by heavy water (D_2O, deuterium oxide). The concentration of heavy water on the Earth's surface constitutes 155 molecules per million. Due to this low concentration, heavy water does not usually attract attention. However, the latest research has demonstrated that the usually ignored deuterium oxide may play a key role in the ageing process.

One of the main factors limiting human lifetime is a natural damage of DNA, both from the direct effect of radiation and from free radicals produced by radiation, mutagens and natural metabolic activities.

The Earth's atmosphere (ozone) shields the bulk of cosmic radiation; however, some amount of radiation still penetrates the atmosphere and reaches the Earth's surface. This radiation, together with Earth's natural radioactive sources, creates an ionizing radiation which negatively affects DNA. In a human body, nearly all DNA damage is eliminated, thanks to an inherent repair mechanism (self-restoration). Nonetheless, in some cases, such mechanisms are disturbed, which results in diseases and stimulates ageing.

Without a doubt, diseases like cancer and a total decrease of the DNA integrity are interrelated. This means that by the end of the life cycle a human "encounters a barrier" when DNA integrity is controversial and there is no method for restoration of all its strands, which would be conducive to prolonging longevity. Consequently, a huge advantage exists in preserving DNA from an early age so as to optimize both the longevity and quality of life.

In papers dealing with the biological impact of low-level radiation, it has been shown that the gradual change of DNA caused by radiation and other mutagens is impossible to be detected by contemporary methods. It was found that an intensive short-time radiation effect results in early aging, but the discussion still continues as to whether mutations are caused by normal radiation levels. Ionizing radiation results in the formation of free radicals, which attack DNA, but this also occurs during natural metabolic processes. Some scientists even assert that low-level radiation triggers an "immune response", which protects DNA from further damages. Natural levels of ionizing radiation produce a weak negative effect, but a very small number of incorrectly restored breaks of DNA strands, building up for decades, can cause ageing effects.

Deuterium, Mitosis and DNA Cell division lies at the core of the growth and sexless propagation of organisms. It is well known that the high level of deuterium delays the mitosis rate (direct cell division), but a precise mechanism of this process is still not clear. In 1989, Jan Lamprecht, Diter Schreter and Nidhart Pavelets conducted research on the impact of deuterium on mitosis at the The Institute of Cell and Tumor Biology and German Cancer Research Center in Heidelberg. In one of the experiments, the cells were subjected to the treatment of heavy water (deuterium oxide) with concentrations of 25, 50 and 75 % for 2 h, while in the other—by the impact of a 75 % deuterium oxide for 2, 6, 12 and 24 h. After that, the number of cells, which were in four phases of mitosis (prophase, metaphase, anaphase and telophase) and interphase (the period of cell preparation for division), was measured.

On average, the mitosis lasts 1–2 h and includes the separation of preliminary doubled material, nucleus division, division of the cell itself, and cytokinesis. The obtained data demonstrated that the unusually large number of cells were in the prophase and metaphase, but especially in the metaphase. If the rate of DNA replication is delayed just during the prophase (the beginning of the formation of the division spindle – the system of protein strands) ionization radiation could break the strands when they are sensitive to damage. However, to verify this influence, it is necessary to carry out in-depth experiments without the presence of deuterium in DNA molecules and enzymes to determine whether the replication of DNA speeds up in this case compared with the DNA containing usual concentrations of deuterium [27].

Hydrogen and Deuterium Bonds in DNA As has already been noted, the DNA molecule is a right-hand twisted double helix consisting of the sugar-phosphate framework and nitrogen compounds (see the insert, Fig. 2.10). It records all information about a specific live organism in the form of alternating cyclic fragments containing carbonyl and amine groups. There are four types of such fragments: adenine, thymine, guanine and cytosine (A, T, G and C). They are arranged in the form of lateral pendants along the whole polymer model of DNA. The order of alternation of these fragments determines the individuality of every live creature. In a paired interaction (A–T and G–C), the hydrogen bond appears between them. It is this bond that holds two DNA molecules in the shape of the widely known double helix [21]. It is believed that deuterium affects biological processes by disturbing the formation of hydrogen bond since the bonds formed by deuterium atoms are stronger that those formed by the usual atom of hydrogen [6, 36].

Hydrogen Bond and DNA In the DNA molecule the hydrogen bond G–C and A–C, which is formed between the strands of the double helix, presents interest. An accurate value of the force of the deuterium bond is difficult to determine. The strength of individual hydrogen bonds of DNA was assessed by D. G. Turner and N. Sagimoto; however, the accuracy of their model is under question. T. R. Griffith believes that deuterium bonds in enzymes affecting DNA is normally by 0.4–1.7 kJ/mol

Table 2.2 Strength of the usual hydrogen bond and formed by a deuterium atom in hydrogen compounds [21]

Organic compound	Usual H-bond, kJ/mol	H-bond at D-substitution, kJ/mol		
		Minimal	Critical	Maximum
Aqueous solution of formaldehyde	6.69	0.25	1.17	2.09
Dimer of formic acid	45.31	0.21	0.42	1.21
Aqueous solution of formic acid	24.95	0.13	1.21	2.38
Formamide dimmer	34.78	0.17	0.67	2.30
Aqueous solution of formamide	18.89	0.13	1.46	2.84

stronger than the usual hydrogen bond. This can be assessed by different methods requiring considerable calculation.

Table 2.2 and 2.3 give the data reflecting the change in the strength of the hydrogen bond in the presence of deuterium atoms in hydrogen bonds.

As can be seen from the data of Table 2.2, the replacement of some hydrogen atoms by deuterium atoms sharply affects the strength of the hydrogen bond, and in other cases the effect, which is more substantial and maximum, can be observed when all hydrogen atoms are replaced with deuterium atoms.

The Length of the Hydrogen Bond In replication and reparation of DNA, the form of the enzyme molecules which control these processes is extremely important. Deuterium somewhat reduces the length of the bond and inhibits the appropriate functioning of the enzymes. However, this effect is most likely quite insufficient, possibly by the order of ~1 %. Generally speaking, the formation of the hydrogen bond in the DNA molecule is a cooperation process which affects stacking-interaction and involves the whole molecule. If enzymes are used as conductors, the strength of the deuterium bond in the DNA molecule is 0.5–2.0 % larger than in an ordinary hydrogen bond. When a small concentration of deuterium exists in nature, and there is a small increase of the force of deuterium bond compared with the usual hydrogen bond, it is tempting to conclude that deuterium does not produce any substantial negative impact on DNA at the concentration of 155 ppm. However, the strengthening of the deuterium hydrogen also produces other potentially harmful effects.

Probable Mechanisms of the Harmful Influence of DNA. Deuterized Enzymes According to a well known theory, deuterium negatively affects the shape of the enzyme molecules involved in DNA-processes. This main concept was put forward by T.G. Griffith in the paper "The possible role of deuterium in stimulation of ageing and other biological mechanisms and processes".

When deuterium is included in a chemical reaction, it is necessary to consider small changes in the induction effect because deuterium is more electrically permeable than hydrogen. The hyperconjugative effect has become apparent since CD_3, for instance, is less delocalized than CH_3, and, what is more important, the effective length of the C–H bond becomes shortened, which affects steric effects. This confirms the opinion that for any stereospecific molecule of the enzyme containing

Table 2.3 Strengthening of D-bond compared with a usual H-bond according to the data of Table 2.3%

Organic compound	H-bond at D-substitution, %		
	Minimal	Critical	Maximum
Aqueous solution of formaldehyde	3.75	17.50	31.25
Formic acid dimmer	0.46	0.92	2.76
Aqueous solution of formic acid	0.50	4.85	9.54
Formamide dimmer	0.48	1.92	6.61
Aqueous solution of formamide	0.66	7.75	15.04

dethrone (deuterium nucleus) on the important section, potential will be involved in erroneous reactions.

Interference of Deuterium into DNA Reparation with Enzymes A large group of enzymes and proteins are involved in the replication and reparation of DNA. Some such enzymes actively use external hydrogen bonds. They are potentially very sensitive to the negative effects of the available deuterium. One very important protein, p53, is responsible for the reparation of DNA. Several different types of damages to DNA may activate p53. For instance, two-strand breakages of DNA after gamma radiation, the emergence of double-strand breaks of DNA caused by gamma radiation, appearance of DNA reparation intermediates after UV-radiation or chemical damage to DNA are all means of p53 activation. It should be noted that more than 50% of human diseases are accompanied by gene mutation forming protein p53 [21].

Delay of DNA Replication Deuterium may also inhibit enzyme DnaB, which is responsible for unfolding and separating DNA during replication. Other enzymes, such primaze and polymeraze, play important roles in synthesizing RNA and the joining of nucleotides to the DNA strand in the course of replication. If the action of one of these enzymes is suppressed, the rate of DNA replication may also be noticeably suppressed, and this is when DNA is the most vulnerable to radiation damage. Thus, deuterium may serve as a catalyst for DNA degradation in a natural level of radiation exposure.

The Inhibition of the Bonding Sections in Replication of DNA The process of DNA replication in some measure is like reparation of a double-strand break of DNA. The complex of enzymes makes it possible to identify and restore the DNA strands located on certain bonding sections. Hydrogen bonds often act on these sections. If deuterium is on them, the synthetic effects and the enhanced strength of the bonds may inhibit DNA reparation. In its turn, if the DNA reparation rate is substantially decreased, ionizing radiation may continue to prevent the reparation process. This is when the strand is destroyed and sensitivity to additional radiation damage is increased.

The Efficiency of Using the Deuterium Depleted Water According to the data given in paper [21], the use of light water differs from various antioxidants, HGH stimulants, vitamins and other means of anti-ageing in one key aspect: light water does not change chemical composition in assimilation. The light water consumed poses direct affects at the cellular level. However, in order to make these effects noticeable, it is necessary to remove deuterium content. By means of exothermal reactions substituting the deuterium, atoms are gradually replaced by normal protium atoms. This process occurs faster when the concentration of deuterium in drinking water is at the lowest possible level. The deuterium constantly enters the organism, which consumed the food that is grown and cooked using light water. Light water may protect DNA from damage and help mechanisms of DNA reparation; however, on its own, light water will not directly restore DNA by itself. Therefore, the question remains as to whether light water can "rejuvenate" organisms and help them function more effectively [21].

References

1. Bachmann K (2008) Unidentified world: the Earth. GEO 1:76–94
2. Mosin OV (2007) Strange water http://www.o8ode.ru/article/water/udivit/waterubnormal.htm. Accessed 10 Mar 2007
3. Chaplin M (2012) Anomalous properties of water. http://www.lsbu.ac.uk/water/anmlies.html. Accessed 28 Nov 2012
4. Antonchenko VYa (1986) Fizika vody (Physics of water). Nauk Dumka, Kiev
5. Kusalik PG, Svishchev IM (1994) The spatial structure in liquid water. Science 265:1219–1221
6. Smirnov ÐN, Syroyeshkin AV (2004) Supramolecular water complexes. Russ Chem J 48(2):125–135
7. Erdey-Grúz T (1974) Transport phenomena in aqueous solutions. Wiley, New York
8. Wilson KR, Cavalleri M, Rude BS et al (2000) Characterization of hydrogen bond acceptor molecules at the water surface using near-edge X-ray absorption fine-structure spectroscopy and density functional theory. J Phys: Condens Matter 14(8):L221–L226
9. Kochnev IN, Vinnichenko MB, Smirnova LV (1989) Temperature anomalies of the absorption spectra and refractive index of water. Res Water Aqueous Syst Phys Methods 6:42–51
10. Vuks MF (1989) Dielectric polarization alcohol in solutions. Res Water Aqueous Syst Phys Methods 73:172–180
11. Voitylov VV, Kakorin SA, Trusov AA (1989) Electroorientation researches electrical conductivity of water-solid. Res Water Aqueous Syst Phys Methods 6:75–83
12. Zhukovsky AP (1981) Motivation of the continuum model of water structure by IR spectroscopy. J Struct Chem 22:56–63
13. Smirnov AN (2001) Acoustic emission during the chemical reactions and physico-chemical processes. Russ Chem J 45:29–34
14. Goncharuk VV, Kazimirov VP (1999) Structure of water. J Water Chem Technol 21(5):5–16
15. Nevzorov AN (2006) Some properties of metastable state of water. Phys Wave Phenom 14(1):45–57
16. Sillinger FH (1980) Water revisited. Science 209(4455):451–457
17. Scientific library (2007) http://sciencerussian.sibenzyme.com/. Accessed 12 Apr 2007
18. Angell CA, Franks F (ed) (1981) Water: a comprehensive treatise, in water and aqueous solutions at subzero temperatures. Plenum Press, New York
19. Grigorovich S (2004) In the beginning was RNA? Sci Life 2:44–55
20. Watson JD, Crick FHC (1953) Genetical implication of the structure of deoxyribose nucleic acid. Nature 17:964–966

21. Goodall KB (2003) In search of the fountain of youth. Preliminary analysis of deuterium's role in DNA degradation. Anti-aging medical news: the official newsletter of the American Academy of Anti-Aging Medicine 7–31
22. Dobretsov NL (2005) Early stages of life origin and evolution. Herald Vavilov Soc Genet Breed Sci 9(1):43–54
23. Kolyasnikov YuA (1993) Polytetrameric model of water structure. Herald Russ Acad Sci 63(8):730–732
24. Le Bihan D, Fukyama H (2011) Water. The forgotten biological molecule. Maiinland Pres Ltd, Singapore
25. Ferronsky VI, Polyakov VA (1983) Izotopiya gidrosfery (Isotopes of the Earth's hydrosphere). Nauka, Moscow
26. Timakov AA (2003) Main effects of light water. In: Abstract of the 8th Russian international scientific conference of Physical-chemical processes during selection of atoms and molecules, RSC Kurchatov Institute, Moscow, 6–10 Nov 2003
27. Zatsepina GN (1998) Fizicheskiye svoystva i struktura vody (Physical properties and structure of water). Izd-vo MGU, Moscow
28. Somlyai G (2001) Let's defeat cancer. Akad Kiado, Budapest
29. Lobyshev VN, Kalinichenko LP (1978) Izotopnyye effekty D_2O v biologicheskikh sistemakh (Isotope effects of D_2O in biological Systems). Nauka, Moscow
30. Sinyak Y, Levinskikh MA, Gaydadimov VV et al (2000) Effect of water with a low concentration of deuterium on the cultivation of higher plants: *Arabidopsis thaliana* and *Brassica rapa*. In: Abstract of the first international forum "Organism and Environment: Life Support and Human Protection in Extreme Conditions", Russian Academy of Sciences, Moscow, 26–29 Sept 2000
31. Sinyak Y, Grigoriev AI, Gaydadimov VV et al (1998) The method for obtaining deuterium-free water and its influence on the physiological status of Japanese quail. In: Abstract of the XI international conference Space Biology and Aerospace Medicine", Russian Academy of Sciences, Moscow, 22–26 June 1998
32. Varnavsky IN (2000) New technology and equipment for production purified biologically active healing drinking water. Dissertation, Moscow Institute of Medical and Biological Problems
33. Sinyak Y, Turusov V, Grigoriev A et al (2002) Possibility of deuterium free water using as antitumoral means with reference to conditions of Martian expedition. In: Abstract of the 34th COSPAR Scientific Assembly. The Second World Space Congress, Houston, USA, 10–19 Oct 2002
34. Somlyai G (2002) Defeating cancer!: the biological effect of deuterium depletion. Akad KÑado, Budapest
35. Somlyai G (1998) The biological effect of deuterium-depleted water, a possible new tool in cancer therapy. J Oncol 30(4): 91–94
36. Smirnov AN, Lapshin VB, Balyshev AB et al (2005) Water structure: giant heterophase clusters of water. J Water Chem Technol 27(2):111–137

Chapter 3
Water Clusters

Abstract The analytical review of existing structure theories, water properties and its clustering has been covered. The latest theoretic researches, related to the build-up of hydrogen bonding and structure of water clusters, have been revealed. A new concept of understanding of water icosahedron has been covered. The interrelation of physical effect on water and clustering has been revealed. The research results of cluster structure of heavy, light and ubiquitous water have been covered. The above-mentioned results are the evidence that water properties basically are dominated by the structural water condition, which depends on deuterium concentration in water.

Keywords Water structure · Water properties · Clustering · Heavy water · Light water · Hydrogen bonding · Water icosahedron

3.1 Water Structures, Properties and Cluster Formation

Recently much attention has been drawn to cluster formation in water. This formation is determined by the involvement of water molecules in the formation of intermolecular hydrogen bonds [1–5]. Advocates of the classical interpretation of the structure of liquids believe that the life cycle of the water clusters is very short (in an order of 10^{-12}); as a result, a cluster of shimmering formations with a short lifespan cannot conspicuously affect water properties. Supporters of the idea of water structuring imply that clusters in the water may be preserved for a long time; therefore, one cannot ignore the presence of associates in a number of cases.

We believe water properties (physical, chemical and physicochemical) are completely determined by the structural state of the water, which depends on its cluster composition (see Fig. 3.1) and on such factors as cluster size and the concentration of the clusters.

The structure of liquid water has been an object of extensive research [6–12]. The study of water structure in the twentieth century gave rise to numerous concepts, most importantly the acknowledgement of the presence of ordered domains. These theories (Samoilov's structural defects, structures of Bernard–Fowler, Poling's "hydrates", shimmering clusters of Frank and Wien and many others) were intensively discussed in a number of studies [6–8] (see also Chap. 2). Paper [13] summarized all these discussions and note: (1) the remote order was neither authentically nor experimentally detected in water; (2) the life-time of the clusters

V. V. Goncharuk, *Drinking Water,* DOI 10.1007/978-3-319-04334-0_3,

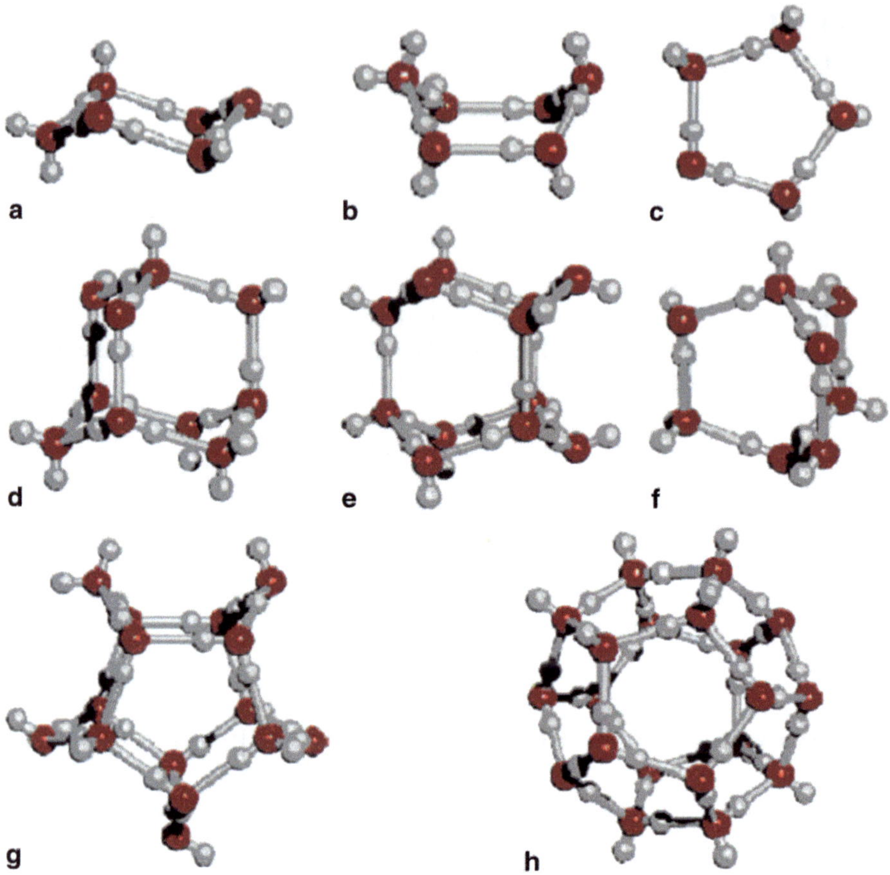

Fig. 3.1 Possible of water clusters

was estimated not to be in excess of 10^{-10}–10^{-9} s; 3) there are reasons to believe the presence of cooperative processes in water are related to the transfer of orientation L- and D-defects of Bierumme [9].

The methods of X–ray scattering or neutrons scattering [14, 15] did not confirm the formation of ordered structures with the dimensions exceeding units of nanometers. This, in turn, does not agree with the data on hysteresis temperature anomalies of the indicator of water refraction [16] or anomalous water scattering in aqueous solutions of organic compounds [17]. Usually, the theoretical calculations of water clusters are limited to tens of molecules or near the phase boundary [6, 16, 18]. M. Chaplin, for example, has calculated and proposed a model of water whose basis is the icosahedron (see Fig. 3.2).

Since a very pronounced ability of self-organization due to the formation of hydrogen bonds is characteristic of water, such properties of water as, for example, fluidity and the capacity of forming gas hydrates implies the formation of expanded

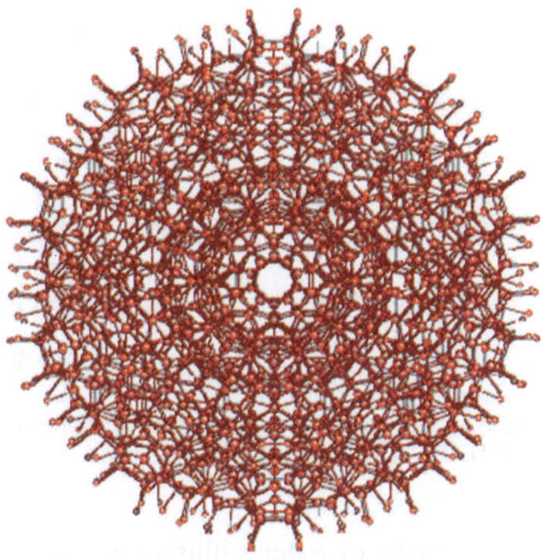

Fig. 3.2 Giant icosahedron water

ordered structures such as the continuum water model [19]. The data on acoustic emission from aqueous solutions in the region from 0.5 Hz to 2.5 kHz imply the presence of a source of generating sound waves with linear dimension larger than dimensions of molecular clusters [20].

At present it has been acknowledged that, owing to the existence of intermolecular hydrogen bonds, water properties differ sharply from properties of other hydrides. For instance, they have an associative structure [7], whose presence on the global scale determines the Earth's climate, as well as the origin and support of life. Anomalous properties of water, just like anomalous properties of oxygen molecules, are the main factors which determine the existence of the Earth's biosphere. Unique, unusual properties of water in the liquid, solid and gaseous states (see Chap. 2) are caused by the presence of strong bonds between hydrogen molecules: two H-bonds in which the H_2O molecule is a proton-donor and two H-bonds with the participation of lone pair of electrons of the oxygen atom. Such a basic structure as that of water organization [21, 22] has been adopted as a basis for consideration of structural models and water properties. The H-bonds determine the formation of rather stable supramolecular structures–clusters, nano- and microdomains, whose values depend on the presence and parameters of interfaces with a solid body, gas or other liquid.

Many anomalous properties of water until today have not been completely explained. One of the reasons for this, from our perspective, is that the phenomenon of the cluster formation of water has been largely ignored by researchers.

One anomalous property of water is that it exhibits the great ability of self-organization due to the formation of hydrogen bonds and cooperative effects [23, 24]. Hence, the primary structures may be dimers, trimers and oligomers. Cyclic structures with the five and six water molecules form various polyhedrons. The most important of these polyhedrons have a symmetry axis of the fifth order (dodecahedron and icosahedron). Such polyhedrons are significant because they can help to

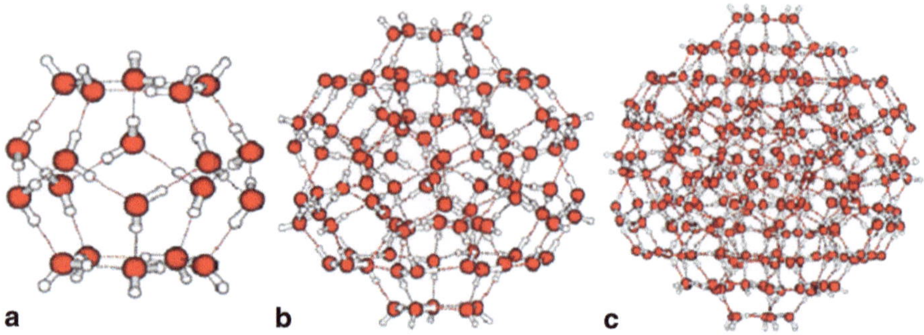

Fig. 3.3 Schematic representation of 5^{12} dodecahedral cluster $(H_2O)_{20}$ (**a**), homological icosahedral $(H_2O)_{100}$ (**b**) and $(H_2O)_{280}$ (**c**) clusters

explain many of water's illusive properties; they provide information, for instance, over water's volatility and its capacity to form gas hydrates.

From such structural elements, long strands and special structures which fill the volume may be formed. The motion of small molecules like water in the liquid state at room temperature, according to contemporary notions, has the nature of rotary fluctuations near the position of equilibrium, which may be conducive to cluster formation. Only at an increase of water temperature to 70–80 °C, according to the Einstein ratio ($\tau = \tau_0\ e^{U/kT}$, where τ is the correlation time of the rotary motion) does the movement of molecules resemble the movement in the gas phase substantially reducing the stability of clusters. Since the reduction of temperature acts on the structure of water as an ordering factor, the formation of clusters is possible from the molecules of dimers, trimers and associates with a large number of molecules. Depending on the research method one can observe from 2–15 up to 50–60 and more water molecules [25–27]. In our opinion, such results are evidence of the fact that the range of the existing sizes is rather wide. Recently, the idea of cluster formation has received the ever broader recognition and experimental confirmation through different methods [28–31].

Due to the rapid development of powerful ab initio technologies, the answers were obtained for most of fundamental questions dealing with the nature of the formation of hydrogen bonds and the structure of water clusters [32, 33]. Therefore, with the use of the ab initio method MP2 geometrical and electron structures of water complexes with ozone were calculated [33]. Also harmonic vibrational frequencies and rotational constants H_2O, HDO and D_2O were determined [32]. The study of various structures of gas hydrates provided us with the possibility of seeing the icosahedral cluster. The research of the structure of water clusters has resulted in the identification of up to 280 water molecules (see Fig. 3.3). The structural and orientation features of various "guest" molecules in the central volume of the cluster icosahedral have also been established. It has been shown that the clusters of water monomers have the highest stabilization energy.

In paper [34], the molecular dynamics method allowed the researchers to calculate thermodynamic properties of pure water clusters as well as water aggregates with inclusion of the CO or CO_2 molecules. For the investigation of the structure of H-bond clusters in methanol and water near the saturation curve and the critical point, the molecular dynamics method was used [35]. The Monte-Carlo methods of mathematical modeling for hydrate clusters with one ion were used when studying the influence of the low-intensive electromagnetic radiation on water clusters with the presence of ions of sodium and potassium [36].

Paper [37] presents the results of experimental and theoretical research of the interrelation between the structure and electrical properties of the clusters.

Clusters containing six, seven, eight and nine water molecules were observed at the Nanotechnology Center in London and Leibniz University in Hanover during an investigation of ice properties at the nanolevel [38]. Apart from the experiments, the paper used theoretical approaches of quantum mechanics. The complex approach showed that, unlike crystalline ice, where the energy of binding is identical between all water molecules, in nanoclusters strong and weak bonds (at the corresponding distances) alternate between individual molecules.

Water clusters attracted a lot of interest because they are formed precisely within a natural temperature interval and are capable of leading to "soft", local variations on water properties, either boosting or slowing down biological and other natural processes. When considering water as a catalyst for biochemical processes, special attention should be paid to the research results dealing with the cluster formation of water. We believe that the study of the mechanisms of cluster catalysis based on the fundamental approaches to the process of catalysis [39] will open hitherto unknown water processes.

3.1.1 *Conceptualizing Boundary Layers of Clusters*

The concept, currently being developed by us, concerns the boundary layers of water clusters; the phase boundary in water (irrespective of the origin of the cluster model structure) is necessary for explaining the link between the water structure and its properties. Such properties include, for example, the ability for overcooling and the redistribution of impurities like deuterium, dissolved gasses and SAS. Properties of the solution and disperse systems are also significant. The use of the water model as a cluster liquid, in which molecules with a variable structure of H-bonds act as a boundary layer, makes it possible to quantitatively describe many phenomena, such as that of overcooling.

The special properties of the boundary layer enable the accumulation of impurities, which compensate for simultaneously occurring polarization charges. The concentration of surface layer substances which are less polarized than H_2O will result in the shielding of cluster charges and, consequently, in the weakening of hydrogen bonds located outside clusters. Consequently, surface layer concentration will result in stabilization of the cluster structure.

In connection with various successfully conducted experiments on the study of gigantic heterophase water clusters (GHWC) [13, 28], the determination of clusters as water fragments with the varied properties required the perfection of the terminology. This made it possible to differentiate the clusters according to the number of water molecules they contained. The GHWC, or "microvolumes" of water, involve HDO molecules in their formation. The existence of GHWC enables the differentiation of ordered associates of smaller sizes, such as microclusters (containing 2–20 molecules) or small clusters (20 to 80–100 molecules). Gigantic heterophase clusters contain 0^{10}–10^{13} of water molecules [28].

The moving force of cluster formation, according to Frenckel [40], is the orientation and polarization of liquid dipole molecules. For the water, which is in the bulk of the sample, the primary ordered forces are structurally similar to the sections of the walls of a glass container. In addition, according to the notions about GHWC that we have developed, the cluster formation is promoted by dissolved deuterium. According to one of the mechanisms proposed, the role of deuterium involves the formation of the volumetric network made of chains containing HDO.

The capacity of oxides to form acidic and basic surface groups with the participation of Lewis and Brensted centers [41] depends on the ordered effect of the structurally related surfaces for instance, silicate. The affinity of the centers of the oxide surface to the electron pairs or to the protons results in the adsorption of water molecules in a certain configuration. Usually, this configuration of surface charges corresponds to the ice structure for silicone oxides. Upon water adsorption, the polylayers near the surface of silicate act as a nucleation factor and form small water clusters. In this case the HDO impurity in water may play a decisive role in the formation and stabilization of GHWC. The energy of ionic hydration for the electrolyte exceeds the energy of water's hydrogen bond by several times (for example, for the ions Na^+—by 5.2, Cl^-—by 4.2 times [42]). In this connection, when adding the first portions of the electrolyte to the water, due to the ingress of a HDO to hydrate shells of ions, it is possible that new clusters are formed, and some changes in the GHWC structure occur.

The water varies in the structure, quantity and composition of microclusters, and small clusters under the polarizing effect of the ions of the electrolyte dissolved in the water, which is in contact with the silicate surface. The authors of paper [43] assumed the possibility of destroying cooperative hydrogen bonds introduced by the electrolyte. In our estimation, the existence of cooperative H-bonds which are capable of encompassing the entire volume of the water sample, in combination with the polarizing effect of the silicate surface, directly points to the possibility of the formation of large size clusters, such as GHWC.

Paper [25, 44–51] described the methods of obtaining the water enriched clusters. Thus, the authors of paper [44] used the cavitation energy of air bubbles for obtaining water containing microclusters. The process was repeated at an increase of water temperature to 60 °C. We believe that the break of the cavitation bubbles is accompanied by the destruction of molecular structures of the liquid to smaller structures, which results in the formation of microclusters. In reference [45], the water clusters were produced by the magnetohydrodynamic treatment. Direct observation of clusters is difficult; therefore, when studying water properties (NMR, IR and UV spectra of laser beam interference, Raman, viscosity, electroconductivity, etc.), their anomalies are referred to the manifestations of clusters.

In paper [25] the initial water was boiled and converted into steam before it was passed through a magnetic field. It was then condensed at a temperature a bit higher than 0 °C, radiating in the region from IR to UV spectrum (according to the conditions of paper [46] the radiation should have a wavelength within the range 610 nm–1 mm). In this case, sodium metasilicate < 1 % was added to the condensate as a template of crystallization. This is the way that microcrystalls of water were obtained.

The presence of clusters in the water was linked with a width of the NMR signal, viscosity and electrical conductance. Their dimension was assessed by the width of NMR line to the oxygen nuclide ^{17}O. For this the water being investigated was poured into a quartz laboratory dish of 2 cm^3 in volume, placed in the resonator of the NMR spectrometer. Water microclusters send a NMR signal < 115 Hz wide. As the standard for the water enriched with the clusters, the authors of paper [25] considered samples with NMR signal 60–70 Hz wide. Electric conduction of the water clusters constituted ~3.7 μS/cm, with a surface tension of less than 61 mH/m. The concentration of microclusters containing 3 to 15 water molecules was assessed by amplitude and broadband of the NMR signal, using the computer software. A substantial error by this method may be introduced magnetohydrodynamic effects. As shown in paper [47], the interaction of the magnetic field of high tension with the wall water layer, in a quartz laboratory dish (especially when using a usual technique of a rotary laboratory dish), results in the emergence of emf, electric current, etc. In this connection, the integrity of the obtained information on clusters in the water under conditions of the influence of magnetohydrodynamic effects causes doubt.

In paper [48] highly-clustered water, possessing an increased affinity to oils was obtained by treating pure water with the electric conduction 0.09 μS/cm. The following methods were used: the magnetic field, a weak electric current, the electric field, ceramics, irradiation, high-frequency radiation, mechanical machining, emf, insonification, a laser beam and natural minerals.

In reference [49], the water was dispersed in the liquid fuel to obtain water clusters. The clusters were then estimated using spectra, which is characteristic of the vibration and oscillation of molecules within the interval of 1.250–1.5 cm. The spectrum of combination scattering was recorded, and was then subjected to mathematical processing. Finally, water clusters with dimensions < 2 nm including 5–300 water molecules were observed in the liquid fuel. The researchers believe that the clusters are partially represented by formations with pentagonal symmetry and, to a lesser degree, with the pentagonal dodecahedral.

The method of the synthesis for cluster water in its contact with the silicate surface is already well known. For example, in paper [50], microclusters consisting of 5–8 water molecules were found. The remote IR spectroscopy method (the spectrum range 50–200 μ) was used to detect the clusters. The IR spectra was recorded for the initial water sample passed through a ceramic membrane (this water had the following composition, (%): SiO_2—50–70; Al_2O_3—10–30; Fe_2O_3—10–20; MnO—0.1–0.3; ZnO—0.01–0.05; CoO—1–1.2). The membrane was prepared with a mixture of powders from these materials, with the diameter of particles 1 to 5 μ. The ratio of the ceramics to the water constituted 20%, and the contact time was at least 12 h. A decrease in the signal intensity of the IR spectrum was used for calculating the amount of clusters. According to the data from paper [51], cluster water with a pH from 5 to 7.5 can have practical applications.

Supporters of clusters [48] believe that after ice thaws the remnants of the ice-like structure preserving close-in orderliness (the tetrahedral ice-like frame) continues to exist in the water within a certain temperature interval.

Some scientists [52] object that water formed from a reticular crystal (ice) is a reticular melt. First of all, it contains clusters with various structures; therefore, ice thawing cannot be identified only by the destruction of a three-dimensional network due to a uniform break of hydrogen bonds. In addition, hydrogen bonds between neighboring water molecules are apt to change their angles, which complicates perceptions about the formation of similar defects in ice thawing.

Having studied the anomalies of precrystallization near the melting point, the researchers came to the conclusion that clusters appear in the course of "conglomerate melting of crystals with hydrogen bonds". At an increase in water temperature of fraction bound into clusters gradually decreases.

The relation between melting and cluster formation in paper [52] was disclosed, although it was perhaps not sufficiently complete. This could be partly because it did not take into account the interaction of the processes accompanying both phenomena. In particular, single molecules without hydrogen bonds may take part not only in intercluster but also in intracluster processes. Accounts of the possible arrangements of single water molecules in the hollows of a tetrahedral ice-like framework have contributed to the development of ideas regarding O.Ya. Samoilov and the Hall's double-structure model. According to the ideas of O.Ya. Samoilov, the anomalies of water properties are linked not with the association of molecules but with the structural features (greater open work pattern of the structure and strongly pronounced close orderliness [8]). GHWC may contain in their structure the fragments of both associated water molecules and the fragments possessing open work patter or strongly pronounced close orderliness.

The authors of paper [53] present a molecular dynamics simulation study. There they unravel the molecular origin of anisotropy in the growth kinetics of hexagonal ice by visualizing the formation of transient water structures in the growing ice interface. During ice growth, the formation of transient structures and their rearrangement to the final ice configuration was observed irrespective of growth direction. However, authors consider that in the direction perpendicular to the basal face of hexagonal ice along which growth occurs most slowly, a two-dimensional transient structure (formed by competing hexagonal and cubic arrangements within the same layer) persists for a significant period of time, in contrast to short-lived transient structures in other directions. This observation of such transient water structures and their rearrangement during ice growth provides a clear explanation of different growth rates on each face of hexagonal ice on a molecular scale.

Factors such as the amphoteric behavior of SiO_2, the surface level presence of both acid and base centers [41] and the proximity between the parameters of molecular water structures and silicone oxide allow water molecules to complete the construction of the SiO_2 structure upon mutual contact; as a result, water monolayers are formed on glass with changed viscosity and density. The build-up of water layers is conducive to the polarization of liquid particles under the impact of electrostatic forces [54]. Water's increased contact time with the silicate surface results in a more secure cluster formation. This is due to the relaxation period of redistribution of water H-bonds in a monolayer. This monolayer enrolls the nearest water layers and is gradually transferred to the rest of the bulk. The impact of the hydrophilic wall on water viscosity may be extended to the 10 nm distances [55].

In our opinion, fragments of the water volume are polarized in the wall regions and constitute clusters 10 nm in size. With an increase of the size and weight to a critical value, such clusters overcome the surface forces holding it, breaking off from the wall and transfering to the bulk of the water, where it exists for some time. According to the estimates of the NMR method on the silicate surface, the thickness of the water layer in the changed structure reaches 30 nm, and a strong impact on the degree of the hydrophilic surface was noted [56]. In the first two layers, the hydrophilic surface water molecules may be dissociated to a greater degree than in the rest of its volume [57]. Actually, the results obtained by the NMR method also confirm the formation of water clusters near the silicate surface.

In water layers larger than 40–50 nm forces the electrostatic repulsion forces are active [58], which under a polarizing effect facilitates formation of cluster structures of up to 40–50 nm in size. The order of such structures depends on the time of the polarizing effect on the electrostatic field and is subject to the regularities of migration polarization [59].

Water's ability to form clusters in regard to its contact with the metallic surface [27, 37] was noted (with the formation of hexamers of the type $(H_2O)_6$ [60]). This corroborates with the properties of water molecules to be in a constant state of linkage to hydrogen bonds.

Nonequivalence of the processes of ice formation and thawing [7, 52] was accepted to explain the accumulation of defects in the solid phase as it approaches the melting temperature [40, 52]. A promising idea expressed in paper [61] viewing clusters as an individual thermodynamic phase which influences the water–ice interface. In fact, it is water's abilities of supercooling and crystallization, rather than its ability to form amorphous ice, confirms its cluster structure.

Processes of polarization affect the properties of water enriched with clusters. Its viscosity (i.e., water viscosity of the local surrounding of water molecules of the EPR probe) varies in continuous polarizing effects [59]. Furthermore, an increase of water microviscosity is categorized according to two polarization processes. Firstly, the structure of small-size clusters is organized according to the mechanism of relaxation orientation polarization. Secondly, the migrational polarization of small clusters can cause them to be restructured into larger clusters, while enlargement of structure increasing also the dipole moment and ability to polarize. Paper [62], for instance, noted that larger clusters are less subjected to thermal disorder and may be oriented around weak electric fields similar to colloid particles.

The thermal disorder of water clusters was also assessed using the Debye formula ($\tau = 4\pi a^3\eta$ kT, where a—is the size of a cluster) [59]. It should be noted that gases adsorbed by the surface may be conducive to cluster stabilization. Moreover, it is well-known [41] that oxygen adsorbed by small particles is bound stronger; hence, it is less active than oxygen adsorbed by large particles. When oxygen is bound to water clusters, it is possible that the size of the cluster is affected; however, this question requires further experimental research. In the electrochemical cell, water clusters are $(1.8\ldots1.7) \times 10^4$ nm in size for the cathode and 1.0×10^4 nm for anodic water; thus, they are slowly destroyed by weak electric fields.

A cluster (as a fragment of the water volume with the changed structure) should be considered as a new phase in contact with the water phase of the surrounding bulk, and, consequently, this region of the solution is characterized by regularities which are intrinsic to the interface. It is known that a surface layer emerges at the interface, delineating a denser phase. No matter what method of cluster formation is considered, the surface layer, in a structural sense, is a transition region between the phases which possesses properties that are different from the intracluster and surrounding water. Such a surface layer has a charge and a potential which is actually the double electric layer.

At present, the information about the properties of the surface layer is absent; therefore, the question calls for more study since by analogy with foams or emulsion the surface layer of the clusters may substantially affect stability of foams and emulsion. Just as for colloids, the surface layers play a role of a structural stabilizer, prolonging the time of cluster existence and influencing its deformability, fluidity, viscosity and the aggregation of clusters of small sizes into gigantic heterophase, among other property changes.

The special properties of the boundary layer may cause it to contain concentrated impurities, compensating for polarizing charges emerging from the formation of clusters. The concentration of substances which are less polarized than H_2O in the surface layer results in the shielding of cluster charges, the weakening of hydrogen bonds and, consequently, the stabilization of the cluster structure.

One of the most important properties of the GHWC boundary layer is the change of the coefficients of passing and refraction of laser light, which makes the clusters visible. Their size and quantity are similar to other disperse impurities which can be determined, as described in papers [63, 64].

Paper [65] describes the technique of investigating GHWC, based on the change of the coefficients of passing and the refraction of laser light. A projection microscope was fitted with a semiconductor laser (radiation wavelength—633 nm, power 3.5 mW) from a stabilized power source of 4.5 V. The system of microscope lenses secured a 20–40-fold magnification. The microscope was equipped with a quartz cell as thick as the liquid layer in the working chamber 0.3–20 mm. Additionally, a video camera, type Video Cam Express, was connected to a PC. As a result of optical refraction, the laser light (see Fig 3.4a and b) is read from a matt screen of the microscope by the video camera and is transferred to the PC. Then Adobe Photoshop computer software is used to produce a difference image (see Fig. 3.4c), and, following this, the amount and the size of GHWC dimensions are calculated in the sample.

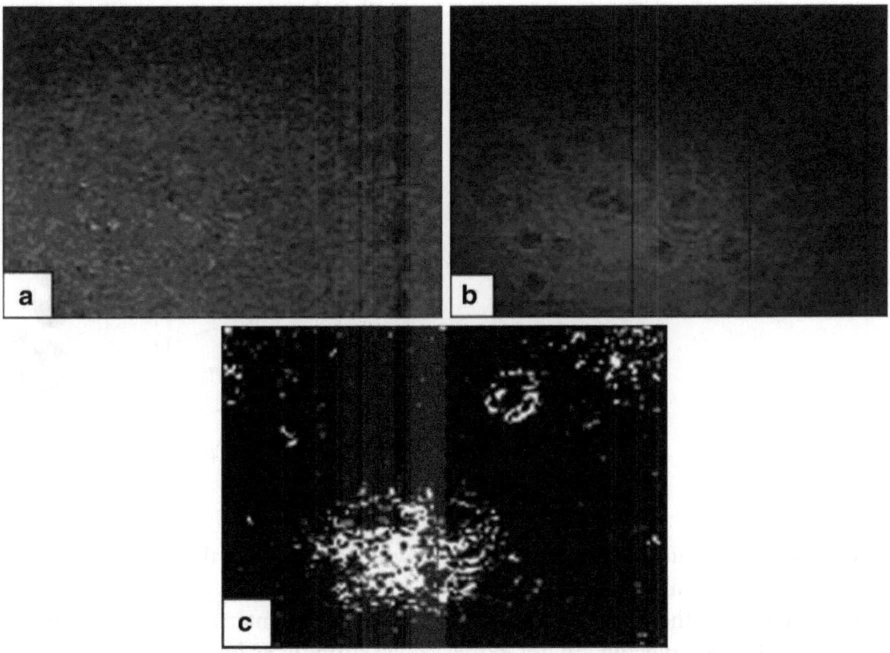

Fig. 3.4 Microscopic image of a standard (**a**), tested (**b**) solutions and the difference image (**c**) [66]

Using the method of laser diffraction, the existence of aquatic microregions which do not contain particles of the disperse phase was established as different from the volume of water according to physical properties like permittivity, which determines the structural heterogeneity of the water.

It was found that water clusters of 0.1 mm in size are responsible for the scattering of monochrome light, and therefore the method of laser diffraction is applicable to the detection of heterophase water clusters.

3.2 Influence of Hydrogen Isotopes on Clustering: Small Water Clusters

Isotope effects on highly concentrated heavy water do not exist under natural conditions. At present these effects have been studied rather well. However, the impact of impurities on heavy molecules of water, mainly deuterium, and on the structure and properties of water is quite a different situation. The results obtained from investigating the role of deuterium in the formation of GHWC [65] is of special interest since they not only expand the field of knowledge about the water structure but are also conducive to the development of hypotheses about water's involvement in the origin of life [66–68].

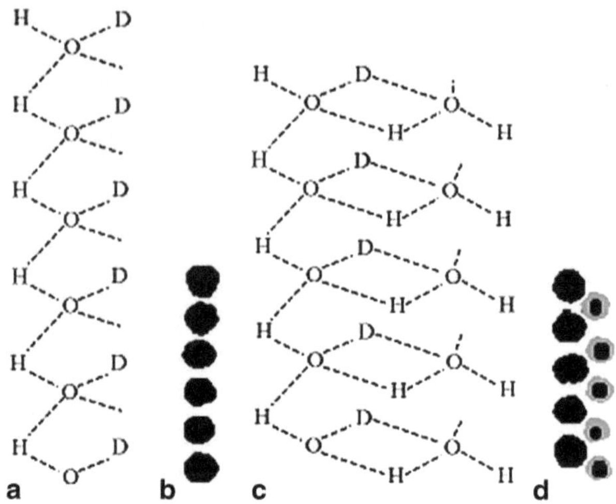

Fig. 3.5 HDO chains in water: (**a**) with one oxygen H bond; (**c**) a chain enforced by participation of deuterium interaction with neighboring water molecules; (**b**, **d**) schematic images

Deuterium in water is completely bound with HDO. Since the D^+ ion is linked with an oxygen atom stronger than the H^+ (the mobility of the D^+ ion constitutes 0.7 of the mobility of the H^+ ion) in an HDO molecule, the intermolecular deuterium links of the HDO molecule are not expressed as strongly as hydrogen bonds of the H_2O molecule.

At the usual concentrations the HDO impurity, possessing low surface tension, may be considered SAS, whose localization in the boundary layer stabilizes the cluster structure.

The non-equivalence of intermolecular H- and D-bonds primarily determines one possible orientation of neighboring HDO molecules in water. Additionally, the influence of the orientation forces and migration polarization result in the formation of chain associates from HDO molecules.

Figure 3.5a shows the simplest chains of the HDO molecules in water; it also presents a more complex and stable formation (see Fig. 3.5c) of the HDO molecules and neighboring H_2O molecules.

The conditions for the formation of the energetically stable network of the hydrogen bonds, as indicated in paper [24], may be realized in the water through the formation of chains and a volume network of HDO molecules. Model fragments of a flat network constructed with HDO molecules in water are illustrated in Fig. 3.6.

In a similar way, the HDO molecules form a three-dimensional network in the bulk of the water. With an increase in deuterium concentration levels, the number of fragments from the cluster network increase, while the dimension of the meshes of the network decreases. In the deuterium depleted water, the integrity of the volumetric network is substantially disturbed, due to the deficiency of the HDO molecules for building chains. To this day, it is not clear how water's physicochemical property changes are subjected to the dramatic weakening of deuterium, i.e., under the condition that the network from HDO cannot be formed (see Fig. 3.6c).

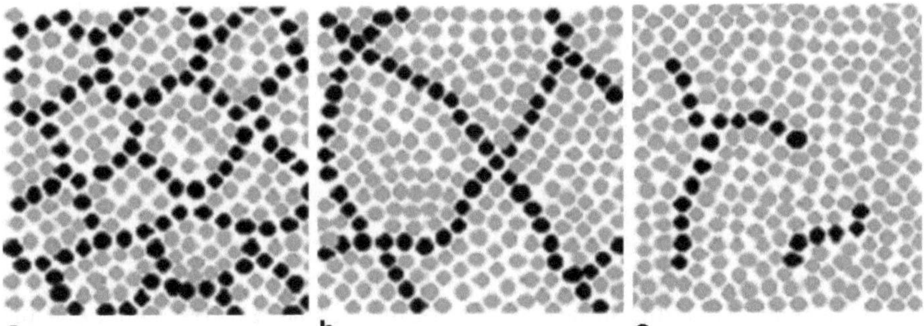

Fig. 3.6 A lattice fragment from the HDO molecules (*red points*), filled with H_2O molecules (*blue points*): (**a**, **b**) is increased content of deuterium in water and usual content of deuterium respectively; (**c**) water strongly depleted by the content of deuterium

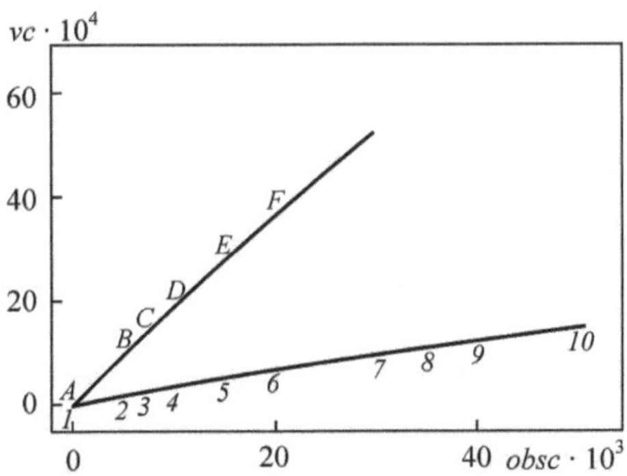

Fig. 3.7 The Bouguer-Lambert-Beer law (*vc*:*obsc*) for scattering the GHWC laser light in water: *A*—bidistilled water; *B–D*—solutions (*B*—NaCl (~0.001 M), *C*—$MgCl_2$ (~0.001 M), *D*—$ZnCl_2$ (~0.001 M); *E*—bidistilled water (indicators of dispersion were determined when used as a water background (4 °C); *F*, *1*—tap water (center of Moscow); *2, 3*—solutions of HCl (pH<4); *4*—heavy water; *5*—protium water; *6*—sea water (the Black Sea); *7–9*—commercially-bottled underground fresh (Aqua Minerale, Bon Aqua, Svyatoi Istochnik); *10*—NaOH solution (~0.001 M)

The investigations have shown that at least two structures are formed in the water. This was corroborated by the data of the experiment given in Fig. 3.7, illustrating the Bouguer-Lambert-Beer law for scattering GHWC laser light of different water. From the figure it can be seen that effectively all tested samples of water in terms of their structural state can be classified by two types.

Due to the non-uniformity of hydrogen and deuterium bonds, chains from the HDO and H_2O molecules are more stable when molecules are packaged in the form of helix-like formations. Their composition becomes more complex, and the functional possibilities increase as the selective adsorption of impurities increases.

According to the GHWC model, in which localized HDO stabilizes the cluster surface, it is necessary to clear up the ratio of deuterium and protium in water. Under these circumstances, the HDO molecules are sufficient for inciting the formation of a film covering the entire surface of the cluster. Calculations have demonstrated that the ratios of water molecules in the bulk of the cluster and on its surface (N_V/N_S) are linked with the linear dimensions of the cluster (α) and particles (*d*) forming it the following way:

$$N_V / N_S = a / 6d \quad (3.1)$$

From the formula, it follows that, as the cluster dimension decreases, the ratio N_V/N_S decreases. When filling the surface layer, for example, the relative amount of molecules increases. In reference to the simplest case (for a cubic associate), under the condition $d = 0.2$ nm and according to the formula (3.1), the following was calculated: in order to locate the HDO molecules in the cluster surface monolayer and maintain the ratio of protium and deuterium 1/6,000, the cluster dimensions should be at least 7.2 μm. It should be noted that the description given is purely geometrical and does not apply to the physical aspect of the process. In other words, it cannot be generalized as the reason for localization of heavy water in the surface layer of water clusters.

In order to fulfill the ratio in the water deuterium:protium = 1:6,500 on the surface of a single mesh, the volumetric network should be $\sim 4.8 \times 10^9$ of the HDO molecules, and in the bulk of a single mesh—about 3.2×10^{13} of the H_2O molecules. The structure of the water within a mesh of the network from the HDO molecules may be described by a two-structure model with allowance for the formation of small water clusters.

Thus, as is known [69], in a "hybrid" water molecule, where one atom of hydrogen is substituted by deuterium, the hydrogen bond is formed from the deuterium atom. An exception is $H_5O_2^{\pm}$, where the hydrogen bond is especially short. Thus, from our perspective, the water structure directly depends on the content of deuterium.

Unfortunately, not enough data have been obtained for reliable determination of the structure of small clusters. In fact, experimental samples contain several isomers, and therefore clusters having different dimensions are often very difficult to separate. Over the last few years, hundreds of models for the formation of water clusters have been proposed (see Fig. 3.1). Despite this, calculations of water properties remain largely inaccurate.

A simple calculation is enough to determine that 1 dm^3 of water contains 125 g of hydrogen, of which 43 mg are deuterium atoms. This provides us with the ground to judge the presence of a substantial concentration of deuterium, which cannot be ignored in the calculations. Accordingly, in heavy water molecules where both hydrogen atoms are substituted with deuterium and tritium (D_2O and T_2O), the hydrogen bond is stronger than in light water (H_2O). It is also highly structured; this is confirmed by large molar volumes [70].

Our paper [32], investigating small water clusters $(H_2O)_{2-6}$, was obtained by means of ab initio methods at the MP2/MP4(SDQ) level within a large range of base sets 6–311++G(d, p)/6–311++G(2d,2p). We analyzed vibrational frequencies using a water monomer and a dimer as examples. We obtained harmonic frequencies and intensity of the IR spectrum, which are in agreement with the experimental data.

The effect of the hydrogen isotope has been considered using the example of symmetrical and asymmetrical oscillations of the O-H bond in the series HOH, HOD, DOD, $(HOH)_2$, $(DOD)_2$, DOD–HOH, HOH–DOD, HOD–HOH. Special attention has been given to the cooperative water effect, which was postulated for the first time by G.S. Frank and W. Wien [71]. According to this effect, the formation of one hydrogen bond is conducive to cluster growth, and, conversely, the cleavage of one bond facilitates the destruction of the whole complex.

Ab initio calculations were conducted by means of the software Gaussian 03 [72]. The structures of water clusters were optimized at the second order Møller–Plesset theoretical level (MP2). The 6–311++G(d, p) base set was chosen both for optimization and determination of frequencies. Vibration frequencies were calculated on the basis of optimized structure by the aforementioned method. Graphic images of water clusters and visualization of IR spectra were then obtained using the Molden program [73]. The energy of cluster formation or stabilization energy E_S and the energy of hydrogen bond E_H were calculated using total energies of the equilibrium state of a cluster and the ground state of a water molecule by the MP4(SDQ) with an enhanced basis set 6–311++G(2d,2p):

$$E_S = E(H_2O)_n - nE(H_2O); \tag{3.2}$$

$$E_H = E_S / N_{HB}, \tag{3.3}$$

where N_{HB}—is the number of hydrogen bonds in a cluster.

The parameter of the conversion criterion for the energy and the gradient of the self-consistent field constituted 10^{-7}.

Water monomer (H_2O) The bond length of the O–H and the valence angle HOH in a water molecule optimized by us constituted 0.09595 nm and 103.47° respectively, which agrees well with experimental data (0.09584 nm, 104.40°) [74]. In a water molecule, three different vibrations are distinguished: ν_1—symmetric vibrations; ν_2—deformation of the valence angle HOH; ν_3—asymmetric vibration. According to our calculations, the values of these frequencies are equal to 3,884.54, 1,628.61 and 400.77 cm^{-1}; this can be compared with the respective vibrational frequencies of the water monomer observed in the matrices of the FTIR neon by means of a spectrometer: 3,660.6; 1,595.6; 3,750.7 cm^{-1} [69].

Unlike H_2O, where vibrational oscillations are combined with the movement of both OH-bonds, in HDO vibration oscillations ν_1 and ν_3 are associated only with one hydroxyl bond and constitute: ν_1=2,864.3204, ν_2=1,427.2593, ν_3=3,946.26 cm^{-1}. As a result, a protium atom is more reactive and easily dissociated than deuterium. The data of the experiment indicate the following HDO: 2,722.9; 1,404.1; 3,699 cm^{-1} [69]. The vibrational frequencies of the IR spectrum of the heavy water monomer D_2O are much smaller, due to an increase in atomic weight. However, similar to H_2O, in asymmetrical and symmetrical oscillations the following modes are involved: ν_1=2,802.03 (2,672.7), ν_2=1,191.21 (1,178.7), ν_3=2,931.60 (2,787) cm^{-1} (the brackets contain experimental data of the frequencies of harmonic

Table 3.1 Rotational constants of water monomer

Water monomer	Rotational constants GHz		
	A	B	C
HOH	799.52	441.78	284.55
HOD	678.54	175.03	195.70
DOD	444.77	221.06	147.67

Table 3.2 Optimized geometrical parameters of water dimer

Bond length (nm)		Angle (degrees)	
O_1H_6	0.1950	$\alpha\ (O_1H_6O_2\ H_5)^a$	180
O–O	0.2941	$\alpha\ (H_5O_2H_6)$	104.0
O_2H_5	0.0959	$\alpha\ (H_3O_2H_4)$	103.5
O_2H_6	0.0966	$\alpha\ (H_6O_2H_1)$	2.1
O_1H_4	0.0961	–	–
O_1H_3	0.0961	–	–

Note: Here and in tables 3.5–3.7, [a] dihedral angle

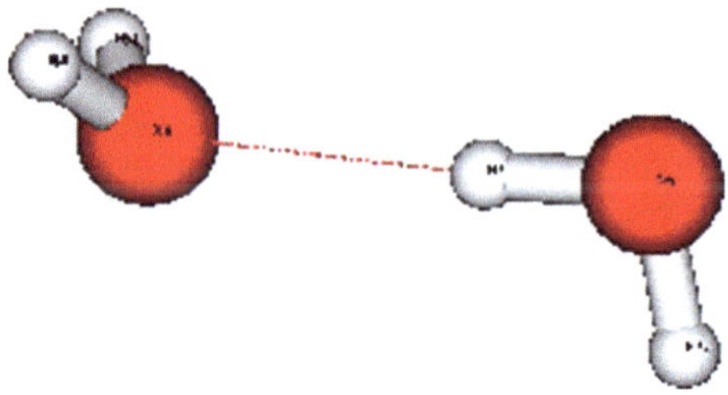

Fig. 3.8 The water dimer $(H_2O)_2$

oscillations of the water monomer isotopes [69]). Based on the IR spectrum of the monomer, the rotational constants of the water monomer isotopes were determined, and are given in Table 3.1.

Water dimer $(H_2O)_2$ The water dimer is its smallest cluster. As a result of optimization, the dimer was found to be characterized by a structure with an ordinary hydrogen bond, belonging to the C_s-symmetry, where both oxygen atoms and an atom of the molecule hydrogen, the donor of the hydrogen bond, are placed. The length of the hydrogen bond in such a structure is equal to 1.950 nm. It should be noted that the hydrogen bond in the structure of the water dimer deviates from linearity by 2.113°, while the distance between oxygen atoms is 0.2914 nm.

The hydrogen atom of the O–H-bond does not participate in the formation of a hydrogen bond and will thus hereafter be referred to as free hydrogen (H_u). A hydrogen atom, on the other hand, is involved in interaction as—bound (H_b). According to the calculation, the stabilization energy of a water dimer constitutes 21.06 J/mol. The data obtained coincide well with such experimental research 22.2±0.033 kJ/mol [75]. Geometrical data of the optimized dimer structure, as well as the shape of the water cluster, are given in Table 3.2 and Fig. 3.8.

Table 3.3 Frequencies of harmonic oscillations of water dimers

Frequency, cm^{-1}	$(HOH)_2$		$(DOD)_2$		DOD–HOH		HOH–DOD		HOD–HOH	
	D	A	D	A	D	A	D	A	D	A
N_1	807	3,876	2,752	2,796	2,753	3,875	3,808	2,795	2,781	3,875
N_2	1,663	1,639	1,214	1,200	1,213	1,614	1,661	1,201	1,428	1,641
N_3	3,973	3,990	2,904	2,922	2,904	3,990	3,973	2,922	3,955	3,990

D donor, *A* acceptor

Fig. 3.9 Water trimer $(H_2O)_3$

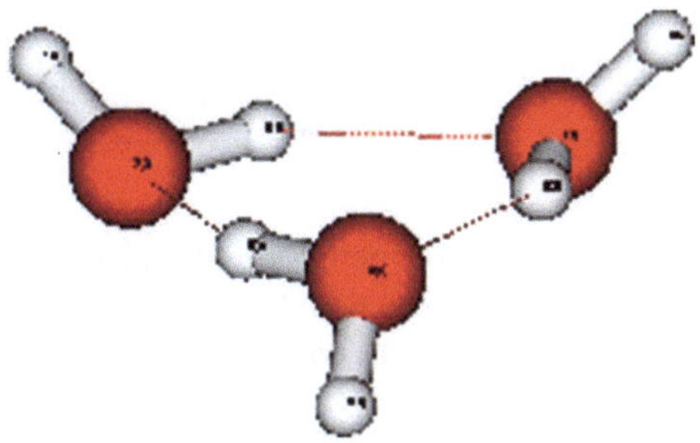

Based on IR spectra, vibrational frequencies were determined for $(HOH)_2$, $(DOD)_2$, DOD–HOH, HOH–DOD, HOD–HOH, and are shown in Table 3.3. From these data, one can see that complexes in which the formation of the hydrogen bonds involves deuterium have lower values of vibrational frequencies. This is evidence of a stronger hydrogen bond.

The analysis of the obtained IR spectra shows that the absorption peak at the highest frequency appears as a result of the asymmetric O–H vibration of the acceptor molecule. The other lowest peak is associated with the oscillation of the O–H donor molecule.

The third stage of absorption is a result of the symmetrical O–H vibration of the acceptor model. Finally, absorption at the lowest frequency corresponds to the stretching vibrations of the bound O–H group of the donor molecule. Such a picture is qualitatively in agreement with the data of experimental research shown in papers [69, 76].

Water trimer$(H_2O)_3$ As for the water trimmer, at present a discussion continues on the essence of its structural conformation. Some researchers point to an open linear chainlike structure [77], while others believe the cyclic structure to be a mostly stable one [78, 79].

The result of our water trimer optimization is evidence of the cyclic structure of the given cluster (see Fig. 3.9). The optimized water trimer is an equilateral triangular structure with three hydrogen bonds. Free atoms of hydrogen are arranged above and under the plane formed by oxygen atoms. Owing to this arrangement, this complex acquires properties of chiral compounds. A water trimer is a unique structure; every water monomer as part of a trimer acts both as a donor and an ac-

Table 3.4 Optimized geometrical parameters of water trimer

Bond length (nm)		Angle (degrees)	
O_1O_2	0.2793	$\alpha\ (O_1O_2O_3)$	60.0
O_1O_3	0.2796	$\alpha\ (H_4O_1H_5)$	104.7
O_3O_2	0.2803	$\alpha\ (H_9O_2H_8)$	105.0
O_1O_4	0.0971	$\alpha\ (H_6O_3H_7)$	105.0
O_1O_7	0.0959	$\alpha\ (H_4O_1H_2)$	20.9
O_2O_4	0.0918	$\alpha\ (H_7O_3H_1)$	22.5
O_2O_8	0.0917	$\alpha\ (H_8O_2H_3)$	20.5
O_2O_9	0.0970	–	–
O_3O_6	0.0959	–	–
O_3O_7	0.0959	–	–
O_3O_8	0.0971	–	–
O_1O_2	0.0943	–	–

ceptor of the hydrogen bond and also contains one free and one bound hydrogen atom. In comparison with the dimer, the water trimer has shorter the hydrogen bond, whose average length constitutes 0.1926 nm.

The distance of O–O (0.2797 nm) also decreases, resulting in the strengthening of the hydrogen bond. The energy of the trimer formation (64.25 kJ/mol) is twice that of the dimer since dimer dissociation involves the breakage of only one hydrogen bond. In order for the trimer to decompose into a monomer, it is necessary to first destroy all three hydrogen bonds.

Geometrical data of the equilibrium state of the trimer are shown in Table 3.4.

Water tetramer $(H_2O)_4$ The tetramer of water just as its trimer has a cyclic structure, where hydrogen free atoms alternate over and under the place of the ring O–O–O–O forming a structure with the elements of the S_4-symmetry [80]. Like the trimer, in the cluster of the tetramer every monomer is simultaneously a donor and an acceptor, but, unlike the trimer, they do not possess the properties of enantiomers since they contain an even number of oxygen atoms as part of the ring. The tetramer also preserves the exponential tendency to a decrease of the O–O bond length (0.2748 nm) and the strengthening of the hydrogen bond (0.1791 nm). We have obtained two optimized structures of the tetramer shown in Fig. 3.10.

In one of the structures, free hydrogen atoms successively alternate above and under the ring plane O_4, and in the other, free atoms alternate in pairs. The formation energy of the latter constitutes 108.27 kJ/mol, which is by 3.85 kJ/mol higher than the stabilization energy of the alternative configuration. Geometrical parameters of these structures are given in Tables 3.5 and 3.6.

Water pentamer $(H_2O)_5$ The water pentamer is one of the most abundant water clusters in nature. It can be found in the solvent shell of hydrophobic clathrates, DNA molecules and proteins [81]. In laboratory conditions, pentamer water clusters may be obtained as a result of the passing of an inert gas through the water in a liquid state, followed by the adiabatic expansion in a vacuum [82]. The stable structure of the pentamer is like the cyclic structure of the trimer, and it also has chiralic properties. The stable structure of the pentamer is like the cyclic structure

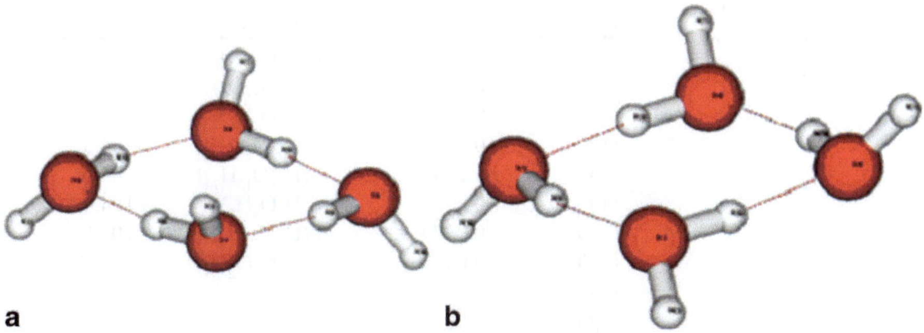

Fig. 3.10 Water tetramers $(H_2O)_4$ *I* (**a**) and *II* (**b**)

Table 3.5 Optimized geometrical parameters of the water tetramer

Bond length (nm)		Angle (degrees)	
O_1O_2	0.2750	$\alpha\ (O_1O_2O_3)$	90.4
O_1O_3	0.2763	$\alpha\ (H_5O_1H_6)$	104.5
O_3O_4	0.2750	$\alpha\ (H_9O_3H_{10})$	104.8
O_4O_2	0.2763	$\alpha\ (H_8O_4H_7)$	104.5
O_1O_5	0.0959	$\alpha\ (H_{12}O_2H_{11})$	104.8
O_1O_6	0.0977	$\alpha\ (H_6O_1O_2)$	8.7
O_3O_{10}	0.0959	$\alpha\ (H_9O_3O_1)$	10.3
O_3O_9	0.0976	$\alpha\ (H_8O_4O_3)$	8.7
O_4O_8	0.0977	$\alpha\ (H_{12}O_2O_4)$	10.3
O_4O_7	0.0959	$\beta\ (O_1O_2O_3O_4)$	0
O_4O_{12}	0.0976	–	–
O_2O_{11}	0.0959	–	–
O_1O_9	0.1811	–	–
O_3O_8	0.1791	–	–
O_4O_{12}	0.1811	–	–
O_2O_6	0.1791	–	–

Table 3.6 Optimized geometrical parameters of the water of the alternative form of the water tetramer

Bond length, nm		Angle, deg	
O–O	0.2748	$\gamma\ (O_1O_2O_4)$	90.0
$O–H_u$	0.0959	$\beta\ (O_1O_2O_4O_3)$[a]	2.5
$O–H_b$	0.0977	$\alpha\ (H_8O_4H_3)$	9.2
H	0.1791	$\alpha\ (H_9O_3H_1)$	9.2
–	–	$\alpha\ (H_6O_1O_2)$	9.2
–	–	$\alpha\ (H_{12}O_2O_4)$	22.5

of the trimer and also has chiralic properties. The average value of the length of the O–O bond in the pentamer, optimized by us, constitutes 0.273 nm, and the deviation from planarity is 16.29°. There are five hydrogen bonds in the cluster of the pentamer, which maximally coincide with the O–O line (angle $OO–H_b=3.5°$). In

Table 3.7 Optimized geometrical parameters of the water pentamer

Bond length, nm		Angle, deg	
O_1O_2	0.2732	$\alpha\ (O_1O_2O_3)$	107.4
O_2O_3	0.2729	$\alpha\ (O_4O_{13}O_2)$	107.0
$O_{13}O_4$	0.2728	$\alpha\ (H_5O_1H_6)$	104.4
O_4O_3	0.2730	$\alpha\ (H_{12}O_2H_{11})$	104.3
O_3O_1	0.2747	$\alpha\ (H_{15}O_{13}H_{14})$	104.6
O_3O_9	0.0978	$\alpha\ (H_7O_4H_8)$	104.4
O_1O_6	0.0979	$\alpha\ (H_{10}O_3H_9)$	104.7
O_2O_{12}	0.0979	$\beta\ (O_4O_3O_2O_1)$[a]	−16.3
O_4O_8	0.0979	–	–
O_1O_5	0.0959	–	–
O_2O_{11}	0.0960	–	–
$O_{13}O_{15}$	0.0959	–	–
O_4O_7	0.0960	–	–
O_3O_{10}	0.0959	–	–

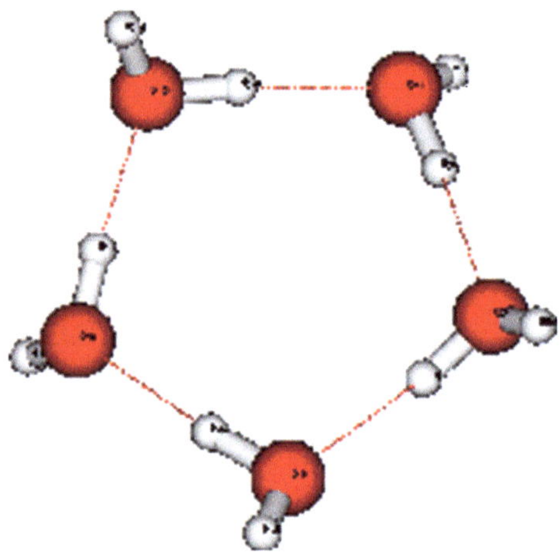

Fig. 3.11 Water pentamer $(H_2O)_6$

the water pentamer, the cooperative effect continues to be preserved; therefore, it is most difficult to break the first hydrogen bond as the following bond breaks easier than the preceding one, and so on. The equilibrium energy of the pentamer constitutes 147.30 kJ/mol.

The geometrical data and the figure of the pentamer equilibrium structure are given in Table 3.7 and in Fig. 3.11.

Water hexamer $(H_2O)_6$ The water hexamer symbolizes the transition from a two- to a three-dimensional dimensional cluster structure. We have optimized a prism-like cyclic hexamer structure (see Fig. 3.12), where every monomer of the water is linked with three hydrogen bonds. This, $(H_2O)_6$ in total, accounts for 9 hydrogen bonds.

In the hexamer, one can distinguish two types of DDA (donor, donor and constitutes donor and acceptor) and AAD (acceptor, acceptor and donor) monomers,

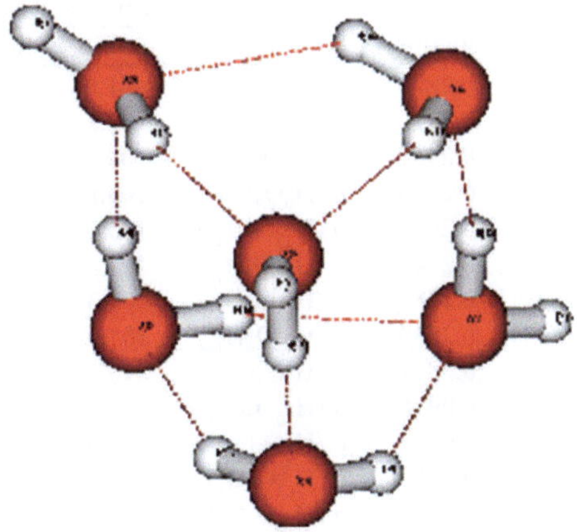

Fig. 3.12 Water hexamer $(H_2O)_6$

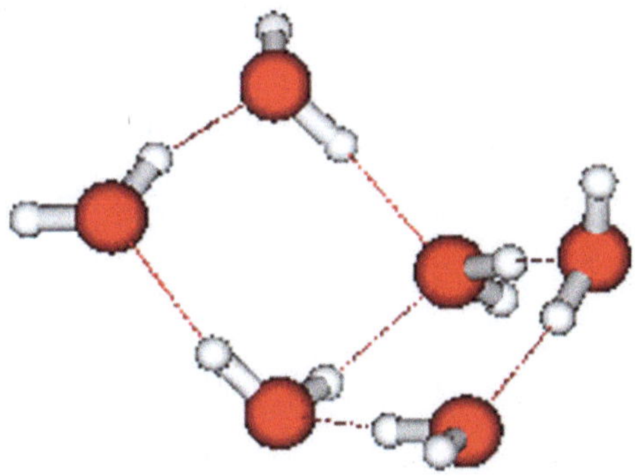

Fig. 3.13 Water isomer $O_3(H_2O)$

whose amount is balanced (3:3). If we look at the hexamer from a different angle, we can see that its cluster is nothing less than a superposition of the dimer with a tetramer, in which free hydrogen atoms are arranged in pairs over the plane of the O_4 ring. The hexamer stabilization energy is highest among the clusters considered and constitutes 183.96 kJ/mol. The other hexamerous structure optimized by us contains only 7 hydrogen bonds and resembles the shape of an "open book" (see Fig. 3.13). In this isomer, four monomers have a DA pair; the other two are connected by three hydrogen bonds and have the configuration DDA and AAD.

It is obvious that the imbalance of the monomer type and a decrease in the number of hydrogen bonds affect the energy component of the cluster. Indeed, the given isomer by 7.89 kJ/mol is less stable than a prism like analog. The geometrical data for the hexamer's stable structure are given in Table 3.8.

The results of investigation have shown that the cyclic ringed cluster structure is more beneficial in terms of energy than linear. The decrease of the O–O bond length, with an increase in cluster size, correlates with a decrease of the hydrogen bond length, which is accompanied by an elongation of O–H_b. A reduction of the

Table 3.8 Optimized geometrical parameters of the water hexamer

Bond length (nm)		Angle (degree)		Bond length (nm)		Angle (degrees)	
O_1O_2	0.2809	$\alpha\ (O_1O_2O_3)$	61.0	O_4H_{13}	0.0968	$\alpha\ (H_9O_2O_5)$	13.7
O_2O_3	0.2952	$\alpha\ (O_4O_6O_5)$	62.2	O_4H_{14}	0.0965	$\alpha\ (H_{12}O_3O_4)$	9.4
O_3O_1	0.2927	$\alpha\ (H_7O_1H_8)$	100.6	O_3H_{17}	0.0960	–	–
O_1O_6	0.2670	$\alpha\ (H_9O_2H_{10})$	103.6	O_5H_{18}	0.0973	–	–
O_2O_5	0.2927	$\alpha\ (H_{12}O_3H_{11})$	105.0	O_6H_{15}	0.0960	–	–
O_3O_4	0.2743	$\alpha\ (H_{15}O_6H_{16})$	105.4	O_6H_{16}	0.0989	–	–
O_6O_5	0.2783	$\alpha\ (H_{17}O_5H_{18})$	105.3	O_2H_7	0.1914	–	–
O_5O_4	0.2934	$\alpha\ (H_{13}O_4H_{14})$	101.4	O_3H_{10}	0.2109	–	–
O_4O_6	0.2891	$\alpha\ (H_7O_1O_2)$	19.1	O_3H_8	0.2133	–	–
O_1H_7	0.0975	$\alpha\ (H_{10}O_2O_3)$	24.2	O_6H_{18}	0.1879	–	–
O_1H_8	0.0967	$\alpha\ (H_8O_1O_3)$	29.0	O_5H_{14}	0.2175	–	–
O_2H_{10}	0.0966	$\alpha\ (H_{13}O_4O_6)$	22.2	O_6H_{13}	0.2028	–	–
O_2H_9	0.0967	$\alpha\ (H_{18}O_5O_6)$	17.8	O_1H_{16}	0.1695	–	–
O_3H_{12}	0.0978	$\alpha\ (H_{14}O_4O_5)$	31.9	O_5H_9	0.2000	–	–
O_3H_{11}	0.0960	$\alpha\ (H_{16}O_6O_1)$	7.7	O_4H_{12}	0.1785	–	–

Table 3.9 Cluster formation energy (E_S) and the hydrogen bound energy (E_H)

Cluster	N_{HB}	Energy, kJ/mol	
		E_S	E_H
$(H_2O)_2$	1	−21.06	−21.06
$(H_2O)_3$	3	−64.25	−21.42
$(H_2O)_4$ [a]	4	−108.21	−27.05
$(H_2O)_4$	4	−112.06	−28.02
$(H_2O)_5$	5	−147.30	−29.46
$(H_2O)_6$ [b]	7	−181.47	−25.92
$(H_2O)_6$	9	−189.36	−21.04

[a] see Fig. 3.10
[b] see Fig. 3.12

O–O results in an increase of the hydrogen bond energy (Table 3.9) in water clusters $(H_2O)_n$ for $n=2\ldots5$.

The greatest deviation of the hydrogen bond from linearity was registered in the trimer cluster (21.3°). In the tetramer, this indicator decreases and reaches nearly complete linearity in the pentamer (3.5°). Hence, it follows that hydrogen bonds in clusters tend to achieve linearity, and the tetragonal structure is similar to the ice structure.

On the basis of the water hexamer, which should be considered as a dimer compound with the tetramer isomer, one may assume that higher clusters, perhaps, completely consist of elementary units (small water clusters). Our calculations have confirmed the cooperative effect of water under experimental observation. The energy of cluster formation increases monotonously with an increase of its size, from 21 in the dimer to 188 kJ/mol in the hexamer.

3.3 Interrelation of the Effects of Physical Action on Water and Clustering

Having analyzed the latest data in the investigation of the water structure, one can reveal the mechanisms and explain unique effects of physical impacts used in water treatment and disinfection processes such as magnetic, ultrasound and ultraviolet treatment.

The mechanism of the magnetic field effect on water systems According to the data from paper [83], the permanent magnetic field (MF) should not affect diamagnetic objects, such as water. At the same time, the results of the practical use of magnetic treatment are evidence of the efficiency of this method in water treatment [84–86].

There are various viewpoints explaining the impact of preliminary magnetic treatment on the processes occurring in water systems. However, the mechanism of MF action on water systems remains insufficiently studied. Experimental research on magnetization of water systems [87–91] is not quite reproducible and does not yet provide an answer to the question of which factors lie behind the MF action and whether they are capable of causing substantial changes in the water itself and its impurities. Paper [47], in particular, considered the issue of magnetic treatment in connection with contemporary notions about the structure of water and solutions. Equally important are such characteristics of MF treatment as the time of the MF aftermath and the time of the system state restoration after the completion of its operation. Such characteristics depend on the ability of the liquids to achieve relaxation of their properties. One can expect that the emf of the electrochemical magnetohydrodynamic (MHD) cell of polarization, as a range of other characteristics of solutions [92] under the influence of kT, decreases over time t according to the exponential law:

$$E = E_0 \exp(-t/\tau_r), \tag{3.4}$$

where τ_r—relaxation time depends on the depolarization of water processes.

Based on the investigation carried out by the method of diffraction X-rays, molecular dynamics and Monte-Carlo, it was shown that the tetrahedral structure of the hydrogen bond characteristic of the water in a free bulk may strongly change in terms of external effects [93]. The capacity of the water to preserve the state characteristic of the preceding external effects was noticed for the first time by biologists and confirmed by spectral methods [94].

Complete relaxation of the state depends on the possibility of rotational processes, translational, orientation and the migrational movement of individual particles and their aggregates. The magnetic field in the moving liquid affects all charged particles simultaneously, and therefore the gradients of the concentration of charged particles appear throughout the whole bulk of the water. After the liquid exits the zone of the MF influence, the diffusion flows are formed, returning the liquid to its initial state.

The relaxation time τ_r of the molecules free cooperative hydrogen bonds according to the Debye formula [95]

$$\tau_r = 4\pi\alpha^3\ \eta\ / kT \tag{3.5}$$

(where α—is the radius of a particle). This formula has the order of 10^{-11}s, in particular for water molecules (at the temperature 293 κ, α = 0.193 nm, η = 10^{-3} Pa s) $\tau_r = 2.2 \times 10^{-11}$ s. However, as shown in paper [96], if the water contains larger particles, the relaxation time may be measured by minutes.

In paper [47], concerning the experimental determination of relaxation time, the emf was measured by means of different time intervals after the liquid exited the magnetic field. In this case, the liquid flowing from the MHD cell placed in the MF entered the cell in a similar manner to the cell placed outside the MF. The electrodes of one cell were electrically connected to the electrodes of the second cell only via the flowing liquid. The experiments have shown that the potential differences in the measuring cell decreased as it was removed from the MHD cell.

The time of achieving the measuring cell by the liquid was calculated based on its rate. Beyond the MF influence (in the given case induction 0.8 T) of MHD, emf decreased, and after 2 s it had a value < 10 % of the initial measurement. A longer MHD emf compared with the expected value. According to calculation, the MHD emf relaxation is evidence of the participation in the polarization process of MF particles, for which migration polarization proceeds at a much slower rate than for particles of molecular dimensions. In the water and solutions observed, such particles may be clusters and ions containing partial hydrate shells, monomolecular water and partially destroyed water clusters. According to the data of papers [94–97], the continuous structure of the water may be violated by the formation of clusters of at least 4–6 to 300 molecules. Under ordinary conditions, forces determined by *kT* and hydrogen bonds of water are proximal, and therefore the clusters are sometimes described as “shimmering” structures with a lifespan in the order of 10 ps. On the contrary, anomalies of water properties, in particular high overcooling [39], point to clusters as more stable formations.

The existence of such clusters has been corroborated with theoretical and numerical methods [97]. After magnetic treatment was applied to the processes of reorientation and polarization of water molecules bound into clusters, a period of relaxation commenced.

Calculations used concurrently with the formula

$$\Delta C = G_E / V_1 \tag{3.6}$$

(where V_1—is the liquid volume) demonstrate the observable relaxation times of MHD emf (1…2, 3 s) which may be achieved through the depolarization of particles with the size of (7…9) × 10^2 nm.

For a long time relaxations of the system have helped us explain the process of magnetic treatment in units representing a permanent magnet. This explanation incorporates the insertion of a pipe of insulation material into the water pipeline. The efficiency of similar units may be associated with the presence of a longer period of relaxation of MHD emf, owing to which the latter acts in the magnetic area. For example, at the speed of 0.5 m/s liquid motion, the residual emf acts in the outgoing flow at the distance of about one meter.

The transformation of water properties under the effect of electrochemical treatment At present, owing to the simplicity of the hardware background of the "live" (cathode) and "dead" (anode) preparation, electrochemically treated water spreads widely. Behind the method lies water electrolysis; however, processes accompanying it have been insufficiently studied to this day.

A process in near-electrode zones is exhausted not only by the variations of the pH but also by the softening of the cathode water. Experiments have shown that after 10–15 min of treatment in a domestic unit with a steel cathode, a graphite anode at the current density 20 mA/cm^2 and a potential gradient of 50 V/cm, hydrogen peroxide is formed in the anode water. This was detected by means of direct titration of anode water with a solution of $KMnO_{4,}$ electron-proton resonance (EPR) and the appearance of a singlet signal of anion-radical atomic oxygen $(O^{-\bullet})$ [98] in a frozen sample of the anode water at UV radiation.

Conditions for the formation of hydrogen peroxide through the reaction

$$O_2 + 2H^+ + 2e^- = H_2O_2$$

in an electrochemical cell are realized near the anode (provided with the presence of protons, the source of electrons and oxygen). The electrons are transferred from free radicals (FR) or via the relay-race mechanism from the water molecules. The formation of hydrogen peroxide and the anion H_2O_{2-} constitute a well-known intermediate stage in the anodic release of oxygen [99]. In the acid, anodic water hydrogen peroxide acts as an oxidant responsible for the source of the oxygen free radical:

$$H_2O_2 = H_2O + O^\bullet.$$

Under the influence of free radicals, the destruction or recombination of the main water elements (OH^-, H^+, $OH^\bullet$, $OH^\bullet$-, O^-, $O^\bullet$, $H_2O^\bullet$, $O_2H^\bullet$-, H_3O^+, $O_3H^\bullet$ etc.) [100] is possible. Short-lived FR catalyze radical processes by means of the chain transfer of an unpaired electron, which results in the formation of cations, anions and FR neutral molecules, whose ingress to organism increases the consumption of antioxidants. Organic impurities of the water are compounds with polycyclic condensed fragments containing the unpaired electron owing to the defects of intermolecular bonds. These impurities are easily converted into stable free radicals [101]. It is undesirable that fragments containing radicals known as stable FR (benzene, aniline, apoxylic, nitroxylic, aminilic, hydrazylic, etc. [102]) are included in drinking. The sensitivity of the water structure to inclusions of foreign particles among various thermodynamic investigations [103] is well documented. It is probable that the spatial lattice of hydrogen bonds, including FR, initiates the formation of anomalous water structures.

The degree that electrolysis influences the state of the water structure was investigated in the current research on the electrochemically treated water by means of the EPR probe, a stable radical whose spectra are sensitive to the variation of the local surrounding of the molecule.

As a result of electrochemical treatment of water, dipole molecules gain the ability to orient themselves to the external electric field, thus creating the possibility of an ordered water structure. The conspicuous molecular orderliness of unmatched molecules may be observed mainly according to the mechanism of relaxation polarization [103].

The two effects of relaxation orientation polarization of water molecules result in the gradual orderliness of the cluster structure; they also lead to the migration polarization of cluster, while the ordered structure contributes to an increase in microviscosity in prolonged water treatment. The longterm effects of the electric field include the formation of small islands (large clusters); they possess increased viscosity, due to the preferential orientation of hydrogen bonds along field power lines. The orderliness of the system increases over the time via the migration polarization of large clusters, which occurs because of the increase in their dipole moment and polarizability.

Of practical interest is the period of preservation of the effect acquired under the influence of electrochemical treatment and in particular, the preservation of the structures with changed orientation.

Based on the obtained results, it is possible to assume that the nature of the decrease in the time of the rotary correlation of the probe after the termination of water treatment is determined by the orientation polarization of the clusters. In the case of cathode water, the dimensions of the clusters constitute $(1.7\ldots1.8)\times10^4$ nm, and for anode water—1.0×10^4 nm.

Absorption of sound in water As has already been noted, one of the reasons for difficult detection of small-size clusters is the impact of the research methods which deforms the closest water structure. It well-known that the absence of hydrogen bonds in water could result in a much lower temperature of melting and boiling than it is known for water. Within the temperature range of the liquid water state, the strength, angle parameters and amount of hydrogen bonds vary. In fact, the value of the energy of H-bonds of the water (18.8 kJ/mol) determines all processes related to the variation of its nearest structure. At temperatures not in excess of ~80 °C the structural formations in the water are partially destroyed. Let us consider, for example, the effect of different research methods on water structure.

One may assume that the destruction of cluster structures is linked to the changes of Q_{cl} in the order of 10% of the energy of the water hydrogen bonds (i.e. $Q_{cl}=1.881$ kJ/mol). The value is therefore quite real for room temperatures, at which water molecules are either structured into primary ice-like aggregates or into secondary clusters. Table 3.10, in comparison, displays radiation energies affecting water in spectral investigations.

As can be seen from these data, most spectral methods produce strong effects on the objects being studied. In the case of water, this may result in the destruction of

Table 3.10 Spectral region and its corresponding radiation energy [104]

Spectral region	Radiation frequency (Hz)	Radiation energy (kJ/mol)
Gamma radiation	10^{19}	3.9×10^{6}
X-ray radiation	10^{17}	3.9×10^{4}
UV-radiation	$(1.5\ldots0.75)\times10^{15}$	597–298
Visual region	3.8×10^{14}	151.3
Close IR-radiation	1.2×10^{14}	50.1
Vibration IR-radiation	2.5×10^{13}	10.0
Remote IR-radiation	10^{12}	0.39
SHF radiation	10^{9}	1.0×10^{-4}
Short radio waves	1.5×10^{6}	1.6×10^{-7}
TV frequencies	5.5×10^{5}	6.1×10^{-8}
Long radio waves	3×10^{3}	1.2×10^{-10}
Radiation of power electric units	3×10^{-1}	1.2×10^{-14}

small clusters or, to be more exact, in the blinking of their structures and the change of their shapes and dimensions.

For example, the impact of the X-ray radiation exceeds the value Q_{cl} by 2×10^{4} times. The transitions of valence electrons corresponding to such energies are implemented irrespective of whether water molecules are included into supermolecular formations. This condition is also accompanied by the destruction of the closed water structure and also of small clusters.

The radiation in the visible region of the spectrum exceeds Q_{cl} in 80.4 times. It is clear that, if small clusters were located in water, then in there illumination a continuous restructuring would occur as a result of the transformation of valence electrons under the effects of light. In combined methods using monochromatic light sources (for instance, laser-acoustic [105]), the action of the last one may be reduced through selection of frequency. The remote IR region is acceptable in terms of minimal effects of the investigation method. In this case, the unit radiation exceeds Q_{cl} only in 2.07 times.

Weaker effects are exerted by methods which use short radio waves (NMR, EPR, DENR, etc.). However, in these methods a water sample apart from electromagnetic radiation (HF and SHF) is subjected to an impact of strong magnetic fields, which exert an unaccounted effect on the water and its impurities.

Energy differences between acoustic research methods As a result of the absorption of the signal by the medium being investigated in the ultrasound methods, the sound intensity being used (I) constitutes at least 10^{-4}–10^{-6} W/cm². In comparison, we point out that the aviation motor creates a sound of the intensity 10^{-4}–10^{-6} W/cm² at a distance of 5 m. Under the effect of a strong vibration, the supermolecular structure in a liquid may be destroyed. For ultrasound methods at the sound velocity in the water $c=1498$ m/s the density of sound energy (I/s) equals $7\times10^{-9}\ldots7\times10^{-11}$ J/cm³. This constitutes 10^{13}–10^{11} kT/cm³ at $T=300$ K ($kT=4.14\times10^{-21}$ J).

We carried out [106] the investigations of the water at a frequency $f=1$ kHz at the signal intensity 10^{-15}–10^{-14} W/cm², i.e. 10–20 dB (threshold of audibility of the signal of the frequency 1 kHz constitutes 10^{-16} W/cm^2). The sound energy density in this case equals $6.6\times(10^{-19}\ldots10^{-18})$ J/cm^3, which constitutes 2.4×10^4–2.4×10^6 kT/cm^3.

The sound wave pressure and amplitude of the sound wave θ in investigations of water systems at sound frequencies is much lower than at ultrasound frequencies. Thus, if at the frequency $f=1$ mHz the value $\theta=1.7\times10^4$ Pa ($\theta=\rho c\omega A$, where ρ—is water density, $\omega=2\pi f$, A is the amplitude of the bias of the particles), then at the frequency 1,000 Hz the pressure amplitude would constitute only 5.5×10^{-4} Pa. It is clear that at such low pressures one should expect a substantially lower disturbing effect of the research method on weakly bound structures, like water clusters. A graphic value is the deformation amplitude δ being developed in the sound wave ($\delta=\omega A/s$). If at the frequency 1 mHz $\delta=7.7\times10^{-7}$ cm, then in the frequency 1 kHz used by us the value δ constitutes only 2.4×10^{-14} cm. Consequently, at sound signals of low frequency and small intensity, the amplitude of the bias and the deformation are substantially smaller than the dimensions of the structural elements of the material being investigated (i.e. water atoms and molecules).

A unit yielding an extremely small density of sound energy (in the order of $6.6\times10^{-19}\ldots10^{-18}$ J/cm^3) was used in the current study. Thus, a minimal perturbation into the structure of the water sample being investigated was introduced. For instance, for heating the water by 1 degree, it is necessary to spend 0.18 J/cm^3, which constitutes 0.42 % of the energy of the breakage of hydrogen bonds. At the radiator intensity 10–15 W/cm^2 (10 dB) the volumetric sound energy (6.6×10^{-19} J/cm^3) is negligibly small compared with either the energy necessary for tangible heating of the water or the cleavage of hydrogen bonds. Due to the infinitesimal amplitude, shear displacements of sound vibrations, which determine structural viscosity, are difficult to achieve.

The sound absorption coefficient α was calculated by the formula [107]

$$\alpha=\left[1/2\left(l_2-l_1\right)\right]\ln\left(Y_1/Y_2\right), \tag{3.7}$$

where Y_1, Y_2 are values of the signals for distances l_1 and l_2.

Figure 3.14a shows the relationship between the signal value Y and the distance to the radiator and the coefficient of sound absorption α, calculated by using these data at various signal intensities of the radiator in distilled water at 20 °C and illuminated with a scattered light of the sample surface ~50 T.

Table 3.11 contains the results determining the value α for water at different types of treatment.

The experiments with water have shown that the value of sound signals with minor intensities and an absence of external effects remain constant for a significant period of time. Under identical experimental conditions (the power of the radiator, the distance from the radiator, temperature, etc.) depend on structural changes in the water.

The value Y decreases with an increase in the distance from the radiator (see Fig. 3.14a). Due to sound attenuation, the absorption coefficient α decreases at an increase of the radiator signal intensity. It is known that sound absorption in liquids

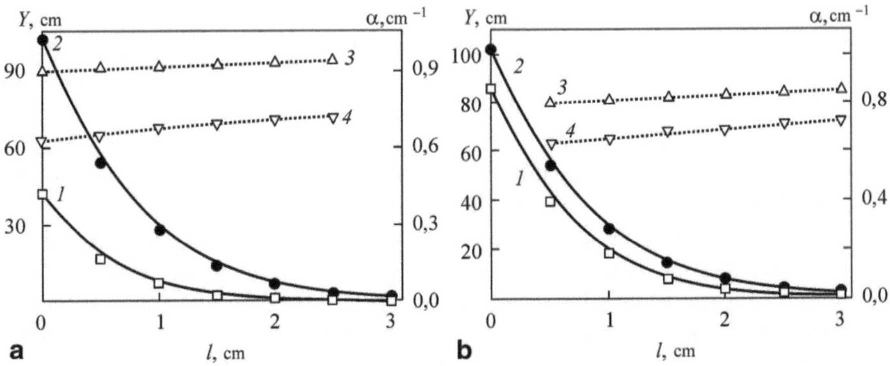

Fig. 3.14 Relationship between the value of the signal *Y* (*1*, *2*), the absorption coefficient *α* (*3*, *4*) and the distance between the radiator and the receiver in distilled water: (**a**) is the intensity of the signal source 1 × 10–15 (*1*, *3*), 2 × 10–15 W/cm² (*2*, *4*), *f* = 1 kHz, *T* is 20 °C (the value α was measured for an illuminated sample); (**b**) the sample was in the dark or 18 h (*1*, *3*), the sample was in the light for 4 h (*2*, *4*), *I*—2×10^{-15} W/cm², *f* = 1 kHz, *T* = 20 °C

Table 3.11 Absorption of sound by water at different types of its treatment

Water sample	Sound absorption coefficient α (cm)		
	Illumination conditions		
	In dark	On light	
	18 h	15 min	4 h
Distillate	0.81	0.80	0.65
Bidistillate	0.86	0.84	0.65
Distillate in the magnetic field 80 mT	0.87	0.85	0.65
Distillate at a higher sound intensity ($I = 3 \times 10^{-14}$ W/cm²)	0.66	0.58	0.51
Distillate, treated by ultrasound 0.1 W/cm for 0.5 min	–	0.64	0.65
Tap water (with electric conduction $5 \times 10^{-5}\ \Omega^{-1} \cdot cm^{-1}$)	0.70	0.69	0.68

is determined mainly by viscosity. For water with the frequency of 1 kHz, the relaxation sound absorption is absent [108], and the value α depends on structural viscosity. According to the Stokes formula, with an allowance for volume viscosity [40], an increase of sound absorption is evidence of an increase in structural viscosity:

$$\alpha = \omega^2 (4\eta_s/3 + \eta_V)/2\rho c^3 \tag{3.8}$$

where η_s—is shear viscosity; η_V—is volume viscosity; $\omega = 2\pi f$, ρ is density.

The concept of volume viscosity as a component of structural viscosity is used in acoustics for explaining the viscosity losses under comprehensive compression, which accompany sound waves in a liquid [107]. The energy of a sound wave is spent on overcoming and increasing volume viscosity η_V, as was shown by the experiments, when the vibration amplitude decreases.

Formula (3.7) is well fulfilled at high frequencies and high sound intensities but does not work in the case of very weak sound intensities. If normal measurements (at the signal intensity 10^{-5}–10^{-4} W/sm^2 and the frequency 10^6–10^9 Hz) for water the ratio $\eta_s \approx 3\eta_V$ [7] is fulfilled, then at the sound intensity 10^{-15} W/cm^2 forces causing longitudinal movement (along the direction of the wave propagation) of the liquid are not capable of activating energy transfer, due to infinitesimal effects and deformation of molecular bonds in the water. An increase of shear volume viscosity at small intensities of the signal may be determined by one and the same mechanism; for instance, the resistance to the change of the orientation of water molecules or water clusters can be observed. This is related to the experimentally observed (see Fig. 3.14a) rapid loss of the signal intensity when removing high values of the absorption coefficient from the radiator. These values are comparable with the value α for liquid crystals [109] and with α for very high frequencies.

It was found that the absorption of low intensity sound in water depends on whether the water sample is in the dark or subjected to light. The determination of sound absorption was carried out after the water sample was in the dark and after 4 h of exposition by light (see Fig. 3.14b). For distilled water kept in the dark, the curve of sound attenuation is lower than for the illuminated sample. This is evidence of the fact that in the water kept in the dark (i.e., under conditions when the water is protected from the influence of electromagnetic radiations), the sound wave loses more intensity than when passing through in the illuminated water. Furthermore, the sound absorption coefficient calculated by formula (3.7) for darkened water samples appeared larger than for illuminated sample (see Fig. 3.14b).

The role of the illumination/obfuscation factor in great measure manifests itself in greater quantities for purer water. As can be seen from the data of Table 3.11, in bidistilled water the rate of change for the value α is greater than in distilled water whereas the rate of change for the illumination mode has no effect on the sound absorption of tap water. Table 3.11 presents the measurement results of sound absorption by the frequency 1 kHz, intensity 2×10^{-15} W/cm^2 at 20 °C. An increase of the signal intensity of the radiator (in the data of the experiment there was an increase of 15 times) results in a decrease of the obfuscating effect and sound absorption by water (see Table 3.11).

Based on the obtained results, one may assume that the macrostructure of water, under conditions of minimum external effects (in the given case the water is protected from light radiation and noise from sound radiation; a metal screen also protects it from the effects of low-frequency electromagnetic radiations), gradually (over several hours) is rebuilt, and the losses of energy in the sound passage becomes greater. Since the mean sample density, in terms of the bulk, in the measuring cell does not change, the effect can be explained by an increase of the liquid viscosity. Owing to the formation of small cluster, the amount of unbound water molecules decreases. This makes difficult to increase viscosity and, according to Eq. (3.8), the passage of sound waves.

To understand the change of the value α depending on the illumination mode, it is necessary to take into certain factors account. First of all, it is known [110] that in terms of compressibility of small mechanical effects, liquids approach solid bodies. This is determined by the fact that there is not enough time for redistribution of molecules under external influence. Molecules, as in solid bodies at small amplitudes of variations almost do not change their positions. The sound energy, in this case, is absorbed to the maximum (3.8) due to the prevailing force of internal friction (i.e., volume acoustic viscosity η_V) whose contribution increases with respect to the force of the sound wave as the sound intensity decreases. Therefore, the values α appear to be high and increase in the absence of illumination due to the formation of cluster structures increasing water viscosity. Having compared the results given in Figs. 3.14a, b, one may conclude that even sound of small power exerts some influence on water macrostructure, though this influence is not so, strong as that of light radiation.

From the data of Table 3.11 it follows that, firstly, tap water has a higher coefficient of absorption of weak sounds than distilled water. Secondly, the effects of the light obfuscation/exposure factor on tap water are much weaker. Both effects, perhaps, are related to the fact that in tap water a higher concentration of impurities of metal ions and water molecules in greater degree are bound to hydrate shells of ions than into clusters, since the energy of hydration of ions is substantially higher than the energy of water H-bonds. The impact of ions is indicative of higher coefficient of the absorption of weak signals by bidistilled water, compared with distilled (for samples not subject to exposure). Therefore, one can conclude that if the water was not subjected to the activating effect of light, cluster formation (or general structuring if other models are taken into consideration) would proceed more strongly in the absence of ions, which reconstruct the water structure into the structure of hydrate shells.

The partial destruction of small clusters during water's transition from a dark to a light environment causes a loose structure to form; in other words, when exposed to light, water adapts a structure that is characteristic of higher temperatures. The notions about the influence of ions on water structure and temperature [43] may lead one to infer that under conditions of light exposure the temperature of water is lower than in the dark. This means that, for instance, at 20 °C in the dark, the water possesses a structure that correlates with a lower temperature. Certainly, the effect of variation of the "structural" water temperature due to the formation of water clusters is substantially lower than due to the formation of hydrate shells of ions since the energy of the polarization effect of ions, judging by the heat effects in solutions, is much larger than that of clusters.

Given the simultaneous obscurity of the water and its magnetization for 16 h by the induction field ~80 mT, one can observe an increase in the absorption of a sound signal (see Table 3.11). After the removal of the magnetic field and the exposure to light, the signal is decreased to the value characteristic of distilled water under conditions of illumination. Consequently, in the case of a long effect, the permanent magnetic field is conducive to an increase of the viscosity of water from a dark environment. Perhaps in the magnetic field the cluster formation is promoted by slowly proceeding to migration polarization [59].

Polarization involves charged clusters, i.e. dipole molecular particles carrying electric charges as a result of their incomplete stoichiometric compensation during the formation of a cluster. The incomplete compensation of the charges of water molecules occurs as a result of deformation of the molecules or due to inequality of the state of molecules inside and on the surface of a water cluster. The formation of the surface charge of a cluster causes its polarization under the effect of external fields.

A cluster may be modeled as a disperse particle with a double electric layer ensuring repulsion forces of neighboring clusters. In this case, the water may be considered as a suspension of clusters and the concentration of the disperse phase (i.e. clusters) in such suspension depends on external conditions (temperature, the presence of impurities, etc.). The external electric or magnetic field polarizes clusters similar to particles of the disperse phase. We believe that such possibilities of the cluster model are advantageous. The continual model of the structured water, most likely, is more suitable for low temperatures; it is best applied, in particular, to conditions of water overcooling [111].

As a result of the experiments, it follows that even a weak permanent magnetic field under a long effect is capable of affecting the structure of distilled water, which is not subjected to the impact of radiations. As the experiments have shown, the three day treatment with the magnetic field did not result in the change of α value. The changes of the structure in the darkened water sample, which had taken place under the effects of the magnetic field, gradually disappeared, and the water structure returned to its initial state. A short-time effect on the water by ultrasound treatment (f=22 kHz, I=0.1 W/cm^2, 0.5 min) sped up the process of decreasing the value α to the value characteristic of the illuminated water sample. This also corroborates the fact that under the effects of light the sample viscosity decreases, due to the destruction of the cluster structure; ultrasound treatment speeds up the process of destroying secondary structures.

Thus, thanks to a very weak disturbing effect of the given method of investigation on the water, the α values were recorded, testifying the changes of viscosity determined by the restructuring of the cluster structure under the impact of different illumination of the sample.

Considering the obtained results one may conclude that the main state of the water structure should be considered as being protected from light, vibrations and various electromagnetic radiations. This state also implies the absence of dissolved substances. In this case, a true equilibrium is fulfilled between water's cluster structure and unbound water molecules. Illumination or other effects result in a gradual destruction of clusters and the transition of the water from the main state to other characteristic states of the given effect.

3.4 The Theoretical Research of Interactions of Water Clusters with Ozone

Ozone molecules, as is known, play an important role in the Earth's atmosphere. The interaction of ozone with atmospheric compounds was studied by many authors [112–114], who noted that ozone complexes with water are key formations in the reactions of the ozone layer.

Under the effect of ultraviolet (UV) radiation, the atmospheric ozone enters into a reaction with various molecules and radicals of the atmosphere. It is known [115] that the solar UV-radiation with the wavelength 185.9 nm causes the ozone formation from oxygen and with the wavelength 200–320 nm—its decomposition. The ozonide anion radical is a primary product of ozone decomposition

$$O_3 + e^- \rightarrow O_3^{\bullet -}$$

The following formation and other free-radical particles are possible with the participation of ozone $O_2^{\bullet -}$, $O^{\bullet -}$, $H_2O^{\bullet -}$, $HO^{\bullet}$ etc. [116–118]. This can be illustrated using singlet oxygen as an example. This oxygen is formed as a result of UV photolysis of ozone. Singlet oxygen, in turn, reacts with water molecules with the formation of the main oxidant—hydroxyl radical. Given reactions are represented by the following scheme:

$$O_3 + h\nu \rightarrow O(D) + O_2;$$
$$O(D) + H_2O \rightarrow 2OH^{\bullet}.$$

It should be noted that ozone may coexist with singlet and molecular oxygen whose ratio in a gas charge equals

$$O : O_{singl} : O_2 = 0.03 : 0.11 : 0.86.$$

In this case, the medium ozone in water cannot be considered independent of the pH because experimental facts confirm the impact of the pH on the preferential activity of ozone by radical or heterolytic mechanism [119–121].

The formation of hydroxyl radical from solvate ozone complexes is also possible:

$$O_3(H_2O) + h\nu \rightarrow 2OH^{\bullet} + O_2.$$

In this connection, the information about the nature of the interaction of ozone complexes with the water is especially important since the hydrate shell may substantially shift the absorption peak and thus affect the formation of free radicals.

Paper [112], dealing with the study of the interaction of ozone with the water, proposed a model in which both terminal oxygen atoms are oriented to the side of one of the water hydrogen atoms. Another study [113] proposed a model in which water hydrogen forms a hydrogen bond with only one of the terminal oxygen atoms of the ozone molecule. Later, a "dipolar" model of the water complex with ozone was proposed, in which a water molecule together with the ozone central atom forms the C_s-symmetry plane [114].

Thus, the information about the structures of the $O_3(H_2O)$-complexes does not provide a single idea about the nature of interaction of ozone molecules with the water.

Ab initio calculations were carried out by means of Gaussian 03 software [72]. The structures of water clusters with ozone were completely optimized at the theoretical level of the Møller-Plesset perturbation theory of the second order. The basic set of wave functions 6–311++G(d, p) of the Gaussian type, with the inclusion of diffusion and polarization functions, was chosen both for optimization and for determination of frequencies. Vibration frequencies were calculated based on optimized structures by the above mentioned method. The graphic representation of water clusters was obtained by means of the Molden software [73].

The energy of forming the water—ozone complex, or the binding energy $E_{S,}$ was calculated based on the energy of the equilibrium state of the cluster and the main state of the ozone molecule by the MP4(SDQ) method with an enhanced basis set 6–311++G(2d,2p):

$$E_S = E\left(O_3\left(H_2O\right)_n\right) - \left(E\left(H_2O\right)_n + E\left(O_3\right)\right) \tag{3.9}$$

The parameter of the convergence criterion for the energy and the gradient of the self-consistent field were set to 10^{-7}.

We have optimized a number of the O_3-H_2O models (see Figs. 3.15–3.19). To confirm the global minimum, the frequency analysis of the water structures with ozone, obtained in optimization, was carried out. As a result, two stable complexes were determined (see Fig. 3.15).

The first model (see Fig. 3.15a) represents a structure in which H_2O and O_3 lie virtually in one plane, and a hydrogen atom is turned to the side of the terminal oxygen atom of ozone. Thus, the hydrogen bond appears between one atom of water hydrogen and the terminal oxygen atom in ozone. However, the length of this bond—0.227 nm—substantially exceeds the length of the hydrogen bond in water clusters [32], and due to this the energy of its formation (9.15 kJ/mol) is inferior to the energy of the hydrogen bond in water clusters.

The second model (see Fig. 3.15b) possesses a large amount of bond energy (9.78 kJ/mol) and has C_s-symmetry. The symmetry axis passes through the ozone central oxygen atom and the water oxygen atom. Based on the results of calculations, such eclipsed configuration, in which hydrogen atoms are opposite to the terminal atoms of ozone is the most stable form of water complexes with ozone at the ratio 1:1.

The *trans*-form [114] stated in the literature (see Fig. 3.17a) was also reproduced in the current study. However, as the calculations have shown, the given configuration is metastable, which is indicated by the presence of the imaginary frequency in

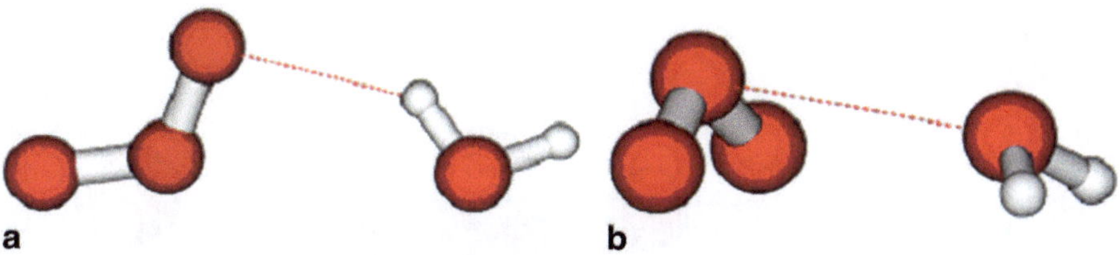

Fig. 3.15 Ozone complex with the water monomer $O_3(H_2O)$ (**a**) and shielded configuration of the isomer $O_3(H_2O)$ (**b**)

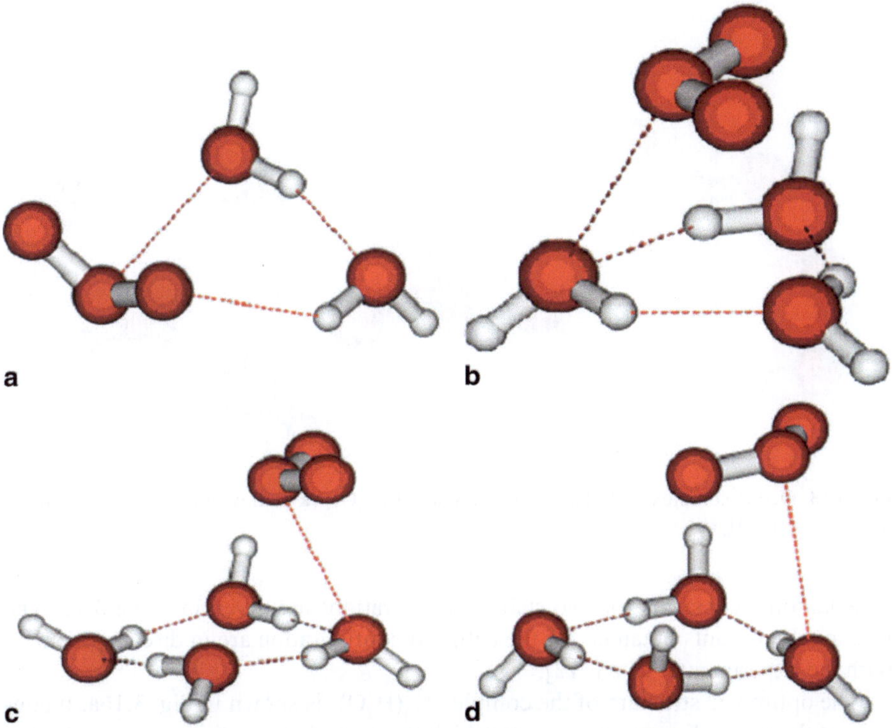

Fig. 3.16 Ozone complex with the water dimer $O_3(H_2O)_2$ (**a**), with the water trimer $O_3(H_2O)_3$ (**b**), with the water tetramer $O_3(H_2O)_4$ (**c**) and isomer $O_3(H_2O)_4$ (**d**)

the vibrational analysis. The energy of the bond of such a complex constitutes only 8.02 kJ/mol.

From our perspective, the dipole orientation (see Fig. 3.17b) is an unstable structure and merely a transitional state to the eclipsed configuration.

The analysis of harmonic frequencies of the dipole structure revealed several frequencies with a negative value. At the smallest perturbation (for example, when removing the symmetry) the given complex acquires the configuration of the eclipsed form.

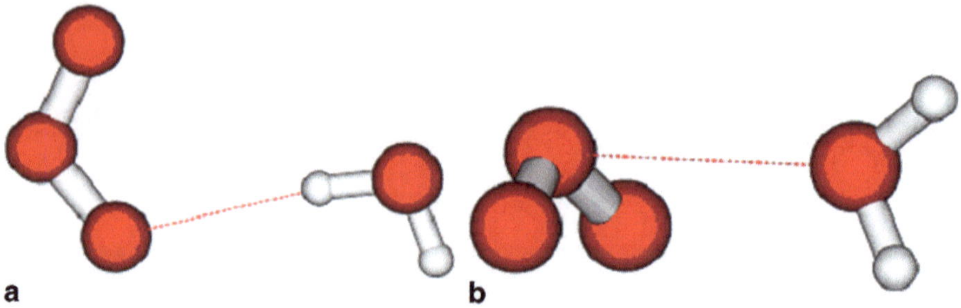

Fig. 3.17 "Trans" configuration (**a**) $O_3(H_2O)$ and "dipole" orientation (**b**) isomer $O_3(H_2O)$

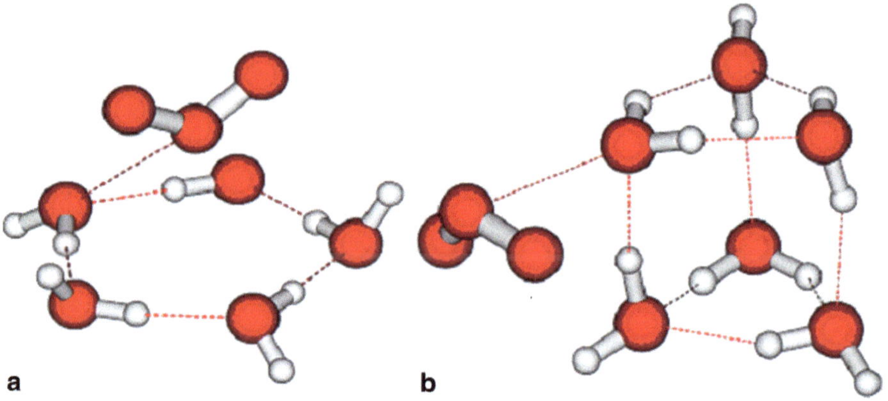

Fig. 3.18 Ozone complex with the water pentamer $O_3(H_2O)_5$ (**a**) and with the cubic form of the hexamer $O_3(H_2O)_6$ (**b**)

Rotational constants for two stable configurations are given in Table 3.12. The rotational constants obtained for the eclipsed configuration are in direct agreement with the experimental data [112].

The optimized structure of the complex $O_3(H_2O)_2$ is shown in Fig. 3.16a. It consists of the water dimer and ozone in which the central atom of oxygen coordinates with the oxygen atom of the water monomer (the donor of the hydrogen bond), where $r_{\text{O-O}} = 0.284$ nm. At the same time, the hydrogen atom, the acceptor monomer, forms a weak hydrogen bond with one of the ozone's oxygen terminal atoms.

The length of the hydrogen bond in the water dimer of a complex with ozone constitutes 0.194 nm, which is comparable with the length of the hydrogen bond in the free dimer 0.195 nm. It should be noted that the water dimer is more strongly bound with ozone than the monomer. Thus, the dissociation energy of $O_3(H_2O)_2$ constitutes 15.49 kJ/mol.

Optimized water–ozone complexes with the number of water monomers 3–6 are compounds in which water molecules have a cyclic form of the earlier studied trimer, pentamer, isomers of tetramer and hexamer [32]. In all enumerated com-

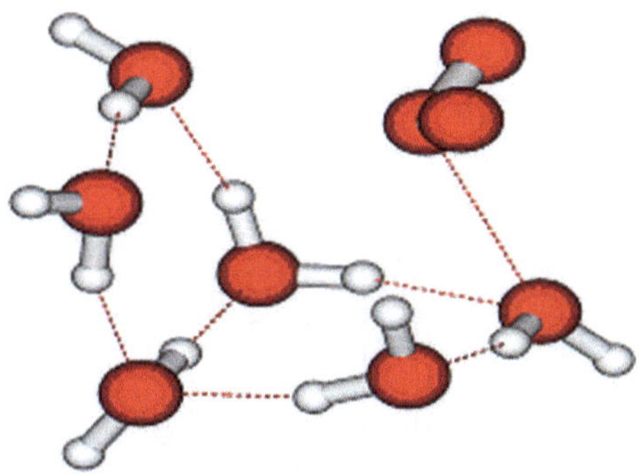

Fig. 3.19 Isomer $O_3(H_2O)_6$

Table 3.12 Rotational constants for the complex O_3-(H_2O)

Complex of ozone with water monomer	Constant (GHz)		
	A	B	C
$O_3(H_2O)$	24.66	2.69	2.43
$O_3(H_2O)^a$	11.42	4.68	3.55
$O_3(H_2O)_{exp}$[1]	11.96	4.17	3.27

[a] isomer

plexes, an ozone molecule is coordinated via the oxygen–oxygen bond by only one monomer from the composition of the water structure (see Figs. 3.16, 3.18, 3.19).

The closest intermolecular distance (0.283 nm) between the binding water molecule and ozone in the complexes $O_3(H_2O)_n$ ($n=1\ldots6$) was obtained for $O_3(H_2O)_2$.

The distribution of the charge on ozone oxygen atoms and the water monomer that interacts with it is given in Table 3.13. As can be seen, the charge on the oxygen central atom in the unbound ozone constitutes $+0.321$, whereas in complexes with water clusters it increases appreciably. According to the obtained data, the greatest positive value for the ozone oxygen central atom is achieved in complexes with the water dimer and tetramer. The obtained distribution of the charge is a result the Coulomb interaction with the binding molecule of the water cluster monomer.

Thus, as a result of theoretical ab initio calculations in studying ozone complexes with water clusters $O_3(H_2O)_n$ for $n=1\ldots6$ the following results were obtained. In the ozone: water ratio $=1{:}1$ two stale configurations with the binding energy -9.15 and -9.78 kJ/mol (Table 3.14) were discovered. It is noteworthy that the structure, in which ozone is bound with the water molecule like the hydrogen bond, possesses a higher total energy value. The highest energy of the $O_3(H_2O)_n$-bond (-90.94 kJ/mol) was obtained in the compound with the isomer of hexamer, and the smallest one (-9.15 kJ/mol) was obtained in the complex with the monomer.

Except for the water dimer in all water clusters studied by us, ozone is bound with one of the closest monomers exclusively via the O–O bond, which was testified from structural data as well as by the charge distribution on atoms.

Table 3.13 Distribution of the charge according to Mulliken on ozone's oxygen atoms and oxygen of the nearest water monomer

Atom	O_3/H_2O	$O_3(H_2O)$	$O_3(H_2O)^a$	$O_3(H_2O)_2$	$O_3(H_2O)_3$	$O_3(H_2O)_4$	$O_3(H_2O)_4{}^a$	$O_3(H_2O)_5$	$O_3(H_2O)_6$	$O_3(H_2O)_6{}^a$
O_t	−0.161	−0.200	−0.226	−0.248	−0.271	−0.224	−0.269	−0.154	−0.209	−0.203
O_c	+0.321	+0.325	+0.475	+0.502	+0.485	+0.503	+0.495	+0.387	+0.449	+0.422
O_t	−0.161	−0.105	−0.226	−0.207	−0.207	−0.254	−0.224	−0.174	−0.207	−0.172
O_w	−0.481	−0.495	−0.518	−0.586	−0.584	−0.650	−0.660	−0.684	−0.600	−0.659

Note: Ot, Oc are respectively terminal and central atom of oxygen in ozone, O_w is an atom of water oxygen bound with ozone

[a] isomer

Table 3.14 Stabilizing energy of $O_3(H_2O)_n$ (kJ/mol)

Method	$O_3(H_2O)$	$O_3(H_2O)^{2a}$	$O_3(H_2O)_2$	$O_3(H_2O)_3$	$O_3(H_2O)_4$	$O_3(H_2O)_4{}^b$	$O_3(H_2O)_5$	$O_3(H_2O)_6$	$O_3(H_2O)_6{}^b$
MP2	−9.61	−15.80	−21.03	−18.94	−18.85	−19.77	−20.77	−22.15	−27.30
MP2(ZPE)[1*]	−5.31	−11.04	−16.47	−15.68	−15.84	−15.84	−18.31	−18.94	−22.03
MP4	−9.15	−9.78	−15.47	−12.46	−13.25	−15.13	−17.05	−16.85	−22.28
$r(O_cO_w)$, nm	–	0.2847	0.2837	0.2921	0.2926	0.3014	0.3127	0.2865	0.2883

[a] energy zero point adjustment

[b] isomer

Thus, the electrostatic O–O interaction is prevailed over the formation of a hydrogen bond in ozone complexes with water clusters. The current study corroborates the hypothesis, put forward earlier [122, 123], concerning ozone transformation involving hydroxyl anions, whose initial stage may be written as

$$O_3 + OH^- \rightarrow O_3^- + OH^{\bullet}.$$

As has already been noted, in complexation of ozone with water a partial positive charge on the ozone oxygen central atom is increased, which in a substantial way is conducive to the interaction with the negatively charged hydroxyl anion. At the same time, the presence of the fractional Coulomb charge on the ozone's oxygen central atom inhibits the formation of the hydrogen bond. From this fact it follows that a relative solubility of ozone in the water, in the first place, is determined by the dipole–dipole interaction of O_c– O_w rather than the presence of hydrogen bonds, which, according to the calculations, have a comparatively small energy of formation.

3.5 Impact of Temperature on Water Clusters

The recent use of computer-aided optical methods of research has made it possible to study the influence of the temperature on the water cluster structure. To this end a microscope with a laser ($\lambda = 633$ nm) has been used. A laser beam after passing a system of lenses with a total 20-fold magnification traveled through a spectrometric tray filled with a sample and then was projected on the screen. Unlike the technique described in paper [28], we used a flow-type rigidly fixed tray placed in a thermostat. The thermostat was used to set the temperature within the range 10–45 °C and maintained at the accuracy ±0.2 °C. The small depth of focus (~1 μ) of the microscope lens made it possible to observe contrast elements in the chosen section of the solution. The image being projected to the screen is a result of light scattering of the laser beam from the phase boundaries of the liquid elements of microscopic dimensions. These elements can be dust particles, air bubbles and others associates of water molecules, which are different from the unassociated water at the interface by the light refraction coefficient.

The image from the screen was read by a video camera and entered into the computer. Then it was converted into a black and white and was transformed into a matrix of a two-dimensional system and was analyzed by means of computer software. The result of analysis was obtained in the form of the total concentration of clusters and the concentration of clusters of certain dimensions (4 to 40 μ). The magnification of the microscope limited the lower size of the objects being observed to the size 2 μm, which made it possible to rule out the count by the computer software of microimpurities like gas microbubbles, colloid particles, etc. The microscope was calibrated using a photomicrography micrometer OMP-2. The scale was then projected to a screen and then by a video camera was sent to a PC. Finally, the calibration of the image elements on the screen was expressed in pixels (in the given case 1 pixel = 2 μ).

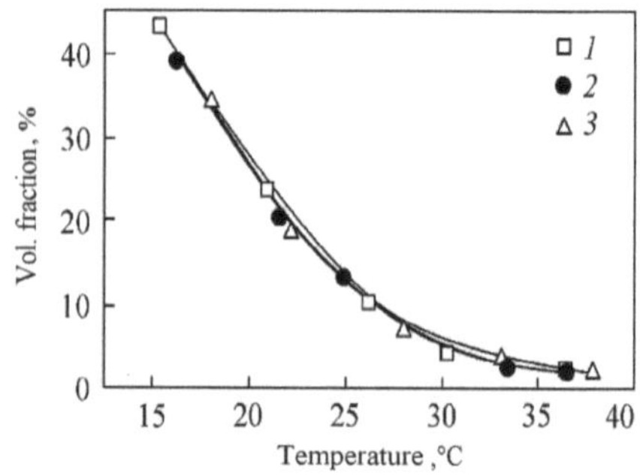

Fig. 3.20 Temperature dependence of the total concentration of GHWC: *1*—distilled water, *2*—magnetized water, *3*—birch tree natural juice

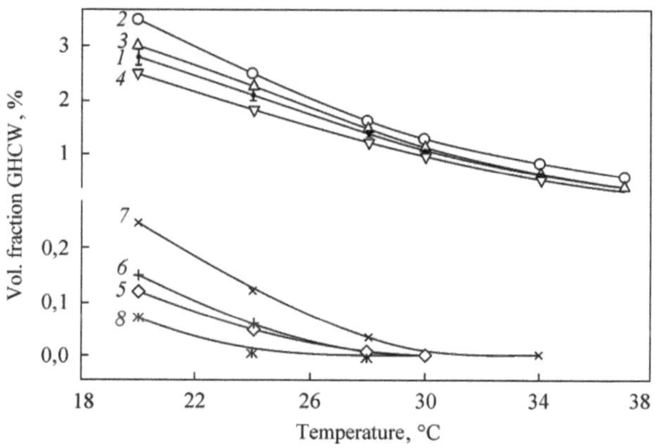

Fig. 3.21 Temperature dependence of the GHWC content of size 4 (*1*—*4*) and 36 μ (*5*—*8*) in the solutions LiCl (*1*, *5*), NaCl (*2*, *6*), KCl (*3*, *7*) and RbCl (*4*, *8*)

Figure 3.20 shows typical curves of the total concentration of clusters (i.e., the total content of GHWC of dimensions 2–40 μ) from the temperature within the interval 15–38 °C. A sample of magnetized distilled water was obtained by passing the water through a column of 2.5 cm in diameter and 10 cm high, and then the column was filled with balls 3 mm in diameter made of iron–nickel magnetic material, creating magnetic induction in an interparticle gap up to 100 mT. Magnetization was carried out under static conditions for a certain period of time. Distilled water was magnetized in a column (in the given case for 27 min) after that the probe was taken and GHWC was determined. The porous liquid of a tree was tested as a natural plant liquid not containing GHWC. For this purpose, sap from a birch tree was taken. Before measuring GHWC, this sap was filtered through a fine porous ceramic filter obtained by sintering mineral particles with sizes 10 to 40 μ. As can be seen from Fig. 3.20, the temperature dependence of the GHWC content in natural birch tree juice by shape is similar to the curve for the water. Thus, one may come to a conclusion that, firstly, GHWC is contained in the porous liquid of plants; secondly, the amount of GHWC in the liquid of plants depends on the temperature.

Figure 3.21 shows temperature dependences for clusters of the size 4 and 36 μ in the 0.01 M solutions of LiCl, NaCl, KCl, and RbCl. The characteristic error of the experiment, estimated in the given case in 5 %, is denoted on curve 1. The concentration was chosen based on the affinity of the 0.01 M solution of NaCl for human organism; therefore it was necessary to evaluate the GHWC content in such a solution at different values of temperature. A number of investigated chlorides correspond to a decrease of the cation's polarizing force, which is determined by an increase of its radius and appears in many properties of aqueous solutions [43]. As can be seen from Figs. 3.20 and 3.21, for all investigated water systems the content of GHWC at an increase of the temperature decreases. The total content of clusters at heating may decrease from 40 at 15°C to 2–2.5 % at 38 °C.

The content of GHWC of various sizes at an increase of the temperature decreases non-uniformly. As can be seen from Fig. 3.20, stronger heating affects large clusters. Actually, at the temperature 30 °C and higher clusters of dimensions 36 μ no longer exist. At the same time, a certain amount of clusters of size 4 μ are preserved, despite heating the solution to 38 °C. It should be noted that at the temperature of the human organism 36–39 °C in aqueous solutions the clusters of the GHWC type are absent. At a lower temperature (30–32 °C) in the solutions GHWC of smaller dimensions are preserved.

A decrease of the GHWC content with an increase of the temperature indicates destruction of clusters under the impact of the energy *kT*. The content of particles in the solution, in the current case referred to as clusters, depends on the energy of the interaction of water molecules like GHWC, capable of cluster formation. The condition for cluster formation is an excess of water molecules of a certain energy level (energy barrier), which results in their interaction with the formation of GHWC. The higher temperature of the solution, the higher energy of the interaction of molecules should be in order to reach and overcome the barrier of cluster formation. At a decrease of the solution temperature, the energy of thermal disordering (*kT*) decreases; therefore, it is easier for water molecules to overcome the energy barrier and GHWC are formed greater. Thus, the formation of GHWC depends on the molecular interaction temperature, which, in turn, is conducive to the change of the content of clusters in the solution. When investigating GHWC, one can observe a final result of such interaction, namely, the change of the concentration of the clusters. The value of the total concentration of clusters shows how much water take part in the formation of GHWC with respect to the amount of water which does not involved in this process.

To determine the activation energy of destruction of clusters the temperature dependences in the Arrhenius' coordinates were constructed [102] (see Figs. 3.22 and 3.23). Curve *1* (see Fig. 3.22) shows the error of determination constituting 5 %. As can be seen from the data given in Figs. 3.22 and 3.23 all points for investigated systems are laid on straight lines, which are individual for each system. This corroborates the applicability of the Arrhenius equation to the given water systems and shows that each of them is characterized by certain energy of activation of destroying clusters of the GHWC type. According to the Arrhenius equation,

$$\ln C = k_0 + E / RT, \tag{3.10}$$

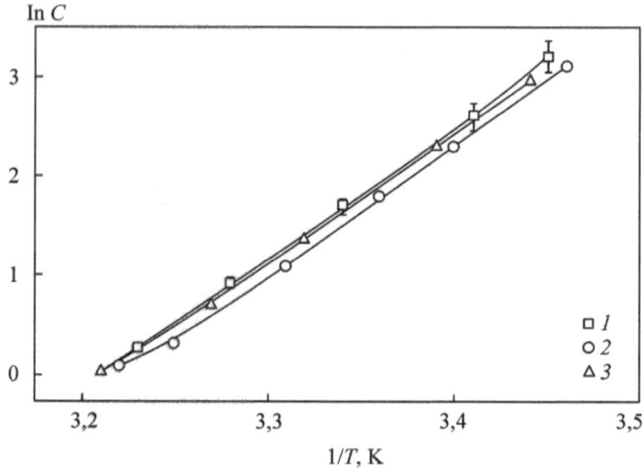

Fig. 3.22 Temperature dependence of the total concentration of GHWC in coordinates of Arrhenius: *1*—distilled water, *2*—magnetized distilled water, *3*—natural birch tree juice

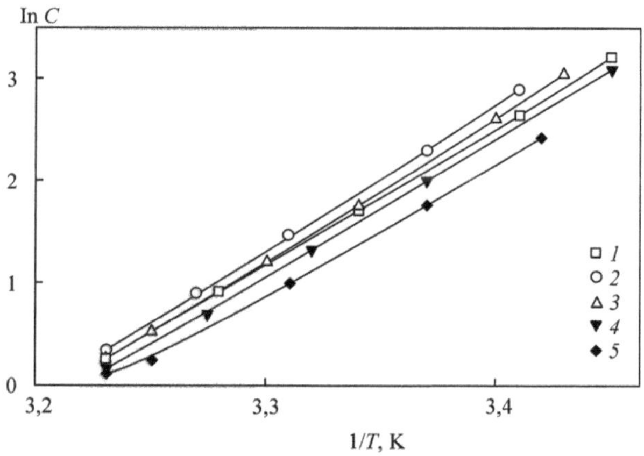

Fig. 3.23 Temperature dependence of the total content of GHWC in solutions of chlorides of alkaline metals (concentration 0.01 mol/dm^3) in Arrhenius' coordinates: *1*—distilled water, *2* – LiCl, *3*—NaCl, *4*—KCl, *5*—RbCl

where *C*—is the GHWC concentration, mol/dm^3; k_0—is the coefficient characterizing the frequency of collision of reacting molecules; *E*—is activation energy, kJ/mol; *R*—is the universal gas constant, J/(mol · K); *T*—is temperature, K.

For calculating the activation energy we will employ graphic relationship ln *C*—1/*T* and following ratios: tg $\alpha = \Delta Y/\Delta X$; $E = R$ tg α ($R = 8.31$ J/(mol · K)). The results of evaluation of the activation energy of destruction of clusters are given in Table 3.15.

From the data of Table 3.15 it follows, that pure water possesses the greatest resistance to heating (i.e., the greatest activation energy of destroying GHWC). In the solutions of the salts NaCl and KCl the activation energy of clusters is lower than in distilled water. Such salts as LiCl and RbCl are most conducive to cluster destruction of the type GHWC. It should be noted that in the destruction of GHWC the known ability of single-charged cations to positive or negative hydration is revealed [43]. Lithium cation as most hydrating and destroys clusters by holding water molecules in the hydrate shell. Contrary to this, Rb^+, which is characterized by negative

Table 3.15 Impact of salts (concentration 0.01 M) or magnification on the activation energy of destruction of water clusters

Indicator	LiCl	NaCl	Distilled water		KCl	RbCl
			Usual	Magnified		
E, kJ/mol	99.7	108.0	141.3	108.0	112.2	95.6

Table 3.16 Total content of GHWC in the water at different treatment modes

Experiment number	Interval of temperature variation, °C	Temperature variation rate, deg/min	Time after achieving temperature 20 °C (min)		
			0.0	10.0	20.0
1	100 → 20	2.7	15.3	15.3	15.3
2	100 → 20	16.0	14.9	15.0	14.9
3	0 → 20	2.7	22.2	21.4	20.2
4	0 → 20	16.0	32.4	25.0	23.8

hydration, destroys clusters, most likely owing to a strong decrease of the average density of the arrangement of water molecules near the cation.

Water magnetization decreases the activation destructive energy of GHWC at the level of the impact of the NaCl additive. Thus, when adding salts to the water or magnetization of the drinking water the content of GHWC in it decreases (in the first place of large dimensions), which is similar to water treatment by heating to >30 °C. To achieve a high degree of water declusterization of the clusters of the GHWC type, the water should be heated to 36 °C and higher. Thus, in a human organism the water is in a declusterized form in terms of GHWC. Unlike this the water in the channels of plants may contain GHWC is substantial concentrations.

In conclusion, we can note the ability of the water to the relaxation of the GHWC content. For this, the following experiments were carried out with the water from pump rooms whose total mineralization constituted 290 mg/dm^3 (see Table 3.16).

The water was heated to boiling and then cooled to 20 °C at different velocity (experiments 1 and 2). In the other case, water samples were cooled down to 0 °C then frozen and heated at various velocities to 20 °C (experiments 3 and 4). After achieving the temperature 20 °C in each of four samples we determined the amount of GHWC through certain time periods (0, 10 and 20 min).

As the results of the experiments have shown (see Table 3.16), in the water samples obtained by its heating from 0 to 20 °C the total content of clusters of the type GHWC is higher than in samples obtained by water cooling from 100 to 20 °C. This agrees with the data obtained above in terms of temperature dependences of the contents of clusters showing that at an increase of temperatures GHWC are destroyed and their concentration falls to zero. At a higher rate of water cooling (experiment 2) the content of clusters is slightly lower than in slow cooling (experiment 1). Consequently, slow cooling of declustered water is conducive to the mechanisms of restoring cluster structures in the water.

The capacity of the water to keep the previous structure in transition of the sample from low temperature to room one is manifested even more conspicuously (experiments 3 and 4). If the cooled water is heated fast (experiment 4) (it in full measure refers to "the melt water"), the destruction of GHWC will be inhibited. The concentration of GHWC for some time remains higher compared with the concentration characteristic of the given concentration. Consequently, the processes of formation and destruction of clusters, type GHWC, take a rather long time (in the order of seconds and minutes), which may be explained by orientation and migration polarization of particles and the presence of relaxation phenomena characteristic of water [61]. The water capacity to diminish the quantity and size of GHWC upon heating, to our mind, is one of its new structural anomalies.

It should also be noted that the studied variations of the water structure expressed in a diminished content of GHWC of various dimensions during heating to the temperature 32–38 °C most likely lie behind a number of other already known water anomalies. In particular, papers [67, 124] found that within the temperature interval 35–40 °C the water is characterized by the minimum of specific heat capacity; hot water (as maximally declusterized) may be frozen faster than cold one; water viscosity at >32 °C depends little on pressure, while within the interval of temperatures 30–50 °C water compressibility is in the minimum. At the temperature >38 °C dimensions the sound velocity in the water sharply decreases, which may be caused by a diminished content of GHWC of large dimensions. This, in turn, makes it possible to infer that GHWC are involved in the transfer of sound waves in a greater degree than the cluster depleted water.

3.6 Cluster Structure of Usual, Heavy and Light Water

In the mainstream of papers [28, 35, 59, 65, 106, 124–127] for the study of clusters in various water systems the authors of [128] considered the impact of changing the temperature for the content of clusters in heavy water (99.99 %th D_2O) and in water samples with the different content of deuterium. For the investigation of the water cluster structure the same technique as in paper [127] was used.

Light, heavy and usual water samples have been investigated. Light water contained 54 ppm or 0.0053997 % of deuterium, heavy one—99.99 % of D_2O (i.e. 9.999×10^9 ppm, was produced at Top-stay Ltd.) usual bidistilled (pump-room water)—154 ppm, which corresponds to 0.015398 % of D_2O. Heavy water was diluted with distilled one in ratios given in Table 3.17. In this case a sample was obtained with a certain content of deuterium, which was placed in a measuring spectrometric laboratory dish and experiments were carried out according to the technique described in paper [126].

In studying the temperature dependence of the contents of clusters, type GHWC, in 99.99 % of heavy water, the mode of its heating or cooling is fulfilled. A thermostat was used to set the temperature within the interval 15 to 37±0.2 °C (heating mode) and 37 to 10±0.2 °C (cooling mode). Diffractograms were measured every three degrees (separately in the heating mode and in the cooling mode). The sample was held at the same temperature for at least 10 min.

Table 3.17 Preparation of sample solutions for the experiment

Experiment number	Volumetric fraction D_2O W (%)	Solution volume V_p (ml)	Volume of distilled water added to sample D_2O V_{H_2O} (ml)
1	99.9900	1	–
2	90.9014	1.1	0.1
3	76.9189	1.3	0.2
4	66.6651	1.5	0.2
5	58.8240	1.7	0.2
6	50.0027	2	0.3
7	40.0052	2.5	0.5
8	25.0090	4	1.5

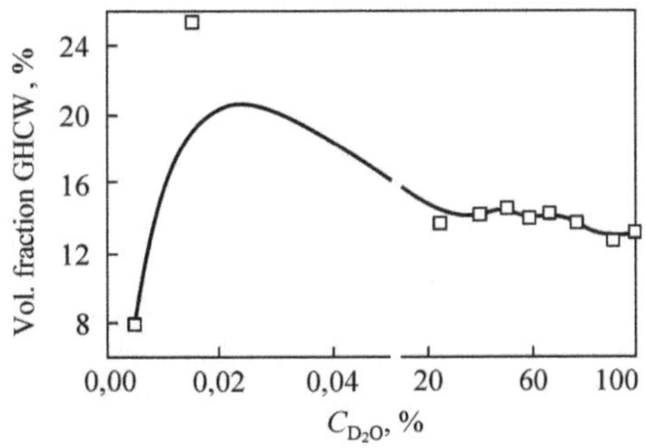

Fig. 3.24 Relationship between the GHWC content in the water and the D_2O concentration in the sample

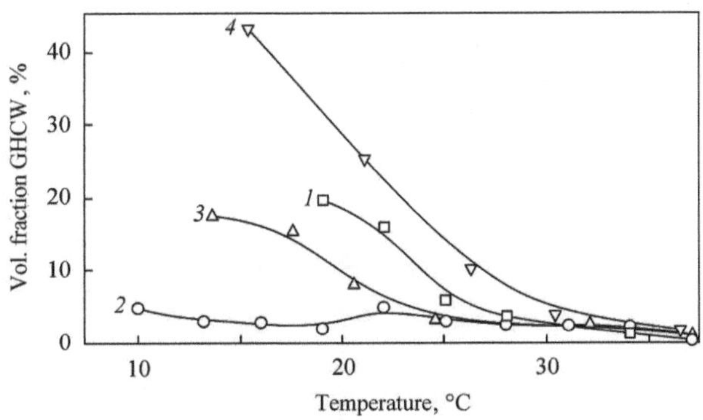

Fig. 3.25 Temperature dependence of the content of the GHWC type clusters in various water: *1*—heavy water, total fraction of D_2O 99.99 % (2×10^6 ppm D_2O), *2*—bidistilled (154 ppm D_2O), *3*—light (54 ppm D_2O), *4*—distilled (154 ppm D_2O)

Figure 3.24 shows the relationship between the content of GHWC in the water and the content of D_2O for a broad region of deuterium concentration. Figure 3.25 shows the temperature dependence of the content of clusters in water samples with different contents of deuterium; Fig. 3.26—the impact of heating and cooling on GHWC in the 99.99 %th D_2O. Design curves for the contents in the water of clusters of the type GHWC depending on water viscosity on temperature are shown in Fig. 3.27.

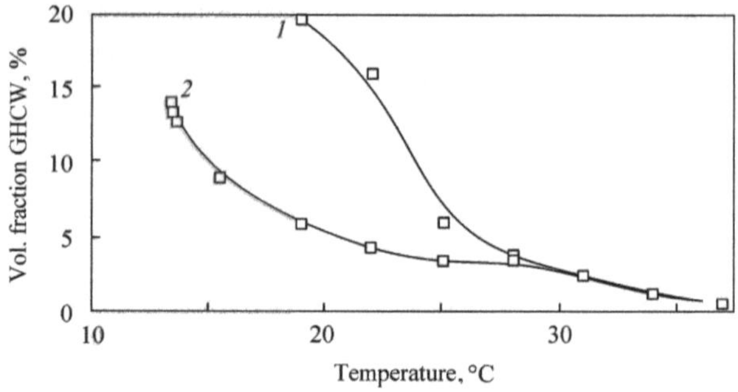

Fig. 3.26 Temperature dependence of the content of clusters in heavy water (99.99 % D_2O): *1*—heating, *2*—cooling

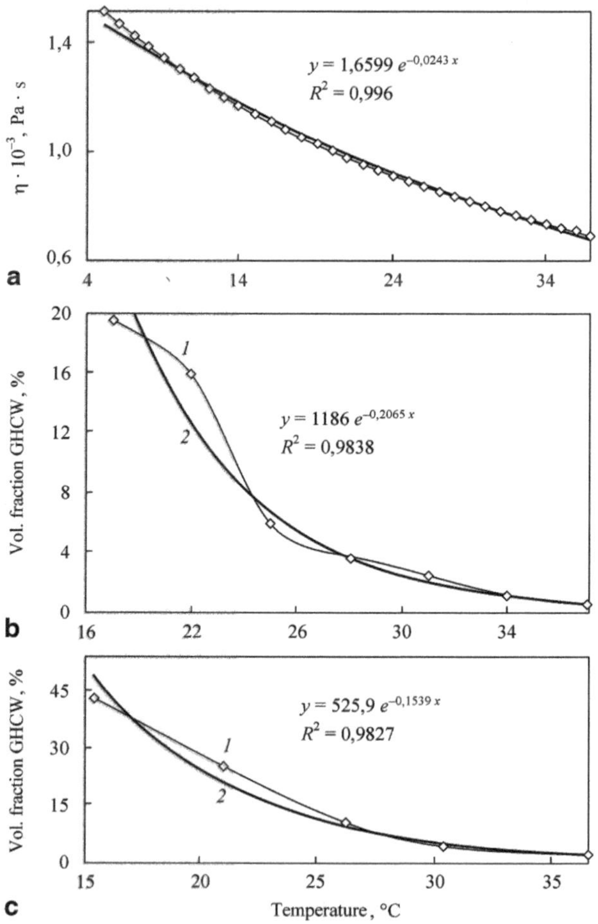

Fig. 3.27 Approximating curves of the relationship between water viscosity and temperature (**a**), experimental data of heating heavy (**b**) and distilled (**c**) water: *1*—experimental data, *2*—trend line

Since deuterium bonds in heavy water are stronger than normal hydrogen bonds [102], one should expect that in heavy water the GHWC concentration or their dimensions will be greater. Experiments showed that such effects are not observed.

It is known that, 99 % of deuterium in usual water is represented in the form of DHO. Simultaneous existence of deuterium and hydrogen bonds in the DHO solutions, to our mind, is conducive to cluster formation due to a higher polarizing effect of mixed dipoles.

As can be seen from the given graphic relationships, the addition of usual water to D_2O effectively does not change the content of gigantic clusters in the region of volumetric from 99.99 to 25 %. D_2O. It refers to GHWC fractions from 4 to 40 μ.

Figure 3.25, shows the temperature dependence of the content of clusters. It can be seen from the Fig. 3.25 that in distilled water one can observe the maximum amount of clusters; in bidistilled water, on the contrary, the amount of clusters is at the very low level and effectively does not depend on the temperature. Such data can be explained by the fact that newly prepared bidistilled water was used, i.e. it was prepared with preliminary thermal treatment. Paper [126] shows that, as a result of thermal treatment, GHWC are destroyed. For restoration of the initial amount of clusters, type GHWC, such water needs certain time to settle down in a calm state. This is required for the slow processes of orientation and migration polarization, which are conducive to self-organization of water molecules into clusters.

From experimental data on the impact of heating and cooling of the heavy water (see Fig. 3.26), one can see that the total volumetric share of GHWC due to heating decreases from 20 to 0.5 %. If one does water cooling from 37 to 28 °C, it can become obvious that effectively one cannot observe any changes, i.e. the happened destruction of clusters and the change of temperature within this interval do not affect restoration of its initial amount.

In cooling curve *2* (see Fig. 3.26) does not coincide with curve *1*, which is an evidence of the formation of clusters in much smaller amount. The presence of the hysteresis loop points to the fact that the formation of clusters follows the other mechanism compared with the process of their destruction (curve *1*). Both curves, most likely, converge into one at lower temperatures. Curves for distilled water (154 ppm) coincide at the temperature 8–10 °C. The obtained results show that the maximum volumetric share of cluster after destruction by preliminary heating constitutes 14 % of GHWC at the temperature 13.4 °C.

One more feature should be noted; this feature is acquired by the water owing to the presence of gigantic clusters. The water containing GHWC may be considered a disperse system, in which water clusters play the role of the disperse phase. Dimensions of GHWC investigated by us were within 2–40 μ and more. Such dimensions of the particles are typical of disperse systems. A very small difference between the densities of the disperse phase and the disperse medium are conducive to sedimentation and aggregative stability of cluster suspensions.

One of important characteristics of water is viscosity. It is known that the temperature dependence of viscosity is described by the exponential equation and is shown in Fig. 3.27a. Based on the quoted data the approximate curve is found.

The exponential nature of the dependence is determined by the fact that in the intermolecular interaction, determining viscosity, hydrogen bonds take part. Since strong cluster formation is observed in the water, it was interesting to compare temperature regularities of cluster formation and water viscosity.

Fig. 3.27b, c shows the approximate dependence, obtained by mathematical methods, for the curves of heating heavy and distilled water.

From design curves and equations corresponding to them it follows that the regularity of the variation of content of GHWC clusters for heavy and distilled water is well described by exponential dependence. This means that the destruction of clusters in heating water is subject to general regularities of the process both for usual and heavy water. Such regularities include the destruction of the structured ice-like framework of water, the existence of free (unbound) water molecules from the hollows of the ice-like framework and the formation as a result of a uniform mass not bound with GHWC. The difference between distilled and heavy water observed in the value of coefficients in the equations of the curves (see Fig. 3.27b, c), indicates a substantial difference of the energy binding water molecules in the case of hydrogen and deuterium bond. From the comparison of the equations of the curves shown in Fig. 3.27b, c with the equation for viscosity (see Fig. 3.27a), one can see that the processes under investigation are well described by exponential curves. Perhaps the phenomena being observed constitute similar regularities, while in the formation of clusters hydrogen bonds play an important role.

Thus, this study on the physics of water using contemporary methods has confirmed the possibility of the formation and existence of clusters of different dimensions. Unique effects of physical impact used in water treatment processes open new ways for perfecting the processes of preparing drinking water of high quality. The changes of the water structural composition, including the amount and concentration of clusters under the external effects, bear information about the evolution of the Earth's atmosphere and hydrosphere which is necessary for stabilization of the Earth's climate and ecological protection.

References

1. Antonchenko VYa (1986) Fizika vody (Physics of water). Nauk Dumka, Kiev
2. Gun'ko VM, Turov VV, Gorbik PP (2009) Voda na mezhfaznoj poverkhnosti (Water at interface). Nauk Dumka, Kiev
3. Castelman AW, Bowen KH (1996) Structure, energetic and dynamics of intermediate states of matter. J Phys Chem 100(31):12911–12944
4. Neilson GW, Mason PE, Ramos S et al (2001) Neutron and X-ray scattering studies of hydration in aqueous solutions. Philos Trans R Soc Lond A 359(8):1575–1591
5. Symons MCR (2001) Water structure: unique but not anomalous. Philos Trans R Soc Lond A 359(6):1631–1646
6. Antonchenko VY, Davydov AS, Ilyin VV (1991) Osnovy fiziki vody (Fundamentals of physics of water). Nauk Dumka, Kiev
7. Zatsepina GN (1998) Fizicheskiye svoystva i struktura vody (Physical properties and structure of water). Izd-vo MGU, Moscow

8. Samoilov OYa (1957) Struktura vodnykh rastvorov elektrolitov i gidratatsiya ionov (The structure of aqueous solutions of electrolytes and hydration of ions). Izd-vo AN SSSR, Moscow
9. Antonchenko VYa, Davydov AS, Zolotariuk AV (1985) Solution and proton motion in ice-like structures. Phys Stat Sol (B) 115(2):631–640
10. Plumridge TH, Waigh RDJ (2002) Water structure theory and some implications for drug design. Pharm Pharmacol 54(9):1155–1179
11. Colson SD, Dunning TH (1994) The structure of nature's solvent: water. Science 265:43–44
12. Kusalik PG, Svishchev IM (1994) The spatial structure in liquid water. Science 265:1219–1221
13. Smirnov VN, Syroyeshkin AV (2004) Supranadmolecular water complexes. Russ Chem J 48(2):125–135
14. Erdey-Gruz T (1974) Transport phenomena in aqueous solutions. Wiley, New York
15. Wilson KR, Cavalleri M, Rude BS et al (2000) Characterization of hydrogen bond acceptor molecules at the water surface using near-edge X-ray absorption fine-structure spectroscopy and density functional theory. J Phys Condens Matter 14(8):L221–L226
16. Kochnev IN, Vinnichenko MB, Smirnova LV (1989) Temperature anomalies of the absorption spectra and refractive index of water. Res Water Aqueous Syst Phys Meth 6:42–51
17. Vuks MF (1989) Dielectric polarization alcohol in solutions. Res Water Aqueous Syst Phys Meth 73:172–180
18. Voitylov VV, Kakorin SA, Trusov AA (1989) Electroorientation researches electrical conductivity of water-solid. Res Water Aqueous Syst Phys Meth 6:75–83
19. Zhukovsky AP (1981) Motivation of the continuum model of water structure by IR spectroscopy. J Struct Chem 22:56–63
20. Smirnov AN (2001) Acoustic emission during the chemical reactions and physico-chemical processes. Russ Chem J 45:29–34
21. Goncharuk VV, Kazimirov VP (1999) Structure of water. J Water Chem Technol 21(5):5–16
22. Anomalous properties of water (2012) http://www.lsbu.ac.uk/water/anmlies.html. Accessed: 28 Nov 2012
23. Nevzorov AN (2009) Some properties of metastable state of water: a new experiment interpretation. Measure World 8(18):38–45
24. Sillinger FH (1980) Water revisited. Science 209(4455):451–457
25. Lorenzen LH (2000) Microclustered water. US Patent 6033678, 7 March 2000
26. Bushuev YuG, Lyashchenko AK (1994) Computer simulation of clusters, cycles, and polyhedrons in water and solutions. J Phys Chem 68(3):470–476
27. Daschbach JL, Peden BM, Smith RS et al (2004) Adsorption, desorption, and clustering H2O on Pt (III). J Chem Phys 120(3):1516–1523
28. Smirnov AN, Lapshin VB, Balyshev AB et al (2005) Water structure: giant heterophase clusters of water. J Water Chem Technol 27(2):111–137
29. Vij JK, Hufnagel F (1989) Millimeter and submillimeter laser spectroscopy of water. Chem Phys Lett 155(2):153–156
30. Fajardo ME, Tam S (2001) Observation of the cyclic water hexamer in solid parahydrogen. J Chem Phys 115(15):6807–6810
31. Yarwood J, Sammon C, Mura C et al (1999) Vibrational spectroscopic studies of the diffusion and perturbation of water in polymeric membranes. J Mol Liquids 80(2):93–115
32. Loboda AA, Goncharuk VV (2009) Ab Initio calculation of water clusters. The vibratory analysis and isotopic effect. J Water Chem Technol 31(2):98–109
33. Loboda AA, Goncharuk VV (2009) Theoretical research into interaction of water clusters with ozone. J Water Chem Technol 31(4):213–219
34. Galashev AE, Pozharskaya GI, Chukanov VN (2002) Physicochemical properties of water clusters in the presence of HCl and HF molecules. Molecular dynamic simulation. J Struct Chem 43(3):458–466
35. Krishtal SP (2004) Structure of H-Bonded clusters in sub- and supercritical methanol and water. Abstract of Ph.Sc Thesis, Institute of Solution Chemistry, Russian Academy of Sciences

36. Hahalin AV (2006) The influence of a low-intensive electromagnetic field on water clusters in the presence of ions. Abst of Ph.Sc Thesis, Moscow State University.
37. Schafer S, Assadollahzadeh B, Mehring M et al (2008) Structure and electric properties of Sn-N clusters (N=6-20) from combined electric deflection experiments and quantum theoretical studies. J Phys Chem A 48(112):12312–12319
38. CNews R&D, Natural Science (2010) http://rnd.cnews.ru/naturs_cience/news/top/index_science.shtml. Accessed 05 Aug 2009
39. Goncharuk VV, Kamalov GL, Kovtun GA et al (2002) Kataliz. Mekhanizmy gomogennogo i geterogennogo kataliza, klasternyye podkhody (Catalysis: mechanisms of homogeneous and heterogeneous catalysis, cluster approaches). Nauk Dumka, Kiev
40. Frenkel J (1946) Kinetic theory of liquids. Oxford University Press, Oxford
41. Morrison SR (1977) The chemical physics of surfaces. Plenum Press, London
42. Mishchenko KP, Ravdel AA (1974) Kratkiy spravochnik fiziko-khimicheskikh velichin (The short reference book of physical-chemical values). Khimiya, Leningrad
43. Mishchenko KP, Poltoratsky GM (1974) Termodinamika i stroyeniye vodnykh i nevodnykh rastvorov elektrolitov (Thermodynamics and structure of aqueous and nanoqueous solution of electrolytes). Khimiya, Leningrad
44. Holloway MA, Holloway JW (2003) Micro-cluster liquids and methods of making and using them. US Patent 6521248, 18 Feb 2003
45. Bogatin JG, Podoyma V (1999) Magnetohydrodynamic device. US Patent 5866010, 2 Feb 1999
46. Lorenzen LH (1998) Process for preparing microclustered water. US Patent 5711950, 27 Jan 1998
47. Goncharuk VV, Malyarenko VV (2003) Mechanism of magnetic fields effects on aqueous systems. J Water Chem Technol 25(3):212–227
48. Toshimitsu H, Kazuhiro M (2003) Water having a superior ability to disperse oil and fats. US Patent Application No 20030035846, 20 Feb 2003
49. Johnson KH, Zhang B, Clark HC (1998) Water clusters and uses therefor. US Patent 5800576, 12 Jan 1998
50. Masakazu U (1999) Water and method for processing thereof. US Patent 5965007, 2 Nov 1999
51. Vainshelboim A, Matravers P, Ostrovskaya A et al (2005) Hair color application using cluster-modified water. US Patent 6905523, 14 July 2005
52. Ubbelohde AR (1978) Molten state of matter: melting and crystal structure. Wiley, New York
53. Seo M, Jang ES, Kim KJ et al (2012) Understanding anisotropic growth behavior of hexagonal ice on a molecular scale: a molecular dynamics simulation study. J Chem Phys 137(15):154503–154511
54. Borisova ME, Koikov SN (1979) Fizika dielektrikov (Physics of dielectric). Izd-vo Leningr un-ta, Leningrad
55. Churaev NV, Sobolev VD, Zorin ZM (1971) Thin liquid films and boundary layers. New York Academy Press, New York
56. Sermon PA (1980) Interaction of water with some silica. J Chem Soc Faraday Trans 76(8):885–888
57. Fripiat JJ (1971) Effects of temperature on apparent intensity of OH valence vibration of inorganic solids. Bull Groupe Fran Argiles 23(2):1–8
58. Deryagin BV (1983) Poverkhnostnyye sily i granichnyye sloi zhidkostey (Surface forces and boundary layers of liquids). Nauka, Moscow
59. Goncharuk VV, Malyarenko VV (2001) Changes in properties of water affected by electrochemical treatment. J Water Chem Technol 23(4):345–353
60. Fajardo ME, Tam S (2001) Observation of the cyclic water hexamer in solid parahydrogen. J Chem Phys 115(15):6807–6810
61. Malyarenko VV (1989) Thermodynamic assessment of the temperature of supercooled water, based on two-structure mode. Ukr Chem J 55(8):810–814

62. Ovcharenko FD, Malyarenko VV, Shilov VN (1977) Study of the electrical conductivity of suspension and of the electrical orientation of colloidal particles. Colloid J 39(1):73–79
63. Demidov AA, Fadeev VV (1980) Possibilities of finding the vertical distribution of impurities in water by laser probing. Rep Soviet Acad Sci 255(4):850–853
64. Glushkov SM, Panchishina IM, Fadeev VV (1989) Raman scattering spectra recorded in the course of the water-ice phase transition and laser diagnostics of heterophase water systems. Soviet J Quant Electron 16(4):843–852
65. Goncharuk VV, Smirnov VN, Syroyeshkin AV et al (2007) Clusters and gigantic heterophase water clusters. J Water Chem Technol 29(1):1–8
66. Timakov AA (2003) Main effects of light water. In abstract of the 8-th Russian international scientific conference of "Physical-chemical processes during selection of atoms and molecules", RSC Kurchatov Institute, Moscow, 6–10 Nov 2003
67. Franks F (2000) Water: a matrix of life. Royal Society of Chemistry, Cambridge
68. Good W (1973) The role of water in the origin of life and its function in the primitive gene. J Theor Biology 39(2):249–276
69. Fomey D, Jacox ME, Thompson WE (1993) The mid- and near-infrared spectra of water and water dimer isolated in solid neon. J Mol Spectrosc 157(2):479–493
70. Kell JS (1977) Effects of isotopic composition, temperature, pressure, and dissolved gases on the density of liquid water. J Phys Chem Ref Data 6:1109–1131
71. Frank HS, Wen W (1967) Structural aspects of ion-solvent inter action in aqueous solutions: a suggested picture of water structure. Discus Faraday Soc 24:133–140
72. Frisch MJ (2003) Gaussian 03. Revision B.05. Gaussian Inc., Pittsburg
73. Schaftenaar G, Noordik JH (2000) Molden: a pre- and post-processing program for molecular and electronic structures. J Comput-Aided Mol Design 14:123–134
74. Hasted JB (1972) Liquid water: dielectric properties. Plenum Press, New York
75. Dyke TR, Mack KM, Muenter JS (1977) The structure of water dimer from molecular beam electric resonance spectroscopy. J Chem Phys 66(2):498–510
76. Huang ZS, Miller RE (1989) High-resolution near-infrared spectroscopy of water dimer. J Chem Phys 91(11):6613–6631
77. Bentwood RM, Barnes AJ, Orville-Thomas WA (1980) Studies of intermolecular interactions by matrix isolation vibrational spectroscopy. Self association of water. J Mol Spectrosc 84(2):391–404
78. Pugliano N, Saykally RJ (1992) Measurement of quantum tunneling between chiral isomers of the cyclic water trimer. Science 257(5078):1937–1940
79. Dyke TR, Muenter JS (1972) Molecular beam electric deflection studies of water polymers. J Chem Phys 57(11):5011–5012
80. Cruzan J, Viant M, Brown M et al (1997) Terahertz laser vibration - rotation tunneling spectroscopy of the water tetramer. J Phys Chem 101(48):9022–9031
81. Csajka FS, Chandler D (1998) Transition pathways in a many-body system: application to hydrogen-bond breaking in water. J Chem Phys 109(3):1125–1133
82. Harker HA, Viant MR, Keutsch FN et al (2005) Water pentamer: characterization of the torsional-puckering manifold by terahertz VRT spectroscopy. J Chem Phys 109(29):6483–6497
83. Feynman R, Leighton R, Sands M (1977) Electricity and magnetism. Addison Wesley Publish Company, Boston
84. Kulskii LA, Dushkin SS (1988) Magnitnoye pole i processy vodoobrabotki (Magnetic field and water treatment processes). Nauk Dumka, Kiev
85. Gehr R, Zhai ZA, Finch JA (1995) Reduction of soluble mineral concentrations in CaSO4 saturated water using a magnetic field. Water Res 29(3):933–940
86. Parsons SA, Judd SJ, Stephenson T et al (1997) Magnetically augmented water treatment. Proc Safety Environ Prot 75(2):98–104
87. Klassen VI (1982) Omagnichivaniye vodnykh sistem (Magnetization of water system). Khimiya, Moscow
88. Bodnarenko NF, Gak YeZ (1984) Elektromagnitnyye yavleniya v prirodnykh vodakh (Electromagnetic phenomena in natural water). Gidrometeoizdat, Leningrad

89. Higashitani K, Oshitani J (1997) Measurements of magnetic effects on electrolyte solutions by atomic force microscope. Proc Safety Environ Prot 75(2):115–119
90. Kronenberg KJ (1985) Experimental evidence for effects of magnetic fields on moving water. IEEE Trans Magn 21(5):2059–2061
91. Lipus L, Krope J, Garbai L (1994) Magnetic water treatment for scale prevention. J Ind Chem 22:239–242
92. Brovchenko I, Paschek D, Geiger A (2000) Gibbs ensemble simulation of water in spherical cavities. J Chem Phys 113(22):5026–5036
93. Bernal JD (1956) The geometric constructions with water molecules. Russ Chem Rev 25:643–660
94. Drost-Hansen W, Clegg J (1979) Cell-associated water. Academic Press, London
95. Sastry S (2001) Water structure: order and oddities (news and views). Nature 409(1):300–301
96. Parcell EM (1984) Electricity and magnetism: Berkeley physics course. McGraw-Hill, New York
97. Liu K, Cruzan JD, Saykally RJ (1996) Water clusters. Science 3:929–993
98. Morkovnik AF, Okhlobystin OYu (1979) Inorganic radical-ions and their organic reactions. Russ Chem Rev 48(11):1968–2006
99. Swallow AJ (1973) Radiation chemistry. Longman, London
100. Malyarenko VV (1993) Nature of free radicals of humine acids. J Water Chem Technol 15(9/10):611–620
101. Rozantsev EG, Sholle VD (1979) Organicheskaya khimiya svobodnykh radikalov (Organic chemistry of free radicals). Khimiya, Moscow
102. Knorre DG, Krylova LF, Muzykantov VS (1990) Fizicheskaya khimiya (Physical chemistry). Vyssh Shkola, Moscow
103. Borisova ME, Koikov SN (1979) Fizika dielektrolitov (Physics of dielectric). Izd-vo Leningr un-ta, Leningrad
104. Gordon A, Ford R (1973) The chemist's companion. Wiley, New York
105. Shen YC, Zhang SY, Liu TZ et al (2001) Detection of laser-generated acoustic waves by an optical transducer operating at guide-wave mode. Prog Nat Sci 11(5):225–229
106. Goncharuk VV, Malyarenko VV (2007) The study of sound absorbtion in water. J Water Chem Technol 29(2):65–71
107. 107.Mason WP (1966) Physical acoustics. Academic Press, New York
108. Rozenberg LD (1968) Moshchnyye ul'trazvukovyye polya (Powerful ultrasonic fields). Nauka, Moscow
109. Kapustin AP (1973) Elektroopticheskiye i akusticheskiye svoystva zhidkikh kristallov (Electrooptical and acoustical properties of liquid crystal). Nauka, Moscow
110. Shakhparonov MI (1980) Mekhanizmy bystrykh protsessov v zhidkostyakh (Mechanisms of fast processes in liquids). Vyssh Shkola, Moscow
111. Qu KY, Jiang Y (2000) Studies on randomness and nucleation rate of freezing from homogeneous nucleation. Acta Phys Sin 49(11):2214–2219
112. Gillies JZ, Gillies CW, Suenram RD et al (1991) A microwave spectral and Ab Initio investigation of O3-H2O. J Mol Spectrosc 146:493–512
113. Zakharov II, Kolbasina OI, Semenyuk TN et al (1993) Molecular ozone-water complex. Ab initio calculation with inclusion of electron correlation. J Str Chem 34(3):359–362
114. Tachikawa H, Abe S (2003) Ozone-water 1:1 complexes O3-H2O: An ab initio study. Inorg Chem 42:2188–2190
115. Bradly RH, Mathieson I (1997) Chemical interactions of ultraviolet-light with wool fiber surfaces. J Colloid Interface Sci 194(2):338–343
116. Minaev BF, Khomenko EM, Loboda AA (2001) A quantum-chemical study of excited states of the HO2 radical. Rus Phys Chem J 75:1158–1162
117. Nelson DD, Zahniser MS (1994) A mechanistic study of the reaction of HO2 radical with ozone. J Phys Chem 98(8):2101–2104

118. Xu ZF, Lin MC (2007) Ab initio studies on the kinetics and mechanisms for O3 reactions with HO2 and HNO. Chem Physics Lett 440(1/3):12–18
119. Staehelln J, Holgne J (1985) Decomposition of ozone in water in the presence of organic solutes acting as promoters and inhibitors of radical chain reactions. Environ Sci Technol 19(12):1206–1213
120. Mvula E, Von Sonntag C (2003) Ozonolysis of phenols in aqueous solution. Org Biomol Chem 1(10):1749–1756
121. Poznyak T, Vivero J (2005) Degradation of aqueous phenol and chlorinated phenols by ozone. Ozone Sci Eng 27(6):447–458
122. Gorlov YuI, Goncharuk VV (2006) Electron structure of the ozone molecule and reactivity of ozone. J Water Chem Technol 28(3):211–221
123. Razumovskii SD (1990) Ozone in the process of restoring water quality. J DI Mendeleev All-Union Chem Soc 35(1):77–88
124. Chaplin MF (2006) Water structuring at colloidal surfaces. In: Blitz J, Gun'ko V (eds) Surface chemistry in biomedical and environmental science. NATO security through science series, Springer, The Netherlands
125. Pollack GH (2013) The fourth phase of water: beyond solid, liquid, and vapor. Ebner, Seattle
126. Goncharuk VV, Bulanova AV, Malyarenko VV (2006) The water as a heterogeneous cluster disperse system. Paper presented at the X Ukrainian-Polish symposium theoretical and experimental studies of interfacial phenomena and their technological applications. Lviv—Uslissia, 26–30 Sept 2006
127. Goncharuk VV, Orehova LA, Malyarenko VV (2008) Influence of temperature on water clusters. J Water Chem Technol 30(2):80–84
128. Goncharuk VV, Orehova LA, Malyarenko VV (2009) The cluster structure of heavy, usual and light water. Ukr Chem J 6:80–85

Chapter 4
Drinking Water: Factors Affecting the Quality of Drinking Water

Abstract The relative critical overview of indicators, which set norms to the drinking water and are used as background for regulatory instruments of the USA, the EU, the WHO, Ukraine and Russia, has been revealed. The analytical review of current problems in drinking water preparation technology at centralized drinking water treatment plants has been delivered. The role of natural organic compounds of natural origin in drinking water quality has been revealed. The application of different technological measures to prepare high quality drinking water has been analysed and explained. The influence of condition of water distribution system on the drinking water quality has been evaluated. It has been cited the examples of modern technological systems of drinking water treatment. A new concept of supply the population with a quality drinking water has been introduced. It is based on representation the need to consume adequate water by a person, which is genetically safe, comprises no man induced alloys, and is characterized by occurrence of such natural substances and microelements that are easiest to digest for a body from water.

Keywords The quality of drinking water · International regulatory instruments · Centralized drinking water treatment plants · Modern water treatment technologies · Man induced alloys, natural microelements · A new concept of supply the population with a quality drinking water

4.1 Ecological and Hygienic Classification of Centralized Drinking Water Supply Sources

In the Ukraine, the goal of creating and adhering to adequate quality and safety standards for drinking water is guaranteed by law. Legal documents such as the Water Code of Ukraine [1] and the Law of Ukraine "On drinking water and drinking water supply" [2] were written to uphold this goal. Codex found in these documents details a range of specific tasks centered around ecological, sanitary, and hygienic contents namely:

- A complex development of interrelated regulatory documents (standards and norms) for ensuring an ecological and sanitary-hygienic safety practice in the field of use and protection of water and water resources;

V. V. Goncharuk, *Drinking Water,* DOI 10.1007/978-3-319-04334-0_4,

- Organization and implementation of the condition of water monitoring with the goal of collecting, processing, saving, and analyzing information on the ecological state and quality of surface water bodies including sources of the centralized water supply, prediction of their changes, and the development of scientifically substantiated recommendations for making managerial decisions in the area of use, conservation, and reproduction of surface water resources;
- Realistic assessment of the ecological state and quality of surface water in regards to the population's centralized water supply based on ecological and hygienic indices and criteria corresponding to the requirements of the standards, techniques, methods, and technologies adopted by the European Union (EU);
- Establishment of ecological rules for water quality pertaining to surface water bodies including sources of drinking water supplies with the aim of assessing their ecological well-being and the determination of a complex web of water conservation measures to maintain and/or to achieve these rules. The ecological standards for water quality should contain scientifically justified permissible values of pollutants and water quality indices (general, physical, biological, chemical, and radiation). Thus, the degree of contamination in water bodies may be determined by relevant categories of water quality;
- Implementation of a complex web of water conservation measures for use, conservation of water, and reproduction of water resources based on scientifically justified state, interstate, and regional programs, and the establishment of three zones housing sanitary sources and facilities catering to the centralized drinking water supply, and certification of water bodies—sources of the centralized drinking water supply including a list of water quality indices corresponding to the state standard.

The aforementioned tasks, ecological and hygienic, are geared to the present-day level of requirements for their implementation and have been ratified by the EU member countries and other developed countries of the world. The requirements of these countries for water-management activities are contained in the UN and EU guidelines found in the Water Management Standards. The political, strategic, and other aspects of water issues [3–11] are reflected in this document. These principles and provisions can be seen from the following perspective.

Water management activities should be based on a systematic ecological approach. This approach is necessary for ensuring the integral, ecologically justifiable use of water and biological resources of water bodies and waterlogged land areas as well as overflow lands adjacent to them together with their biota. Additionally, human beings are regarded as a pivotal factor in the well-being of this natural environment. The integrity of humans' normal functions depends on their communities use age within these targeted areas. This usage can be monitored through the conservation of natural ecological systems and water resources, comprehensive protection of the population's health through ensuring a reliable drinking water supply, proper biological maintenance of recreational areas, and a consorted effort toward meeting the needs of industry, agriculture, fisheries, transportation, and energy. With a continued examination of a systematic ecological approach, conservation and restoration cannot be ignored.

One of the main goals of the water-management policy should be conservation and restoration of water ecological systems coupled with the achievement of an ecological-quality rating comparable to an unadulterated, natural condition. Due to the effects of hydroengineering and pollution found in industrial-municipal wastewater, the natural state of the surrounding ecology cannot be maintained and is difficult to restore. The crucial functions and natural properties intrinsic to water systems, generally referred to as basic ecological quality, should be the objective of this water-management policy. An exploration of the breadth and depth of water management and current policies is further revealed with the following information.

The draft of the EU Directive on Water Management Policy, and the Directive 2000/60/EC [6–8] both contain the classification of surface water categorizing the levels of ecological quality with excellent, high, good, moderate, poor, bad and the verbal characteristics of the biological, hydromorphological, and physicochemical components of ecological quality labeled with the first three levels of this classification. The directive for the achievement of the high and good ecological categories for these water bodies should be combined with the directive for the achievement of excellent and good quality of surface water based on the determination of its target values including hydrological conditions. The target values of the water quality indices represent those of their threshold values which are necessary for establishing "ecological naturalness", i.e. background quality of surface water which can be achieved over a certain time for intended bodies of water. Water quality should be estimated by abiotic (physico-chemical) and biotic indices in the aggregate while taking into account the hydrological conditions of these water bodies and features of their economic activities. The establishment of the target values of water quality indexes implies the complex use of surface water, in particular, for drinking water irrigate ion, watering livestock, fisheries, recreation, and aesthetic purposes. Water consumers place different requirements on surface water quality, therefore indexes and criteria of water quality in many aspects differ [4, 5]. These consumers are found around the world.

Since 1972 the water policy in the United States has been regulated by the Federal Clean Water Act of the Environmental Protection Agency. The statutory regulations found in this act detail the effluent policies and procedures for the restoration and maintenance of the country's surface water involving the hydrophysical, hydrochemical, and hydrobiological conditions intrinsic to them in the natural environment as well as attainment, as far as possible, of background water quality ensuring the existence of wild flora and fauna, shell-fish, and fish combined with a proper drinking water supply and human recreation on these marked water. Included in this act are statutes to regulate individual state governments in the development of "water quality standards" establishing the "target use" of individual bodies of water and the compilation of relevant quantitative criteria and descriptive definitions of water quality. Dovetailing these definitions is the scope and sequence of waterbodies classified by their applications. For instance, the values of water quality criteria not used for a public water supply differ from those used to supply the population with purified water since public health in these cases needs protection. As a rule, the locations of water suitable for a supply of drinking water are well known and are classified as such. Many water are used not only for supplying the population with potable water, but also for pisciculture (raising fish in ponds) and for catching fish and other hydrobionta [10]. The US is highly regulated in its water use.

The US has developed methods of calculating the ecological risk from toxic substances contained in water and affecting human health. This method takes into account the safety of several factors. First, potable water quality, purified water, is secured by the requirements set forth by the U.S. Safe Drinking Water Act. A significant regulation is to control the results of water purification of the contents of more than 80 substances of low concentrations—"levels of maximum pollution" [4]. Second, is the consumption of fish and mollusks potentially contaminated with toxic substances. When developing water quality criteria for public health protection, methods are used which make it possible to determine finite concentrations of pollutants distinguished by their biological effect. Such finite concentrations correspond to nonthreshold (carcinogenicity) and threshold (acute, subacute, and chronic toxicity) effects [4].

Third, is contact with water during passive or active recreational use. The criteria lies in the assessment of background concentrations of specific substances having toxic effect which do not present a substantial risk for human health as a result of prolonged action.

The above data indicates that the imperative of the water policy of the EU and the US is in close link with the goals of achieving a good and high status of water ecosystems for the formation of water of good and high quality suitable for the existence and reproduction of fish and aquatic invertebrates requiring water purity and at the same time fit for a public drinking water supply. Concern for the well-being of water ecosystems i.e., maintenance or restoration of their normal state is in alignment with the concern for the population's health and its supply of pure water as well as furnishing good recreational conditions including angling. This data is useful for the Ukraine.

The paradigm which gained a firm foothold in countries of western civilization, "ecologization" of water management policy including a strategy of the secure drinking water supply for the population, is also sufficiently fruitful and important for the Ukraine. At present, Ukraine, pursuant to government resolution [12], is carrying out harmonization of national regulatory conservation documents in conformity with relevant regulatory documents of highly-developed countries of Europe and the rest of the world. Being mindful of this circumstance, supporters came close to scientific justification of the Ukraine's important state standard "Sources of Centralized Drinking Water Supply" (in terms of surface water) developed on the initiative and with the support of the A.V. Dumanskii Institute of Colloid and Water Chemistry, National Academy of Sciences of Ukraine. Several advances have been made.

First, the basis of regulatory documents for the assessment of water quality in surface water bodies is the regulated system of indices and criteria for the composition and properties of water, i.e., the classification of surface water quality. The indexes of water quality are the components and properties of physical, chemical, and biological natures distinguished among other components and properties by having features and indexes characteristic only of them and having a generally-accepted scientific title and dimension. Second, for the ecological estimate of water quality in surface water bodies, hydrophysical, hydrochemical, and bacteriological indices are used as well as indicators characterizing the contents in water of specific substances of toxic and radiation effects. In addition, hydrophysical, hydrochemical, and hydrobacteriological indexes characterize the contents of specific substances used for the hygienic assessment of water quality in the same surface water bodies at the sections used as sources of the centralized drinking water supply for special sanitary-hygienic (health-related) indexes of water quality such as organoleptic, sanitary-microbiological, parasitological, and virological. Furthermore, criteria of water quality of surface water include such things as quality characteristics of water composition and properties in the form of their numerical values relating to the values of indexes. Numerical values of individual indexes in any surface water body including those used for centralized drinking water supply vary within a sufficiently broad range of values (from minimal to maximum) reflecting their regular variability during the functioning of water systems and during the formation of water quality under the effect of natural and anthropogenic factors. Keeping numerical values in mind, in order to judge the purity or contamination of water in surface water bodies including sources of centralized drinking water supply, scientists need a comparison scale that represents an arranged series of values of individual water quality indexes thus criteria. The ranks represent sections within boundaries of the whole range of criteria reflecting the degree of water purity and contamination and are referred to as classes, categories, grades, and indexes. Such integral characteristics of water quality as saprogenicity and trophicity reflecting the state of water ecosystems in their structural-functional integrity [13–16] are determined based on elementary criteria. By keeping these advances paramount, the Ukraine should find a thrust forward in its water quality management.

Believing that the classification of water quality should be an ambition for the future of the Ukrainian state water standard, the task is to construct a modern classification for the quality of surface bodies of water indicating acceptable sources of centralized water supplies. Taking into consideration the common nature of the main goal being the creation and adherence to adequate quality and safety standards for drinking water then the classification of surface water quality developed should meet certain ecological, hygienic, and technological requirements. By its very content, it may be that only ecological and hygienic values in a well-ordered system of ecological and hygienic indexes of the water composition and properties thereof will be taken to the technological stations for conditioning into drinking water. Thus, water conditioning is approached through the following methods.

To date two substantially different methodological approaches, monovariant and polyvariant, have been formed to assess water quality in surface water bodies and to resolve the issue on its suitability for centralized drinking water. First, given the monovariant condition, two possibilities arise. These include whether or not the water is pure and safe or whether or not it is dirty and dangerous. Additionally, when given the polyvariant multi-valued assessment of water quality, one may form a different idea about the degree of purity (or impurity) and make a judgment about different possibilities of its practical use. The following can be taken into consideration:

With the monovariant approach, criteria for water quality assessment are used as standards of ecological safety of water consumption seen as:

- maximum permissible concentrations (MPC) of substances in water bodies whose water is used to meet drinking, household, and other needs of the population—sanitary-hygienic (health-related) MPC;
- maximum permissible concentrations in water bodies whose water is used for fishery needs—fishery-economic MPC, and;
- maximum permissible concentrations of radioactive substances in water bodies whose water is used for meeting drinking, household, and other needs of the population—radiohygicnic MPC.

With the polyvariant approach, when employing its assessment of water quality, one uses classifications of their degree of purity (or pollution) in terms of ecological and hygienic indexes and criteria. Different types of such classifications are designed for unbiased characterization of water composition and properties under conditions of broad variability of their values in surface water bodies. Based on the differential characterization of surface water, a conclusion is drawn about the suitability of water for different purposes of water management. A peek into Ukraine's past is helpful in understanding this presentation.

The systems of health-related and fishery-economic MPC were approved in the former USSR as the primary criteria for assessing surface water quality. A characteristic feature of fishery MPC was the fact that they were much stricter than MPC of harmful substances established for bodies of water designed for drinking and household purposes owing much to the higher sensitivity of toxic water pollution in hydrobiota than humans [17]. A critical analysis of the system of fishery-economic MPC coupled with the explanation of a whole range of drawbacks was seen in certain studies [17]. The main drawbacks to consider as scientists solve Ukraine's water quality issues are as follows:

- ecological violations in bodies of water including the worsening of water quality occurs even when observing MPC because they are monitored only in places where the dumping of concentrated wastewater takes place and they do not extend to unconcentrated diffusion of polluted sources of water bodies (wash-off from agricultural lands, build-up areas, storage sites for fertilizers and pesticides, etc.);
- in most cases MPC cannot really be achieved given the existing technologies and financial capabilities of enterprises and therefore are generally violated;
- for water a great number of MPC were established (more than one thousand fishery and about 1,500 health-related ones), however, not more than 10% can be monitored of the total number of existing adverse substances, and;
- the development of a new MPC does not meet the penetration rate of new toxic substances into water bodies since this requires great organizational efforts and financial costs.

Long-term practice of using fishery and health-related MPC indicates that in the overwhelming majority of points when monitoring surface water quality the established MPC are broken. These violations differ from each other only by the composition of indices, frequency, and the magnitude of deviation from the standard. Thus both MPC systems do not ensure reliable protection of water bodies, but instead create an illusion as if most of the specific substances having toxic effects are controlled by nature-conservation agencies. In addition to the above drawbacks, interestingly both MPC systems have an advantage. The advantages are lists of experimentally established safety criteria for human health and hydrobionta provided they are found in minimal concentrations in the water. For health-related MPC, there are such concentrations of harmful substances at which they do not produce any direct or indirect effect on the population's health (when the organism is exposed to them in the course of a whole lifetime) and do not worsen hygienic conditions of water management [18]. Therefore, MPC values of harmful substances in the regulatory documents of the former Soviet Union, Ukraine, and the Russian Federation dealing with requirements for drinking water and water in sources of the centralized water supply system combined with regulatory documents displaying similar limiting values from the US, the EU, and the World Health Organization (WHO) may serve as chief orientations when establishing lower permissible levels of harmful substances for a class (category) of surface water of high quality. This may be seen in a rating such as 'very pure' which corresponds to conditions of drinking water. So, as more work is dedicated to the main goal, supporters can use present and past regulatory information to formulate the desired advancement.

In order to obtain an idea about such limiting values and use them further in this author work for the first edition of The Ecological-Hygienic Classification of Surface Water Quality—Sources of the Centralized Drinking Water Supply System. Tables 4.1 and 4.2 are composed with lists of limiting values of those indexes of the composition and properties of drinking water. Included in these tables are water sources of centralized drinking supplies in which information is integrated into the relevant regulatory documents of the former USSR, Ukraine, Russian, the US, and two international organizations. In these two tables, the limiting values are broken down into 8 groups of indexes with tables specific to 513 limiting values relating to 158 indexes. The greatest number of the limiting values (97) is specified in the WHO regulatory documents for drinking water, in SaNPiN standards of the Russian Federation 2.1.4.559-96 (94), and in the standards of the US Environmental Protection Agency (81). Ukrainian SaNPiN standards number 383 from December 23, 1996 includes only 36 limiting substances. The greatest number of limiting values for harmful inorganic substances (priority) is given in Russian SaNPiN standards 2.1.4.559-96 (41), and for harmful organic substances (priority) in the WHO standards for drinking water (61) (see Table 4.3). A further look at the policies of surrounding countries is useful.

Since 1975 the governments of the member states of the European Union in their activities to secure safe drinking water for the population have been guided by Directive 75/440/EEC dealing with the requirements for surface water designed for drinking water intake in member countries of the Community [26]. Annex 2 of this Directive is referred to as "Quality Characteristics of Surface Water" which are used for obtaining quality drinking water. This characterization in fact represents

Table 4.1 List of limiting values for indexes of composition and properties of potable water and water in sources of the central water supply system according to standards of the USSR, Ukraine, Russian Federation, and the USA, and International Agencies (all groups of indexes, except indexes for the content of harmful organic substances—priory)

Water quality-indexes	Measuring units	The USSR		Ukraine	Russian Federation		Environmental Protection Agency (EPA)		EU	WHO
		GOST Standard 2874-82 [19]	SaNPiN №4630-88 [18]	SaNPiN №383 of Dec. 23, 1996 [20]	SaNPiN 2.1.4.559-96 [21]	List of fishery MPC, 1995 [22]	MCLG—Maximum contaminant level goal-desirable [23]	MCL—Maximum contaminant level-mandatory [23]	EU Directive 98/83 of Nov. 3, 1998 [24]	WHO standards for potable water [25]
Organoleptic indexes										
Smell	Point	≤2	≤1	≤2	≤2	–	Values for odor, and after-taste are normalized, but units are not compared with units enforced in Ukraine		–	–
Taste	Point, DI[a]	≤2	–	≤2	≤2	–			–	–
Color	Degrees, Pt-Co-scale	20	Should not show in water column 20 cm	20[c](35)	20	–		15	20	15
Content of suspended substances	NF[b] (mg/dm³) at 20 °C	–	–	0.5 (1.5)	2.6	–	–	0.5–1.0	4	5(1)
General physico-chemical indicators										
(pH)[c] index	pH units	6.0–9.0	6.5–8.5	6.5–8.5	6.0–9.0	–	–	6.5	6.5–8.5	6.5–8.5
Total mineralization (contents of salts)	mg/dm³	1,000	1,000	1,000(1,500)[c]	1,000	–	–	500	1,500	1,000

Table 4.1 (continued)

Water quality-indexes	Measuring units	The USSR		Ukraine	Russian Federation		Environmental Protection Agency (EPA)		EU	WHO
		GOST Standard 2874-82 [19]	SaNPiN №4630-88 [18]	SaNPiN №383 of Dec. 23, 1996 [20]	SaNPiN 2.1.4.559-96 [21]	List of fishery MPC, 1995 [22]	MCLG—Maximum contaminant level goal-desirable [23]	MCL—Maximum contaminant level-mandatory [23]	EU Directive 98/83 of Nov. 3, 1998 [24]	WHO standards for potable water [25]
Total hardness	mmol/dm^3	7.0	–	7.0(10.0)[c]	7.0	–	–	–	1.2	–
Sulfates	mg/dm^3	500.0	500	250(500)[c]	500	–	–	250	250	250
Chlorides	mg/dm	350	350	250(3 50)[c]	350	–	–	250	250	250
Permanganate oxidizability	mgO/dm^3	–	–	4.0	5.0	–	–	–	5.0	–
Alkalinity	$mgHCO_3^-$/dm^3	–	–	–	–	–	–	–	30.0	–
Oxygen content[c]	mgO_2/dm^3	–	>4.0	–	–	–	–	–	–	–
BOD_{tot}	mgO_2/dm^3	–	>3.0	–	–	–	–	–	–	–
Total organic carbon	mg/dm^3	–	–	3.0	–	–	–	–	–	–
COD	mgO/dm^3	–	>15.0	–	–	–	–	–	–	–
Microbiological indexes										
Total number of bacteria	Number of colony-forming bacteria, (CFB/cm^3)	≤100	Water should not contain pathogens	≤100	50	–	–	500	10 (at 22 °C) 100 (at 27 °C)	

Table 4.1 (continued)

Water quality-indexes	Measuring units	The USSR		Ukraine	Russian Federation		Environmental Protection Agency (EPA)		EU	WHO
		GOST Standard 2874-82 [19]	SaNPiN №4630-88 [18]	SaNPiN №383 of Dec. 23, 1996 [20]	SaNPiN 2.1.4.559-96 [21]	List of fishery MPC, 1995 [22]	MCLG—Maximum contaminant level goal-desirable [23]	MCL—Maximum contaminant level-mandatory [23]	EU Directive 98/83 of Nov. 3, 1998 [24]	WHO standards for potable water [25]
Number of bacteria of a group of colon bacilli (coliform microorganisms)—BGCB index	CFB/dm^3	≤3	–	≤3	None	–	–	5 % smpl/month	None	None
Number cf thermostable colon bacilli (fecal coliforms—FC index)	CFB/100 cm^3	–	–	None	None	–	–	–	None	None
Number of pathogenic microorganisms	CFB/dm^3	–	–	None	–	–	–	None	None	

Table 4.1 (continued)

Water quality-indexes	Measuring units	The USSR		Ukraine	Russian Federation		Environmental Protection Agency (EPA)		EU	WHO
		GOST Standard 2874-82 [19]	SaNPiN №4630-88 [18]	SaNPiN №383 of Dec. 23, 1996 [20]	SaNPiN 2.1.4.559-96 [21]	List of fishery MPC, 1995 [22]	MCLG—Maximum contaminant level goal-desirable [23]	MCL—Maximum contaminant level-mandatory [23]	EU Directive 98/83 of Nov. 3, 1998 [24]	WHO standards for potable water [25]
Number of fecal Streptococci	In 100 dm^3	–	–	–	–	–	–	–		–
Number of sulfatereducing clostridia	In 100 dm^3	–	–	None		–	–	–	<1	–
Number of enterococci	In 100 mL	–	–	–	–	–	–	20	1,000	10,000
Lactopositive colon bacillus bacteria	In 1 dm^3	–	≤10000	–	–	–	–	–	–	–
Virological indexes										
Numbers of coliphages	In 1 dm^3	–	≤100	None	None	–	–	–	–	–
Number of enteroviruses	In 10 dm^3	–	–	None						

Table 4.1 (continued)

Water quality-indexes	Measuring units	The USSR		Ukraine	Russian Federation		Environmental Protection Agency (EPA)		EU	WHO
		GOST Standard 2874-82 [19]	SaNPiN №4630-88 [18]	SaNPiN №383 of Dec. 23, 1996 [20]	SaNPiN 2.1.4.559-96 [21]	List of fishery MPC, 1995 [22]	MCLG—Maximum contaminant level goal-desirable [23]	MCL—Maximum contaminant level-mandatory [23]	EU Directive 98/83 of Nov. 3, 1998 [24]	WHO standards for potable water [25]
Parasitilogical indexes										
Number of pathogenic colon protozoa (except cysts of lamblia, Cryptosporidia. etc)	In 25 dm^3	–	Water should not contain pathogens	None	None	–	–	None	–	–
Number of colon helminthes (eggs and larvae)	In 25 dm^3	–		None	None	–	–	–	–	–
Radioactivity indexes										
Total activity of β-emitters	Bq/dm^3	–	–	1.0	–	–	–	–		1.0

Table 4.1 (continued)

Water quality-indexes	Measuring units	The USSR		Ukraine	Russian Federation		Environmental Protection Agency (EPA)		EU	WHO
		GOST Standard 2874-82 [19]	SaNPiN №4630-88 [18]	SaNPiN №383 of Dec. 23, 1996 [20]	SaNPiN 2.1.4.559-96 [21]	List of fishery MPC, 1995 [22]	MCLG—Maximum contaminant level goal-desirable [23]	MCL—Maximum contaminant level-mandatory [23]	EU Directive 98/83 of Nov. 3, 1998 [24]	WHO standards for potable water [25]
Annual dose of individual β -emitters for one person	mSv	–	–	–	–			–		–
Tritium (I^3)	Bq/dm^3	–	–	–	–	–	–	–	100	–
Cobalt-60	Bq/dm^3	–	–	–	–	–	–	0.005		–
Strontium-89	Bq/dm^3	–	–	–	–	–	–	0.003		–
Strontium-90	Bq/dm^3	–	–	–	–	–	–	0.020		–
Iodine-129	Bq/dm^3	–	–	–	–	–	–	0.080		–
Iodine-131	Bq/dm^3	–	–	–	–	–	–	0.016		–
Cesium-134	Bq/dm^3	–	–	–	–	–	–	0.014		–
Cesium-137	Bq/dm^3	–	–	–	–	–	–	0.009		–
Lead-210	Bq/dm^3	–	–	–	–	–	–	0.95		–
Radium-228	Bq/dm^3	–	–	–	–	–	–	0.20		–
Total indication dose	rβv per year	–	–	–	–	–	–	–	0.1	–
Total activity of α-emitters	Bq/dm^3	–	–	0.1	–	–	–	–		0.1
Harmful inorganic substances (priority)										
Aluminum[d]	mg/dm^3	0.5	0.5	0.2(0.5)	0.5	–	–	0.2	0.2	0.2

Table 4.1 (continued)

Water quality-indexes	Measuring units	The USSR		Ukraine	Russian Federation		Environmental Protection Agency (EPA)		EU	WHO
		GOST Standard 2874-82 [19]	SaNPiN №4630-88 [18]	SaNPiN №383 of Dec. 23, 1996 [20]	SaNPiN 2.1.4.559-96 [21]	List of fishery MPC, 1995 [22]	MCLG—Maximum contaminant level goal-desirable [23]	MCL—Maximum contaminant level-mandatory [23]	EU Directive 98/83 of Nov. 3, 1998 [24]	WHO standards for potable water [25]
Ammonium nitrogen (NH_3 and NH_4)	mg/dm^3	–	2.0 (ammonia for nitrogen)	–	–	0.05 (ammonia)	–	–	0.5	1.5
Barium	mg/dm^3	–	0.1	0.1	0.1	0.74	–	2.0	0.1	0.7
Beryllium	mg/dm^3	0.0002	0.0002	–	0.0002	0.0003	–	0.004	–	–
Boron	mg/dm^3	–	0.5	–	0.5	0.5	–	–	1.0	0.3
Vanadium	mg/dm^3	–	0.1	–	0.1	0.001	–	–	–	–
Bismuth	mg/dm^3	–	0.1	–	0.1	–	–	–	–	–
Tungsten	mg/dm^3	–	0.05	–	0.05	0.0008	–	–	–	–
Europium	mg/dm^3	–	–	–	0.3	–	–	–	–	–
Iron[d]	mg/dm^3	0.3	0.3	0.3	0.3	0.1	0.3	–	0.2	0.3
Cadmium	mg/dm^3	–	0.001	–	0.001	0.005	–	0.005	0.005	0.003
Potassium	mg/dm^3	–	–	–	–	50.0	–	–	12.0	–
Calcium	mg/dm^3	–	–	–	–	180.0	–	–	100.0	–
Cobalt	mg/dm^3	–	0.1		0.1	0.01		–		–
Silicon	mg/dm^3	–	10.0		10.0			–		–
Lithium	mg/dm^3	–	0.03	–	0.03	–	–	–	–	–
Magnesium	mg/dm^3	–	–	–	–	40.0	–	–	50.0	–
Manganese[d]	mg/dm^3	0.1	0.1	0.1	0.1	0.01	–	0.05	0.05	0.5(0.1)
Copper[d]	mg/dm^3	1.0	1.0	1.0	1.0	0.001	–	1.0–1.3	2.0	2.0(1.0)
Molybdenum	mg/dm^3	0.25	0.25	–	0.25	0.0012	–	–	–	0.07

Table 4.1 (continued)

Water quality-indexes	Measuring units	The USSR		Ukraine	Russian Federation		Environmental Protection Agency (EPA)		EU	WHO
		GOST Standard 2874-82 [19]	SaNPiN №4630-88 [18]	SaNPiN №383 of Dec. 23, 1996 [20]	SaNPiN 2.1.4.559-96 [21]	List of fishery MPC, 1995 [22]	MCLG—Maximum contaminant level goal-desirable [23]	MCL—Maximum contaminant level-mandatory [23]	EU Directive 98/83 of Nov. 3, 1998 [24]	WHO standards for potable water [25]
Arsenic	mg/dm^3	0.05	0.05	0.01	0.05	0.05	–	0.05	0.01	0.01
Sodium[d]	mg/dm^3	–	200.0	–	200.0	120.0	–	–	200.0	200.0
Nickel	mg/dm^3	–	0.1	0.1	0.1	0.01	–	–	0.02	0.02
Niobium	mg/dm^3	–	0.01	–	0.01	–	–	–	–	–
Nitrates	mg/dm^3	45.0	45.0	45.0	45.0	400.0	–	44.0	50.0	50.0
Nitrites	mg/dm^3	–	3.3	–	3.0	0.08	–	3.3	0.5	3.0
Mercury	mg/dm^3		0.0005		0.0005	None (0.00001)	–	0.002	0.001	0.001
Rubidium	mg/dm^3	–	0.1	–	0.1	0.1	–	–	–	–
Samarium	mg/dm^3	–	–	–	0.024	–	–	–	–	–
Lead	mg/dm^3	0.03	0.03	0.01	0.03	0.1	–	0.015	0.01	0.01
Selenium	mg/dm^3	0.001	0.01	0.01	0.01	0.0016	–	0.05	0.01	0.01
Silver	mg/dm^3	–	0.05	–	0.05	–	0.1	–	0.01	–
Carbohydrate[d]	mg/dm^3	–	–	–	0.03	–	–	–	–	0.05
Strontium (stable)	mg/dm^3	7.0	7.0	–	7.0	10.0	–	–	–	–
Antimony	mg/dm^3	–	0.05	–	0.05	–	–	0.006	0.006	0.005
Thallium	mg/dm^3	–	0.0001	–	0.0001	–	–	0.002	–	–
Tellurium	mg/dm^3	–	0.01	–	0.01	0.0028	–	–	–	–
Phosphorus elementary	mg/dm^3	–	0.0001	–	0.0001	-	–	–	–	–

Table 4.1 (continued)

Water quality-indexes	Measuring units	The USSR		Ukraine	Russian Federation		Environmental Protection Agency (EPA)		EU	WHO
		GOST Standard 2874-82 [19]	SaNPiN №4630-88 [18]	SaNPiN №383 of Dec. 23, 1996 [20]	SaNPiN 2.1.4.559-96 [21]	List of fishery MPC, 1995 [22]	MCLG—Maximum contaminant level goal-desirable [23]	MCL—Maximum contaminant level-mandatory [23]	EU Directive 98/83 of Nov. 3, 1998 [24]	WHO standards for potable water [25]
Fluorides	mg/dm^3	0.7–1.5	0.7–1.2	1.5	1.5	+0.05 to background but not higher 0.75 mg/dm^3	–	2.02–4.0	1.5	1.5
Chloride										
Active	mg/dm^3	–	None	–	–	–	–	–	–	0.5–5.0
residual free	mg/dm^3	0.3–0.5	–	0.3–0.5	–	–	–	–	–	–
residual bound	mg/dm^3	0.8–1.2	–	0.8–1.2	–	–	–	–	–	–
Chromium (III)	mg/dm^3	–	0.5	–	0.5	0.07	–	0.1	–	–
Chromium (VI)	mg/dm^3	–	0.05	–	0.05	0.02	–	–	0.05	0.05
Cyanides	mg/dm^3	–	0.1	–	0.035	0.05	–	0.2	0.05	0.07
Zinc[d]	mg/dm^3	5.0	1.0	–	5.0	0.01	5.0	–	5.0	3.0

[a] DI—dilution index (until disappearance of odor and aftertaste)

[b] NF—nephelometric turbidity units

[c] Values, shown in brackets, are permitted allowing for a specific situation

[d] Indexes, which under certain range of values have organoleptic properties

Table 4.2 List of limiting values of composition and properties of drinking water and water in sources of centralized water supply in compliance with regulatory documents of the USSR, Ukraine, Russian Federation, the USA, and International Organization (indexes of the contents of harmful organic substances—priority)

Substances	Measuring units	USSR	Ukraine	Russian Federation		US-EPA	EU	WHO
		SaNPiN N4630-88 [18]	SaNPiN N 383 of Dec. 23, 1996 [20]	SaNPiN 2.1.4.559-96 [21]	List of fishery MPC, 1995 [22]	MCL [23]	EU Directive 98/83 EU of Nov 03, 1998 [27]	WHO standards for drinking water [25]
Harmful organic substances								
Chlorinated alkanes								
Carbon tetrachloride	μg/dm^3	–	–	6	None (0.000014)	5	–	2
Dichloromethane	μg/dm^3	–	–	7.5	–	5	–	20
1,2-Dichloroethane	μg/dm^3	20	–	–	100	5	3	30
1,1,1-Dichloro-ethane	μg/dm^3	–	–	10,000	–	200	–	2,000
1,1,2-Dichloro-ethane	μg/dm^3	–	–	–	–	5	–	–
Chlorinated ethylenes								
Vinyl chlorides	μg/dm^3	50	–	50	None (0.000008)	5	0.5	5
1,1-dichloroeth-ylene	μg/dm^3	–	–	–	100	7	–	30
1,2-dichloroeth-ylene	μg/dm^3	–	–	–	–	170	–	50
Trichloroethylene	μg/dm^3	60	–	–	10	5	10	70
Tetrachlorethylene	μg/dm^3	20	–	–	160	5	10	40
Aromatic hydrocarbons								
Benzene	μg/dm^3	50	–	10	50	5	–	0.7; 10
Toluene[a]	μg/dm^3	50	–	500	50	1,000	–	700(24–700)
Xylenes[a]	μg/dm^3	–	–	50	50	10,000	–	500(20–1,800)

Table 4.2 (continued)

Substances	Measuring units	USSR	Ukraine	Russian Federation		US-EPA	EU	WHO
		SaNPiN N4630-88 [18]	SaNPiN N 383 of Dec. 23, 1996 [20]	SaNPiN 2.1.4.559-96 [21]	List of fishery MPC, 1995 [22]	MCL [23]	EU Directive 98/83 EU of Nov 03, 1998 [27]	WHO standards for drinking water [25]
Harmful organic substances								
Ethylbenzene[b]	μg/dm^3	10	–	10	1	700	–	300(2.4–200)
Styrene[b]	μg/dm^3	100	–	100	100	100	–	20
Polycyclic aromatic carbohydrates								
Benz (a) pyrene	μg/dm^3	0.005	–	0–5	–	0.2	0.01	0.6
Chlorinated benzenes								
Monochlorobenzene[a]	μg/dm^3	–	–	20	–	100	–	300(10–120)
1,2-Dichlorobenzene[a]	μg/dm^3	2	–	2	1	600	–	1,000(1–10)
1,4- Dichlorobenzene[a]	μg/dm^3	–	–	–	–	–	–	300(0.3–30)
Trichlorobenzene	μg/dm^3	30	–	30	1	70	–	20(5–50)
Other compounds								
Acrylaldehyde	μg/dm^3	–	–	20	–	–	–	–
Di (2-ethylhexyl) adilat	μg/dm^3	–	–	–	–	400	–	80
Di (2-ethylhexyl) phthalate	μg/dm^3	–	–	–	–	6	–	8
Acrylamide[b]	μg/dm^3	10	–	10	None	TT[b]	–	0.5
Epichlorohydrin[b]	μg/dm^3	10	–	10	–	TT[b]	0.1	0.4
Hexachlorobutadiene	μg/dm^3	10	–	10	–	–	–	0.6
Hexachlorocyclopentadiene	μg/dm^3	1	–	1	–	50	–	–

Table 4.2 (continued)

Substances	Measuring units	USSR	Ukraine	Russian Federation		US-EPA	EU	WHO
		SaNPiN N4630-88 [18]	SaNPiN N 383 of Dec. 23, 1996 [20]	SaNPiN 2.1.4.559-96 [21]	List of fishery MPC, 1995 [22]	MCL [23]	EU Directive 98/83 EU of Nov 03, 1998 [27]	WHO standards for drinking water [25]
Harmful organic substances								
Ethylenediamine-tetraacetic acid, EDTA	µg/dm^3	–	–	–	–	–	–	200
Trinitroloacetic acid	µg/dm^3	–	–	–	–	–	–	200
SAS[a]	µg/dm^3	–	–	500	–	–	–	–
Heteroorganic compounds								
Dialkyl tin (compounds)	µg/dm^3	–	–	2	–	–	–	–
Tributyl tin oxide	µg/dm^3	–	–	4	–	–	–	2
Bis (tributyl tin) oxide	µg/dm^3	–	–	0.2	–	–	–	–
Tetraethyl tin	µg/dm^3	0.2	–	0.2	–	–	–	–
Tributil tin methacrylate	µg/dm^3	0.2	–	0.2	–	–	–	–
Tetraethyl lead	µg/dm^3	None	–	None	–	–	–	–
Pesticides								
Individual pesticides[c]	µg/dm^3	–	–	–	–	–	0.1	–
Pesticides (total content)[d]	µg/dm^3	–	0.1	–	–	–	0.5	–
Alachlor	µg/dm^3	–	–	–	–	2	–	20
Aldicarb	µg/dm^3	–	–	–	–	–	–	10

Table 4.2 (continued)

Substances	Measuring units	USSR	Ukraine	Russian Federation		US-EPA	EU	WHO
		SaNPiN N4630-88 [18]	SaNPiN N 383 of Dec. 23, 1996 [20]	SaNPiN 2.1.4.559-96 [21]	List of fishery MPC, 1995 [22]	MCL [23]	EU Directive 98/83 EU of Nov 03, 1998 [27]	WHO standards for drinking water [25]
Harmful organic substances								
Aldrin (dieldrin)	μg/dm³	–	–	–	None	–	0.03	0.03
Atrazine	μg/dm³	–	–	–	5	3	–	2
Bentazon	μg/dm³	–	–	–	–	–	–	30
Carbofuran	μg/dm³	–	–	–	–	40	–	5
Chlorotoluron	μg/dm³	–	–	–	–	–	–	30
Chlordane	μg/dm³	–	–	–	–	2	–	0.2
DDT	μg/dm³	2	–	2	None	–	–	2
1,2-Dibromo-3-chloropropane	μg/dm³	–	–	10	–	0.2	–	1
2,4-Dichlorophenoxy acetic acid butyl ester	μg/dm³	–	–	30	–	70	–	30
1,2-Dichloropropane	μg/dm³	400	–	400	–	5	–	20
1,3-Dichloropropane	μg/dm³	400	–	400	–	–	–	20
Ethylenedibromide	μg/dm³	–	–	–	–	0.05	–	–
Heptachlor and heptachloroepoxide	μg/dm³	–	–	50	0.5	0.6	0.03	0.03
Hexachlorobenzene	μg/dm³	50	–	–	–	1	–	1
Dalapon (sodium 2,2-dichloropropionic acid)	μg/dm³	–	–	2,000	3,000	200	–	–

Table 4.2 (continued)

Substances	Measuring units	USSR	Ukraine	Russian Federation		US-EPA	EU	WHO
		SaNPiN N4630-88 [18]	SaNPiN N 383 of Dec. 23, 1996 [20]	SaNPiN 2.1.4.559-96 [21]	List of fishery MPC, 1995 [22]	MCL [23]	EU Directive 98/83 EU of Nov 03, 1998 [27]	WHO standards for drinking water [25]
Harmful organic substances								
Dinosob (2(1) methyl propyl-4, 6 dinitrophenol	μg/dm^3	–	–	100	–	7	–	–
Isoproturon	μg/dm^3	–	–	–	–	–	–	9
Lindane	μg/dm^3	–	–	2	–	0.2	–	2
2-Methyl-4-chloro-phenoxy acetic acid (MCPA)	μg/dm^3	–	–	–	–	–	–	2
Methoxychlor	μg/dm^3	–	–	–	–	40	–	20
Metolachlor	μg/dm^3	–	–	–	–	–	–	10
Molinate	μg/dm^3	–	–	–	–	–	–	6
Pendimethaline	μg/dm^3	–	–	–	–	–	–	20
Pentachlorophenol	μg/dm^3	–	–	–	–	1	–	9
Permetrine	μg/dm^3	–	–	–	–	–	–	–
Propanil	μg/dm^3	–	–	–	30	–	–	–
Pyridate	μg/dm^3	–	–	–	–	–	–	–
Simazine	μg/dm^3	–	–	None	2.4	4	–	2
Trifluoralin	μg/dm^3	–	–	–	–	–	–	20
Chlorophenoxyherbicides (except 2.4 D)								
2,4 DV	μg/dm^3	–	–	500	–	–	–	90
Dichlorprop	μg/dm^3	–	–	–	–	–	–	100
Phenoprop	μg/dm^3	–	–	–	–	–	–	9
Mecoprop	μg/dm^3	–	–	–	–	–	–	10

Table 4.2 (continued)

Substances	Measuring units	USSR	Ukraine	Russian Federation		US-EPA	EU	WHO
		SaNPiN N4630-88 [18]	SaNPiN N 383 of Dec. 23, 1996 [20]	SaNPiN 2.1.4.559-96 [21]	List of fishery MPC, 1995 [22]	MCL [23]	EU Directive 98/83 EU of Nov 03, 1998 [27]	WHO standards for drinking water [25]
Harmful organic substances								
Silvex (2. 4. 5 TP)	μg/dm³	–	–	–	–	50	–	9
2,4,5-T	μg/dm³	–	–	–	–	–	–	9

[a] Substances, which given a certain range of value, have organoleptic properties
[b] TT—method used under conditions of a relationship between the contents and the value of initial dosage, which should not exceed 0.05 % for acryl amide at the dose 1 mg/dn^3 and 0.01 % for epichlorohydrin at the dose of 20 mg/dm^3
[c] The parametric value of individual pesticides -except aldrin, dieldrin, heptachlor, and heptachloroepoxide
[d] Total content of pesticides—the sum of individual pesticides

Table 4.3 Number of indexes of quality of drinking water and water in sources of centralized drinking water supply and their values as part of the regulatory documents of the USSR, Ukraine, Russian Federation, the USA, and International Organizations

Water quality-indexes	Number of water quality-indices	Number of standard values								
		USSR		Ukraine	Russian Federation		USA(Environmental Protection Agency -EPA)		European Union (EU)	World Health Organization (WHO)
		GOST standard 2874-82	SaNPiN №4630-88 [18]	SaNPiN №383 of Dec. 23, 1996 [20]	SaNPiN 2.1.4.559-96 [21]	List of fishery MPC, 1995 [22]	MCLG—Maximum contaminant level goal-desirable [23]	MCL—Maximum contaminant level-mandatory [23]	EU Directive 98/83 of Nov. 3,1998 [24]	WHO standarts for potable water [25]
Organoleptic	4	3	2	4	4	–	–	2	2	2
General physico-chemical	11	5	7	7	6	–	–	4	7	4
Microbiological	8	2	2	5	4	–	–	4	7	4
Virological	2	–	1	2	1	–	–	–	–	–
Parasitological	2	–	1	2	2	–	–	1	–	–
Radioactivity	13	–	–	2	–	–	–	9	2	2
Harmfulin-organic substances (prioritty)	44	15	38	13	41	32	3	17	25	24
Harmfu-lorganic substances (priority)	74	–	22	1	36	22	–	41	10	61
Total	158	25	73	36	94	54	3	78	53	97

the classification of the quality of surface water consisting of three categories with two subcategories in each, i.e. it has six ranks (classes) of quality (see Table 4.4 and 4.5). The categories were identified by means of differentiation using a technological principle based on the determination of those standard methods of treating surface water which are necessary to convert them into drinking water. For category A1—this is a simple physical treatment, for example, fast filtering and disinfection: for category A2—normal physical and chemical treatment and disinfection, for example, coagulation, flocculation, filtering, and chlorination, and for category A3—intensive physical and chemical treatment, for example, coagulation, flocculation, decantation, filtering, absorption on layered carbon, ozonation, and residual chlorination. Each of three categories may have two values such as G—"guide" and J—"mandatory" [26]. Such type of classification in principle may be sufficiently flexible and convenient. The EU classification for quality of surface water includes 46 indexes (parameters) such as physical, chemical, and microbiological. This is comparatively more than in the above-discussed analogous from the USSR classification. However, this common European classification of natural water in surface water bodies used for the preparation of drinking water is not without drawbacks. In many cases, values G and J in categories A1, A2 and A3 are not filled, and the absence of any values of some important indexes (beryllium, cobalt, nickel, vanadium, total and residual carbon) in all three categories is seen, and the absence of a variation range in the values of most indices within the subcategories except for pH and the content of fluorides is present, and the absence of strictly hydrobiological indexes of water quality, for example, biomass of phytoplankton, which is always present in water bodies is also prevalent. Forthwith, the present classification for the quality of surface water as part of Directive 75/440/EEC [26] may no longer correspond to the latest requirements of Directive 2000/60/EC for water policy in the EU member countries [8] and perhaps requires revision and improvement. To substantiate this claim, the following data is offered.

The national and foreign standards considered above for limiting the values of indexes of surface water and health-related classifications of their quality as related to the centralized drinking water supply reflect a water-consumers' interpretation of "water quality" in the world and the Ukraine. Within this study of water-consumption, the understanding that surface water vary and may be suitable or unsuitable in their compositions and properties for individual types of water consumption and water management is important for serving as a basis for standards and classification of water quality as seen in industrial and agricultural water supplies, fishery water, and drinking and health-related water supplies. Within this understanding is the quality of the ecosystems including but not limited habitats for water plants and animals. The ultimate intent being ecological standards and classifications of water quality is their use for protection of water ecosystems against anthropogenic effects, improvement of their state by conducting water conservation measures, and preservation of biological diversity in water bodies. A continued look into the ecosystem is necessary.

With all the specificity of water consumption and ecological standards and classifications of surface water quality, some of them approach each other either in

Table 4.4 Comparison of classifications of water quality of surface sources of centralized drinking water supply in regulatory documents of the USSR and EU

Water quality-indexes in surface water bodies	Measuring units	USSR Standard 2761-84 [27]			EU Directives 75/4401/EEC of June, 16, 1975 [26]					
		1st class	2nd class	3rd class	category A_1		category A_2		category A_3	
					G optimal value	I, mandatory value	G optimal value	I, mandatory value	G optimal value	I, mandatory value
Organoleptic indexes[a]										
Odor	Points and dilution indicator (DI) at 25°C	2	3	4	3	–	10	–	20	–
Color	Degrees, Pt-Co-scale	35	120	200	10	20[b]	50	100[b]	–	–
Weighted substances	mg/dm^3	20	1500	10,000	25	–	–	–	–	–
Indexes of salt composition										
Total mineralization[a]	mg/dm^3	<1000	–	–	–	–	–	–	–	–
Sulfates[a]	mg/dm^3	<500	–	–	–	–	–	–	–	–
Chlorides[a]	mg/dm^3	<350	–	–	200		200	–	200	–
Tropho-saprobiological indexes										
Chemical										
pH indicator[a]	pH units	6.5–8.5	6.5–8.5	6.5–8.5	6.5–8.5	–	5.5–9.0	–	5.5–9.0	–
Ammonia	mg NH_4^+/dm	–	–	–	0.05	–	1.0	1.5	2.0	4.0[b]
Nitrite	mg NO_3/dm^3				25.0	50[b]	–	50[b]	–	50[b] 8 8
Phosphates (P_2O_5)	mg P_2O_5/dm^3	–	–	–	0.4	–	0.7	–	0.7	–
Oxygen-saturated water	%	–	–	–	>70	–	>50	–	>30	–

Table 4.4 (continued)

Water quality-indexes in surface water bodies	Measuring units	USSR Standard 2761-84 [27]			EU Directives 75/4401/EEC of June, 16, 1975 [26]					
		1st class	2nd class	3rd class	category A_1		category A_2		category A_3	
					G optimal value	I, mandatory value	G optimal value	I, mandatory value	G optimal value	I, mandatory value
Permanganate oxidizability	mg O/dm^3	7	15	20	–	–	–	–	–	–
Bichromatic oxidizability	mg O/dm^3	–	–	–	–	–	–	–	30	–
BOD_{tot}	mg O_2/dm^3	3	5	7	–	–	–	–	–	–
BOD_5	mg O_2/dm^3	–	–	–	<3	–	<5	–	<7	–
Hydrobiological										
Phytoplankton[c]	mg/dm^3	1	5	50	–	–	–	–	–	–
	cell/cm^3	1,000	10,000	100,000						
Bacteriological										
Total coli-forms at 37 C in 100 $см^3$	Th.cell/dm^3	–	–	–	50	–	5,000	–	50,000	–
Inorganic substances of toxic action[b]										
Priority										
Barium	μg/dm^3	–	–	–	–	100	–	1,000	–	1,000
Boron	μg/dm^3	–	–	–	1,000	–	1,000	–	1,000	–
Iron[a]	μg/dm^3	1,000	3,000	5,000	100	300	1,000	2,000	1,000	–
Cadmium	μg/dm^3	–	–	–	1	5	1	5	1	5
Manganese[a]	μg/dm^3	100	1,000	2,000	50	–	100	–	1,000	–
Arsenic	μg/dm^3	–	–	–	10	50	–	50	50	100
Copper[a]	μg/dm^3	–	–	–	20	50	50	–	1,000	–
Mercury	μg/dm^3	–	–	–	0.5	1	0.5	1	0.5	1
Lead	μg/dm^3	–	–	–	–	50	–	50	–	50
Selenium	μg/dm^3	–	–	–	–	10	–	10	–	10

Table 4.4 (continued)

Water quality-indexes in surface water bodies	Measuring units	USSR Standard 2761-84 [27]			EU Directives 75/4401/EEC of June, 16, 1975 [26]					
		1st class	2nd class	3rd class	category A_1		category A_2		category A_3	
					G optimal value	I, mandatory value	G optimal value	I, mandatory value	G optimal value	I, mandatory value
Total chromium	μg/dm^3	–	–	–	–	50	–	50	–	50
Zinc[a]	μg/dm^3	–	–	–	500	3,000	1,000	5,000	1,000	5,000
Cyanides	μg/dm^3	–	–	–	–	50	–	50	–	50
Fluorides	mg F/dm^3	–	–	–	700–1,000	1,500	700–1,700	–	700–1,700	–
Beryllium, vanadium, cobalt, nickel, in compliance with WHO standards for drinking water are also priority metals in terms of toxicity: however they were not included in the USSR and EU classifications.										
Facultative										
Aluminum[a]	μg/dm^3	–	–	–	–	–	–	–	–	–
Molybdenum	μg/dm^3	–	–	–	–	–	–	–	–	–
Tungsten	μg/dm^3	–	–	–	–	–	–	–	–	–
Lithium	μg/dm^3	–	–	–	–	–	–	–	–	–
Antimony	μg/dm^3	–	–	–	–	–	–	–	–	–
Thallium	μg/dm^3	–	–	–	–	–	–	–	–	–
In the USSR and EU classifications these components are absent; however they were included in WHO standards for drinking water										
Organic substances of toxic action[d]										
Priority (A)										
Hydrocarbons polycyclic aromatic (benz(a) pyrene and others)	μg/dm^3	–	–	–	–	0.2	–	0.2	–	1
total pesticides (parathion, BHC, dieldrin)	μg/dm^3	–	–	–	–	1	–	2.5	–	5.0

Table 4.4 (continued)

Water quality-indexes in surface water bodies	Measuring units	USSR Standard 2761-84 [27]			EU Directives 75/4401/EEC of June, 16, 1975 [26]					
		1st class	2nd class	3rd class	category A_1		category A_2		category A_3	
					G optimal value	I, mandatory value	G optimal value	I, mandatory value	G optimal value	I, mandatory value
Phenols (volatile)	μg/dm^3	–	–	–	–	1	1	5	10	100
Facultative (B)										
List of MPC of substances of this subgroup is given in	–	–	–	–	–	–	–	–	–	–

A—priority substances of toxic action are the most important in terms of adverse effects and spreading and require paramount attention

B—facultative substances of toxic action are less spread and harmful, their determination is desirable, but is done selectively

[a] In addition to indexes of this group, other substances (in certain concentrations) which are designated in the subsequent group of indexes for quality of surface water, have organoleptic properties

[b] Exceptional climatic and geographic conditions

[c] With respect to depths exceeding 1 m from the water surface

[d] Total toxicity of water in surface water bodies is determined by the methods of biotesting

Table 4.5 Comparison of MPC of facultative organic substances of toxic action in water of surface water bodies—sources of centralized drinking water supply in regulatory documents of the USSR, the USA, and the European Union

Substances	Measurement units	MPC				
		Fishery (according to standards of the RF [22]	health-related			
			SaNPiN 2.1.4.559-96 [21]	EPA USA, 2000 [23]	EU Directive 98/83 EU of Nov. 3, 1998 [24]	WHO standards related to drinking water [25]
Chlorinated alkanes						
Carbon tetrachloride	μg/dm^3	0.014	6	5	–	2
Dichloromethane[a]	μg/dm^3	–	7.5	5	–	20
1,2-Dichloroethane	μg/dm^3	100	–	5	3	30
1,1,1-Dichloroethane	μg/dm^3	–	10,000	200	–	2,000
1,1,2-Dichloroethane	μg/dm^3	–	–	5	–	–
Trihalomethanes (THM, sum)	μg/dm^3	–	100	–	–	–
Dibromochloro-methane	μg/dm^3	–	10	–	–	–
Chlorinated ethylenes						
Vinyl chloride	μg/dm^3	None	50	5	0.5	5
1,1-dichloroethylcne	μg/dm^3	100	–	7	–	30
1,2-dichloroethylene	μg/dm^3	–	–	170	–	50
Trichloroethylene	μg/dm^3	10	–	5	10	70
Tctrachloroethylene	μg/dm^3	160	–	5	10	40
Aromatic hydrocarbons						
Benzene	μg/dm^3	50	10	5	–	10(24–170)
Toluene	μg/dm^3	50	500	1,000	–	700(4–2,600)
Xylenes[a]	μg/dm^3	50	50	10,000	–	500(20–1,800)
Ethylbenzene	μg/dm^3	10	10	700	–	300(2.4–200)|
Styrene[a]	μg/dm^3	100	100	100	–	20
Polycyclic aromatic hydrocarbons						
Benz(a)pyrene	μg/dm^3	–	0–5	0.2	0.01	0.7

Table 4.5 (continued)

Substances	Measurement units	MPC				
		Fishery (according to standards of the RF [22]	health-related			
			SaNPiN 2.1.4.559-96 [21]	EPA USA, 2000 [23]	EU Directive 98/83 EU of Nov. 3, 1998 [24]	WHO standards related to drinking water [25]
Chlorinated benzenes						
Monochlorobenzene[a]	μg/dm³	–	20	100	–	300(10–120)
1,2-Dichlorobenzene[a]	μg/dm³	1	2	600	–	1,000(140)
1,4- Dichlorobenzene[a]	μg/dm³	–	–	–	–	300(0.3–30)
Trichlorobenzene	μg/dm³	1	30	70	–	20(5–50)
Heteroorganic compounds						
Dialkyl tin	μg/dm³	–	2	–	–	–
Tributyl tin oxide	μg/dm³	–	4	–	–	2
Bis (tributyl tin) oxide	μg/dm³	–	0.2	–	–	–
Tetraethyl tin	μg/dm³	–	0.2	–	–	–
Tributil tin methacrylate	μg/dm³	–	0.2	–	–	–
Tetraethyl lead	μg/dm³	–	None	–	–	–
Other compounds						
Acrylaldehyde	μg/dm³	–	20	–	–	–
Di (2-ethylhexyl) adilat	μg/dm³	–	–	400	–	80
Di (2-ethylhexyl) phthalate	μg/dm³	–	–	6	–	8
Acrylamide	μg/dm³	None	10	TT[b]	–	0.5
Epichlorohydrin	μg/dm³	–	10	TT[b]	0.1	0.4
Hexachlorobutadiene	μg/dm³	–	10	–	–	0.6
Hexachlorocyclopentadiene	μg/dm³	–	1	50	–	–
Ethylenediaminetetraacetic acid, EDTA	μg/dm³	–	–	–	–	200
Trinitroloacetic acid	μg/dm³	–	500	–	–	–

[a] Given specific concentrations have organoleptic properties

[b] TT—method used under conditions of a relationship between the contents and the value of initial dosage, which should not exceed 0,05 % for acryl amide at the dose 1 mg/dn³ and 0.01 % for epichlorohydrin at the dose of 20 mg/dm³

their substance or purpose. Ecological and fishery standards and classifications of surface water quality are very close since all fish are hydrobiota while the water ecosystems are their habitat. Ecological, health-related, and drinking water standards and classifications are very close to each other by their main purpose. It is in water ecosystems that water of a specific quality is formed which is used to ensure people's needs without causing harm to their health owing to health-related monitoring and relevant technologies of water conditioning at drinking water supply facilities. Such a broader view of the significance of water quality in surface water bodies is reflected in the definition of this concept in a new regulatory document [16] where water quality is characterized by its composition and properties. This characterization is evident as a component of the water ecosystem and living habitat of hydrobiota and from the viewpoint of suitability for specific purposes of human use. Nevertheless, the ecological assessment of water quality in surface water bodies in many ways differs from water consumption sectoral approaches to the assessment related both to different goals of using surface water and the features of ecological classifications of their quality. Taking ecological approaches into consideration, interconnected principals are important. In our opinion the data from national and foreign experience in establishing requirements for drinking water utilizing classification of surface water quality and analyzing sources of centralized drinking water supplies as mentioned above as well as the ecologo-hygienic classification of water quality in surface water should be based on three interconnected principles: ecological, hygienic, and technological (water conditioning).

To begin, the ecological principle is predetermined by processes involving the formation of water quality. This quality is seen in sources of centralized drinking water supplies formed under natural conditions having in mind anthropogenic effects of the ecosystem and termed ecologo-hygienic classification. This classification is represented by the indexes of the salt composition, hydrochemical, hydrobiological, and bacteriological indexes as well as specific indexes of toxic and radiation elements.

Next, the hygienic (health-related) principle is predetermined by the fact that the water taken for subsequent water treatment should be safe in respects to hygienic concerns. The hygienic component of the ecologo-hygienic classification of water quality in a surface source of drinking water supply should be represented by organoleptic indexes as well as indices of salt composition important for health, hydrochemical, hydrobiological, bacteriological indexes, and indexes of toxic and radiation elements. With a detailed examination of the above two principles, the indexes of the ecological and hygienic components of the ecologo-hygienic classification of water quality of surface water intended for treatment into drinking water are rather similar except for organoleptic ones. The difference between the ecological and the hygienic positions may reveal itself in the ratios of quantitative values (criteria) of the total indexes of water quality. In the case of such differences, it may be preferable to use stricter hydroecological criteria since hygienic requirements in this case are met only with a greater safety margin.

Last, the technological (water-conditioning) principle is predetermined by the fact that the water taken from the surface water source should be a suitable source for production of drinking water having full physiological value. The ecologo-hygienic classification should contain the following criteria: salt composition, hydrochemical, hydrobiological, and bacteriological indexes including indexes of specific substances of toxic and radiation elements, especially in the 3rd class of quality (polluted water) which characterizes water suitable for conversion into drinking water under given water treatment technologies existing in the Ukraine.

Being guided by the above principles, we have developed a draft report titled Ecologo-Hygienic Classification of Water Quality for Ukraine's Surface Water—Sources of Centralized Drinking Water Supply. In comparison to the classification for surface water intended for drinking water supply approved by the EU [29], we suggests six quality categories (I–VI) as part of four quality classes (1–4) of surface water. These categories are expounded upon in the forthcoming paragraphs.

To start, water of class 1, category I ("very pure"), correspond to an advisable designation. Unfortunately, a few of these types of water sources accessible for a drinking water supply exist. It is possible that sources pertaining to this class 1 exist in the mountainous areas of Crimea and the Ukrainian Carpathian Mountains. Water from these areas corresponds in many indexes to drinking water quality and does not require much if any purification. The criteria of all water quality indexes of standard 1st class, category I, are established taking into account the requirements for drinking water quality (see Tables 4.1, 4.2).

Another way, water of class 4, category VI ("dirty" and "very dirty") correspond to numerous sections of water bodies predominately found in industrial regions of the country. Water located in these regions is unsuitable as a source for drinking water even with conditioning measures in technological, ecological, and hygienic terms. It is important to identify this type of surface water in order to conscientiously and justifiably exclude these sources from the stock of modern and potential sources of central drinking water supply in the Ukraine.

Following this, surface water of class 2 ("pure") with two categories—II ("pure") and III ("rather pure"), and of class 3 ("polluted") with two categories—IV ("weakly polluted") and V ("polluted"), are truly suitable for use in centralized drinking water supply systems. They encompass the overwhelming number of sections of Ukraine's surface water. Therefore, the values of the composition and properties of water quality in individual categories of all groups of indexes except organoleptic ones are established taking into account the criteria of the system of ecological classifications of quality for Ukraine's land surface water [15, 16]. In this case one should take into consideration the fact that criteria of tropho-saprobiological indexes are closely correlated to each other. To compute the water saprobity index according to Pantle-Buck (in Sladecek's modification) [28] it advisable to use the most complete list of general indicators of water quality (from fungi, algae, and protozoa to fish) which allow for Ukraine's conditions [29].

4.2 Water Quality Requirements to the Drinking Water Supply Sources and Rules of Their Selection

Sources of the Centralized Drinking Water Supply. Hygienic Requirements for the Water Quality and Rules of Selection This standard applies to sources of drinking water supplies and sets up hygienic requirements for the selection of a new appraisal of existing sources of the centralized drinking water supply system. This standard comprises 54 regulatory documents. Some of these are summarized below.

Main Provisions Water sites where water quality meets existing hygienic, epidemiologic, ecological, and technological requirements are used or can be used for centralized drinking water supply. This corresponds to water site requirements established for drinking water supply sources as determined on the basis of the following caveats:

- hygienic and ecological assessment of formation conditions and level of protection of underground water sources within the sanitary protection area;
- hygienic and ecological assessment of the water supply sources and neighboring areas upstream and downstream of the diversion facility within the sanitary protection area;
- qualitative assessment based on the examination of water samples taken monthly over the last three years and qualitative assessments of the amount of water in the water supply sources;
- sanitary assessment of the water diversion site, and;
- forecasting of sanitary and ecological conditions of the water supply sources.

Quality classification of surface water which are sources of the centralized drinking water supply by hygienic and ecological criteria (see Table 4.6) includes 80 indicators used for the assessment of drinking water quality in compliance with the sanitary legislation and consists of seven different groups: I group—4 organoleptic indicators, and II group—17 total sanitary indicators of water chemical composition, and III group—6 hydrobiological indicators, and IV group—6 microbiological indicators, and V group—2 parasitologic indicators, and VI group—9 radiation safety indicators, and VII group—36 priority toxicological indicators of the chemical water composition (of them 25—non-organic and 11—organic components).

Quality classification of underground water which are sources of the drinking water supply by hygienic and ecological criteria (see Table 4.7) includes 71 indicators used for assessment of drinking water quality in compliance with the sanitary legislation and consists of seven different groups comprising: I group—4 organoleptic indicators, and II group—14 total sanitary indicators of water chemical composition, and III group—2 hydrobiological indicators, and IV group—6 microbiological indicators, and V group—2 parasitologic indicators, and VI group—9 radiation safety indicators, and VII group—34 priority toxicological indicators of water chemical composition (of them 27—non-organic and 7—organic components). Indicator value range (criteria) of water quality of two classifications is split into four classes: 1st class—excellent desired quality of water, and 2nd class—good, accept-

Table 4.6 Quality classification of surface water which are sources of the centralized drinking water supply by the hygienic and ecological criteria[a]

Water quality indexes in surface water bodies	Measurement units	Water class quality			
		1	2	3	4
I Organoleptic indicators[b]					
Odor	Points and dilution	<1	1–2	3–4	>4
	indicator(DI) at 25 °C	<2	2–16	17–23	>23
Aftertaste	DI	<1	1–2	3–4	>4
Color	Degrees Pt-Co-scale	<20	20–80	81–120	>120
Transparency	mg/dm³	<20	20–1,500	1,501–500	>5,000
II General chemical indicators					
Dry residue	mg/dm³	<400	400–650	651–1,000	>1,000
Sulfates[b]	mg/dm³	<40	40–120	121–250	>250
Chlorides[b]	mg/dm³	<30	30–100	101–250	>250
Magnesium	mg/dm³	<10	10–30	31–80	>80
Hardness total[b]	mmol/dm³	<10	10–30	5.1–7.0	>7.0
Alkalinity[b]	mol/dm³	<3	3.0–5.0	4.1–6.5	>6.5
pH index[b]	pH units	6.9–7.0	6.8–6.5	6.4–6.1	>6.1
		7.1–7.5	7.6–8.1	8.2–8.5	>8.5
Ammonium nitrogen[b]	mg N/dm³	<0.10	0.10–0.30	0.31–1.00	>1.00
Nitrite nitrogen[b]	mg N/dm³	<0.002	0.002–0.01	0.011–0.050	>0.050
Nitrate nitrogen[b]	mg N/dm³	<0.20	0.20–0.50	0.51–1.00	>1.00
Phosphate phosphorus[b]	mg P/dm³	<0.015	0.015–0.050	0.051–0.200	>2.000
Dissolved O_2	mg O_2/dm³	>8.0	8.0–7.1	7.0–5.0	<5.0
Saturation O_2	%	96–100	95–81	80–60	<60
		101–105	106–120	121–140	>140
Permanganate oxidizability	mg O/dm³	<3.0	3.0–10.0	10.1–15.0	>15.0
Bichromatic oxidizability	mg O/dm³	<9.0	9.0–30.0	31.0–40.0	>40.0
BODTotal	mgO_2/dm³	<1.3	1.3–3.0	3.1–7.0	>7.0
Total organic carbon	mg C/dm³	<5.0	5.0–15.0	15.1–25.0	>25.0

Table 4.6 (continued)

Water quality indexes in surface water bodies	Measurement units	Water class quality			
		1	2	3	4
III Hydrobiological indicators					
Phytoplankton dominance of blue-green algae, mainly in water bodies (reservoirs, estuaries, lakes): —quantity	th. Cel/dm^3	<10	10–40	50–100	>100
—biomass	mg/dm^3	<1	1–4	5–10	>10
The dominance of diatoms mainly in water bodies (rivers, canals): —quantity	th. Cel/dm^3	<1	1–4	5–10	>10
—biomass	mg/dm^3	<1	1–4	5–10	>10
Total level of chronic toxity	Chronic unit	<1	1–2	3–4	>4
Microscopic (imperfect) fungi	cel/dm^3	None	None	None	None
IV Microbiological indicators					
Total microbial number	CFU/cm^3	Scores	Hundreds	Thousands	Tens of thousands
Total coliforms (lactopositive coliform bacillus bacteria), CBGB index, no more than	CFU/m^3	100	1,000	10,000	50,000
Thermostable coliform bacteria (TCB), index	CFU/100 cm^3	None	50	500	>1,000
Presence of pathogenic nitrobacteria (Salmonella, Shigella)	Number/dm^3	None	None	None	Number/None
Coliphages, index	Plaque-forming units/dm^3	None	10	100	1,000
Enteroviruses, adenoviruses and antigen of rotaviruses, reoviruses and viruses of hepatitis A	Number/dm^3	None	None	None	Number/None
V Parasitological indicators					
Number of pathogenic coliform protozoans	cells, cysts/25 dm^3	None	None	None	None
Number of coliform helminthes	cells, eggs, larvae/50 dm^3	None	None	None	None
VI Radiation safety indicators					
Total activity of α- emitters (Σ α- activity)	Bq/dm^3	<0.1	<0.1	<0.1	<0.1

Table 4.6 (continued)

Water quality indexes in surface water bodies	Measurement units	Water class quality			
		1	2	3	4
Total activity of β- emitters (Σ β- activity)	Bq/dm^3	<1	<1	<1	<1
Strontium-90 (^{90}Sr)	Bq/dm^3	<2	<2	<2	<10
Cesium-137 (^{137}Cs)	Bq/dm^3	<2	<2	<2	<100
Uranium (summary activity/concentration of the natural isotopes mixture)	Bq/dm^3 (mg/dm^3)	<1 (0.04)	<1 (0.04)	<1 (0.04)	<1 (0.04)
Radium-226 (^{226}Ra)	Bq/dm^3	<1	<1	<1	<1
Radium-228 (^{228}Ra)	Bq/dm^3	<1	<1	<1	<1
Radon-222 (^{222}Rn)	Bq/dm^3	<100	<100	<100	<100
Tritium (H-3)	Bq/dm^3	$<3 \times 10^4$	$<3 \times 10^4$	$<3 \times 10^4$	$<3 \times 10^4$
VI Toxicological indexes of chemical composition of water (priority) Inorganic					
Aluminum (Al)[b]	$\mu g/dm^3$	<50	50–200	210–500	>500
Barium (Ba)	$\mu g/dm^3$	<100	100–1,000	1,001–2,000	>2,000
Beryllium (Be)	$\mu g/dm^3$	<0.2	0.2–2.0	2.1–4.0	>4.
Boron (B)	$\mu g/dm^3$	<100	100–200	201–400	>4,000
Bromides (Br)	μg/dm3	<100	100–200	201–500	>500
Vanadium (V)	$\mu g/dm^3$	<2	2–10	11–20	>20
Total iron (Fe)[b]	$\mu g/dm^3$	<50	50—100	101–1,000	>1,000
Cadmium (Cd)	$\mu g/dm^3$	<0.1	0.1–0.5	0.6–5.0	>5.0
Cobalt (Co)	$\mu g/dm^3$	<10	10–20	21–50	>50
Lithium (Li)	$\mu g/dm^3$	<10	10–50	51–100	>100
Manganese (Mn)[b]	$\mu g/dm^3$	<10	10–100	101–1,000	>1,000
Arsenic (As)	$\mu g/dm^3$	<1	1–10	11–50	>50
Copper (Cu)[b]	$\mu g/dm^3$	<1	1–25	26–50	>50
Molybdenum (Mo)	$\mu g/dm^3$	<1	1–25	26–200	>200
Nickel (Ni)	$\mu g/dm^3$	<20	20–50	51–100	>100
Mercury (Hg)	$\mu g/dm^3$	<0.20	0.20–0.50	0.51–2.5	>2.5

Table 4.6 (continued)

Water quality indexes in surface water bodies	Measurement units	Water class quality			
		1	2	3	4
Lead (Pb)	µg/dm^3	<5	5–20	21–100	>100
Selenium (Se)	µg/dm^3	<1.5	1.5–5.0	5.1–10.0	>10.0
Antimony (Sb)	µg/dm^3	<0.1	0.1–0.5	0.6–1.0	>1.0
Thallium (T)	µg/dm^3	<0.1	0.1–0.5	0.6–0.2	>1.0
Fluorides (F^-)	µg/dm^3	<700	700–1,000	1,001–1,500	>1,500
Chromium (III), Cr (III)[b]	µg/dm^3	<100	100–250	251–500	>500
Chromium (VI), Cr (VI)[b]	µg/dm^3	<4	4–10	11–50	>50
Zinc (Zn)	µg/dm^3	<10	10–100	101–1,000	>1,000
Cyanides (CN^-)	µg/dm^3	<1	1–10	11–50	>50
Organic					
Benz(a)pyrene	µg/dm^3	<0.01	0.01–0.70	0.71–5.00	>5.00
Benzene[b], xylene[b], toluene[b]	µg/dm^3	<5	5–30	31–70	>70
Ethylbenzene	µg/dm^3	<0.5	0.5–2.0	2.1–5.0	>5.0
Oil products (general, carbons)[b]	µg/dm^3	<10	10–50	51–200	>200
Pesticides organochlorine (sum)	µg/dm^3	<0.1	0.1–1.0	1.1–5.0	>5.0
Synthetic surface active substances (SSAS)	µg/dm^3	<10	10–50	51–250	>250
Tetrachlorobenzene	µg/dm^3	<0.5	0.5–2.0	2.1–5.0	>5.0
Carbon tetrachloride	µg/dm^3	<0.5	0.5–2.0	2.1–6.0	>6.0
Trihalomethanes (TGM) chloroform, dibromhlor-metan, dihlorbrommetan (sum)	µg/dm^3	<50	50–100	101–200	>200
Phenols volatile[b]	µg/dm^3	<1	1–10	11–50	>50
Chlorophenols[b]	µg/dm^3	<0.3	0.3–0.5	0.6–1.0	>1.0

Here and in the Table 4.7:
None—indicator is not available
[a] Indicators of I, II, IV, V, VI, VII groups attribute to the hygienic, indicators of II, III, V, VI, VII (ecological)
[b] Except for group I, certain substances of groups II and VII have organoleptic properties (under certain conditions)

Table 4.7 Quality classification of underground water which are sources of the centralized drinking water supply by the hygienic and ecological criteria[a]

Water quality indexes in surface water bodies	Measurement units	Water class quality			
		1	2	3	4
I Organoleptic indexes[b]					
Odor	Points and dilution indicator(DI) at 25 °C	<1	1–2	3–4	>4
		<2	2–16	17–23	>23
Aftertaste	DI	<1	1	2	3
Color	Degrees Pt-Co-scale	<15	15–20	21–35	>35
Transparency	mg/dm^3	<0.5	0.5–1.5	1.6–5.0	>5.0
II General chemical characteristics					
Dry residue (mineralization)	mg/dm^3	<500	500–1,000	1,001–1,500	>1,500
Sulfates[b]	mg/dm^3	<250	250–350	351–500	>500
Chlorides[b]	mg/dm^3	<250	250–300	301–350	>350
Magnesium	mg/dm^3	<10	10–20	21–30	>30
Hardness total[b]	mmol/dm^3	<4	4–7	8–10	>10
Alkalinity[b]	mol/dm^3	<1.5	1.5–4.0	4.1–6.5	>6.5
pH index[b]	pH units	6.5–7.0	6.0–8.0	6.0–8.5	>8.5
Ammonium nitrogen[b]	mg N/dm^3	None	0.05–0.50	0.51–2.00	>2.00
Nitrite nitrogen[b]	mg N/dm^3	<0.05	0.05–0.50	0.51–1.00	>1.00
Nitrate nitrogen[b]	mg N/dm^3	<5.0	5.0–7.0	7.1–10.0	>10.0
Phosphate phosphorus[b]	mg P/dm^3	<0.3	0.3–0.5	0.6–1.0	>1.0
Permanganate oxidizability	mg O/dm^3	<4.0	4.0–5.0	5.1–6.0	>6.0
Bichromatic oxidizability	mg O/dm^3	<4.0	4.0–6.0	6.1–10.0	>10.0
Total organic carbon	mg C/dm^3	<2.0	2.0–3.0	3.1–4.0	>4.0
III Hydrobiological indexes					
Total level of chronic toxity	Chronic unit	<1	1–2	3–4	>4
Microscopic (imperfect) fungi	cel/dm^3	None	None	None	None

Table 4.7 (continued)

Water quality indexes in surface water bodies	Measurement units	Water class quality			
		1	2	3	4
IV Microbiological indexes					
Total microbial number (TMN)	CFU/cm^3	One	Scores	Hundreds	Thousands
Total coliforms (lactopositive coliform bacillus bacteria), CBGB index, no more than	CFU/m^3	None	None	1–10	100
Thermostable coliform bacteria (TCB), index	CFU/100cm^3	None	None	None	None
Presence of pathogenic nitrobacteria (Salmonella, Shigella)	Number/dm^3	None	None	None	None
Coliphages, index	Plaque-forming units/dm^3	None	None	None	None
Enteroviruses, adenoviruses and antigen of rotaviruses, reoviruses and viruses of hepatitis A	Number/dm^3	None	None	None	None
V Parasitological indexes					
Number of pathogenic coliform protozoans	Cells, cysts/25 dm^3	None	None	None	None
Number of coliform helminthes	cells, eggs, larvae/50 dm^3	None	None	None	None
VI Radiation safety indexes					
Total activity of α- emitters (Σ α- activity)	Bq/dm^3	<0.1	<0.1	<0.1	<0.1
Total activity of β- emitters (Σ β- activity)	Bq/dm^3	<1	<1	<1	<1
Strontium-90 (^{90}Sr)	Bq/dm^3	<2	<2	<2	<10
Cesium-137 (^{137}Cs)	Bq/dm^3	<2	<2	<2	<100
Uranium (summary activity/concentration of the natural isotopes mixture)	Bq/dm^3 (mg/dm^3)	<1 (0.04)	<1 (0.04)	<1 (0.04)	<1 (0.04)
Radium-226 (^{226}Ra)	Bq/dm^3	<1	<1	<1	<1
Radium-228 (^{228}Ra)	Bq/dm^3	<1	<1	<1	<1
Radon-222 (^{222}Rn)	Bq/dm^3	<100	<100	<100	<100
Tritium (H-3)	Bq/dm^3	$<3\times10^4$	$<3\times10^4$	$<3\times10^4$	$<3\times10^4$
VI Toxicological indexes of chemical composition of water (priority) Inorganic					
Aluminium (Al)[b]	μg/dm^3	None	<500	501–2,000	>2,000

Table 4.7 (continued)

Water quality indexes in surface water bodies	Measurement units	Water class quality			
		1	2	3	4
Barium (Ba)	μg/dm^3	<100	100–200	201–1,000	>1,000
Beryllium (Be)	μg/dm^3	<0.2	0.2–1.0	1.1–2.0	>2.0
Boron (B)	μg/dm^3	<200	200–500	501–1,000	>1,000
Bromides (Br)	μg/dm^3	<10	10–25	26–100	>100
Vanadium (V)	μg/dm^3	<10	10–50	51–100	>100
Total iron (Fe)[b]	μg/dm^3	<300	300–1,000	1,001–2,000	>2,000
Cadmium (Cd)	μg/dm^3	<1	1–2	3–4	>4.0
Cobalt (Co)	μg/dm^3	<10	10–20	21–50	>50
Litium (Li)	μg/dm^3	<10	10–50	51–100	>100
Manganese (Mn)[b]	μg/dm^3	<50	50–100	101–500	>500
Arsenic (As)	μg/dm^3	<10	10–20	21–50	>50
Copper (Cu)[b]	μg/dm^3	<1	1–2	3	>3
Molybdenum (Mo)	μg/dm^3	<200	200–300	301–500	>500
Nickel (Ni)	μg/dm^3	<20	20–50	51–100	>100
Mercury (Hg)	μg/dm^3	<0.5	0.5–10	1.1–2.0	>2.0
Lead (Pb)	μg/dm^3	<10	10–30	31–100	>100
Carbohydrate (H_2S)	μg/dm^3	None	<5	5–10	>10
Selenium (Se)	μg/dm^3	<1	1–10	11–15	>15
Antimony (Sb)	μg/dm^3	None	<10	10–20	>20
Strontium (stable) (Sr)	μg/dm^3	2,000–7,000	2,000–7,000	2,000–7,000	2,000–7,000
Thallium (T)	μg/dm^3	None	<0.5	0.5–1.0	>1.0
Fluorides (F^-)	μg/dm^3	<700	700–1,000	1,001–1,500	>1,500
Chromium (III), Cr (III)[b]	μg/dm^3	<100	100—200	201–500	>500
Chromium (VI), Cr (VI)[b]	μg/dm^3	<10	10–20	21–50	>50
Zinc (Zn)	μg/dm^3	<100	100–500	501–1,000	>1,000
Cyanides (CN^-)	μg/dm^3	None	<10	11–50	>50

Table 4.7 (continued)

Water quality indexes in surface water bodies	Measurement units	Water class quality			
		1	2	3	4
Organic					
Benz(a)pyrene	μg/dm³	<0.01	0.01–0.70	0.71–5.00	>5.0
Oil products (general, carbons)	μg/dm³	None	<20	20–50	>50
Synthetic surface active substances (SSAS)	μg/dm³	None	<10	10–50	>50
Pesticides organochlorine (sum)	μg/dm³	<0.1	0.1–0.2	0.3–0.5	>0.5
Carbon tetrachloride	μg/dm³	<0.1	0.1–10	1.1–2.0	>2.0
Trihalomethanes (TGM) chloroform, dibromhlor-metan, dihlorbrommetan (sum)	μg/dm³	<10	10–20	21–50	>50
Phenols volatile	μg/dm³	None	<1	1–2	>2
Chlorophenols	μg/dm³	None	None	None	None

able quality of water, and 3rd class—satisfactory, acceptable quality of water, and 4th class—mediocre, limited acceptability, undesired water quality. Furthermore, a classification of surface water in the Ukraine is used as a source of identifying centralized drinking water supplies utilizing the facultative toxic compound (see Table 4.8) for periodic monitoring of sanitary conditions and for possible presence of such toxic substances which may be harmful for the population (necessity of the monitoring, periodicity, and seasonal prevalence are determined in each particular case). Characteristics of water composition by the surface and underground water classification indicators for the water from the centralized drinking water supply systems are determined by research and assessment of water quality indicators which were standardized and well approbated [30]. This standard takes into account domestic and foreign methodologies used in the world practice to monitor certain indicators of surface and underground water developed for centralized drinking water supply in compliance with the EU Directive and recommendations of the World Health Organization, and requirements of the International Organization for Standardization (ISO).

Rules for Selection of New and Monitoring of the Existing Surface and Underground Sources of the Centralized Drinking Water Supply For the centralized drinking water supply, the surface and underground sources with the 2nd and 3rd class water quality are assessed by utilizing the hygienic and ecological criteria (see Tables 4.6, 4.7). Water treatment procedures and equipment required for this purpose are selected for each particular water source based on technological examination or experience of the plant's operation [30].

Assessment of the Results of Drinking Water Quality Examination of the Existing and Designed Water Supply Sources Results of water quality examination by the epidemiological safety indexes are assessed in the following sequence. First, in the event of detecting bacterial contamination such as TMN, saprophytes, microorganisms, CBGB in amounts above the permitted rates in the water, repeated sampling should be done and additional examination employing the use of TCB indicator, pathogenic microorganisms, and coliphages tests. Should repeated detection of any bacterial contamination in two water samples taken in a sequence at the same point of examination be found, it is necessary to perform additional monitoring of water quality of the surface water sites by sanitary and chemical indexes. In case of non-correspondence to the standards (detection of an industrial impact on the water quality and the condition of its formation in the water supply sources) in the water diversion sites, it is necessary to tighten control over observance of rules in the sanitary protection area, improve water cleaning and decontamination technology, and carry out special control of the epidemiological safety of the drinking water before its reaching external distribution systems and internal drinking water systems. For the purpose of an objective assessment and forecasting of epidemiological situations regarding drinking water supply sources, the development of substantiated complete plans of antiepidemic and prophylactic measures and the determination of their priority should be enacted. In part this action should include national and departmental systems of ecological monitoring with participation of the local

Table 4.8 Quality classification of surface water which are sources of the drinking centralized water supply by the facultative toxicological indexes[a]

Water quality indexes in surface water bodies	Measurement units	Water class quality			
		1	2	3	4
Toxicological indexes (facultative)[a] *Organic*					
1.2-Dichloroethane	μg/dm³	<3	3–25	26–100	>100
1,1,1-Dichloroethane	μg/dm³	<200	200–300	301–1,000	>1,000
1,1,2-Dichloroethane	μg/dm³	<1	1–3	4–5	>5
Chlorinated ethylenes					
Vinyl chlorides	μg/dm³	<0.5	0.5–15	16–50	>5
1,1-dichloroethylene	μg/dm³	<7	7–30	31–100	>100
1,2-dichloroethylene	μg/dm³	<50	50–110	111–170	>170
Trichloroethylene	μg/dm³	<5	5–30	31–70	>70
Tetrachlorethylene	μg/dm³	<5	5–50	51–160	>160
Heteroorganic compounds					
Bis (tributyl tin) oxide	μg/dm³	<0.01	0.01–0.10	0.11–0.20	>0.02
Dialkyl tin	μg/dm³	<0.1	0.1–1.0	1.1–2.0	>2.0
Tributyl tin oxide	μg/dm³	<0.2	0.2–2.0	2.1–4.0	>4.0
Tetraethyl tin	μg/dm³	<0.01	0.01–0.10	0.11–0.20	>0.20
Tributil tin methacrylate	μg/dm³	<0.01	0.01–0.10	0.11–0.20	>0.20
Other compounds					
Acrylamide	μg/dm³	<0.5	0.5–3.0	3.1–10.0	>10.0
Acrolein	μg/dm³	<1	1–10	11–20	>20
Hexachlorobutadiene	μg/dm³	<0.6	0,.6–3.0	3.1–10.0	>10.0
Hexachlorocyclopentadiene	μg/dm³	<1	1–20	21–50	>50
Di (2-ethylhexyl) adilat	μg/dm³	<80	80–200	201–400	>400
Di (2-ethylhexyl) phthalate	μg/dm³	<6.0	6.0–7.0	7.1–8.0	>8.0
Epichlorohydrin	μg/dm³	<0.1	0.1–3.0	3.1–10.0	>10.0
Ethylenediaminetetraacetic acid (EDTA)	μg/dm³	<10	10–60	61–200	>200
Trinitroloacetic acid	μg/dm³	<50	50–200	201–500	>500

[a] Facultative toxicological indexes are less harmful and less widely spread toxic than priority substances of toxic action. Their identification is expedient but this is done selectively and less often, depending on the availability of analyses methodologies and laboratory equipment

authorities and the national sanitary and epidemiological control authorities in the collection and record keeping of corresponding information. Further sources for assessment are designed with the upcoming mindset.

New sources of centralized drinking water supply are selected in the following order based on their reliability:

- Interstratal pressure water;
- Interstratal non-pressure water;
- Ground water (first aquifer from the surface), and;
- Surface water (rivers, water reservoirs, lakes, and channels).

In this selection process, preferences are given to the most reliable sources. If the water reserve is not sufficient or it cannot be used from a technical and/or ecological standpoint, the water deficit should be replenished from less reliable water sources taking into consideration the water quality after its mixing and before its inflow into the distribution system. Selection of new sources of centralized drinking water supply systems with several sources available is determined by a technical and an economic comparison of possibilities to obtain qualitative drinking water in alignment with sanitary legislation. If development of new sources and/or updated design of existing facilities is undertaken, reserves of underground water should be taken into account.

Technological Requirements for Water Treatment Depending on its Quality Class Technological approaches to the conditioning of surface and underground water depend on the physical, chemical and microbiological nature of contaminants. These contaminants are common for surface and underground water.

1st Class—Very Good, Desired Water Quality Treatment of the category of 1st class water requires its disinfection by one of the following reagents: chlorine, hypochlorite, chlorine dioxide, and chloramine. Disinfection with ultraviolet radiation in combination with O_2, H_2O_2 is also necessary along with treatment with making use of ozone and filtration with coagulation. In the future, cleaning by filtering through active carbon or slow filters, cleaning and disinfection with other reagents, and methods permitted by the National Sanitary and Epidemiological Service of the Ministry of Health of Ukraine should be employed.

2nd Class—Good, Acceptable Water Quality, 3rd Class—Satisfactory, Acceptable Water Quality The following processes are required for the water treatment of the 2nd and 3rd quality class.

- *Conditioning by the organoleptic indexes*:

After-taste, smell: aeration, oxidation, and adsorption by activated carbon.

Coloring: oxidation; coagulation—flocculation, sedimentation, filtering, ultrafiltration.

Suspended substances: sedimentation, microfiltration, ultrafiltration, filtering through an additional alluvial layer, coagulation, flocculation, sedimentation or flotation, filtering, and contact coagulation.

- *Conditioning by the water chemical composition index:*

Ammonia nitrogen, nitric nitrogen, and nitrate nitrogen: biological cleaning by filters with fixed heterotrophic biomass and biosorption, and ion exchange in filtering through ionites (anionites for nitrates, cationites for ammonia ions), and nanofiltration.

Phosphorus of phosphates: phosphate removal, filtering through activated aluminum oxide, and treatment with lime.

Permanganate oxidation characteristic, dichromate oxidation characteristic (COD), standard biochemical oxygen demand BOD_{total}, common organic carbon: preliminary biological cleaning in natural conditions, supported biological cleaning with immobilized microbial flora, preliminary oxidation, and coagulation—flocculation with following flotation or sedimentation and filtering, carbon treatment, contact coagulation, ozone treatment with following biosorption by biological activated carbon, slow filtering, disinfection, and membrane filtering.

- *Conditioning by the microbiological, parasitologic and hydrobiological indexes:*

Phytoplankton: microfiltering, preliminary chlorination with following coagulation—flocculation, preliminary chlorination and pressure reagent flotation, filtering through fast filters, and filters with biological activated carbon.

Microbiological and parasitologic indexes: disinfection with one of the reagents such as chlorine, hypochlorite, chlorine dioxide or chloramines, bactericidal irradiation, coagulation, ultrafiltration, and nanofiltration. It is recommended to treat water with disinfectants permitted for use in the Ukraine.

- *Conditioning by the non-organic toxic action content indexes:*

Aluminum (for the 3rd quality class), barium, and beryllium (for the 3rd quality class), cadmium, arsenic, nickel, and mercury (for the 3rd quality class), lead, antimony, and thallium for coagulation—flocculation, sedimentation, filtering, contact coagulation, nanofiltration, and ion exchange on selective sorbents.

Boron: fast filter water clearing, filtering through ionite filters with boron-selective resin in OH-form, and after treatment by biological activated carbon and disinfection.

Iron, manganese: use of strong oxidants to produce hydroxides, coagulation, filtration, filtering through modified sorbents, nanofiltration and sorption by biological activated carbon, silica gel, granite, and marble chips.

Fluoride: fluorination by adding fluoride compound powder or its solution to water, defluorination: ion exchange by fluoride-selective ionites, sorption on the surface of newly formed aluminum or magnesium hydroxides, on hydroxylapatite, and on modified clinoptilolite.

- *Conditioning by the organic toxic action content indexes:*

Preliminary mechanical and chemical cleaning, preliminary biological supported cleaning with immobilized microbiosis or with sand dunes, artificial water sources, oxidation with chlorine dioxide, ozone, hydrogen peroxide, and UV-irradiation with filtration through activated carbon, slow filtration, and nanofiltration.

- *Conditioning by the radiation safety indexes:*

Sorption by natural natrium-based sorbents (clinoptilolite, vermiculite), treatment with bentonite and following coagulation—flocculation, treatment with hydrogen peroxide with bivalent iron and following coagulation, treatment with a mixture of powdery sorbents (bentonite, clinoptilolite, and lime) and following coagulation, sorption by activated modified carbon, sorption by selective ionites, sorption by mixed composite sorbents (selective for radionuclides), and aeration with volatile components (Radon-222).

4th Class—Mediocre, Limited Acceptability, Undesired Water Quality If no other water supply sources are available, and in the event of economic expediency, the 4th quality class water will be treated with all the above mentioned methods. Meanwhile, the consumption of reagents for the duration of treatment increases in compliance with the technological requirements and possibility of usage of the 4th class water.

4.3 Current Problems and Potentialities of Drinking Water Preparation Technology at Centralized Water Treatment Facilities

The supply of quality drinking water is best achieved with stellar management practices incorporating a broad spectrum of various activities including but not limited to a positive community outreach [31]. Consequently, water management can be an area of conflict between government, the scientific district, and the surrounding populace. Therefore, the creation and development of new technologies for drinking water conditioning should take into account the effective combination of the above communities. Quality drinking water in the Ukraine is of paramount importance. This is due to this eastern European country's deterioration of its ecological state noticeable in the centralized system of water sources and its economic difficulties which severely constrain budgeting monies into new technologies. Now for an inquiry into the technical side of current problems and potentialities.

Role of Natural Organic Compounds in the Formation of the Quality of Natural and Drinking Water In surface sources of water supply the presence of natural organic compounds (NOC) determines the nature of many processes of both self-purification and water conditioning [32–35]. NOC components play an important role in the complexation with traces of heavy metals [36, 37], migrations of hydrophilic organic compounds in sources of water supply and also in the kinetics of aggregation of colloid particles [38]. Consequently, the state of natural water and NOC substantially affect the technology of water conditioning and water quality. Therefore, in the research of recent years sufficiently large attention is paid to clarification of the NOC nature and their interaction with other components of natural water. It is known that the NOC reactivity depends on such physico-chemical properties such as molecular weight, the amount of aromatic rings in the structure, the elementary composition, and presence of functional groups [39]. A more in depth look at natural organic compounds is reflected in the following data.

Modern notions about the fractional composition of the NOC refer to their classification by different hydrophobic and hydrophilic fractions [38–42] (see Table 4.9). The NOC fractional composition to a great extent depends on the geoclimatic conditions for the formation of water quality in surface sources of water supply (see Table 4.10). Molecular weights of humic compounds also substantially differ for different water sources. The study of river water in many countries shows that the values of the molecular weight constitute 0.3 to 2.15 and in lake water it is <17.8 kamu. For ground water this index equals 0.64–1.0 kamu [43]. Information about NOC fraction composition is one of the central problems for the technology of drinking water treatment because these compounds determine many features of the water conditioning processes. A number of researchers focus attention on the fact that not only hydrophobic-hydrophilic NOC properties and their acid-basic properties substantially affect the quality of natural water determining the transport of heavy metals, NOC reactivity, and biological accessibility of the traces of heavy metals for microorganisms (phytoplankton, zooplankton, algae, etc) [36, 44–47]. As researchers give thought to this information, they continue to define possibilities utilizing this data.

Researchers notice that the NOC acid-basic properties are helpful in determining the complexing of metals and working on a definition for the possibility of removing the latter from water during water treatment [35, 36, 48]. This also refers to such metals often encountered in natural water as iron and manganese [48]. For instance, in [49] it is shown that Fe(III) in surface water bodies is mainly in the form of complex compounds with natural organic ligands of different chemical nature and molecular weight. Among these complexes fulvate complexes prevail [50]. The increase of the NOC contents in natural water and their complexing with iron and manganese may lead to the increase in the concentration of heavy metals in drinking water due to their release from the composition with NOC during chlorination [51]. Thus, deep removal of NOC during water treatment makes it possible not only to avoid the formation of toxic chlorination products at the final stage of drinking water treatment, secondary microbial water pollution in distribution systems, but also to prevent penetration of ions of heavy metals into the drinking water [51–54]. All this determines the necessity of obtaining complete information on the quality and quantity of NOC in specific sources of water supply which is extremely important for the adequate prediction of technological models of water treatment. As a result, the degree of complete removal of NOC at all stages of drinking water conditioning determines is biological stability [55, 56].

The purpose of [34] did not make consideration toward the analytical possibilities of determining the fractional composition and other characteristics of NOC as their goal. Nevertheless it is expedient to guide the reader in the appropriate literature references on this issue. A determination of the NOC acidic-hydrophilic properties, makes it possible to imply their anthropogenic or soil origin, and is given in [35, 57]. The approach to the separate determination of NOC hydrophobic-hydrophilic fractions both in material and after oxidative treatment of water is described in [38, 40, 58–60]. The analytical methods when considering NOC as a complex heterogeneous mixture consisting of humic and fulvic acids, low-molecular organic acids, proteins, hydrocarbons and other classes of compounds are investigated in paper [61].

Table 4.9 Composition of NOC individual fractions as per [38–42, 44, 59, 60]

Fraction	Designation	Class of organic compounds (NOC)
Humic acids	HA	Portion of humic matter, which precipitates at pH 1
Hydrophobic acids	HOA	Soil fiilvic acids, C_5–C_9 aliphatic carboxylic acids, 1- and 2-ring aromatic carboxylic acids, 1- and 2-ring phenols
Hydrophobic bases	HOB	Portion of humic matter, which is retained by resin KhAD −8 at pH ~7 and which may be washed off by HC1; 1- and 2- ring aromatic amines, but excluding pyridine, protein like matter
Neutral hydrophobic compounds	HON	Mixture of hydrocarbons; $>C_5$ aliphatic alcohols; amides; ethers; ketones; aldehydes; long-chain ($>C_9$) aliphatic carboxylic acids and amines; >3-ring aromatic carboxylic acids and amines
Hydrophilic acids	HIA	$<C_5$ aliphatic carboxylic acids; polyfunctional carboxylic acids; mixture of different hydroxyacids
Hydrophilic bases	HIB	Amphoteric protein-like materials, which contain aliphatic carboxylic acids, aminosugars, peptides and proteins; $<C_9$ aliphatic amines, pyridine
Hydrophilic neutral compounds	HIN	Short-chain aliphatic amines; alcohols; aldehydes; ethers; ketones; $<C_5$ aliphatic amides; polyfunctional alcohols; hydrocarbons; cyclic amides; polysaccharides

Table 4.10 Percent contents of fractions of natural organic matter isolated from different sources of water supply

Fraction	Content, %		
	[38]	[59]	[45]
HA	19	–	–
HOB	Traces	0–6	0–22
HOA	54	8–12	19–68
HON	12	13–22	0–25
HIA	7	44–55	8–50
HIB	5	4–6	1.5–10
HIN	3	9–25	1–35

In the technology of water treatment of natural water determination of the share of NOC or diluted organic carbon (DOC), being biodegradable organic carbon (BDOC), is of substantial importance [62–65]. Biological accessibility of BDOC is an important factor determining water quality in the distribution systems of water supply. The gist of determining BDOC lies in the fact that a water sample is inoculated with a weighted amount of biologically active sand within several days during aeration. The BDOC concentration is calculated as a difference between the initial DOC and the minimal concentration of the latter observable during the incubation period [63]. It should be noted that some researchers underline the importance of separate identification of assimilated organic carbon (AOC) in which BDOC may be considered as part of AOC [55, 65–68]. In addition, these researchers believe that it is the content of AOC which correlates with the secondary bacterial growth and is a measure of the bacterial mass, while BDOC is a DOC fraction, which may be assimilated and mineralized by heterotrophic microorganisms [68]. In line with the data of [69] BDOC is a fraction of organic carbon, which may be mineralized by heterotrophic microorganisms. AOC is the fraction of total organic carbon (TOC) converted into the cellular mass by specific strains or the bacteria population. For determining AOC a water sample is incubated with the specific bacteria population and the maximum level of bacteria composition (in the stationary phase of growth) is taken proportional to the limiting content of the nutrient matter (biogene) found as carbon with the redundant amount of other nutrient substances. The limiting concentration of biogene is converted into an equivalent of carbon by using the empirical coefficient determining the relationship between the amount of the produced cellular mass of bacteria and the substrate. AOC, in general, includes the most easily degradable BDOC fraction.

The secondary products appearing in the water and related with natural processes produce a substantial effect on the water quality in natural sources [70–72]. Thus apart from NOC, many inorganic compounds may appear in the water as a result of the weathering of soil minerals or decay of vegetation. The increase in the concentration of humic compounds in the water is inevitably accompanied by the increase in the content of iron [49]. For all intents and purposes all dissolved iron is represented by organic complex compounds of different molecular weight determining its migration mobility. The presence of NOC in the water produces a little effect on the content of manganese since this metal in small amounts is bound with it into complexes [49].

One of these undesirable elements is boron whose natural origin in water may be detected also by the mineral composition of soils and physiology of the vegetation. Since boron is a substantial trace element in the nourishment of higher vegetation, it can be extracted from the soil and diffused into the leaves, where its highest concentrations can be detected. When the higher vegetation undergoes biological decay, boron returns to the soil together with decayed natural organic matter. Owing to this life cycle of the plants one can observe certain correlation between the boron contents and organic carbon in the soil. Moreover, the investigations demonstrated much affinity between boron and humic compounds. Given the high content of the latter boron is sorbed in the soil in great amounts [73, 74]. Hence it could be expected that the washing out of boron to the underground and surface water will be correlated with NOC content in the water.

For a surface water source effectively not subjected to anthropogenic effects, it is established that changing boron concentration from 20 to 50 mg/dm^3 reads to the DOC content 1.5 to 3 mg/dm^3. In this case, the value of ratio boron DOC equals 36 if both parameters are expressed respectively in mg/dm^3 or μg/dm^3. The statistical analysis of surface water showed that there is a linear relationship between the content of organic matter and boron in the water [75]. A deviation from the discovered correlation may be used for establishing the anomaly in water chemical properties or its pollution as a result of anthropogenic effects.

NOC plays a substantial role in the formation of water quality in a water body when affected by sunlight. In paper [76] is evidence that under sunlight, radiation causes destruction concurrently with the formation of acids having a low molecular weight. Microorganisms also take part in this process. Finite products of the said processes are formic, acetic, pyruvic, oxalic and succinic acids. Formation of such hydrophilic low-molecular compounds invariably will affect the efficiency of water treatment processes and will require special techniques. A hypothesis is expressed that decay of organic compounds in fresh water exposed to the sunlight may be initiated at the expense of the mechanism of Fenton's photoreaction in the presence of amorphous iron oxide and fulvic acids [77].

There are also other viewpoints with respect to the NOC role in the formation of the water quality of water bodies. It is believed that an increase in the content of NOC is one of the reasons for inhibition of the development of aqueous plants and animals since in the sections of water bodies having high water color the oxygen concentration is always low [49, 78]. Perhaps this is determined by the deterioration of the photosynthesis processes which naturally leads to the decrease of bioproductivity of the water bodies binding biogenic elements into inaccessible forms. A number of researchers believe that the presence of NOC in water may be considered a favorable factor for detoxication of ecologically hazardous inorganic and organic compounds. In papers [79, 80] is shown that toxicity of heavy metals, polyaromatic hydrocarbons, pesticides, and herbicides in the presence of NOC sharply decrease or is not revealed at all [49, 80]. Thus, the monitoring of the contents of natural organic matter in sources of water supply is a useful and simple tool of water quality when producing drinking water. The reason for this lies in the formation of secondary products of decontamination when using ozone or chlorine compounds. In addition, the content of NOC has large significance for monitoring the emergence and growth of a biofilm in the distribution systems and as a result affects the biological stability of water. It should be recognized that large waterworks for water conditioning should conduct the monitoring of (TOC/BDOC) and adapt the technology of water conditioning to the change of this index.

Pretreatment of Natural Water by Filtration Through Natural Barriers (dunes, dams, embankments, underground horizons) The WHO concept of providing the population with high-quality drinking water implies microbiological safety with minimal use of chemical treatments. With the objective of reducing the chemical makeup, special attention should be attached to biological methods to minimize the formation of necessary doses of chemicals and to reduce the formation of decontamination by-products [81]. One example of reduction is to pump water from surface sources of drinking water to surface sand horizons such as dams and bankings for filtration as a pretreatment from organic and microbial pollution [82–85]. This pretreatement is advantageous due to cost effectiveness and minimizing the use of chlorine which forms a toxic compound during the reaction with NOC [86]. With this promising method, scrutiny is necessary.

It is known that biological stability of water is determined by the presence of compounds such as BDOC and AOC both contributing to the growth of microorganisms and affecting the growth of heterotrophic bacteria [87, 88]. In addition, a substantial role is played also by the presence of other biogenic compounds and water temperature [89]. Humic compounds representing the main function of natural organic compounds are present in all surface water and are poor in response to biological decay. Insufficient biodecomposition runs very slowly though may be accelerated by water pretreatment or by the use of chemical oxidation [90–95]. So, the adherence to filtration through sand is a likely alternative.

Microbiological and physico-chemical processes determining the removal of microorganisms and other impurities from water when filtrating it through sand, plays a substantial role in the removal of natural organic carbon in such a system [96]. Processes of purification of NOC normally run inside the microbiological consortium localized in the sand matrix such as in the biofilm. The bacterial cells in the biofilm are in the matrix of extracellular compounds mainly representing proteins and carbons [97]. The presence in the biofilm of sand through which water is filtered ensures a large interphase surface for the whole water volume where biochemical processes occur thanks to numerous channels, cracks, voids, and bacteriological clusters [98]. When the water passes through sand with a biofilm, NOC are sorted out, metabolized and, removed from the water to a certain degree. More enlightening details about this process can be seen in the following.

The efficiency of natural water treatment when it passes through sand horizons has been investigated in detail in [99]. Despite the fact that the mechanisms of many phenomena involved in this process to date have remained unclear, some regularities of natural self-purification of water in sand horizons still have been revealed. Surface sand horizons perform an important barrier function which is confirmed by substantial reduction in the filtered water of AOC content, the amount of bacterial mass determined by the methods of direct calculation, or by the method of seeding on the surface of agar. Hydrophilic and hydrophobic acidic fractions of NOC are bioresistant and are still destroyed to some extent determined by their presence in sand layers of aboriginal microbiota. In near-surface treatment of water plankton microbiota takes part. The aboriginal microflora of the horizon displaying low activity in the oligothrophic medium of the sandy horizon when pumping water demonstrates the metabolic potential for degradation of bioresistant fractions of humic acids. However, such a metabolic potential by itself was insufficient for substantial reduction of the TOC and AOC concentrations and in the long run insufficient to obtain biologically stable drinking water. However, all hope is not lost.

Despite this, the obtained results of research concerning hydraulic, physico-chemical and chemical-biological processes demonstrated a good and stable sustainable effect of purification when filtrating through dams and banks. Such filtration ensures the removal of suspended particles of pathogenic microorganisms of some organic matter and most of trace amounts of polluting compounds except some polar organic compounds which are persistent ones [100, 101]. Microorganisms make a substantial contribution to water treatment in the case of its underground filtration due to enzymatic degradation or partial metabolism of water impurities and owing to simultaneous action with physico-chemical processes (e.g. adsorption). Biofilms formed in such ecosystems during underground filtration may contain algae, bacteria, fungi, and other eukaryotic microorganisms filtering through sand particles then passing into the system of water distribution. Since the biofilms represent a predominant form of life in the water medium, it is very important to have information on the structure and functions of these biocenoses as explored below.

Restrictions in the presence of biogenic elements and stress situations in the environment may lead to physiological and morphological changes in many species of aqueous bacteria. Paper [102] deals with the study of the state of the biofilms in the systems when filtrating water through banks. It was shown that biofilms investigated at various sections of one and the same system differed from each other and consisted of different populations. The index of similarity in identical water media was higher than such between different aqueous sections inside the single medium. This was evidence of the fact that identical biological niches were populated with identical bacterial populations. Most identified bacteria belonged to *beta- Proteobacteria.* In addition, saprophytic microbacteria and legionellas were identified in the biofilms. The same microorganisms were detected also inside the water-distribution system which pointed out the possible transfer of the bacteria from the surface water to drinking water. *Legionella pneumophila* and indicators of faecal pollution, *Enterococcus falcium,* and *Efaecalis* were not detected in the biofilms of this research. So, this data was not enough to make a conclusion on the presence or absence of pathogenic microorganisms in the biofilms. The impacts of toxic compounds present in the surface water were not detected in the studied samples of the biofilm. This indicated an increased tolerance of the populations of the biofilm with respect to toxic compounds. In other research, this author finds more of interest concerning natural filtration data.

Research of the extracellular enzyme activity of biofilms demonstrated its reduction during transfer from the surface water to the drinking water. This reflects improvement in the trophic status indicating that the purification process, in principle, performed its function. Application of pre-treatment of water by its filtration through dunes, banks, dams, and sand strata is widely used not only in Central Europe [85, 102–105], but also in many countries of the third world where surface water are heavily polluted.

It is also necessary to note that filtration (or infiltration through soils) is used also for artesian water. Thus, in 1998, in Finland they developed and started to implement a project referred to as "The reuse of artesian water: technique of infiltration processes in soil and water quality". As a result of the implementation of this project, they managed to optimize the process of water treatment allowing for its quantity, quality and the ecological effect of the soils [104]. Thus, the use of the processes involving water pre-purification of bacterial pollution, NOC, and other pollutants by pumping and filtering it through natural or artificial sand barriers is an accessible and cheap method of water pretreatment before its further purification in obtaining biologically stable drinking water. The Finnish method necessitates the following two procedures: adsorption on the horizon rock through which water passes, and NOC biodegeneration in the biofilm on the grains of the rock.

First, during sorption of dissolved organic substances (DOS) on the rock of the horizon there occurs their distribution between sorbed and volume phases [105–107]. In this case, both the properties of the rock mineral-water surface interface and the water phase in terms of their reactivity with respect to various effects vary. When assessing adsorption purification ability of the rocks, the researchers propose to take account seasonal changes of the concentration and composition of DOS in natural water which substantially change when snows melt and in the period of floods [108–110]. During these periods NOC peak concentration is achieved and then after some time it reduces exponentially. To account for such a phenomenon a mechanism of the intensive washing of rocks with large volume of river and ground water was proposed [108, 109]. However, the authors of paper [107] propose another view of this phenomenon. In the basin of a certain river the water samples of the spring floods from a surface water source and the so-called "perched water" (i.e. underground water of shallow location before the first waterproof horizon) were investigated. It is shown that the level of DOS contents and their molecular weight are identical in both cases. However, in the precipitation period the "perched water" samples taken from the same places differ substantially by characteristics and properties of DOS from the spring samples. These differences are referred to as biogeological changes in the DOS with a total characteristic within the summer period and during the periods of precipitations under arid conditions.

Second, NOC biodegeneration in the biofilm on the grains of the rock is evident in the biogeological conditions implied from the adsorption through the ground rocks. Methods of chromatography and spectroscopy were used to demonstrate that the total characteristic of DOS by components both in surface and underground water samples varied due to fractionation in sorption on the rocks. This is expressed in the preferential removal from water of compounds with a larger molecular weight of more aromatic and more hydrophobic components. Thus, it is confirmed that the processes in the ground substantially affect water quality more than other phenomena. Since sorption from processes during underground water purification of biochemical processes play a large role in this topic, the author continued to consider other papers dealing with the stability of organic matter in soil [111–113]. This consideration is explored in the next few paragraphs.

Resistance of humus matter to biological degradation is a direct result of internal factors intrinsic to the system such as the presence of condensed three-dimensional structures with a large amount of aromatic blocks resistant to microbiological effects. External factors (together with the role of organic matter penetration to the soil) mainly are attributable to microisolation of humic fractions [113]. This implies physical occlusion of the latter to microaggregates resistant to biodegradation which prevent penetration of soil enzymes and interaction with the surface of clays and amorphous oxides.

In paper [112] is information about the difficulty of isolating one limiting factor responsible for bioresistance of humic acids. Structural classic characteristics such as the degree of aromatic property, polydispersity, or the amount of functional groups containing oxygen produce their limiting influence only. The aliphatic property has a positive impact on mineralization of organic carbon, but it is true only for half of ortho-alkyls. Alkyl carbonic chains in peat humus acids manifest the same bioresistance as aromatic ones. Resistance to biodegradation, as is shown in [112], depends on complex interactions of structural factors in which unordered cross-linked macromolecular structures of humic acids produce a principal impact on limiting effects of soil enzymes.

An example of effective use of water pretreatment by filtration through the dam is described in [114]. This technology was used for improving the quality of drinking water in Berlin. The most substantial factors were assessed which affect the efficiency of the artificial pumping of the surface water through banks at two different modes. Every month a specially run analytical program gave information on the changes of the contents in natural water before and after the pumping of DOS, the values of UV-absorption in water samples (UV A_{254}), adsorbed organic halogens, and any model organic compound. The monitoring was run for more than 1 year. The modes of carrying out infiltration differed by different oxidizing-reducing conditions, time, and distance of water transportation. In one case, anaerobic conditions dominated (transportation time constituted 4–5 min), while in another case there were predominantly aerobic conditions (transportation time—up to 50 days). All these factors substantially affected the kinetics of DOS degradation and the degree of removing organic halogens as well as the traces of the model organic matter. Aerobic conditions turned out to be more effective.

Thus, the use of the process of water pretreatment by pumping into underground horizons should be preceded by the study of geochemical characteristics of soils from the viewpoint of their sorptivity and the research into the NOC structural and chemical characteristics with respect to the ability of the latter for biodegradation.

State-of-the-art Approaches to Justify the Technology of Conditioning Quality Drinking Water As is shown in [32], the main goal of modern technologies for conditioning quality drinking water is through obtaining biologically stable water having very low levels in the contents of organic compounds, nitrogen and phosphorus for preventing secondary water pollution as a result of microbial growth in the distribution systems.

In addition, the presence of NOC in the drinking water treated determines the formation of trihalomethanes (THM) and other halogen-organic compounds when decontaminating water with chlorine which also determines the necessity of deep removal of NOC during water conditioning.

In 1988, the US Environmental Protection Agency (EPA) published rules which set minimal levels for the contents of disinfection products for drinking water constituting 80 μg/dm^3 for general THM and 60 μg/dm^3 for five haloidacetic acids (HAA) [115]. In 2000, the EPA toughened these requirements to 40 μg/dm^3 for THM and 20 μg/dm^3 for HAA. In addition, the EPA set special requirements to completeness of TOC removal during water treatment to prevent the formation of secondary products of decontamination that are higher than standard requirements. These requirements were set based on the analysis of the experience of many years showing that identified THM and HAA involve 50 % of the total content of haloid-organic products that are formed during chlorination [116–118]. Other haloid-containing compounds that remain unidentified are hazardous for human health. Therefore, in-depth removal of TOC during water conditioning prior to chlorination leads to the reduced formation of unidentified compounds which reduces the risk related with hazards for human health. Specific requirements for the completeness of removing THM set in [115] depend on the quality of initial natural water and their content is the function of TOC concentration and water alkalinity (see Table 4.11) [119]. As is seen from the given table the maximum degree of removing TOC, the bulk of which is represented by NOC, constitutes 50 % with the TOC initial concentration less than 8 μg/dm^3 and alkalinity roughly 1.2 mg-eq/dm^3.

Table 4.11 Removal degree of total organic carbon in the coagulation process with different water alkalinity in terms of $CaCO_3$

Concentration of TOC, mg/dm³	Alkalinity, mg/dm³		
	0–60	>60–120	>120
	TOC removal degree, %		
>2.0–4.0	35	25	15
>4.0–8.0	45	35	25
>8.0	50	40	30

It should be noted that in Ukraine the content of TOC in the sources of central water supply reaches 50 mg/dm³ with water alkalinity >2.5 mg-eq/dm³. In paper [119] also is found data that given an accurately chosen dosage of the coagulant, which has the gradient of reducing the TOC content 0.3 mg/dm³ per each 10 mg/dm³ of $A1_2(SO_4)_3$, one can increase the degree of TOC removal to 64–66 % with water alkalinity >1.2–2.4 mg-eq/dm³. Nonetheless, even given such a high degree of TOC removal and preventing the formation of THM not to exceed 65 and stabilizing HAA to 78 % is preferable. Consequently, it is necessary to use special techniques which allow one to increase coagulation efficiency in order to enhance its action. As this process is continued to be explored, more literature is necessary for a proper understanding.

In literature such modification of the coagulation process is referred to as "enhanced coagulation" [120]. The stage of water treatment with coagulants is essential effectively for all technologies of water treatment at the centralized stations of water supply. One of the techniques of "enhanced coagulation" may be water pretreatment by such oxidants as chlorine, ozone, and chlorine dioxide with the goal of oxidizing reduced forms of iron and manganese and the destruction of organic substances imparting tang and smell, water decoloration, and prevention of the microbial growth in the facilities [121]. However, for some categories of natural water oxidants may act as an additional factor during coagulation for removal from hardness [122]. It is known that most often chlorine is used at the preliminary stage as an oxidant although it is not always advisable due to reasons considered earlier. The use of preozonation instead of chlorine makes it possible to use it after coagulation through precipitation. Then the probability of forming secondary products of chlorination stands in the way of removing TOC [123]. However, in some cases, observed that preozonation before coagulation may prevent the removal of TOC [124, 125]. In water after ozonation and coagulation higher concentrations of TOC may emerge than in the case of simple coagulation owing to partial oxidation of NOC leading to the production of secondary products which are less effectively removed by coagulation. This is attributable to the fact that organic compounds formed during ozonation are more hydrophilic and have a lower molecular weight compared with initial NOC and accordingly they are more difficult to be removed by coagulation [126].

Thus water preozonation before coagulation needs to be carried out at such doses whereby oxidation products formed may be partially polar than initial NOC. This requires determination of technological parameters of ozonation for each specific water supplies depending on qualitative and quantitative composition of NOC and accompanying anthropogenic pollutants in the initial water. The use of intermediate water ozonation after coagulation through precipitation sets quite a different problem when determining effective modes of oxidation. Ozonation is advisable to be carried out in such a way that the secondary products formed are more hydrophilic and have lower molecular weight. Then hydrophilization and reduction of their molecular weight leads to larger biodegradation positively affecting the situation when using biodegradation and biosorption at the subsequent stages of purification.

Taking into account the above it is expedient to assess the effective degree of oxidation of organic substances before coagulation and biofiltration. In paper [127] is found the method of assessing NOC biodegradability in different water sources based on the concept of determining the kinetics of running the "fast" and "slow" stages of NOC biodegradation before or after oxidation. The proposed approach may be applied for assessing the efficiency of combining the ozonation/biopurification processes and control of the NOC composition in natural water. The investigations showed that ozonation of water containing high concentrations of TOC leads to the increase of the concentrations of NOC fractions both fast biodegradable and slowly decomposing. The NOC fraction (slowly biodegradable) makes difficult its removal during biofiltration. Process efficiency may be increased by adding to the water easily degradable organic carbon which will be conducive to metabolization of slowly decomposing TOC fractions and increase the efficiency of biodegradation of ozonized NOC [128].

In paper [129] one more approach is proposed for assessing the effective degree of oxidizing organic substances in water before the subsequent stages of its treatment. It is known that the efficiency of biosorption of organic matter on activated carbon (AC) depends on the value of changes of Gibb's free energy of adsorption ($-\Delta G^{\circ}_{a}$) with the idea that the higher this value the smaller the contribution of the biodestructive component and vise versa [130]. This relationship is related to solubility of the organic compounds for example to its hydrophilic nature since hydrophilic compounds biodegrade easier and is less effectively sorbed on the AC hydrophobic surface. It is proposed to use the value $-\Delta G^{\circ}_{a}$ as the criterion of assessing the rational degree of oxidizing organic compounds before biosorption or biofiltration. It should be noted that such an approach may be used also when combining the oxidation through coagulation processes. If preoxidation increases the changes of Gibb's free energy of adsorption in relation to oxidation products, then it is determined by the increase in the degree of hydrophobic nature of the oxidizing products formed. The decrease of $-\Delta G^{\circ}_{a}$ on the contrary indicates the accumulation of the products with a larger degree of a hydrophilic nature compared with the initial compound. Naturally, the decrease of $-\Delta G^{\circ}_{a}$ after ozonation will increase the efficiency of biosorption or biolfiltration while the increase of $-\Delta G^{\circ}_{a}$ will lead to the better effect of coagulation.

One of the areas in water conditioning of drinking water that is currently practiced is a simplified schematic of removing turbidity and bacterial pollution of water by the so-called "in-line filtration" and "direct filtration" methods. The "in-line filtration" method implies contact with coagulation in the layer of the filtration medium such as the combination of coagulation and filtration. The term "direct filtration" denotes successive implementation of the processes of coagulation, flocculation, and filtration without the clarification stage [131]. These methods are characterized by reduced doses of coagulants, fewer numbers of the strains formed, less areas, and a reduction of capital costs. However, since filtration is only a physical barrier both these methods have total efficiency somewhat lower than the technologies with the clarification stage before filtration [132]. Nonetheless, in certain situations such a simplified pattern of water conditioning is justified and therefore the US Environmental Protection Agency (US EPA) introduced the above methods into the Enhanced Surface Water Treatment Rule [133, 134].

Papers [135–139] deal with the in-line filtration and direct filtration methods. Generalizing the results of these investigations one may note that use of the direct filtration method (at the initial water temperature of 14–20 °C, filtration rate 12.0–14.4 m/h) and the aluminum coagulant with anionic flocculant make it possible to reduce bacterial pollution by the order of magnitude 1.3–3.8. Replacement of aluminum by polyaluminum chloride and anionic polymer by the cationic polymer on the average secured the reduction of the bacterial pollution by the order of magnitude 2.6–2.9 [135, 136]. More effective purification is observed when using in-line filtration. At the water temperature of 6.9–16 °C, the use of $Al_2(SO_4)_3 \bullet 14H_2O$, of anionic and cationic flocculants, and the filtration rate of 12.5–20.0 m/h made it possible to reduce the value of bacterial pollution by the order of magnitude 1.5–4 [132]. Replacement of $Al_2(SO_4)_3 \bullet 14H_2O$ by iron chloride and the use of cationic flocculant at the filtration rate of 7.3 m/h contributed to achieving the reduction of bacterial pollution by the order of magnitude 4.7 at water temperature (− 5 °C) [137, 138]. The increase of the filtration rate to 29–38 m/h, but at water temperature 18–19 °C, and with the same type of the coagulant and flocculant ensures the decrease of the degree of bacterial pollution by the order of magnitude 3.0–4.4 [139]. So, a through examination of in-line coagulation is helpful.

Thus, the use of in-line coagulation concurrently with the use of the iron coagulant and the cationic flocculant even at the lower temperature of water may ensure its effective purification. It should be noted that the coagulant dose in these investigations was 20 mg/dm³ for $A1_2(SO_4)_3$ and 10 mg/dm³ for $Fe_2(SO_4)_3$. However, the relationship between the coagulant dosage and water temperature, titration rate, and concentration of NOC was not found out. Efficiency of using in-line coagulation in combination with preliminary volume coagulation at water temperature 3–5 °C was studied in paper [140]. In-line coagulation as the second stage of water treatment at low temperature of the latter made it possible, with initial color 58 deg., to reduce it to 11.5–14.0 deg., turbidity—to 0.33–0.44 mg/dm³, the content of residual aluminum—to 0.046–0.056 mg/dm³ and iron—to 0.007–0.008 mg/dm³. Thus, after in-line coagulation the indicators of water quality were more effective than standard ones which exist at present.

The use of potassium ferrate for coagulation water treatment is an interesting area. In potassium ferrate iron is hexavalent (FeO_4^{2-}), i.e. it is the compound like potassium permanganate [141]. An advantage for using such a reagent lies in that it incorporates simultaneously properties of an oxidant and a coagulant. Potassium ferrate differs from other inorganic salts by that ferrate-anion (FeO_4^{2-}) easily dispergate and dissolves in aqueous solutions (~ 15 g/dm^3). In doing so, it quickly decomposes in acid solutions, but its stability increases as the solution pH increases [142]. FeO_4^{2-} is decomposed under the action of reducing reagents and the composition of the reducing products depends on the solution pH. In an acid medium:

$$Fe_4^{2-} + 8H^+ + 3e^- \rightarrow Fe^{3+} + 4H_2O;$$

Weak, neutral, and weak alkaline media

$$FeO_4^{2-} + 4H^+ + 3e^- \rightarrow Fe(OH)_3 + OH^-;$$

Alkaline medium

$$FeO_4^{2-} + 2H_2O + 3e^- \rightarrow FeO_2^- + 4OH^-.$$

With the dose of potassium ferrate of 10 mg/dm^3 (pH 3.39) $Fe(OH)_3$ is the reduction product. Ferrate may decompose a broad spectrum of aqueous impurities thanks to its strong oxidizing ability [143–145]. The research conducted demonstrated that potassium ferrate may simultaneously reduce by 50 % the content in the water of alcohol and benzene and part of some biologically resistant compounds such as phenyl benzene [146]. Owing to a high oxidizing ability, potassium ferrate may act as an effective disinfecting agent. Its disinfecting ability is the action of monochloramine within a wide range of pH values [147, 148]. For example, when treating household wastewater after the secondary sedimentation tank with the dose of 10 mg/dm^3 of K_2FeO_4 for 15 min 85 % BOD_5, 10^4/cm^3 of fecal coliform bacteria, and about 10^4/cm^3 of the total content of bacteria were removed from water. When treating surface water with potassium ferrate with the dosage of 50 mg/dm^3 about 99 % of suspended matter and 94 % of turbidity were removed. Thus, potassium ferrate turned out to be a more effective coagulant than $A1_2(SO_4)_3$ and $FeCl_3 \bullet 6H_2O$ at the dose of 275 mg/dm^3 [141].

Fulvic acids and natural organic matter also may be effectively removed from water when using potassium ferrate [141]. About 90 % of NOC absorbed in the UV area are removed at the ratio of the concentration of ferrate ions to fulvic acid 12:1 (by weight). The oxidizing ability of ferrate is enhanced when pH increases in a certain series (optimal pH value 8–9). The joint use of ferrate, the coagulant and decrease of turbidity (polyaluminum chloride) were more effective when removing fulvic acids and turbidity reduction if ferrate was used at the stage of preoxidizing. Fulvic acids with concentration 2 mg/dm^3 are removed from water completely with the joint use of 8 mg/dm^3 of ferrate and 0.8 mg/dm^3 of polyaluminum chloride. Since Fe^{3+} accelerates decomposition of ferrate the joint use of ferrate and Fe^{3+} promotes the effect of reducing the content of fulvic acids. Potassium ferrate may be synthesized by the Tian method [149] which includes oxidation of iron nitrate with hypochlorite in a strongly alkaline medium. Then potassium ferrate is precipitated by adding potassium hydroxide. The sediment is filtered and dried out.

It is known that ozone may be used at the beginning (preozonation), and the middle and end of the flow chart [150]. During ozonation the main goal is transformation of large organic molecules for enhancing the efficiency of the subsequent stage of treatment such as coagulation, flocculation and adsorption [151, 152]. At the end of the flow chart ozone is used in high dosages for preventing the development of microbes in the distribution networks [153]. However, a number of papers noted an ambiguous change of the properties of secondary products by oxidizing organic matter in water. Very often one can observe the inadequate bacterial growth in the distribution systems or in the case of biological water treatment [122, 154–156]. As a result of such ambiguous behavior of secondary products of oxidizing the production of biologically stable water cannot be guaranteed [127].

In paper [157] gel-chromatography methods are used in combination with the changes in the intensity of the ultraviolet region of absorption and sensitive method of determining DOC to study the changes, which take place in the DOC structure as a result of ozonation and chlorination. Concurrently with the chemical analysis the researchers determined also the degree of biological degradability of oxidizing products by monitoring the changes in the AOC content in water in compliance with the methods described in [158, 159]. In paper [159] used a sample of fulvic acid, which contained five main fractions: a humus fraction with hydrophilic macromolecules of acid compounds; structural units representing precursors or products of destruction of the humus fraction; the low-molecular organic acid fraction, which is represented by oxalyl acid, rumaric acid, etc.; the amphiphilic fraction with a broad choice of compounds; hydrophobic compounds.

It was found out that ozonation of the aqueous solution of this sample of fulvic acid only slightly changes its composition. When the ozone dose is increased from 7 to 20 mg/dm^3 one could observe a small decrease of DOC content. A macromolecular fraction of humic compounds remains in the solution and manifests absorption intensify in the ultraviolet region $\lambda = 254$ nm.

After chlorination the low-molecular fraction grows significantly when determining DOC and when assessing the intensity of UV-absorption by dissolved organic matter. In this case the macromolecular humic fraction is not subjected to biological destruction. The aromatic acidic fraction of fulvic acid does not biodegrade either, whilst some portion of low-molecular organic acids in the corresponding fraction was biodegradable.

As a result of ozonation there occurs decrease of the molecular weight of ozonized fulvic acid [160]. In this case the degree of biodegradability and the amount of DOC varied from 13 in the initial sample to ~17 % in the ozonation one. Similar results were obtained in [161], where it is shown that chlorination reduces the DOC content by 18 %, while addition of chlorination to ozonation increases the biodegradability degree to 24 against 18 % in chlorination. According to [162] the improved biodegradability is achieved as a result of destruction of bioresistant molecules of dissolved organic compounds into small molecules, which are better assimilated by microorganisms. The smaller molecules of fulvic acids are the most suitable substrate for bacteria because they contain a broad variety of degradable functional groups. Both ozonation and chlorination of fulvic acids lead to the change of molecular dimensions to smaller values and the appearance of more polar compounds, which may be a reason of better biodegradability [163–165].

Thus, a detailed characteristic of humic compounds after their oxidation treatment is an extremely complicated task due to a complex structure and polydispersity of their composition. Determination of DOC content and gel-chromatographic analysis in combination with the change of UV-absorption intensity make it possible to characterize humic compounds and predict their behavior in physical, chemical and biological processes [166]. But still the approach proposed in [129] appear to be more attractive since it makes possible to obtain more quickly necessary information by assessing the change of Gibb's free energy of absorption of NOC oxidation products. It is found out that preliminary NOC oxidation may either decrease or increase the potentials of THM and HAA formation. The THM formation normally decreases as the ozone dose increases. The potential of THM formation initially increases as the ozone dose increases and reaches a peak and then decreases. In paper [167] it is show that the value of this peak for different NOC fractions varies from 1.4 to 3.5 mg O_3/mg of DOC. In this case one observes the value of the HAA formation potential after oxidation of such initial compound as *n*-hydroxybenzoic acid (~10 times higher than for the initial solution i.e. without ozonation). It is clear that secondary products of this acid ozonation are target compounds for the HAA formation and in the given case ozonation leads to the increase of their content in water after chlorination. Research into the HAA formation in water with the increased content of humic matter and high color showed that after chlorination one can see the growth of the HAA concentration (up to 120 mg/dm^3). From this amount 44 % accounts for dichloroacetic acid and 4 % for trichloroacetic one. The HAA highest concentration was noted in the period of water maximum temperature and the highest content in it of NOC [168].

Thus, the use of preliminary ozonation of natural water before further stages of its treatment requires thorough research of the composition and properties of NOC of a specific source of water supply and specific interaction with ozone.

In paper [169] noted that ozonation converts DOC from hydrophobic into hydrophilic organic carbon without substantial removal of DOC. The positive role of such transformation manifests itself especially effectively during further filtration of water through BAC [170].

However, as can be seen from the above analysis there is some optimal limit for oxidizing NOC before subsequent biosorption. Thus it was established [171, 172] that the maximum of BDOC is formed with the degree of TOC destruction in ozonation is ~30 % and does not increase even though the ozone dose or the reaction time increase. It is possible that TOC transformation into BDOC may be inhibited by the BDOC products themselves formed in the ozonation process since the latter are capable of absorbing ozone. The positive effect of preozonation of water before biosorption on AC was noted also in [173–176].

In ozonation of NOC the amount of organic acids formed by ~10 times exceeds the total amount of aldehydes and ketones [177]. In this case ozonation increases the level of AOC content in the investigated natural water from 30 to 585 μg/dm^3 which requires its deep removal during further water treatment.

As was noted earlier, one of the effective and promising methods of removal of AOC from water is biological water treatment following preoxidation especially on granulated activated carbon (GAC). Comparative testing of GAC and standard filtration media such as anthracite and sand showed that a large population of microorganisms is held on GAC and a dense biofilm is formed and this film may remove AOC and other secondary ozonation products more effectively with relatively short time of contact with carbon [177, 178]. In ozonation DOC are converted from humic compounds into nonhumic fractions and there occurs transition of macromolecular fractions to low-molecular ones [179, 180]. On the one hand, such transformation increases the degree of adsorption cleaning and on the other hand—during the ozonation process some intermediate products may also be produced, which are undesirable during adsorption. Therefore, when combining preozonation and adsorption on GAC it is important to establish a clear-cut interrelationship between the ozone dose and GAC adsorption ability [167].

It was shown that efficiency of adsorption increases substantially when using sufficiently high ozone doses (6.0 mg O_3/mg of DOC). However, such influence is related to, most likely, the development of biological activity in the carbon layer. It is known, aliphatic aldehydes, hydrogen peroxide and organic peroxides saturated with carboxylic acids are typical secondary products of DOC ozonation [181, 182]. On the one hand, the appearance of hydrophilic secondary products of high solubility whose amount increases with the increase of ozone dose [183] reduces efficiency of physical adsorption, but on the other hand biosorption efficiency increases [131].

Problems of successful use of AC at different stages of the flow chart are discussed in the world literature within the last 5–7 years. Paper [184] notes that GAC and membranes should be used at the water treatment stations if the contents of TOC in surface sources are greater than 4 mg/dm^3 and a station services 100,000 of population. In the event of using underground water for drinking water supply this requirement is implemented even at the TOC content more than 2 mg/dm^3 and the size of more than 50,000 people. This is determined, as was noted, by the necessity of the deepest removal from water of organic compounds—precursors for the formation of toxic haloid organic substances during disinfection.

No doubt, the stages of adsorption on AC in most flow charts precede the stages of coagulation-flocculation, precipitation and mechanical filtration. When cleaning of NOC such definite sequence of operations [185] affects both the water quality and operational costs. With the optimal sequence the removal of more than 70 % of all NOC contained in water may be achieved. However, paper [186] showed that AC may be successfully used also at the preliminary stage of water treatment to coagulation for effective removal of humic compounds if water has little turbidity.

AC in the technology of water conditioning is also used for the removal of organochloric pollutants formed after disinfection [187]. In the case of removing NOC from natural water after the stage of coagulation-precipitation the operational time of the adsorption filter till the reduction of absorptivity may achieve 1,000 days. After that period carbon becomes like sand and removes organic compounds only at the expense of "the effect of the biofilm". In this case it is advisable to regenerate carbon by treating with an alkaline solution [188]. Losses of carbon as a result of such regeneration are insignificant and may constitute ~2–3 % at the expense of removal during filtration (friction and other mechanical factors) [189].

In the case of thermal regeneration the carbon losses reach 18–20% and it is expedient to charge the filter with 21–22% of GAC of the initial medium [189]. Efficiency of drinking water conditioning by the sorption methods may be enhanced by using a mixture of different sorbents. Thus, paper [190] proposes as the components of the mixture to use GAC and AC in the form of fibers. In paper [34] the necessary GAC specific surface is 300–500, while that of carbon fibers is up to 750 m^2/g. Efficiency of the combination of these sorbents is determined, perhaps, by the different porous structure and chemistry of the surface of the sorbents used, which predetermined successive selectivity of their sorption with respect to pollutants of natural water of different chemical origin. Such combination of adsorbents appeared to be especially successful when removing phenol polluting substances (up to 100%) and ions of heavy metals (up to 90%) from water.

The other multicomponent mixture for conditioning drinking water is proposed in [191]. The mixture consists of AC modified with ions of rare-earth metals, silicagel, granite chips, marble chips, sand, and cation exchanger. When using such a mixture the degree of removing organic compounds reaches 90%, while that of Ca, Mg and Fe ions—99.9%. The combination of anion exchangers and AC yields a good effect of purifying natural water of humic compounds. It was shown that during filtration through anion exchanger there occurs chromatographic separation of components as a result of more selective adsorption of humic acids. Since competition on the part of macromolecular humic substances decreases the AC sorption capacity with respect to organic low-molecular non-electrolytes increases [192]. The same effect can be observed also when removing fulvic acids from water [193].

In any case the selection and combination of adsorbents with different properties should be carried out depending on the composition of impurities of natural water of each specific source of water supply. In doing so it is necessary to take into consideration the fact that in natural water along with NOC microimpurities of anthropogenic origin are present effectively always. NOC compete with microimpurities in adsorption determining insufficient efficiency of their removal in adsorption treatment of water [194, 195]. This phenomenon is extremely important because the NOC concentration (n—level, mg/dm^3) by the several orders of magnitude exceeds the concentration of pollutants (n—level, $\mu g/dm^3$), which is characteristic of many pesticides and chlorinated hydrocarbons. For some compounds imparting taste and odor the level of pollutants constitutes nanograms/dm^3. The NOC nature (distribution in terms of the molecular weight, chemical composition) also affects the degree of competition [196, 197]. In the event of simultaneous competitive adsorption of NOC and microimpurities the NOC fractions whose molecular dimensions are comparable with micropolluntants lead to the greatest reduction of the adsorption of the latter [198, 199]. And on the contrary the NOC fractions of large molecular dimensions affect less adsorptivity of small organic pollutants [198, 199]. Similar behavior in adsorption of ingredients of NOC of different molecular dimensions was noted also for AC preliminary saturated with NOC [194, 199, 200]. However, it was noted, that NOC components of large molecular dimensions may substantially reduce the adsorption rate of microimpurities [199, 200].

From the viewpoint of AC physical characteristics the NOC impact on the reduction of adsorption of microimpurities is less pronounced when using carbon containing small micropores. This is related with blocking pores or reduction of accessibility to them as a result of NOC adsorption near entrance to the micropores. For AC having wide distribution of the pores by dimensions such influence is much smaller [198–204]. However, the presence of wider pores in carbon is not the only decisive factor when assessing competition in adsorption of NOC and microimpurities. The authors of the paper [198] show that given identical distribution of pores by dimensions, but with different chemistry of the AC surfaces then the mutual impact of NOC and microimpurities will be different. Thus, adsorption of trichlororethane on AC, presaturated with NOC, became smaller as the amount of acid centers on the carbon surface increased [205]. It is shown that in the general case hydrophobic adsorbents (with a lower content of acid centers on the surface) are the most effective with respect to microimpurities when adsorption is done on AC with both presaturated and nonsaturated NOC [205, 206].

A detailed research into the impact of the pore dimensions and the chemistry of AC surface on the competitive adsorption of NOC and such representatives of microimpurities as relative hydrophobic trichloethylene (TCE) and hydrophilic methyltetrabutyl ether (MTBE) showed the following [197]. To prevent the blocking of AC pores the dimensions of the pores should be approximately two times larger than the effective diameter of the molecules of the target adsorbate. Apart from the factor of pore distribution by dimension, a substantial role is played by sorbent hydrophilic property expressed as a sum of the contents of oxygen and nitrogen (mmol/g) in AC. This index may be used as a criterion of selecting suitable grades of carbon for simultaneous removal from water of NOC and microimpurities. As carbon wetting ability increases or the oxygen and nitrogen contents increase, then the adsorption of microimpurities decreases. This is determined by the enhancement of water adsorption on polar groups of the AC surface, which leads to the formation of aqueous clusters blocking the access of microimpurities to adsorption centers. The given effect is observed both for water-hydrophobic TCE and wettable MTBE. Based on the research conducted it was proposed to use for simultaneous water treatment of NOC and microimpurities of carbons with the content (oxygen + nitrogen) < 2–3 mmol/g.

In addition to anthropogenic organic micorimpurities AC effectively removes from water pollutants produced by aqueous microorganisms. Thus, microcysts (MU), which belong to the class of hepatoxins and possess extremely high toxicity, are synthesized by cyanobacteria present in surface sources of water supply. They cause liver cancer. When the filters with AC are used in water conditioning then the MU content in water is reduced from 0.15 μg/cm^3 to complete absence. However, it should be noted that in the flow chart water ozonation was used before a carbon filter. Exclusion from this flow chart of an ozonisator did not allow removing microorganisms from water [207].

Over the last decade ozonation in combination with AC in the technology of water treatment is one of the most popular techniques. In this case carbon was mainly used in the mode of biofiltration [208]. Dissolved organic compounds both of natural and technogenic origin present in water even in trace concentrations are the sources of energy and carbon in synthesis of the biomass of microorganisms. The special overview [209] deals with the issues of biosorption of organic compounds, therefore we will briefly consider only technological aspects of natural water biofiltration through AC.

Paper [210] quotes results of commercial tests in the course of which organic compounds in the process of biofiltration were removed by 80–90 % and absorbable organic haloid derivatives by 95 %. The water filtration rate in the system under consideration constituted 10 to 30 m/h.

Paper [211] gives the results of pilot production research into the efficiency of drinking water purification of organic pollutants in the filters with GAC at different modes of their operation. The nonregenerated peat GAC of the Norit grade and the regeneratable coal, Filtrasorb grade 400 were tested. Norit, as used both in slow filters with the bed height of 0.15 m and in fast filters of the bed height of 2 m. Filtrasorb 400 was used only in rapid filters with the bed height of 2 m. The filtration rate in the slow filter is 0.625 and the rapid one—8.3 m/h. Thus, the time of the water contact with carbon was the same in both cases and constituted 14.4 min. Concurrently the water was cleaned in the slow and rapid filters with sand both without preozonation and with ozonation. To the rapid filters with GAC the water was fed only after ozonation. To filters (both sand and carbon ones) this water was subjected to usual physicochemical treatment. The variation of the content of organic matter in water was assessed by such integral indexes as TOC permanganate oxidizability, optical density at 254 nm. As a result of the 12-month tests it was found out that owing to bioactivity in the slow filters with sand on the average 10 % of organic pollutants (in terms by TOC) were removed, while in the slow filters with GAC—25 %. The difference was determined mainly by adsorption of the indicated substances on GAC. The degree and the rate of biooxidation in the filters with sand and GAC are of the same magnitude. Ozonation ensures additional reduction of the concentration of organic pollutants in water by 5 %. The rapid filters with GAC remove by 10 % more TOC than the slow ones. On the average during one year the degree of surface water purification of organic matter on the rapid filters with GAC constituted ~40 %.

Analysis of the data on the operation of commercial plants combining ozonation and filtration through GAC shows that efficiency of removing TOC depends on the quality of the initial water. Thus, at upon ozonation the TOC initial concentration of 60–156 mg/dm^3 can be reduced by 11 and in the case of two-stage treatment on filters with AC this value may be reduced by 85 %. The TOC content in the outflowing water varied from 1.9 to 13.9 mg/dm^3 [212]. This value so far is not in line with the state of water biological stability, but makes it possible to avoid the formation of extraordinary concentrations of haloid-organic compounds during disinfection. In paper [213] the other composition of the initial water ozonation and biofiltration through GAC leads to reduction of the content of organic matter by ~50 %. Ozonation in combination with the AC treatment at the second stage leads to the removal of nearly 60 % of organic compounds. As was stated above AC may be used effectively also for removing haloid-organic compounds that form during preliminary and disinfecting chlorination of water. In traditional processes of treatment there occurs the removal of HAA is achieved only by 20 %, however in filtration through GAC—49–86 % of these acids whose initial amount depends on the initial content of TOC in the water of the source [214].

For removing haloid-organic compounds from water one also used the combination of ozonation and biofiltration. Thus, carcinogenic hardly-decomposed compounds (chloroform, bromdichloromethane, dibromochloromethane and bromoform) at first are destroyed by ozone action and then the formed nontoxic products are removed by biofiltration through GAC in the fluidized-bed state [215].

In [177, 216, 217] the processes of removing such biologically assimilating compounds as aldehydes, ketones, and glyoxals were studied. It was noted that aldehydes are removed well on filters with GAC even in that case if the TOC removal substantially deteriorates. For effective implementation of biosorption it is recommended to maintain the water filtration rate within 2.4–12 m/h and the time of the solution contact with carbon at least 5–10 or even 25 min. In this case at the start of filtration the degree of removing biologically assimilable carbon increases and then goes to the plateau and remains at the same level for a long time. When using as a medium a fresh GAC, biological activity in the layer of carbon with respect to aldehydes sets within a week. In this case 90 % of aldehydes are removed. At the same time it was necessary to have 80 days in order to achieve a 50 % removal of AOC [217].

Comparative research of biosorption efficiency with respect to secondary ozonation products of fresh GAC and GAC saturated in the process of a three-year operation, i.e. the carbon already populated with microorganisms, is interesting. During contact of water with carbon within 4.2 min the latter removed on the very first day of filtration all formaldehydes and glycol and 88 % of methylglyoxal. Fresh GAC initially did not remove the mentioned above substances and started to operate effectively only on the 7th to 14th day after filtration. Saturated GAC at a smaller time of contact (up to 1.3–1.4 min) still removed up to 80 % glyoxal on the 7th and 17th days, while it took 52 days for GAC to achieve the same effect [177].

Thus, the quoted data show that BAC is an effective sorbent for removing from water a broad spectrum of wettable low-molecular biologically assimilable organic compounds. It is especially effective, as is shown in paper [153] to use multistage processes of ozonation–biological treatment for removing DOC. Multistage treatment compared with the traditional single stage one makes it possible to remove DOC much more effectively and with smaller expenses. The ten-fold one-minute ozonation and subsequent treatment make it possible to remove 95 % of DOC. At the same time prolonged ozonation for 64 min followed by biotreatment removes only 70 %. Similar results aimed at substantial increase of the degree of water purification of DOC in the multistage technology of ozonation–bioxidation compared with the single stage one were described in [218].

Authors of the papers [219–222] showed the possibility of using as an oxidizing agent chlorine dioxide before adsorption on GAC. In oxidizing NOC chlorine dioxide manifests itself as a sufficiently effective oxidant and effectively does not lead to the formation of THM. A shortcoming of using chlorine dioxide is the formation of inorganic secondary products—chlorites and chlorates. Usually 1 mg of ClO_2 produces about 0.7 mg of chlorite and different amounts (normally lower) of chlorate. The maximum level of pollution with secondary products varies depending on the source of water supply, while the dose of ClO_2 does not exceed 0.24–1.0 mg/dm^3 [223]. Consequently, it is expedient to consider factors affecting the value of ClO_2 demand for treating natural water. This value depends on the presence in natural water of Fe(II), Mn(II), nitrites and NOC. The latter are most significant since metals may be effectively removed during oxidation and filtration through a quartz filter.

From the economic point of view the use of ClO_2 is more preferential than ozonation. However, in this case also it is expedient after the stage of acidification to use filtration through GAC for removing the products of oxidizing NOC and inorganic secondary products. It is shown [224] that filtration through GAC makes it possible in addition to oxidized organic compounds to effectively remove chlorites. However, the literature contains limited information on the efficiency of the combination of oxidizing NOC by chlorine dioxide with subsequent sorption on GAC. Most likely, in the given cases it is predominantly physical adsorption of organic compounds that takes place since the presence of residual amounts of chlorine compounds prevents normal functioning of the biofilm. Paper [223] shows that the impact of ClO_2 on NOC does not substantially change the NOC fraction identified by the absorption intensity in the UV-region. However, small changes in the distribution of molecular dimensions of the NOC fractions after oxidation take place. Absorption in the UV-region by fractions with a large molecular weight (1,900–1,000 Da) decreased, while as absorption by fractions with the molecular weight <500 Da after oxidation somewhat increased. The research demonstrated that the demand in ClO_2 is determined, mainly, by the presence in water of macromolecular fractions of NOC ($>1{,}000$ Da).

It should be noted that pretreatment by ClO_2 substantially enhances the degree of adsorption removal of NOC from water. After passage through the layer of GAC of 1,600 volumes of water with respect to the carbon volume the content of TOC in the untreated water constitutes 2 mg/dm^3, while for the treated water this value does not exceed 1 mg/dm^3. During the filtration process the concentration of chlorites and chlorates in water does not exceed 0.05 mg/dm^3.

Magnetic ionic exchange may be used as a definite alternative technology of removing NOC from natural water in place of coagulation or adsorption on AC. Magnetic ion-exchange resins (MIER) possess traditional anion-exchange properties. MIER has the polyacryl matrix in the chloride form, macroporous structure and strongly basic functional groups. In contrast to traditional anion-exchange resins MIER contains magnetic iron oxide included into the polymer matrix. The magnetic component is conducive to agglomeration and precipitation of the resin and determines a smaller dimension of resin balls so that they may be used for water treatment in the pseudo-liquefied state. The MIER balls have the diameter ~ 180 μm, which are 2–5 times smaller than traditional ion-exchange resins. This leads to the increase of the interphase and the decrease of mass-transfer resistance in the liquid and solid phases. Since the kinetics of ion exchange is determined by mass transfer in the liquid and solid phases MIER should remove NOC much faster than traditional ion-exchange resins [225].

The tests conducted showed that the removal of NOC occurs sufficiently fast. In this case UV adsorbing substances are removed within first 10–20 min. Water treatment with MIER makes it possible to remove these substances in a greater degree than the total dissolved organic carbon. In addition, when using MIER it is possible to remove bromides [226–228]. However, when increasing water alkalinity the removal degree is reduced. In paper [227] was shown that MIER remove hydrophobic and hydrophilic fractions of organic acids. Water treatment with MIER is more effective compared with increased coagulation in relation to the removal of UV adsorbing organic matter, DOC, precursors of forming THM and HAA [229].

MIER also removes a wider series of fractions of organic acids and fractions with a different molecular weight than coagulation [230]. In addition, when treating water with aluminum coagulant after MIER the increase of the degree of removing UV adsorbing organic substances or DOC does not take place. However, the use of coagulation is more effective for raw water with low alkalinity and high electrical conductivity. For both types of water treàtment (MIER and coagulation) the efficiency series of removing NOC components was roughly identical: UV-adsorbing substances > precursors for the formation of HAA > DOC > precursors for the formation of THM.

The undoubted advantage of using MIER is the possibility of removing bromides from all studied types of natural water [230]. The degree of removing bromides increases as the water alkalinity and concentration of bromide ions decrease, in the process of coagulation bromides are not removed.

In the issue of preventing the formation of disinfection products in water there are two approaches. The first approach lies in monitoring the content of precursors, which may react with the disinfecting agent and then introduce the technological stage of their removal.

Monitoring and removal of the precursors is the main factor for removing NOC present in the water. The strategy of monitoring and removing the precursors for the formation of disinfection products in the general form are given in Table 4.12.

Quantitative and qualitative compositions of disinfection toxic products depend both on the type of the disinfectant and the water quality in the water source. It is known that NOC include hydrophilic acids, hydrophobic bases, wettable neutral compounds, hydrophobic acids and neutral hydrophobic compounds. All these fractions during chlorination may form haloid-organic compounds, but the most problematic fraction in terms of forming THM and HAA is the fraction of hydrophilic acids.

Next data's summarized possible technological processes of water treatment for five main fractions of NOC (see Table 4.13).

Membrane Processes in the Technology of Drinking Water Treatment Membrane technologies compete in terms of efficiency and economic efficiency with other methods of water treatment owing to their ability of removing a broad spectrum of organic and inorganic impurities and pathogenic organisms. However, a wide use of membranes and economic indicators of the process substantially depend on possibilities of preventing precipitation of membranes. When conditioning water from natural surface sources membranes are clogged, mainly, with NOC, polysaccharides, proteins, which cause a substantial reduction of the volume flow [231–233].

In this connection the use of membranes in the technology of water treatment is preceded with coagulation/flocculation, ozonation, adsorption on AC and other methods of water conditioning.

Table 4.12 Strategy of monitoring and removing precursors for formation of disinfection toxic products

Strategy	Details
Monitoring water supply source	Includes control of water uptake site with the aim of lowering concentration of precursors
Monitoring water treatment	Use of alternative disinfectant-oxidizers, which do not form THM such as ozone, chlorine dioxide potassium permanganate, chloramines
	Movement of chlorination point along the production line of water treatment with the aim of lowering the concentration of precursors, which form smaller concentration of secondary products
	Chlorination at pH low values, which reduce the chlorine dose owing to the presence of more active hypochlorite acid or free chlorine suitable for disinfection
Physicochemical removal of precursors	Includes removal of precursors by such techniques as enhanced coagulation, enhanced softening, sedimentation, filtration, adsorption on AC and membrane separation
Oxidation/transformation	Includes processes, which vary chemical structure of precursors

Table 4.13 Technological processes of removing different NOC fractions from natural water

NOC fraction	Water treatment processes
Nonhumic compounds	Membranes, biodegradation on BAC
Humic compounds	Membranes, coagulation, adsorption
Macromolecular NOC (2,000–5,000 Da)	Membranes (ultrafiltration), coagulation
Medium-molecular NOC (500–2,000 Da)	Ultrafiltration or nanofiltration, adsorption
Low-molecular NOC (<500 Da)	Nanofiltration, ozonation, biosorption

First of all, it should be noted, that the methods of water pretreatment, certainly depend on the type of the membrane processes used. At first consider the specificity of using microfiltration and ultrafiltration whose use for drinking water treatment grows fast [234]. The latter is determined, by the ability of micro- and ultrafiltration systems to effectively remove turbidity and the bulk of pathogenic microorganisms present in natural water [235–237]. However, the ability of these membranes to remove dissolved organic compounds and precursors for the formation of haloid-organic substances is limited [238–240]. This is determined by the fact that the dimensions of the molecules of organic pollutants normally are smaller than the dimensions of the pores of microfiltration and some of ultrafiltration membranes used in water treatment. The molecular weight of the pollutants removed for such membranes is limited by a number 10 to 1,000 kDa [241]. On the other hand, molecular weights for many humic aqueous substances normally are less than 10 kDa. Such fractions cannot be removed from water by ultra- and microfiltration to such a degree as to prevent in the future the formation of haloid-organic substances.

Use of the multibarrier approach to the organization of the technology of using micro- and ultrafiltration makes it possible to solve successfully the problem of removing organic matter and precursors for the formation of haloid-organic compounds. Thus, paper [242] noted that the use of pulverized activated carbon (PAC) increases the ability of membranes to remove organic impurities from natural surface water. This is determined by the fact that low-molecular organic compounds are adsorbed on PAC and then may be removed from water in membrane treatment concurrently with carbon. In this connection preliminary adsorption on PAC before the membranes is a more effective method of water treatment than simply membrane filtration for the removal of large molecules of humic mater from natural water [243].

Pretreatment together with coagulation also enhances the ability of membrane systems to remove dissolve organic compounds [244, 245]. According to the opinion of researchers [244, 245], low-molecular organic compounds in the coagulation process aggregate into a larger particles and then may be removed from water in membrane treatment. In paper [35] authors suggested that the mechanism of their removal from natural water is determined by adsorption of organic matter on the freshly formed surface of micelles made of aluminum or iron hydroxides and their subsequent removal during membrane filtration.

Paper [234] demonstrated that micro- and ultrafiltration all alone cannot sufficiently fully remove from water TOC and precursors for the formation of THM defined as the potential for the formation of chloroform. The comparative research of water pretreatment by coagulation and adsorption showed the advantage of the coagulation treatment over membrane filtration. Such combination made it possible to remove ~75 % of organic matter and the precursors for the formation of THM from raw water. For this purpose it was enough to use a coagulate dose of only 0.3 mg/dm^3 (in terms of aluminum). Such a small dose of coagulant is determined by the fact that the concentration of organic matter in water is ~2.5 mg/dm^3, while the concentration of the precursors for the formation of THM varied 100 to 250 $\mu g/dm^3$. Organic compounds and the precursors for the formation of THM mainly had small dimensions. More than 75 % of organic matter contained in raw water and more than 95 % of the precursors for the formation of THM had the molecular weight < 100 kDa.

It was demonstrated [244] that for effective removal of such type of compounds from water using membrane methods without pretreatment it was necessary to use membranes with the pore nominal size < 0.01 μm. Ordinary ultra- and microfiltration membranes in the absence of any water conditioning removed respectively 8.6 and 15.4 % of organic matter and 23.1, 23.4 % of the precursors for the formation of THM [245]. The use, prior to feed to the membranes, of aluminum-chloride coagulant increased the degree of removing organic compounds and the precursors for the formation of THM respectively up to 74.6 and 73.9 %. Dimensions of the aggregates of particles, which were formed in the process of coagulation, were relatively small and varied in the series roughly from 0.5 to 10 μm. These aggregates were well retained by membranes.

Other methods of separating weighted amounts such as filtration through sand filters and precipitation were less effective when removing particles within the limits of the said dimensional interval. This accounts for the striving to obtain, in water conditioning using traditional methods, larger aggregates of particles by using higher doses of coagulants and corresponding technologies of removing weighted amounts [246].

As a working hypothesis it was proposed a mechanism confirmed by experimental data, which indicate the advantages of using low doses of coagulants in membrane filtration [234]. According to this hypothesis it is assumed that given low doses of the coagulant (~ 0.3 mg/dm^3 in terms of aluminum) organic pollutants tend to form organoaluminum complexes. Complexing reactions run very fast literally within seconds. In contrast to the complexing processes the use of high concentrations of the coagulant water treatment is controlled by adsorption of pollutants on amorphous hydroxides of metals. In this case the process is relatively slow and runs within minutes [247]. The high rate of the complexing reaction explains why there is no need in the stage of flocculation for attaining a relatively high degree of removing pollutants in membrane filtration [234]. The stage of filtration is normally needed in order to maximally remove pollutants at high doses of coagulants used.

In paper [248] the use of PAC in the process of water pretreatment, made it possible to achieve the degree of removing organic pollutant and precursors for the formation of THM is 27.19 and 29.9 % respectively. This is determined by the fact that PAC failed to effectively remove organic compounds with the relatively high molecular weight, which were present in the river water under study. In addition, the contact time between water and PAC was a bit less than three minutes while for achieving the equilibrium in adsorption, for instance, atrazine and other organic pesticides present in the surface water of water sources, ~80 min are needed [248].

Thus, it should be noted that in the choice of the method of water pretreatment, type of concentration of the coagulant, type of adsorption and other factors in micro- and ultrafiltration it is necessary in the first turn to take into attention the characteristic of water in a water source [246, 249, 250].

Other approaches to the flow chart of the multistage technology of water treatment may be in the case of using nanofiltration or reverse osmosis. As a rule, before such processes coagulation-flocculation, ozonation and adsorption are used, especially, if water in a water source has a higher concentration of NOC [233, 251]. By means of micro- and ultrafiltration large NOC fractions are not always removed from water. To raise the efficiency of membrane separation, as is shown in paper [252, 253] it is expedient to use oxidation of impurities with ozone or H_2O_2/UV before nanofiltration. Oxidation of H_2O_2/UV may lead to complete destruction of synthetic organic matter and transformation of humic or hydrophobic substances into nonionic or less hydrophobic compounds. The use of H_2O_2/UV for water pretreatment simultaneously has a double advantage: prevention of the reduction of the volume flow due to limitation of the possibility of membrane clogging with NOC; effectively complete removal of synthetic organic pollutants and hydrogen sulfide.

In addition, nonselective responses of hydroxyradicals in the course of pretreatment prevent the formation of undesirable nonhumic compounds (polysaccharides and biodegradable oxidation products). Nonetheless, such pretreatment of the water from the source may completely prevent precipitation on the membranes. Therefore, in order to remove different classes of impurities and restore technological characteristics of the membrane their effective chemical purification is needed [254]. In choosing the corresponding chemical purifying agent one should take into account not only the composition and structure of the sedimentation formed, but also the chemical nature of the membranes since it determines the recovery of membrane properties and economic indicators of the process [254, 255].

The choice of the agent for purification of the membranes depends on the chemical properties of the pollutants and, accordingly, on the type of water pretreatment prior to membrane filtration. In paper [35] some aspects of this issue are considered. Add-on oxidation processes referred to as "AOP" and using ozone, hydrogen peroxide, UV-radiation, photocatalysis or a combination of these methods may change the functional groups, the molecular structure, distribution of the molecular weight, physicochemical and biological characteristics of NOC. Depending on the mode of oxidative pretreatment of water NOC molecules undergo the following transformations: small changes of the functional groups without substantial destruction of NOC macromolecules; significant destruction of large aromatic skeletons into macromolecular compounds (for instance, aliphatic organic acids). These two transformations affect membrane processes both positively and negatively in terms of membrane clogging and their ability to be cleaned. When assessing the effect of insignificant changes of the functional groups one should take into consideration the adsorption of organic polluting compounds determined: by physical adsorption of organic molecules on the membrane surface; electrostatic interaction between organic matter and the membranes and the formation of hydrogen bonds between the membranes and NOC molecules. Physical adsorption determined by Van der Waals forces is reverse and is characterized by a small binding energy. Hydrogen bonds posses high energy of interaction, which 2–4 times exceeds the energy of Van der Waals interaction, while the electrostatic bonds have the energy 8.3–12.5 kJ/mol [256].

Phenol and peptide groups of NOC in humic acids may form hydrogen bonds with carbonyl and polyamide groups on membranes made from cross-linked aromatic polyamides [257]. In addition, peptide blocks may also form hydrogen bonds with phenol groups of other NOC molecules thus enlarging the dimensions of the aggregates. All this leads to substantial clogging of the membranes. However, under the effect of H_2O_2/UV oxidizing the phenol groups in the NOC aromatic skeleton undergo transformation into quinone groups [258].

The quinone groups are less inclined for the formation of hydrogen bonds with carbonyl and amide groups on the surface of a membrane and form predominantly hydrogen bonds with water molecules [232, 258]. In addition, quinone groups do not form hydrogen bonds with the peptide skeleton of other NOC molecules, which prevents the formation of NOC polymolecular layers on the membrane surface in adsorption [253]. Thus, the technological mode of preliminary oxidation of water impurities before membrane filtration in the case of shallow oxidation should maximally ensure the transformation of phenol groups into quinone ones. Given the substantial destruction of NOC macromolecules in the AOP process the relative content of humic, aromatic and resin-like fractions of NOC decreases [232, 259, 260].

In paper [261] NOC transformation under the impact of ozonation is studied and it has been shown that in this case the C=C double bonds were broken, simple organic acids and aldehydes were formed. It was noted that among 27 simple organic acids which were formed, the oxalic acid is dominated in the presence of acetic, lactonic and formic acids. In trace amounts gallic, phthalic, fumaric, maleic, valeric and glycolic acids were present. Oxidation of H_2O_2/UV as is shown in [258], leads to the formation of oxalic, acetic, maleic and *n*-butyric acids. Propionic, caproic and butyric acids were present is small amounts. These organic acids due to their molecular dimensions may lead to some pollution—chemical poisoning of the external surface of nanofiltration or reverse osmosis membranes and in the case of ultrafiltration or microfiltration membranes—poisoning of the pore space of the latter.

It is advisable to discuss briefly the possibilities for membrane cleaning upon conditioning drinking water. The elimination of clogging is based on the combination of two factors: the effect of dissolution and desorption by the cleaning agent and the effect of the hydrodynamic shock to the polluting layer [254, 262]. This approach implies that the initial polluting layer is destroyed by the chemical agent and then easily removed by a hydrodynamic flow. It is proposed [254, 255, 262] to use for cleaning membranes double cleaning with alkaline and acid solutions, which makes it possible to remove both acidic and basic NOC fractions. The given method is sufficiently practical, effective from the economic point of view and makes it possible to recover membrane properties. However, in paper [253] the preference is given to alkaline washing due to the following reasons. Cleaning with the alkaline solution is advisable in the case of the high content of humic and fulvic acids, which are well soluble in alkaline solutions. At high pH values NOC molecules become conformationally more linear [231, 263] and consequently the polluting layer becomes more spongy and therefore the membrane lends itself easier to cleaning.

A hypothesis was put forward that hydroxylic ions in the alkaline solution (0.1 M NaOH) may promote destruction of the polluting layer along three avenues: increase of the ionic force; increase of the solubility of the NOC fractions; the reduction of the negative charge of NOC and the membrane as a result of Na^+ adsorption [264]. Preoxidation treatment of raw water by H_2O_2/UV leads to a more effective washing of the membranes off polluting compounds with the alkaline solutions.

Apart from chemical acidification of organic matter in natural water, biological methods are also used in the practice of water treatment before membrane filtration. The main goal of biological pretreatment of water before membrane filtration is prevention of membrane clogging. It is known that the removal of organic matter and biogenic compounds from water before the membrane system may minimize both biological and organic pollution [265]. Though humic compounds are weakly-degradable compounds, a bioreactor or a biofilter are capable of removing only a certain percent of low-molecular biodegradable compounds [266]. In addition, as it is shown in [266], different versions of the combination in the position of the biofilter with the membrane system leads to the purified drinking water of nonuniform quality and prevents secondary microbial growth in distribution networks. The results of biofilter location before the membrane system and after it were compared. It was established that TOC and organic acids are removed in the greater measure in the biofilter–membrane system than the membrane–biofilter system. In the series of experiments under study it was determined that biological pollution does not make a tangible contribution to the total effect of pollution. Membrane location after the biofilter increases the time of its stable operation. This is related to the fact that the biofilter removes compounds, which lead to pollution with organic matter. The reduction of the precipitation rate of pollutants also decreases when water turbidity increases [267]. This turbidity was created by introducing kaolin to water. This effect of preventing membrane clogging is quite predictable and is often used in practice of water treatment. However, the configuration of the membrane–biofilter system leads to a much greater total removal of organic matter from water, the organic matter being determined by specific absorption in the UV region at 245 nm.

Nanofiltration has been the most widely used technology of membrane water treatment in recent years. Thus, in Norway the formation of the market of membrane filtration equipment dates back to 1990. At present, the number of water treatment facilities based on the methods of ultrafiltration and nanofiltration exceeds 100; in this case preference is given to the methods of nanofiltration. On the average, the capacity of the facilities constitutes 2,500, while the maximum one is 26,000 m^3/day. When purifying water with nanofiltration the bacteria and viruses are removed completely, the concentration of organic polluting substances is reduced by 90–95 %, while Fe and Mn decrease respectively by 95 and 30–40 % [268].

In Paris in one of the water treatment stations servicing 800,000 people the nanofiltration method was used. The station capacity—140,000 m^3/day, the number of membrane modules operated concurrently exceeds 9,000. The flow chart used at this station is the multistage one. The water taken from the Seine at first is treated by the methods of coagulation–flocculation, precipitation, ozonation, filtration on fast filters. Then after adjusting the pH values the water is fed to the membrane modules after which it is disinfected with UV-radiation [269]. A similar experience of using the membrane technology for drinking water treatment is also presented in Russia [270]. Their technology incorporates a stage of simplified aeration, two-stage filtration, oxidation with the subsequent purification on adsorption filters and at the final stage -membrane filtration. As a result of water treatment the iron content in water is 0.2 mg/dm^3, color—18 deg, oxidizability—2.4 mg O/dm^3.

The water of water sources used for drinking water supply in addition to suspended particles and NOC, contained pollutants of anthropogenic origin, for instance, polycyclic aromatic hydrocarbons (PAC) that were in microquantities. The concentration of such PAC as benzopyrene, phenanthrene should not exceed 0.2 μg/dm^3 in drinking water. Benzopyrene, whose concentration in drinking water possesses the highest carcinogenic activity. Treatment of water containing microquantities of PAC by UV radiation followed by nanofiltration makes it possible to remove individual groups of PAC with efficiency of up to 80–90 % [271].

The technology of successive combination of different methods of membrane filtration for obtaining water of high quality acquires popularity. Thus, in the Netherlands tested the technology of water treatment on ultrafiltration modules for removing suspended substances, bacteria and viruses and then tested on reverse osmosis modules, which makes it possible to remove dissolved compounds, anthropogenic pollutants. Since the resultant water is effectively completely demineralized it was proposed to mix it with ground water treated using an ordinary flow chart [272].

A similar approach was proposed by the Axiva GmbH Company, which developed the technology of deep treatment of the water of surface water bodies comprising three stages of purification. At the first stage the water is sent for preliminary microfiltration, while at the second stage—for ultrafiltration and at the final stage—for reverse osmosis filtration [273]. No doubt, such especially clear water requires further conditioning in terms of mineral components.

Thus, the quoted short overview of using, membrane technologies for drinking water treatment demonstrates their efficiency, economic performance and prospects given certain limitations of capacity depending on the needs of the consumers. A common feature in all cases is the combination of the membrane technology with other methods of water treatment and conditioning.

Removal of specific pollutants from drinking water Nitrates. Pollution of underground and surface water with nitrates due to ingress of nitrate fertilizers, livestock wastes, wastes from processing enterprises and treatment facilities leads to serious eutrophication of the water supply source and hazards to human health. The presence of nitrates in drinking water is especially dangerous for children and pregnant women [274]. Blood cancer of women and congenital anomalies of children are also related to the presence of nitrate in drinking water [275, 276].

At present, the best and most acceptable methods of removing nitrates from drinking water are ion exchange and reverse osmosis whereby membranes of ultra-low pressure are used [277]. Paper [278] reveals the technology of purifying underground water of nitrates with the account of background natural pollutants. This technology provides for three successive stages of filtration: mechanical purification on a sand medium, ion exchange on a high-basic anion exchanger and sorption on AC. Regeneration of the medium of the anion-exchange filter is carried out with a 10% solution of hydrochloric or sulfuric acids followed by washing and conditioning with a 4% solution of alkali (NaOH).

However, both methods lead to the formation of concentrated waster products, which require further processing. In addition, they are not selective with respect to nitrate-ions against the background of other ion impurities. Due to high specificity of denitrifying bacteria, low cost and high rate of denitrifaction the biological denitrification from this point of view is an attractive alternative method of removing NO_3^- [279, 280].

Biological denitrification is based on the use of facultative bacteria, which consume NO_3^- as an acceptor of electrons under anaerobic conditions. Organic compounds (ethanol, methanol and acetate), reduced sulfurous compounds and hydrogen are donors of electrons used in biological denitrification [281]. The process of biological denitrification is normally carried out in bioreactors with a compact-packed layer of the medium. However, usually after this a post-cleaning of the system and water are needed in order to remove organic carbon and redundant biomass. Capital costs for facilities, operational expenses and the necessity of the presence of highly-skilled personnel also downgrade the attractiveness of the traditional biological denitrification. In these terms, it seems that the improved process of nitrification of water intake in situ by adding biogenic carbon either to porous partitions of the facility directly during water intake or to the matrix of the water horizon is more effective. The first approach is promising however it requires the cut of the ground.

It is limited by the depth of the water-bearing horizon and therefore is rather costly [282]. As far as the second approach is concerned the enhanced in situ biodenitrification (EISBD) process lies in injection to the underground well of organic carbon with the goal of enhancing the operation of denitrifying bacteria, which reduce nitrates to nitrogen and at the same time simultaneously oxidize organic carbon to CO_2. The EISBD technology is sufficiently effective and ecologically acceptable, does not require large costs for solving the problem of disinfection of nitrates [283].

Until recently the use of this technology was limited due to fear of the biofouling of the injection wells [284]. However, detailed research of the given process makes it possible to show that bio-growth may be controlled [284, 285]. This is achieved by the correct choosing of the carbon source, which produces less biomass than the other sources [276]. It is proposed to use acetate as such source [286]. The denitrification reaction when using acetate for the needs of oxygen cell breathing and synthesis may be described by the equation

$$17NO_3^- + 25CH_3COO^- + 17H^+ \rightarrow 6N_2 + 5C_5H_7O_2N + 25HCO_3^- + 16H_2O.$$

Paper [283] presents the experience of 41 cities in which a pilot project was implemented for this technology for the analysis of economic and innovation aspects of using the technology in municipal systems of water treatment. Paper [276] demonstrated that capital costs and equipment costs were in the amount of 75,000 USD. Acetate costs constituted \$ 0.05 per 1,000 dm^3 for 12 mg/dm^3 of NO_3^-—N mg/dm^3. The costs for cleaning the well were US\$ 0.06 per 1,000 dm^3. Operational costs including wages and depreciation deductions were assessed as US\$ 0.16 per 1,000 dm^3 of water. The quoted costs are comparable and even lower than the operational costs for ion exchange, which constitute respectively ~US\$ 0.14 and 0.55. However, it should be borne in mind that the cost of decontaminating waste that form as a result of ion exchange and reverse osmosis may be very substantial and not included into the figures. The tests conducted showed that the process proceeded rather fast without the silting of the injection wells within the same three warmest months of the year. For cleaning wells—1,200 dm^3 of the cleaning solution were used, which contained 5 % of hydrogen peroxide. The solution was injected under pressure for four hours [276]. A similar cleaning procedure was described in paper [284].

A new approach to do denitrification of drinking water was proposed in [287]. The possibility of using a new extraction membrane bioreactor (MBR) for the removal of nitrates from drinking water was investigated. The conceptual model of the extraction membrane bioreactor lies in the following: water containing nitrates is fed to a pipe membrane and NO_3^- ions diffuse through the membrane pores. The population of denitrifying bacteria circulates along the external side of the pipe membrane creating the motive force for mass transfer of NO_3^-. Biogenic substances and carbon are fed there for securing vital functions of the microbial population. The membranes have high penetrability in terms of NO_3^- and at the same time separate the microbial mass from water. The advantage of such an extraction membrane bioreactor lies in the following:

- the final product (water) is separated from the microbial biomass by the membrane and consequently water pollution with the biomass is ruled out. Extraction of organic carbon may be minimized by limitation of the load in terms of carbon and maintaining high density of the biomass;
- heterotrophic bacteria secure the high rate of denitrification. Several types of donors (ethanol, methanol, acetic acid) of electrons may be used for controlling the growth of the biomass;
- the reactors may be created sufficiently cheap using pipe or fiber membranes. The pressure in these modules is relatively low and the effective monitoring over the biological growth is possible [288].

The membrane extraction bioreactor was also investigated in [289]. The authors of this paper described the laboratory system in which two chambers are separated with a half-penetrable membrane. The denitrification rate in the given experiment reached 5.8 g of NO_3^-—N/(m^2 per day). However, in the given paper one could observe substantial concentration of methanol and microbial pollution in the eluate. In paper [290] the impact of the membrane material on the process efficiency in MBR was investigated and it was noted that in the case of acetate-cellulose membranes the denitrification rate of 1.2 g of NO_3^-—N/(m^2 per day) was achieved. In this case the presence of ethanol and microbial pollution in the eluate was recorded. In paper [291, 292] on the same type of the membranes the degree of NO_3^- removal reached 90 % and the denitrification rate was 4 g of NO_3^-—N/(m^2 per day). However, methanol pollution was found in the eluate at the COD value of 16 mg O/dm^3. The described shortcomings were overcome when used akrylonitrile fiber membranes [287]. In this case the extraction membrane bioreactor ensured the 99 % degree in removing nitrates at the average denitrification rate 6.1 g of NO_3^-—N/(m^2 per day) at the NO_3^- initial concentration of 200 mg/dm^3, however in this case ~ 8 % of added methanol transferred to water through the membrane. Thus, further research of the process is necessary for its effective practical use.

In paper [293] the method of regulating the efficiency of the technology of biological denitrification by automatic monitoring over the addition of organic carbon to the system is proposed. In the given technology the values of oxidation-reduction potential and pH respectively in the reduction and oxidation phases were used as the monitoring parameters. If in the water treated the C/N relationship was low and the deficit of organic carbon was created the denitrification process slowed down. The proposed system of regulation in real time made it possible to add organic carbon depending on the variation of C/N relationship. The average degree of removing OC and nitrates in the process was respectively 94 and 96 %.

Chemical recovery of nitrates is a certain alternative to the biological denitrification [294]. It is shown that the former proceeds at a higher rate compared with biological denitrification and relatively more selectively with respect to nitrates than reverse osmosis and ion exchange [295, 296]. Metallic iron, which is a strong and ecologically acceptable reagent, was used in the quoted investigations as a reducing agent. In paper [297] it is shown that iron powder treated with a 10 % hydrogen may reduce nitrates and nitrites at pH 7 in the ratio of N_2:H_2 = 90:10 %. Reduction of nitrate in the oxygen atmosphere occurs in the interval of pH values 2–5 [298].

However, the obtaining of ammonium ions in the quoted experiments is a negative moment. Therefore in paper [299] investigated the conditions for the recovery of nitrates with metallic iron in acidic and alkaline conditions in order to find the optimal mode of their transition to ammonia and ammonium ions to gaseous nitrogen. It is shown that nitrates are mainly reduced to gaseous nitrogen at the beginning of the reduction reaction, while ammonia/ammonium—at the end of the reaction. At low values of pH the main product of reduction is ammonium, while at pH high values N_2 is formed. In the reduction products N_2 was found (40.1; 56.0; 68.0; 77.8 and 81.0 % after a 150-min reduction at pH respectively 2.0; 3.5; 5.0; 7.0 and 8.5). The described method of water denitrification is still in the stage of experimental research and requires further development.

Manganese and Iron Manganese compounds are often found in the sources of drinking water supply in the dissolved form as Mn^{2+}. Although manganese at concentration < 500 μg/dm³ does not harmfully affect human health, its presence in drinking water at concentration > 100 μg/dm³ is not advisable due to increased water color [300].

The European Environmental Commission (80/778/UUC) recommends the top limit for the manganese content in drinking water of 50 μg/dm³.

The most traditional method of removing manganese and iron from water up to date is the widely used technique of oxidation-filtration. The development of this method is aimed at its perfection and selection of a more effective material for filter media. Thus, in paper [301] ozone and standard rapid filters were used as the oxidizers for joint removal of iron and manganese. It is shown that at the pH value of the initial water of 6.8–7.6; water temperature 8–12 °C; filtration rate 5–10 m/h for achieving normalized water quality indexes for the ozone dose of 1.5 m/h is needed and the time of water contact with the ozone-air mixture is of 12 min. At such technological parameters the residual ozone concentration in water constituted 0.1–0.3 μg/dm³, while the content of iron and manganese decreased to standard indexes.

Paper [302] carried the rehabilitation of the water treatment station for increasing the degree of removing iron and manganese by installing a two-stage filtration circuit with reagent treatment. In this case aerated sand filters were used: 10 parallel connected filters at the first stage and 8 filters at the second stage with the station capacity of 5,000 m³/day. The initial water contained 5 μg/dm³ of iron and 1.5 μg/dm³ of manganese, and up to 2 μg/dm³ of ammonium nitrogen. After putting into operation of two stages of aerated filters the content of iron in drinking water constituted 0.2, that of manganese 0.05 and ammonium nitrogen—0.12 μg/dm³. For ensuring a higher quality of water purification of iron it is proposed to rehabilitate filters by replacing the granular medium with the foampolystyrene one. This makes it possible to secure the high quality of water purification, to reduce water discharge of washing water by ~20 % and costs for washing the filters [302].

Built in Hamburg station with the capacity of 20,000 m³/day for treatment drinking water from artesian wells is focused on the removal of iron, manganese and accompanying water pollutants. The water from the wells 300 m deep is directed to a saturator for saturation of it with technical oxygen. Then follow biofilters in which iron and manganese are oxidized to undissolved oxihydrates. The water treatment flow chart is a multi-stage one. After water purification the content of carbonic acid is 3.2–5.0; that of iron—1.3–1.8; that of manganese—0.13–1.18; of ammonium nitrogen—0.08–0.17; of humic substances—not more than 0.2 μg/dm^3 [303].

A promising area of removing dissolved manganese is the use of biological filters with adapted microflora [304, 305]. Adaptation of the biological filter capable of removing manganese from water is achieved by passing aerated water through a column charged with sand in the absence of chlorine. This makes it possible for benthos microorganisms to form a biofilm in the interporous space of the medium [306]. It was noted that biological treatment with the goal of manganese bioaccumulation may reduce the concentration of the latter from 350 to 20 μg/dm^3 [307]. Time necessary for population of the filter with bacteria may vary from weeks to months even after the seeding of the filter medium with flora from other adapted filters [308]. Thus, the water treatment station in Kent (Great Britain), using underground water with manganese concentration 500–600 μg/dm^3, took three months to secure the removal initiation of manganese during the biological process [306]. The biofilter maturing time for successful removal of manganese from water may substantially differ for two different water sources even though the chemical and biological compositions of water are very similar [309]. It is possible to speed up the process of filter maturation by using the special monoculture *Leptothrix discophora*. This culture is one of the predominant microorganisms in the Mn-biofilters [310] and effectively oxidizes both Mn^{2+} and Fe^{2+} to oxides. This strain attracts the greatest attention of the researchers due to its rather fast growth kinetics and metabolism efficiency in oxidizing iron and manganese [311].

The investigations carried out in [309] showed that the biofilter seeded with the culture *Leptothrix discophora* SP-6 the most effectively work within the concentrations of 2,000–3,000 μg/dm^3. The transfer from a laboratory unit to a pilot one may be carried out without any problems whatsoever. For seeding a large pilot unit a culture from the laboratory unit is used. The data obtained at the large scale pilot unit showed that the biofilm formed in the interporous space of the granular filter medium removes Mn^{2+} better compared with the laboratory model and the frequency of the filter reverse washing decreases. The main advantage is the fact that the seeding of the filter with the above-described strain makes it possible to reduce time of filter maturation from one month to several days.

Arsenic The problem of water pollution with arsenic and especially of underground water, which is used for drinking water supply, has been intensively discussed lately owing to its extremely harmful impact of human health.

Paper [312] shows that consumption of drinking water polluted with arsenic at the concentration of ~ 100 μg/dm^3 causes carcinogenic effect. As(V) may substitute phosphate in some biochemical reactions, while As(III) is capable of reacting with thiols in proteins and inhibit their activity [313].

The maximum level of drinking water pollution with arsenic was reviewed and established by the directive of the European Commission in 2003 for all systems of drinking water supply in the EU at the level of 10 μg/dm^3 against 50 μg/dm^3 in past years [314]. This standard was recommended by the WHO [315].

Distribution of various forms of arsenic As(V), As(III) in natural water depends mainly on the value of the redox potential and pH [316]. Under the conditions of oxidation the pentavalent arsenic—$H_2AsO_4^-$, $HasO^{2-}{}_4$ (pK_a=2.19, pK_E=6.94) is a prevailing form in surface water. On the other hand, under moderate reducing conditions characteristic of underground water, As(III) is a thermodynamically stable form. In such water at pH values normal for natural water arsenic is in the neutral state form the arsenic acid (H_3AsO_3, pK_a=9.22) [317].

As(III) in a very small degree is adsorbed by a solid surface, which creates certain difficulties when it is removed from water using usual technological techniques [314]. For extracting arsenic compounds from water it was proposed to use the coagulation–direct filtration process [318], which made it possible to reduce the coagulant dose for water treatment. In addition to this it was proposed to use ionic exchange, softening with lime with concurrent removal of arsenic, adsorption on iron oxide or activated aluminum oxide and reverse osmosis [319].

However, all types of technologies considered are not sufficiently effective when removing As(III). The preliminary stage of oxidation by this form of arsenic to the pentavalent one is necessary in order to increase the degree of extracting As(III). For oxidizing As(III) different reagents are used such as potassium permanganate, chlorine, ozone, hydrogen peroxide or manganese oxides [320, 321]. The technique of As(III) preoxidation before the removal is rather effective, however, as was noted earlier, has certain shortcomings. This is related to the formation of toxic secondary products and substantial increase of water treatment costs.

The use of biological oxidizing of iron by microorganisms for simultaneous extraction of As(III) and As(V) is a certain alternative to the methods considered; this technique makes it possible to simultaneously remove arsenic compounds [314]. The gist of the method lies in the following. Iron oxidation is carried out by using the microorganisms *Gallionella ferruginea* and *Leptothrix ochracea.* During this process iron oxides build up on the filter medium together with the microorganisms, which creates favorable conditions for arsenic absorption from the water flow. Under optimal conditions As(III) is oxidized by these microorganisms, which increases the total effect of its removal up to 95 % even at the initial concentration of 200 μg/dm^3. In addition, the As(V) concentration decreases more effectively under the same conditions. The arsenic residual concentration in water processed in such a way may be reached below the level of 10 μg/dm^3.

The considered technology has a number of advantages over the earlier investigated physicochemical processes. First of all, it allows to avoid the use of chemical reagents for oxidizing As(III), it is more cost effective and ecologically acceptable. There is no need in control by the quantity of sorbent for removing arsenic since the sorbent (iron oxides) is formed constantly in the medium layer. In addition, the proposed combined process (biological oxidation—filtration—sorption) may be used for simultaneous removal of iron, manganese and arsenic.

Oil products, pesticides In surface and underground sources of drinking water supply oil products, aromatic polycyclic hydrocarbons, pesticides and other xenobiotics are often widely common types of anthropogenic micropollution.

Recently in extraordinary cases the introduction of PAC to the technology of sorption purification has been used for purification of drinking water. An advantage of the carbonization process is fastness of its introduction and the possibility of varying the doze and the site of introducing carbon. Thus, the introduction of carbon pulp was used at the water treatment facilities of the Ufa-Vodokanal company in emergency pollution of water sources with oil products. In this case the feed of the latter to the metering points from the time of receiving a signal about pollution did not exceed 30–40 min. The use of the carbonization process in the case of pollution with oil products $<0.32\ \mu g/dm^3$, with benzopyrene <20 of MAC, with a broad fraction of organic compounds with the number of carbon atoms C_{10}–$C_{32}<0.23\ \mu g/dm^3$ made it possible to obtain drinking water of standard quality [322]. It should be noted that a MAC dosage and the time of contact should be determined for each specific water source and different seasons of the year.

At some France's water systems the use of PAC is determined by the presence in the initial river water of pesticides exceeding the sanitary standards. Paper [323] proposed the method of two-stage water treatment in several versions. In one of them the water was pretreated with reagents. The other stage was sorption on PAC and at the last stage ultrafiltration was used. The initial river water contained atrazine (at the concentration of 2 $\mu g/dm^3$), which during treatment was effectively removed.

Paper [324] quotes the results in the operation unit for purifying river water of atrazine in a somewhat different version. The unit incorporated an input tank with a mixer to which PAC was metered, two flocculation/coagulation stages, a thin-walled sedimentation tank and a sand filter. It was found out that the activated carbon sorption stage removed 84 % of atrazine given the carbon dosage of 25 g/m^3. At the output from the sand filter the removal degree constituted 98 % while the initial concentration of atrazine in the inflowing water was 1 $\mu g/dm^3$.

Fuel oxygenates, which are added to gasolene with the goal of reducing CO emissions are a rather widely used anthropogenic source of polluting underground water. Methyltributyl ether (MTBE) is one of the most widely used fuel oxygenates and is used, for example, in the USA in ~70 % of all brands of gasoline [325]. This leads to the situation that MTBE contaminates both underground and surface sources of water supply due to the spillage of gasoline. MTBE is resistant to biochemical decomposition and sufficiently well soluble in water (43.0–54.3 $\mu g/dm^3$). The most common processes of removing MTBE from water are air scavenging, adsorption on GAC, deep oxidation (AOP). The efficiency and cost efficiency of each of these processes in great measure depends on water quality indexes and the capacity of the water treatment facility [326, 327]. The comparative research of all above-mentioned methods of water purification of MTBE showed the following [325]. The scavenging technique is the most economically efficient method at the sufficiently high capacity of the unit (~3,800 dm^3/min). However, for raising the purification degree from 80 to 99.5 % it is necessary to increase the height of the tower, where the process takes place, respectively from 2.0 to 12–20 m. In addition, the scavenging towers may be fouled with iron floes, deposits of carbonates and biological fouling. Air scavenging is ineffective either at low temperature.

The use of AC is the most expensive process under all conditions except cases when water has sufficiently high value of COD (~61 mg O/dm^3). Filters with GAC also may be fouled with iron, carbonates and biological deposits. In addition, GAC sorptivity with respect to MTBE depends on the presence of competing impurities and the use of GAC is needed at relatively high concentration of MTBE (>1 mg/dm^3).

Al low capacity of the water treatment facility (~38 dm^3/min) the lowest cost of purification is characteristic of O_3/H_2O_2-treatment (AOP-process). But the given process is ineffective for high ratios of COD/total content of dissolved substances/alkalinity (the research of the ratio constituted: 61 mg O/dm^3/424 mg/dm^3/432 μg/dm^3). Removing from water such a xenobiotic as MTBE and similar contaminants requires the establishment of technological and economic parameters for each specific water source.

It is important to mention the method of removing organic xenobiotics from drinking water described in [328] by using the combination of photocatalysis with flocculation and adsorption on AC. It was shown that the use of photocatalysis only with the use as a catalyst of TiO_2 and PAC independently does not lead to deep destruction or complete removal of organic matter. The removal of the latter becomes more effective with simultaneous introduction to water of TiO_2 and PAC. The process becomes even more effective when the photocatalysis stage together with PAC precedes water treatment with $FeCl_3$. Perhaps, the presence of a positive effect of combining PAC with photocatalytic oxidation is determined by the same reason considered in [329], which investigated the impact of AC on ozone transformation into the $OH^•$-radicals.

The most effective are sorbents with a high content of alkaline surface groups and with the extensively developed surface. The introduction of carbon increased the relationship between the concentration of $OH^•$-radicals and ozone 3–5 times. It is shown that AC acts rather not as a catalyst, but as an initiator or a promoter of ozone transformation into $OH^•$-radicals.

Ozonation of the water of the Zurich lake at the ozone dose equal to 1 mg/dm^3 in the presence of 0.5 and 1.0 g/dm^3 of AC led to the increase of ozone decomposition constant K_D in 10 times. In this case, preferentially compounds resistant to oxidation with ozone are removed. The presence of AC during the process of water ozonation in the Zurich Lake led to the reduction in the content of dissolved organic carbon by 40 % within the first 60 min of treatment. Adsorption of low concentrations of dissolved organic matter on AC does not change its ability to promote ozone transformation into $OH^•$-radicals.

The certain interest presents biological methods dealing with removal of organic matter of artificial origin. Thus, in paper [330] the process of water treatment with substantial quantities of organic compounds and ammonium nitrogen is investigated. For water treatment at the first stage flush-mounted immersed biofilters were used. As a medium they used anthracite chips and granules of porous ceramics with the size of the granules on the average ~ 4 mm and porosity at least 40 %. Water temperature was 5–6 °C. This method ensured the degree of removing organic pollutants at least 70, while that of ammonium nitrogen—79–85 %.

The use of bioreactor charged with the carrier suspension with a biofilm and operating in the airlift mode (BAC-reactor) was proposed for the removal of organic sulfur-containing compounds in natural underground water. Active biomass growing on polluting organic xenobiotics is obtained by seeding the particles of the carrier of a mixed microbial culture taken from the active sludge of the water treatment facility. Despite the slow growth of the biomass and delayed kinetics of degradation a high rate of the process in the BAC-reactor was attained owing to the high concentration of the biomass (< 12 g/dm^3). Degradation ability constituted 8.7 kg of COD (m^3/day) at the degradation degree of 70% of the total quantity of organic matter in terms of COD [331].

The biological method of purifying underground water of such compounds as ethers and alcohols, alkylated ethers and alcohols was patented in the USA [332]. In the given method water treatment was carried out on GAC with a fixed biofilm. Carbon was in the fluidization state due to water circulation. At the content of 2.21 mg/dm^3 of methylrributyl ether the adaptation period of the microflora attached to carbon constituted 110 days, after which MTBE was destroyed almost completely.

The biological method of removing trace amounts of pesticides (benthazione, isoproturone, dichloroprone, etc.) in aerobic conditions was investigated in [333]. The unit comprised columns connected in series, the height of the columns being 105 cm, charged with a mixture of pebble, sand and crushed coal. The initial concentration of the pesticides was 2–6 μg/(dm$^3 \times$ day).

Thus, a short overview of the method of purifying natural water of specific natural substances demonstrates that in the modern technology of water conditioning methods of enhanced oxidation (AOP), sorption and biochemical oxidation are most prevalent ones. Depending on the quality of the initial water the requirements to the degree of its purification and economic factors the modern technologies widely incorporate the combination of these methods with traditional methods, which makes it possible to ensure the high quality of drinking water.

Effect of the Condition of Distribution Systems on Drinking Water Quality Today, one of the central problems in supplying population with high-quality drinking water is the condition of water supply distribution systems. According to the directives of the European Union, drinking water must conform to the consumer requirements as it leaves the tap [334]. Consequently, it is necessary to maintain the high water quality over the entire distribution system including the plumbing installations of individual residential buildings.

As it leaves the pure water tank of a centralized supply plant, drinking water is mainly of satisfactory quality and is free of microbiological contamination. But in the process of conveyance to the consumer's tap its microbiological quality may significantly degrade. Loss of water quality in distribution systems is largely determined by the degree of the water's biological stability. Bacterial growth, i.e., formation of a biofilm, occurs mostly in water that contain organic or inorganic substrates, such as humic and fulvic acids, polymeric hydrocarbons, proteins, carboxylic acids, nitrogen and phosphorus compounds, etc., and also depends on the pipe material, presence of disinfectants, supply of bacteria with the water, and the hydraulic parameters [335–337]. Factors that favor survival of bacteria in chlorinated water include their attachment to the pipe walls, aggregation of bacteria, age of the biofilm, encapsulation, and the type of disinfectant (free chlorine, monochloramine, etc.) [338]. Addition of chlorine, even in quantities of less than 3 mg/dm^3, has a limited effect on the existing biofilm and does not prevent formation of a biofilm on a clean surface. Fecal bacteria penetrating into water are immobilized and develop on existing biofilms [339]. Given the vast quantities of water produced by centralized water treatment plants, the disinfection process cannot always be relied upon to guarantee a total absence in the water of pathogens or indicator microorganisms [340–342]. It was demonstrated in a number of studies that potential pathogens are capable of surviving [343–345], reproducing [346], and forming biofilms [347] even in properly conditioned drinking water.

Development of heterogeneous biofilms on pipe surfaces was explained by investigators on the basis of a variety of approaches, such as modeling of mass transfer constraints [348–350] or a biological approach based on the cell-to-cell link [351–353]. It is demonstrated in [47] that by using the mass transfer constraints model it is possible to account for the morphology of a heterogeneous biofilm. In cases where the substrate concentration in water is low or where bacteria are growing fast, mass transfer restricts the bacterial growth in the lower (bottom) parts of the biofilm, which makes for more heterogeneous biofilm morphology. In biofilms that are not limited in the consumption of the substrate (i.e., when the substrate concentration in the water is low and microorganisms grow slowly); bacteria grow at the same rate throughout the thickness of the biofilm, which makes its morphology more homogeneous. Later [350], the proposed approaches were supplemented with a factor correcting for detachment of bacteria under the effect of moving water on a stable, but still growing, biofilm. Thus, development of a biofilm is basically the product of the balance between bacterial growth and detachment of bacteria. On the one hand, bacteria may break away as a result of prolonged processes of erosion and abrasion that mainly affect the biofilm surface. But on the other hand, there operates a mechanism similar to that of a precoat filter, with particles breaking away from a submerged biofilm.

Determination of the predominant role of one or the other of these mechanisms plays a significant role in the water treatment technology [354]. This is due to the following reasons. Particles that break away from a biofilm as a result of detachment are larger than its thickness by approximately one order of magnitude, while erosion and abrasion removes smaller particles. The type of biofilm detachment, which is largely determined by the film morphology [350], has a significant effect on the water quality since the residual quantities of a disinfectant will act more efficiently on pathogens in small aggregates than in large ones. At the same time, erosion can help remove surface irregularities which will result in a smoother biofilm while the detachment process will strip away large sections of the biofilm that tend to increase heterogeneity of its morphology. A smooth biofilm surface will help reduce the local and general breakaway forces acting on it, while the effect of a rough surface will be to increase both the general breakaway force and the localized forces that tend to tear away those parts of the biofilm which exceed its average thickness [354]. Therefore in order to preserve the water quality in distribution systems it would be desirable to maintain conditions, which promote formation of a smooth biofilm surface, even though in stable operation the general quantitative growth of a biofilm is equal to the total amount of breakaway losses. This equilibrium can be upset by altering the hydrodynamic conditions. Excess biomass can be removed from a distribution system by adjusting the rate of the water flow or reversing its direction—methods that are actually used in practical operation [355]. Thus, to a certain degree the biofilm growth in a distribution system can be controlled.

One significant factor that influences the initiation of deposition and the biofilm growth in distribution systems is the pipe material. Most often this material is iron. In distribution systems made of iron-based metals one can find deposits on the internal surfaces of pipes, which are the complex composition of biomass, iron compounds, hardness salts, and organic compounds serving as the nutrient medium for microorganisms [356]. In their mechanical properties they are soft deposits that can be effectively removed from the pipes. Deposition of iron corrosion products reduces the hydraulic carrying capacity of pipes and increases the consumption of electric power. Another negative effect of the deposition of corrosion products is the appearance after distribution systems of "red" or "dyed" water because of the washing out of iron from the deposits.

According to paper [357, 358], deposits formed in old iron or steel water distribution systems have a porous and laminar structure. Compounds usually found in deposits of iron corrosion products include α-FeOOH; γ-FeOOH; Fe_3O_4; $Fe(OH)_3$; $5Fe_2O_3 \cdot 9H_2O$; $Fe_4^{II}Fe_2^{III}(OH)_{12}(CO_3)$, and $CaCO_3$ [359].

Iron washed out of iron or steel pipes is most often responsible for a dyed tap water. For example, in the USA it became necessary to regulate the secondary maximum iron content in tap water at a level of 0.3 mg/dm^3 [360]. Iron is mainly washed out of the pipes into bulk water in ferriferous forms. It was found that the washing out of iron compounds from corroded pipes may be influenced by such water quality parameters as dissolved oxygen, pH, alkalinity, buffer capacity of the water, hydrodynamic characteristics, temperature, special features of the water treatment processes, use of inhibitors, and fluctuation of the water quality [358].

As a rule, the rate of corrosion of a clean iron surface increases with rising oxygen concentration in the water. However, a laminar film of corrosion deposits affects the rate of diffusion of oxygen to the metal and, consequently, the corrosion process. It was found that increase of the concentration of dissolved oxygen in a water resulted in a decrease of its iron content under stagnant conditions. But with the same amount of the oxidant less iron was washed out into flowing water than under stationary conditions. Also, the rate of decrease of the content of oxygen (or another oxidant) in flowing water was higher than in stationary water. It had been demonstrated that by increasing the oxidant concentration in water and maintaining stable hydrodynamic parameters it was possible to reduce the washout of iron from corroded pipes. The composition and structure of corrosive films may affect the quantity of iron compounds washed out into the water, and the oxidants present in water oxidize Fe(II) both in the films and in the water [358].

In addition to altering the color, taste, and turbidity of water, iron corrosion products may also induce chemical decomposition of residual chlorine in pipes, which will eventually affect the microbiological characteristics of the water quality [338,

361]. This raises the problem of cleaning the pipelines for improving the drinking water quality. One method of removing soft deposits from a distribution system is discussed in [356]. The proposed cleaning process consisted in flushing pipes with compressed air and water. The water's bacterium content, iron concentration, and turbidity were the highest in the morning (around 9:00 a.m.), when the water consumption was at its peak. The cleaning reduced the diurnal fluctuation of the water quality but had only a short-term positive effect as far as iron was concerned. On the other hand, removal of soft deposits which contained large amounts of nutrients reduced the bacterial growth in the distribution system in the summer season when the temperature conditions were favorable for bacteria. Deposits removed by this method were not found to contain viruses (of the NLV Norwalk-like type) or coliform bacteria, but heteroform bacteria were present in large numbers. Immediately after the purging of the pipes the content of biologically consumable phosphorus compounds in the tap water decreased below the level achieved at the waterworks—an indication of accumulation of phosphorus compounds in soft deposits in the pipes. This accumulation, even in small quantities, may sharply increase the microbial growth [356, 362, 363].

In contrast to uncleaned pipes, purged pipelines showed no diurnal changes in the concentration of iron. Microbial growth decreased substantially immediately after purging. In the course of the summer season the concentration of heterotrophic bacteria in drinking water rose to a level that was 5 times as high as in water from previously cleaned pipelines and 11 times higher than in uncleaned distribution systems. After the pipe purging the water quality remained stable during approximately three months. The growth of biological deposits on the surface of pipes in the summer period was partially due to the rise in temperature, but the main cause was the presence of biogenic substances. After removal of deposit the microbial growth decreased significantly.

A detailed analysis of causes of different nature that are responsible for deterioration of water quality in pipelines was made in [364, 365] with the municipal distribution system of Athens as an example. Without going into an extensive review of all the causes analyzed and referring the reader directly to the reports cited we will outline the problems studied in them. All negative factors associated with the operation of a distribution system were divided into four groups: water losses, structural limitations, shortcomings caused by reduced carrying capacity, and deterioration of the water quality characteristics. Water losses result from corrosion holes, pipe coupling leakage, high nighttime water consumption, etc. Structural limitations are associated with failures of bimetallic couplings, structural damage below the groundwater level, disintegration of pipes in high-alkalinity soils, etc. Factors that affect the carrying capacity include increased water turbidity, pipe corrosion, insufficient pressure of the pumping equipment, deposits in pipelines, and high fluctuation of the volume of water in accumulator reservoirs. Deterioration of the water quality is caused by the intrusion of pollutants through holes in pipes under certain conditions, washout of foreign impurities from pipes, inferior quality of the initial water, and insufficient purification. The authors of [364, 365] analyze the principal and secondary causes, as well as the above-mentioned factors, and suggest remedies that can be summed up as follows:

- preventive measures to guard against stagnation at dead ends of pipelines where low flow velocities can be avoided by creating loops in which water will circulate;
- regulation of the pH value and prevention of corrosion through addition of special reagents: this will help to reduce the internal corrosion, leaks and ruptures, and also the rate of leaching metals or other chemicals from pipe walls;
- improved control of the operation of pumps for achieving optimum efficiency;
- improved control of pressure waves to form self-sustaining and controllable pressure zones.

Another useful measure that will help to improve the water quality is periodic purging of pipes to remove deposits. A concept of creating a self-cleaning distribution system is presented in [366]. It is proposed to maintain a constant water flow velocity at a level of 0.4 m/s which is sufficient for a daily removal of deposits from a distribution system. Theoretical studies have shown that although the actual flow velocities may be lower, fluctuation of velocities produces a self-cleaning effect. Similar approaches are also applicable to pulsation purging of pipelines. Here the recommended flow velocity is 1.5 m/s. The main distinguishing feature of a self-cleaning distribution network is use of branched system instead of closed-loop circuits. This makes it somewhat different from the concept proposed in [364, 365], but the validity of this approach was confirmed by the monitoring of a model system over a period of five years. Branched self-cleaning systems have proved to be much more efficient than closed ones.

Since potential pathogenic microorganisms can survive, procreate, and form biofilms in pipelines under normal conditions of conveyance of drinking water [343–347], a number of studies focused on the effect of UV-radiation on the state of biofilms in drinking water pipelines [367–369]. In paper [370], assessment of the state of a biofilm with the aid of a scanning electron microscope failed to reveal any significant differences in the bacterial morphology of biofilms in systems with and without ultraviolet. A practically identical result was obtained in [368], although here it was also demonstrated that reduction of exposure to UV-irradiation from 12 to 2 h enhanced the assimilating capacity of a biofilm with respect to the growing amount of biogenic substances. However, ultraviolet had no significant effect on the composition of the biofilm, in contrast to its effect on microorganisms in bulk water. This suggests that ultraviolet had no effect on specific microorganism strains in biofilms. According to the authors of [368], this phenomenon is possible because of the shielding of the biofilm by depositing particles or their aggregates, which makes a considerable contribution toward the survival of bacteria susceptible to ultraviolet. This would also account for survival of pathogenic microorganisms in water distribution systems and was corroborated by a series of special tests in which pathogens were introduced into a system.

Thus, UV-irradiation cannot be recommended for disinfection of water distribution systems. In this connection we must also mention the study of the effect on a biofilm of free chlorine in quantities of 1 to 3 mg/dm^3 or polyhexamethylenebiguanidine in a dosage of 10 mg/dm^3 in [371]. Such disinfection substantially reduced the biofilm density as compared to the control figure. However, the best result was obtained by shock treatment. The largest decreases of the biofilm density were obtained with a weekly free chlorine dosage of 10 mg/dm^3.

The pipe material is one of the factors that significantly affect the growth of biofilms in pipes. While external water distribution systems are mainly made of iron or steel, the copper and plastics are often used for indoor pipelines. The fouling of these materials with biofilms progresses in different ways [337, 344, 372, 373]. Thus, for example, although the formation of biofilms in copper pipes proceeded more slowly than in plastic ones, pipes of the two materials showed no differences in the numbers of microbes after 200 days of operation. Copper ions were responsible for the lower microbial number during the first 200 days, but after that the microbial number began to increase.

In addition, the numbers of viral particles was lower in water leaving copper pipes and in biofilms on these pipes. Also, microorganism communities in these two types of pipes differed in structure. The content of cyclopropane fatty acids was higher in copper pipes than in those made of polyethylene, which is indicative of higher numbers of gram-negative bacteria in the stationary phase of the bacteria's growth. The content of saturated fatty acids typical for gram-negative bacteria was also lower in copper pipes than in polyethylene ones ($<3\,\%$ as against 5–6 %). Conversely, the content of eukaryotic cells was higher in films on copper pipes. There were also differences in the community structure for both types of pipes depending on pipe length and age. Also, a certain amount of phosphorus compounds was washed out of new plastic pipes [373].

In examining the properties of biofilms on the surfaces of distribution system pipes it is profitable to consider the possibilities of modeling their initiation and growth. Modeling of this process will enable producers of drinking water to select an improved strategy of water quality control and investment management [374].

Although a detailed discussion of the feasibility of modeling is outside the scope of this review, we will elaborate on a few general approaches to this problem.

Basically, modeling of the quality of water in distribution systems is done for a stationary (stable) or a dynamic environment. In the steady-state modeling the external conditions in a distribution system are constant in time, and the nodal concentrations of the components that will be obtained upon achievement of the equilibrium are defined. These methods can provide the general information on the spatial distribution of the water quality in a system. In dynamic models the external conditions change in time, and the time intervals of the principal component concentrations are defined. These approaches were elaborated for the steady state in [375–377] and for dynamic models in [378–381]. The dynamic model and numerous existing methods for its solution are described in paper [382].

The next step in the development of water quality modeling for distribution systems and the understanding of regularities in biofilm growth was creation of models that made it possible to include the content of organic carbon and phenomena involved in bacterial growth in addition to the kinetic regularities. This is the so-called "single-component" model. The term "component" used in this model applies to the water quality constituents, such as organic substrate, biomass, and free chlorine. A comprehensive model is necessary for the identification and monitoring of potential local sites of bacterial growth in a system [383].

In paper [384], Lagrange's method was demonstrated to be the most effective approach in water quality modeling. Two methods, the Lagrangian Time-Driven Method (TDM) and the Event-Driven Method (EDM), were compared for varying admissible concentration deviations and calculated time variant water quality indexes. Based on these methods the authors of [382] proposed a new hybrid method (EDMNET) that improved the accuracy of the previously developed Langrangian methods. All the above-mentioned methods were integrated in an existing hydraulic similarity model. The integrated model was tested to examine different problems arising in a distribution system under varying conditions. It was demonstrated that the proposed method could accurately model the nodal concentrations in a maximum number of segments of a distribution system and involved fewer calculations than the other methods.

Thus, the above analysis of the state of water distribution systems and its effect on the drinking water quality indicates that the reason of pipe's ill-condition includes incomplete removal of suspended particles in the process of water treatment, washout of fine impurities from filter media, deposition of metallic oxides or calcium carbonates, postflocculation, biological activity, and corrosion. The build up of new deposits appearing in pipes after cleaning progresses is quite fast.

In paper [341] was found that in one year after the cleaning the new deposits were almost as thick as the old ones which had accumulated over many years. Further, it was confirmed that the meliorative effect appeared to be transient (especially for iron concentrations and turbidity), because of fast growth of new deposits. New soft deposits usually accumulate in certain places of the distribution system (at low night flow velocities in narrow ends of the network and in reservoirs). Cleaning is not the only solution of the problem of excluding deposits from distribution systems. This can also be achieved by selecting and observing appropriate hydraulic parameters, using modeling methods to control the water quality in a system, and improving water treatment prior to distribution, especially with respect to biologically assimilable organic carbon. Cleaning and disinfection of pipelines are advisable and necessary in case of contamination of drinking water. The pipe material does not have a decisive role in the formation of a biofilm but only influences the kinetics of this process.

Among of the disinfectants the most effective action on biofilms is produced by monochloramine and on suspended bacteria by chlorine. Membrane filtration is believed to hold considerable promise in preventing formation of biofilms in drinking water supply systems [339].

Some examples of modern technological systems for production of high-quality drinking water A drinking water supply technology must fully ensure safety of the drinking water for the consumer by removing or reducing to minimal concentrations all components that may represent potential health hazards both in initial and treated water. A technology for production of high-quality drinking water is usually designed with regard to practical possibilities of the producer and the water quality in the supply source.

As noted in paper [81], the health hazard associated with the presence of toxic substances in drinking water differs from hazards caused by microbiological contamination. Except for cases of heavy contamination of a water supply system, few chemical constituents present in water are capable of causing acute disorders. This is why chemical pollutants are assigned a lower priority than microbial ones whose action is usually stronger and more wide-ranging. However, the use of chemical disinfectants in water treatment usually results in the formation of by-products that are potential health hazards. Thus, a water treatment technology is faced with two conflicting problems: on the one hand, the need for a strong disinfecting effect, and, on the other hand, minimization or exclusion of a side effect of disinfection by-products on human health.

Disinfection represents the final stage of protection of drinking water from external contamination and a secondary growth of microorganisms in the water distribution system. In the last analysis, the entire sequence of treatment processes can be viewed as preparation of the water for an effective and reliable disinfection. According to the WHO guidelines, the present trend in the water treatment technologies is composed of optimization of chemical reagents used in water treatment and development of physical or biological treatment methods designed to reduce the dosages of such reagents and, consequently, to decrease the formation of disinfection by-products. This approach is assumed to be based on the principle of a multibarrier treatment procedure with maximum exploitation of natural water treatment processes on various filter devices or screens.

One excellent illustrative example of the production of high-quality water is a multibarrier technology used in the Netherlands [86]. The water quality in the supply source is constantly monitored. If it is not quite satisfactory, the river water is mixed with underground water pumped out from a depth of 120 m. If the river water quality is not acceptable at all, the river intake is disconnected and the water supply, although at a lower rate, is provided from the underground source solely.

The basic components of the technological system are presented in Fig. 4.1 and include the following elements. At the first stage, the initial water is treated with the coagulant $FeCl_3$ to remove suspended matter, including phosphates, partially dissolved organic substances, viruses, and heavy metals. After the pretreatment the water is collected in two reservoirs from which it flows, via canals, to five ponds (975 ha). Here it percolates into a sandy soil (dunes). To achieve a water quality comparable with groundwater, the retention period must be as long as possible. In this particular case the approximate duration of this infiltration is two months.

During this time the water quality continues to improve: nitrates are broken up and trace quantities of organic compounds and >99 % of fecal bacteria and viruses are removed.

After that the water is conveyed to a collecting pond via a special water collection system and channels. The largest amount of water comes from a collector located above the confining well. If the water has an increased content of mineral salts, it is diluted with water pumped out from a deep well reaching below the confining well. After infiltration the water is enriched with oxygen in a cascade-type facility, filtered, and ozonized. This treatment partially destroys organic matter and pesticides and kills pathogenic viruses and bacteria which are then removed by the subsequent filtration process. After ozonation, the water is softened, if necessary,

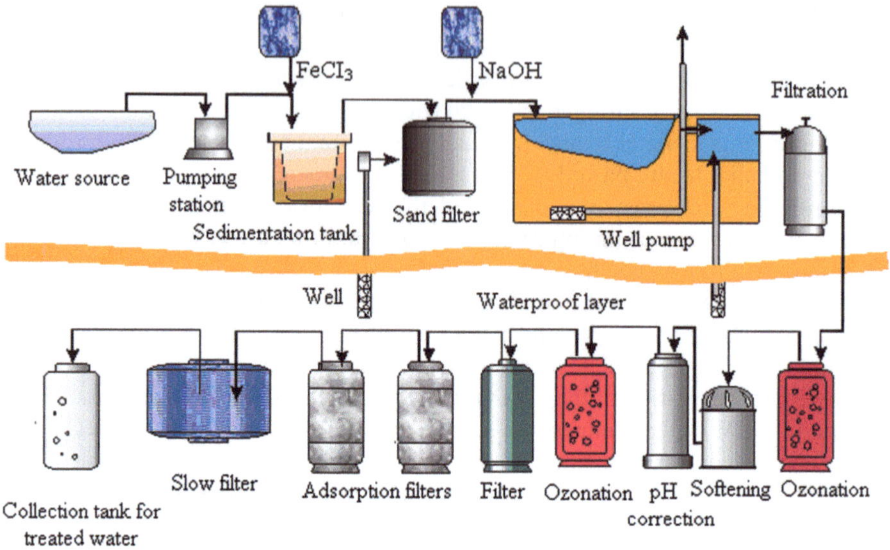

Fig. 4.1 Schematic of drinking water treatment in the Netherlands

and its pH value is adjusted. At the next stage, the water, after a second ozonation, undergoes a two-phase adsorptive treatment for a thorough removal of trace quantities of organic substances by physical adsorption and biological oxidation. In summertime the filtrate is enriched with oxygen. After a certain time the activated carbon (AC) is removed for regeneration and the filters are recharged. After the carbon filters water proceeds to slow sand filter. The main purpose of this stage—is to further disinfect the water by removing residual bacteria and traces of organic compounds. The water after the slow filters is transparent, colorless, tasteless and safe for consumption. The dosage of a disinfectant needed for its decontamination can be reduced to a minimum, and the designers of this system believe that such water, being biologically stable, requires no further disinfection. In other words, the above-described procedure can serve as an example of a chlorine-free technology.

In this technology attention is drawn to the location of the slow filter in the flow diagram. The current trend in modern treatment procedures is to place slow filters at the final stages of multibarrier systems [90, 385]. This reduces the organic impurity load on the medium biofilm of the slow filter by several orders of magnitude and makes its operation more efficient by minimizing the frequency of backwashes (normally more than two flushings a year). With correctly loaded filter media, low-rate filtration through sand produces a greater improvement in the water quality than any other single conventional treatment process [386–389]. Obviously, slow filters can only be used at waterworks that have sufficiently large areas [81].

The efficiency of slow filters in the treatment of water from the Lake of Zurich by the procedure presented in Fig. 4.2 was studied in [388]. The water was treated by ozonation followed by filtration through a rapid filter, then a second ozonation process, and filtration through activated carbon. After the above processes it was conveyed to slow filters. The first ozonation (1.1 mg $O_3/dm^3 \pm 25\,\%$) was designed to

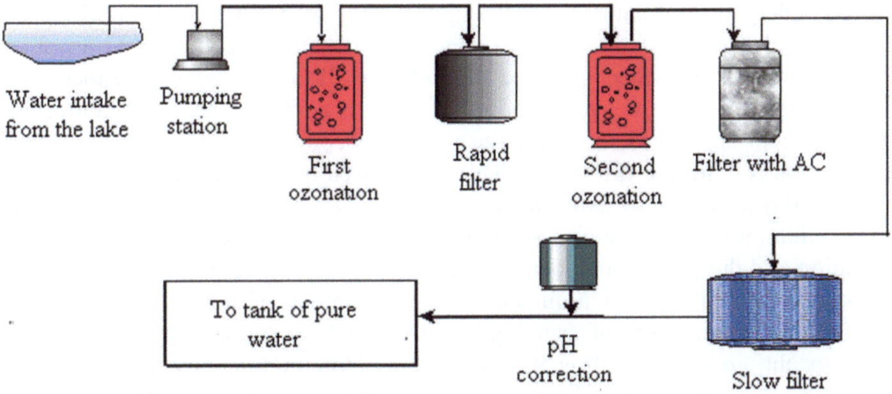

Fig. 4.2 Schematic of the conditioning of drinking water from the lake of Zurich

disinfect the water and facilitate flocculation before the rapid filters. These sand and pumice filters removed upward 95 % of suspended matter. The second ozonation before the AC filters was done with an O_3 dosage of 0.5 mg O_3/dm^3. For further purification the water was passed through slow filters. One such filter contained four beds of grains that increased in size in the direction of filtration: fine sand (50–85 cm), coarse sand (5 cm), fine gravel (5 cm), and coarser gravel (5 cm). Thus, the total height of the sand filter medium varied between 100 and 65 cm. After 10–15 years of operation the sand remaining in the filter must be removed. The Zurich waterworks operated 14 low-rate sand filters. The procedure provided for operation in the dark at 4–8 °C. Each filter had an area of 1.12 m^2 and a maximum specific capacity of 16 m^3/m^2. The whole system had a daily treatment capacity of 250 m^3.

The main problem encountered in the operation of slow filters is their silting. Two operative modes were tested for the technology in question over several years: at average filtration velocities of 0.5 and 4–7 m/h. In the former case all filters worked for more than 9 years with periodic rapid cleaning every two–three years. In the latter case a filter remained in continuous operation for nearly 8 years without a single cleaning. That filter was only slightly more silted than the former ones. In the course of operation the permeability of the filters gradually decreased by 0.5 m/h indicating an increasing silting of the system. This is why the filters were washed up and cleaned up by replacing the topmost 5 cm of the sand bed. This increased the hydraulic permeability by 1–2 m/h for a period of three years. However, this method does not restore the initial percolating capacity. Silting of slow filters is primarily due to microbiological activity of the medium, with biomass filling at least 10 % of the pore space of the upper layer. Different modes of filter operation, such as were described above, also produce changes in the quantity of biomass, especially extracellular polysaccharides and proteins. Differences were also observed in the diversity of microbiological populations. Despite the continued interest in slow filters, the problem of their silting calls for further study.

A technology patented in Great Britain [389] can serve as an example of production of drinking quality water from heavily polluted surface sources. A simplified schematic of this technology is presented in Fig. 4.3. Water from a polluted surface source undergoes pretreatment involving processes of coagulation and precipitation. Then the water is collected in a pond from which it is conveyed to a coarse filter. After that the water is passed through two filtration stages with media of different grain sizes. A coagulant or a flocculant is added to the water before the first filtration stage. Thus, the first stage practically combines filtration with contact coagulation (or "in line filtration"). This is followed by several stages of sorption on AC and disinfection by exposure to ultraviolet. The final stage is a thorough fine purification on two reverse osmosis units connected in parallel one of which is in operation and the other undergoes regeneration. Naturally, after the final stage the drinking water must be conditioned to adjust the contents of mineral components.

In Namibia producers of drinking water use mostly surface sources and to a lesser extent, groundwater. It is also planned to obtain high-quality drinking water by realizing the technological system shown in Fig. 4.4 which comprises the following processes. At the first stage the water is ozonized and fed to a coagulatio n/flocculation facility. Then the water is conveyed to a flotation tank and passed through a two-bed filter. After this the water is ozonized for a second time and fed on to a BAC filter. This is followed by filtration through two stages of AC adsorption filters and UV-irradiation. Chlorine is used for a final disinfection. The quality of the water obtained by this procedure is better than that prescribed by the drinking water standards [390].

A multistage water treatment procedure proposed in [391] includes oxidation of water impurities by ozone at the first stage to be followed by filtration through a biologically activated medium. Then the water is to be softened in filters with an ion-exchange resin, deodorized on AC, and passed through a reverse osmosis unit. The disinfection is to be done by ultraviolet and a second ozonation.

Construction of a waterworks with a capacity of 24,000 m^3 per day designed to supply 200,000 people at a rate of 120 dm^3 per person per day is reported in [392]. Here, water drawn from lakes and other sources is collected in an intake reservoir (Fig. 4.5). Coarse mechanical filters are installed immediately after the pump plant and are followed by several more filtration stages. Next the water reaches a reagent flotation stage to be treated with coagulants and flocculants, passes through a flotation reservoir, and proceeds to biological treatment facilities. Further treatment, also conducted in several stages, is designed to produce a thorough disinfection. First, the water is ozonized and then passed through two stages of filtration through AC. Here, organic compounds of a certain molecular weight interval are removed at stage 1 and trace quantities of pesticides and other toxic substances at stage 2. A final removal of pathogenic microflora and viruses is effected at a reverse osmosis stage.

A drinking water plant in Ivry-sur-Seine (France) also uses a multistage comprehensive technology (Fig. 4.6) [393]. The first water treatment stage includes preozonation. Then the water reaches a stage of coagulation with $FeCl_3$. This stage also provides for addition of PAC in emergency situations. Treatment at the next two stages is based on in-line filtration with $FeCl_3$ and is carried out on two different types of media: a special biologically active medium "Biolite" and quartz sand. After filtration through a sand filter the water is ozonized and fed on to an AC

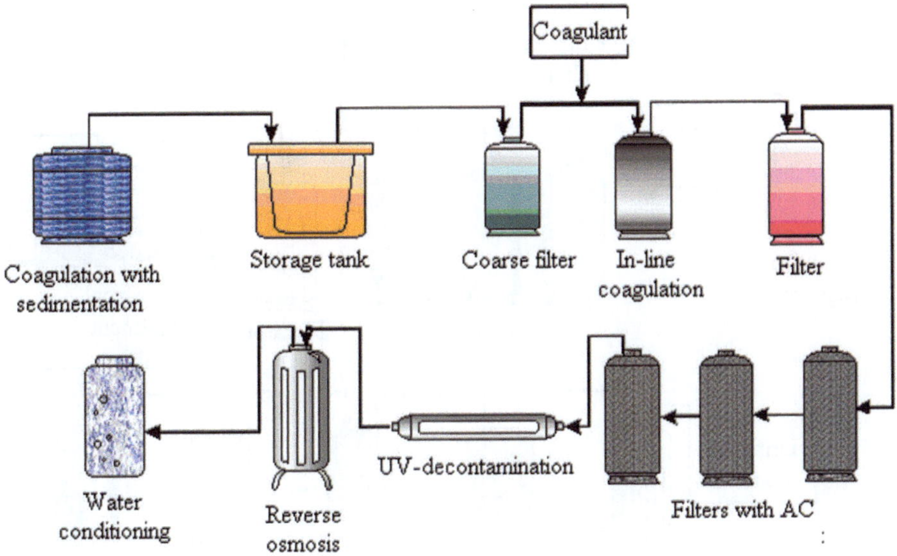

Fig. 4.3 Schematic of drinking water treatment in Great Britain

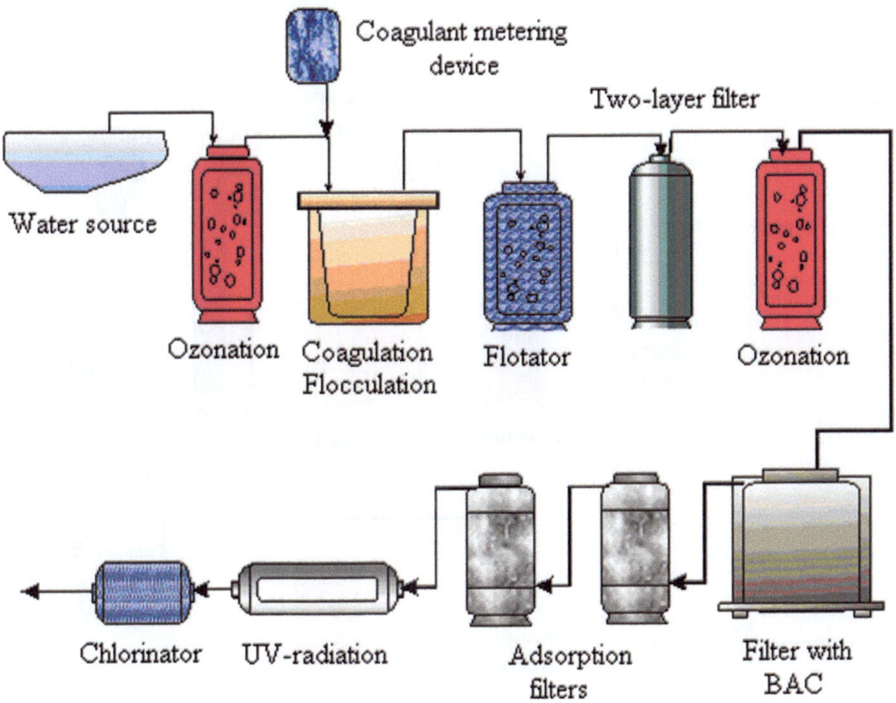

Fig. 4.4 Schematic of drinking water treatment in Namibia

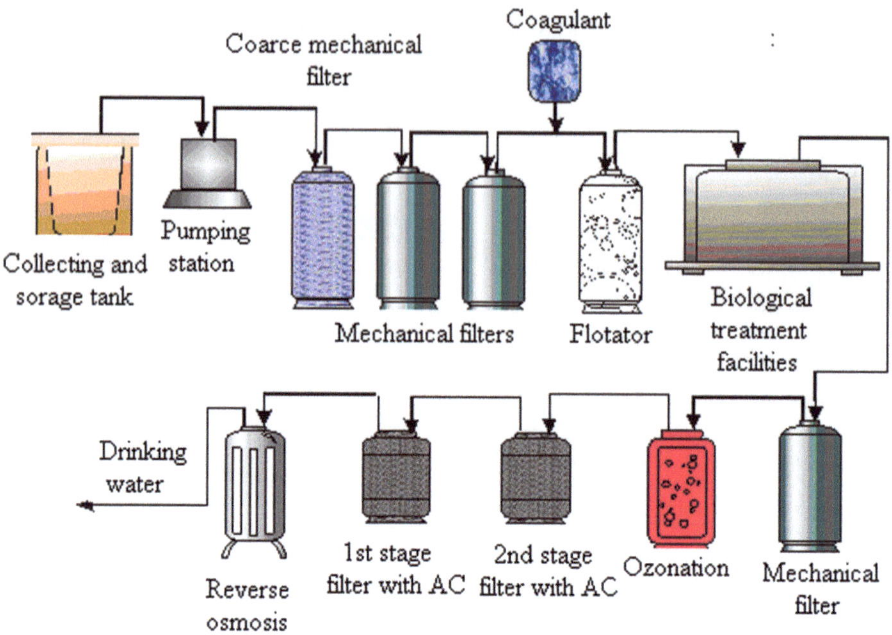

Fig. 4.5 Schematic of drinking water treatment with using biological processes

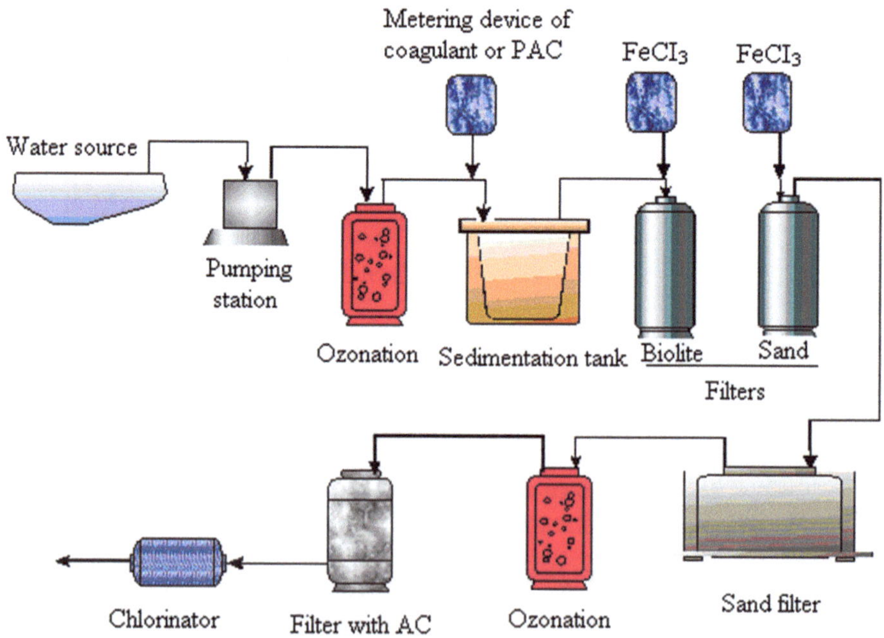

Fig. 4.6 Schematic of drinking water treatment in Ivry-sur-Seine (Paris)

filter. The final disinfection is done with chlorine. Water purified by this technology has the following characteristics: TOC—2.5; NH_4^+—0.01; NO_2—0.01; NO_3—22; Fe—0.01; Mn < 0.02 mg/dm^3; pH—7.7.

Valuable experience in the design and operation of high-capacity water treatment plants has been gained by the Ecoholding Co. in Russia. This company also uses a comprehensive approach in the production of drinking water. First water undergoes coagulation treatment and then sedimentation in high-efficiency thin-layer sedimentation reservoirs followed by adsorption on AC. For a final stage, the water is passed through membrane filters. The volume of concentrate formed in the process of treatment does not exceed 15 % of the total volume of treated water. Waterworks of the above design have been built or are under construction in many Russian cities (Mozhaysk, Shchekino, Borovichi, Taganrog) [394].

In Yaroslavl, in addition to the conventional drinking water technology of coagulation–sedimentation–filtration the following stages were incorporated in the technological system for the treatment of the Volga water: aeration, flocculation mixers and chambers with elements of thin-layer sedimentation, ozonation, and two-bed and carbon filters. This multibarrier technology significantly improved the drinking water quality, as compared to the conventional procedure, and also offered added flexibility in adjusting the treatment parameters depending on the quality of the initial water [395].

Another example of a multibarrier drinking water technology is described in [396] where three waterworks using heavily polluted river water were studied. One of these plants comprised stages of coagulation/flocculation, flotation, ozonation, filtration, and disinfection. AC filters were additionally used at another one, and at the third plant treatment was done only by coagulation/flocculation, filtration, and disinfection. The results of these studies showed that the best water quality characteristics had been achieved by the technology that included the multiple stages [396].

The reconstruction of water treatment facilities in Rostock (Germany) provided partial replacement of chlorine with ozone as the disinfecting agent. This resulted in an appreciable decrease of the content of halogenous organic substances in treated water. Ozone was added at several water treatment stages, and after ozonation the water was purified on AC. After the reconstruction the water quality conformed to Germany's most stringent standards [397].

Advantages of multistage drinking water treatment procedures incorporating ozonation and filtration through AC are described in [398–400]. In all cases multistage procedures produced higher water quality than the conventional processes of coagulation and filtration.

Integrated purification of drinking water by using ozone in conjunction with BAC in the process of water treatment began to be implemented in Tokyo (Japan) at several waterworks in 1992, 1999, and 2004 years. These projects helped to solve the problem of removing ill-smelling compounds, organochloric substances, and ammonium nitrogen [401].

Final disinfection is the central issue of all drinking water technologies. This problem must be the subject of a separate extensive review. This is why in the present chapter will touch upon it only in terms of the use of the various water disinfection methods in modern water treatment technologies.

Obviously, chlorination remains the most widespread disinfection method used in the overwhelming majority of technological procedures [402]. However, as was already repeatedly noted in paper [35], epidemiological studies have shown that consumption of chlorinated water is associated with an increased carcinogenic hazard and may impair the reproductive functions in humans [403]. Analysis of data on the effect of long-term consumption of chlorinated water on human health, together with data from epidemiological, chemical, and genotoxicological studies, indicated that alternative disinfectants had to be found to reduce the potential health hazards. The proposed alternatives include ozone, chlorine dioxide and chloramine [404], and ultraviolet [405–407].

Studies carried out over eight years on the identification of disinfection by-products (DB) showed that disinfection with chlorine resulted in the formation of the largest amount of halogenated DBs, and with chloramine of the same types of DBs but in lesser quantities and lower concentrations. Ozonation also produced dibromoacetonitrile, in addition to nonhalogenated DBs, if the water undergoing treatment had an increased content of bromine. With low bromine levels, chlorine dioxide induced the formation of a relatively small quantity of DBs, as compared to the other disinfectants [408]. In addition, chlorine dioxide had a stronger bactericidal effect than chlorine or sodium hypochlorite.

Another advantage of this technology is that effectively no chlorine-containing vapors or gases are released in the area where chlorine dioxide is obtained by the membrane electrolysis method [409]. The above-mentioned CIO_2 production method provides for the use of a unique ejector in the electrolytic procedure. Analysis of the operation of disinfection facilities at the waterworks of Monchegorsk (Russia) showed that the membrane electrolysis technology was highly efficient and ecologically safe for the city.

Experience of disinfection of water from the Moskva River with CIO_2 instead of CI_2 indicates that while the former disinfectant produced significantly lower concentrations of THM its by-products included chlorite ions in quantities that were several times larger than the maximum permissible concentration (MPC). To remove such products it is necessary to include an additional AC sorption stage in the technological chain. Then in overall the water treatment costs are increased in three–four times [410]. We believe that it would be wrong to attribute the rise in costs solely to the change of disinfectant, because addition of an AC treatment stage helps solve a wider range of issues related to the water quality. Obviously, a substantial improvement in quality should significantly increase the producer's costs.

The problem of reducing the contents of organochloric secondary products of disinfection can be solved by using CI_2. Thus, for example, in [411] reagents containing chlorine were added after water had been treated by coagulation flocculation, sedimentation, and filtration. It was found that in this way the disinfectant dosage and the quantity of toxic disinfection by-products could be significantly reduced.

Exposure to ultraviolet for destroying microbiological impurities has been steadily gaining ground in modern disinfection technologies [412]. It is known that

UV-radiation at a wave length of 253.7 nm destroys DNA in the cells of bacteria and other microorganisms thereby preventing their propagation. A UV radiation dosage of the order of 10,000 $\mu W \times s/cm^2$ is sufficient to kill 99.9 % of microbiological impurities. Although UV-sterilizers have been used in practical disinfection for quite a long time, they have gained widespread acceptance in centralized water supply only over the last few years.

For example, since 2001 in the north of Sweden a drinking water disinfection system using ultraviolet sources with a radiation intensity of 200 J/m^2 has been implemented at a waterworks with a daily capacity of 144,000 m^3. There, disinfection is performed on two technological lines with a combined total capacity of 6,000 m^3/h which can operate simultaneously or alternately [413]. Also in Sweden, the leading manufacturer of disinfection systems WEDECO AG Water Technology built a drinking water disinfection facility at the waterworks in Gallivare. The waterworks supplies 450,000 residents operating at a daily capacity of 200,000 m^3. Use of ultraviolet instead of chlorination improved disinfection of the water and significantly reduced the needed quantities and stocks of chlorine-containing reagents [414].

A three-year experiment in drinking water disinfection by ultraviolet was conducted in Germany. Pretreated and filtered water was caused to flow around a source of UV-radiation and then tested for quality. Despite positive results, the main conclusion drawn from these tests was that ultraviolet disinfection units must be constantly checked and monitored during operation [415].

In Switzerland 70 % of the drinking water supply is disinfected with ultraviolet, chlorine and chlorine dioxide ranking next in the quantities of water treated. It should be noted that in Switzerland the drinking water demand is satisfied by 43 % by artesian wells and by 39 % by springs. Thus, here disinfection by UV-radiation is used predominantly for groundwater [416].

In Russia, the groundwater is the traditional sphere of application of UV-radiation as a disinfectant too. Today, more than 700 plants equipped with ultraviolet units of Russia's domestic manufacturer KIT Co. are in operation [417, 418].

After this brief discussion of the various means of drinking water disinfection it should be emphasized that in all cases the choice of the disinfection method will be governed by the quality of the initial water, the technology of its treatment ensuring a certain level of biological stability, and the condition of the water distribution system. In view of the above considerations, a conclusion can be made that use of reagents containing chlorine is often necessary to make water microbiologically safe.

Thus, the above review of some examples of modern technological systems for the production of drinking water demonstrates that in addition to the traditional classical technology of preoxidation—coagulation—flocculation -sedimentation—filtration, new water treatment techniques are now used on an increasingly wide scale. First of all, we should note the use of ozonation combined with AC which has now become almost universal in the countries of Europe and America. Also, we cannot fail to notice the increasingly widespread use in drinking water production of biological methods and primary ozonation instead of chlorination, changes in the place of addition of chlorine-containing reagents to the water undergoing treatment, and the absence, in some cases, of a final chlorination.

Integrated membrane technologies for drinking water treatment followed by conditioning have received wide acceptance. Another promising trend that deserves mentioning is the use of slow filters at the final stage of a multistage water treatment procedure. Finally, the principal distinctive feature of modern technological systems is multistage and multistep procedures designed to produce high-quality water.

The choice of the drinking water technology and the economic feasibility of the project based on it are determined by a large number of factors that include assessment of the water source supply resources, the available funding, etc. In this regard the computer programs proposed for economic and financial feasibility assessment of drinking water supply projects (EVAPRO) become an attractive tool. The EVAPRO system is designed to facilitate feasibility studies for different investment and technological scenarios and suggest modifications of the initial technological systems. This software incorporates automation of the feasibility studies for water projects on the basis of the available information, such as the population, the expected demand for water, potential water resources, investment possibilities, monetary resources, and water price policies. It provides a set of tools for linear scaling programming, numerical methods, and financial calculations, and also combines technological optimization with financial analysis in the feasibility assessment of water supply projects [419]. Use of this approach in the development of modern water supply systems appears to be quite promising.

4.4 A New Concept of Supplying the Population with a Quality Drinking Water

The question regarding the quality of drinking water has long ago stepped over national frontiers and has taken on a nature of global responsibility. In particular, UNESCO has endorsed the extensive International Hydrological Program. More than 100 countries in the world have taken part in the development of this program. And the period of 1981 through 1990 was announced as decade of drinking water. These measures have stimulated research in the field of drinking water supply, which was conducive to revealing the main factors affecting the quality of drinking water and recognizing the acute necessity of taking urgent measures at the legislative level on both national and international scales. In response, the European Union Council accepted the Resolution 98/83/EC regulating the quality of drinking water.

Practical implementation of these legislations has not yielded any substantial results for a number of reasons, mainly because specifically developed technical measures have not taken into consideration the amassed knowledge about the factors affecting the quality of drinking water. Among these factors, the following should be taken as priority: inequality in regional supply of quality sources of water, which results in the necessity of finding alternative water sources; a strong tendency for the worsening of water quality in traditional sources of drinking water supply; secondary contamination of water at water treatment facilities and in the water system; the necessity of rapid response to the quality of the initial water at the time of obtaining drinking water; new possibilities of obtaining drinking water due to the

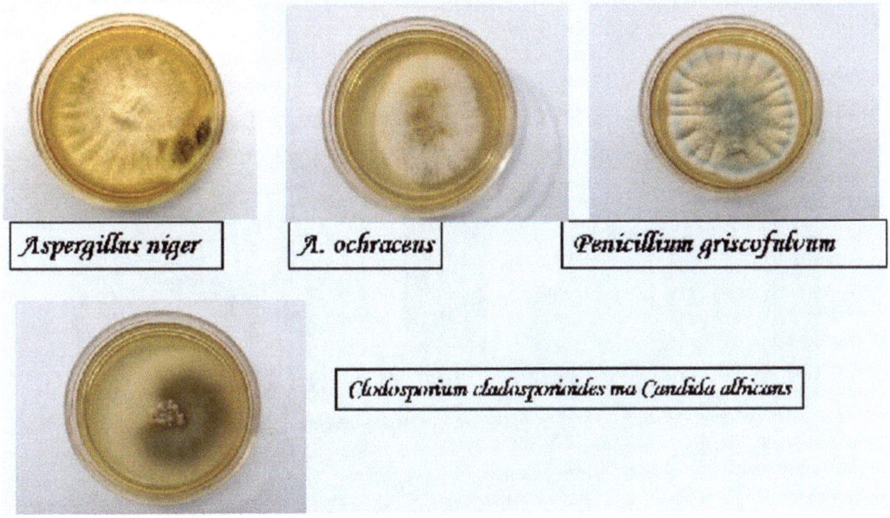

Fig. 4.7 Micromycetes isolated from tap water of the city of Kiev

use of modern technologies; the impact of the conditions of water supply on water quality and, certainly, the current state of the water system.

In the case of transportation of high-quality drinking water in the water system, one can observe its secondary pollution by the products of pipe corrosion and microflora. The employees of the Dumanskii Institute of Colloid and Water Chemistry, National Academy of Sciences of Ukraine (ICWC) have recently detected the growth of micromycetes in the water system of the city of Kiev (see Fig. 4.7).

The presence of these microscopic fungi results in the deterioration of organoleptic indexes of drinking water quality and causes a number of serious human diseases (aspergulleses, bronchitises, allegries, dermatomycoses, etc.). Thus, regarding the central system of water supply, it is in principle impossible to secure the customer with high quality drinking water due to its secondary pollution in the distribution system. This problem becomes increasingly difficult due to the unsatisfactory technology of water treatment, the technical state of water piping networks and water supply in a number of populated areas is provided by schedule. Thus, for instance, the use of waterworks unsuitable for deep removal of organic and other nutrient matter for microorganisms results in the formation of carcinogenic substances during water decontamination and, conversely, in the propagation of the microflora including pathogenic microflorain the pipelines. This occurs especially when water supply is provided according to the schedule.

The solution for the urgent issues regarding drinking water supply is only possible when based on the fundamentally new concept of providing the population with quality drinking water. Such a concept provides for the following. Water safety in terms of toxicological and microbiological indexes is prepared at waterworks. This water is suitable for household or domestic use and is fed to the distribution networks. In this case it is obtained mainly from the water of surface sources.

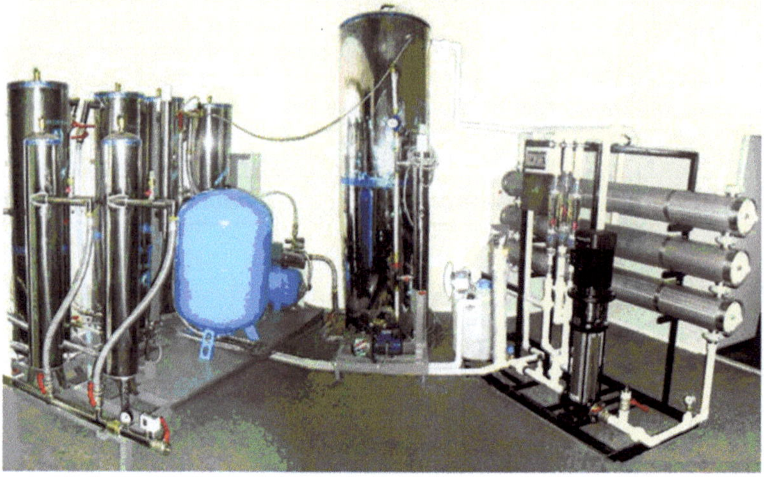

Fig. 4.8 Multifunctional module unit Vega-3-UM

High quality drinking water in the amount sufficient for satisfying human physiological needs is prepared at the consumption site on the plant of water pump rooms. It is then not fed to the distribution network but is supplied to the consumer by other methods (by gravitation flow or bottling). Such drinking water is obtained mainly from underground water protected against anthropogenic influence or from tap water. In the places where the underground sources of water supply met the requirements for quality drinking water, simplified plants were used. The initial water was disinfected by a safe reagentless method. Such a concept was implemented at the plants in water pump well rooms with strictly defined and regularly controlled quality indexes [420]. In other cases, quality drinking water was obtained in water pump well rooms on module units of various complexities, depending on impurities of the initial water. The capacity of each unit is determined by the number of consumers and, as a rule, does not exceed 5 m^3 of treated water per hour. The results of employing module units proved the efficiency of solving the problem of drinking water supply by means of water pump well rooms (see Fig. 4.8, 4.9).

4.4.1 The Concept of Choosing a List of Indexes and Their Regulatory Values for Determination of Hygienic Requirements and Control over the Drinking Water Quality in Ukraine

A conceptual approach to the assessment of the quality of drinking water proposed in [421] is aimed at the notion man's necessity to consume physiologically wholesome water. This implies not only its purity, safety and absence of any anthropogenic im-

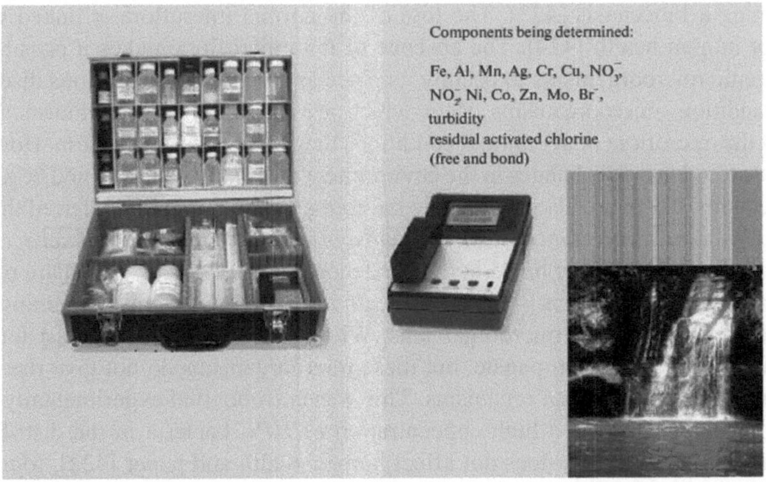

Fig. 4.9 Portable laboratory Aqua-Test for monitoring water quality (developed with Polyus Company Ltd.). Methodological support including methodologies for photometric analysis according to GOST Standards and the technique developed at the ICWC of Ukraine's Academy of Sciences. Governing batch (40 pcs) for the first time issued in Ukraine (in 2004)

purities, but also the presence of necessary natural substances and microelements, which are most easily assimilated by an organism from water. It is known that the quality of physiologically wholesome water was formed in the absence of an anthropogenic influence on natural sources in a specific area and is optimally balanced by nature for obtaining by organism of salts and microelements necessary for it in the given climatic conditions. In compliance with this requirement, first regulatory requirements to water quality were formed historically.

The indexes of the quality of physiologically wholesome water at present cannot always be met at centralized stations of drinking water supply for largely populated areas [421]. Therefore, we would like to underscore the importance of achieving the indexes of water quality considered below when using new advanced technologies, both with centralized systems of drinking water supply and local ones.

Water surface and underground sources of water supply contain both natural impurities, determined by geographical, climatic and geological factors and anthropogenic ones. The state of a body of water with an intrinsic background of chemical and biological components without anthropogenic impurities is the most suitable for obtaining drinking water in a specific geographical region. Natural impurities cannot produce a negative impact on human health owing to their genetic proclivity to consumption of this water. Exceptions to this include such hazardous natural impurities as, for example, radioactive compounds, heavy metals, arsenic, fluorides, etc.

Sterile water in nature is an extremely rare case. Even in water obtained from deep wells and springs, some quantity of bacteria can be found. However, surface water always contain microorganisms in a large or small number. These organisms are identified as a normal microflora; therefore, they are always present in water [422]. The mankind developed simultaneously together with microbes intrinsic in it forming a biocenosis [423]. The loss of the normal microflora is linked with a risk for human health [424]. The absence of this microflora makes it possible for pathogenic microorganisms to occupy its place and thus cause infectious diseases.

In addition, microorganisms occur which are identified as opportunists. Being part of the microflora under normal conditions they do not cause any harm. However, if there are substantial changes in the environment they may propagate and be a cause of infections. There are also pathogenic microorganisms that are transferred through water (*Mycobacterium tuberculosis, Cryptosporium parvum, Vibrio cholera*, etc.).

Natural or saprophytic bacteria (a rule denoted as *HPC, SPC, TMN*) are present in all natural water systems [425]. After treatment drinking water still contains some non-pathogenic level of microorganisms. When optimal conditions exist for their vital activity they may propagate, but these microorganisms do not give rise to infectious diseases in human organisms. This was corroborated experimentally [426, 427]. Paper [428] notes a high concentration of *HPC* bacteria in the distribution system, which, as a rule, does not affect human health and paper [422], found out that they do not cause of infectious diseases even among animals with a suppressed immune system.

Thus, one may conclude that water in a natural state without the anthropogenic impact is generally safe for consumption. Natural impurities present in it are in balance equilibrium with natural microorganisms residing in the water. However, the ingress to such water of impurities of an organic or inorganic nature, which are not typical of the given water ecosystem, may lead to the disturbance of this equilibrium and the emergence of both chemical and microbiological contaminants determined by opportunists bacteria.

Anthropogenic contamination of water supply sources (especially surface ones) hazardous for the population's health occurs as a result of ingress to the water of substantial amounts of nitrites, pesticides, heavy metals, oil products and chlorinated hydrocarbons. Surface water supply sources, unlike underground sources, contain a large number of organic matters, which may be conducive to the formation of compounds in water treatment processes.

Scientific approaches to the choice of an advisable list and regulatory values of water quality indexes were most fully presented in the latest edition of the Guidelines for Drinking Water Quality, Recommendations of the World Health Organization, 2004 [429]. Its fundamental provisions are used when preparing national standards in many countries. However, in each specific case they are characterized by a different set of indexes and philosophy of their use. The WHO Guidelines recommend such concentrations of water components which do not lead to a substantial risk for human risk throughout a lifetime.

We offer the following approaches as a conceptual foundation for the indexes in the assessment of the quality of physiologically wholesome drinking water [421]. First of all, based on the above, one should single out a number of water quality indexes which are characteristic of Ukraine's natural water without the account of the anthropogenic load on bodies of water. For assessing the natural ecological state of water sources, it is advisable to use approaches developed in the EU Water Framework Guidelines adopted in 2003 and the experience of the ecological classification and regulating the quality of Ukraine's surface water [15, 430]. In line with these documents, water resources may be classified by several levels:

- high ecological state and maximum ecological potential;
- good ecological state and good ecological potential;
- satisfactory ecological state and satisfactory ecological potential;
- unsatisfactory ecological state and unsatisfactory ecological potential;
- poor ecological state and low ecological potential.

Thus, the initial natural state of a water object mostly precise is reflected by the level of a high ecological state and the maximum ecological potential. In natural water, the characteristics of biological components in the high ecological state completely or almost completely correspond to undisturbed conditions.

The characteristics of physical-chemical components also completely or virtually correspond to undisturbed conditions.

With the account of the world experience Ukraine saw the creation of its own methods of ecological assessment of water quality, which is a basis for assessing human activity impact on the surrounding environment [15]. Guided by these methods, we introduced a new conservation standard, "an ecological standard" for the quality of surface water. This standard is necessary for assessing ecological well-being of water objects in terms of water quality and determination of a complex of water protection measures. This standard was developed in compliance with Clause 37 of Ukraine's Water Code [1]. The notion of "the ecological standard of water quality" implies the characteristic based on scientifically substantiated quantitative values of water quality indexes (hydrophysical, hydrochemical, hydrobiological, bacteriological, and specific matter) which reflect the natural state of a water object and the objective of water conservation activities aimed at improvement or conservation of its ecological well-being [1, 16, 430]. In this case it is implied that the natural ecological state of surface water is such a state of natural water objects, which existed or may exist provided the absence or minor impact of human activity.

A conceptual approach to the composition of physiologically wholesome water should take into consideration apart from the macrocomponent composition only those microcomponents, which constitute a natural background and indispensable for organism.

It should be noted that at the present moment the developmental level of science, technology and economy in the country does not make it possible to meet all characteristics of the quality of physiologically wholesome water. Therefore, along with the targeted directives for obtaining such water we consider also the maximum limit levels for the content of macro- and microcomponents of water, which have been determined based on the critical analysis of the latest recommendation of the WHO, EU Guidelines, developed countries of Europe, the USA, and Ukraine's experience. In the [421] we are going to consider some commentaries to setting maximum possible levels in the content of various components in water in the context of adopted recommendations (experience of Ukraine, the EU, the USA, the WHO, and Russia).

Taste and smell in the water may have natural origin from inorganic and organic impurities and biological processes, from contamination with synthetic chemical compounds, as a result of corrosion of pipelines or water treatment. In addition, taste and smell may occur during storage and transportation of the water due to microbiological activity. The setting of the quantitative values of these indexes is determined both by the requirements for water safety and by the necessity of ensuring a high level of organoleptic indexes of drinking water. Requirements as to the safety of water consumption are ensured by achieving the required level of extracting components imparting taste and odour to the water in the course of its treatment. This refers to such contaminations of biological origin as products of metabolism of actinomycetes and fungi, invertebrate animals, cyanobacteria, algae and iron bacteria.

Chemical substances affecting taste and smell include aluminum, ammonium, chlorides, chlorine, chlorophenols, copper, dichlorobenzene, dissolved oxygen, ethyl benzene, hydrogen sulfate, iron, manganese, monochloramine, monochlorobenzene, oil contaminants, sodium, styrene, sulfates and surface-active substances (SAS). Limitations of the quantitative content of these components, in addition to ensuring the safety of drinking water consumption also guarantee the attainment of acceptable levels of taste and smell. The WHO Guidelines for Drinking Water Quality do not contain specific quantitative values of taste and smell of water and they recommend to be based on local factors, social and cultural circumstances and conditions of the environment. In Ukraine this value traditionally constitutes less than 2 points.

As to *color* it is implied that most people visually sense it at the level more than 15 standard units (degrees). The color level less than 15 units may be acceptable for consumers; however, this index is varied depending on local conditions. There are medical justification for color, however, one should bear in mind that intense color may indicate the presence of stained organic matter, which produce toxic secondary products during decontamination. Requirements to reduce the content of organic hydrocarbon determine the necessity of reducing water color less than 15 deg.

Recommended values for *turbidity* have not been established. However, the authors of [427] believe that turbidity less than 5 nephelometric units (EM/dm^3) are usually acceptable for consumers although it may vary depending on local conditions. In this case one should remember that particles determining turbidity may protect microorganisms during decontamination and thus stimulate further bacteriological growth. Therefore, the last edition of the WHO Guidelines for Drinking Water Quality indicates that for effective disinfection of water average turbidity should be <0.1 EM/dm^3. Variations of water turbidity are the most important controlled index of water quality.

The question of the optimal content of *salts of calcium and magnesium* in drinking water has been discussed. According to the opinion of the authors of [431], the most scientifically substantial research, carried out in Sweden [432], was based an extensive study of cardio-vascular diseases. In this case, any interaction between water hardness, the level of the contents of calcium or magnesium and heart disease or mortality was not detected. However, a range of people involved in the given research was rather limited. Only 7 % of the total quantity of examined people consumed drinking water with the magnesium level of more than 8 mg/dm^3 and 16 %—hard water with the content of calcium 28.5 mg/dm^3 or more.

As is noted in [431] at present there is no proof that a low level of the content of calcium and magnesium in drinking water is substantial factor, which cause cardio-vascular diseases. Nonetheless, there is noted a certain positive effect in epidemiological research about the role of hard water in reducing by 10–15 % total mortality caused by cardio-vascular diseases. If this effect takes place in reality, the minimal recommended values for the content of calcium and magnesium in drinking water could lower the risk factor in cardio-vascular diseases. However, this issue calls for further research.

By the data of [433] the optimal composition of drinking water is determined by maintaining carbon dioxide equilibrium and acceptable values of the *pH and alkalinity*. Water companies in the Netherlands recommend the following values:

- total content of inorganic carbon > 1 mmol/dm^3 is proposed so as to provide a possibility of forming a protective carbonate layer on pipe material;
- range of pH variations in the series 7.8 < pH < 8.3 is proposed in order to maintain the dissolving properties of water to be sufficiently low for such metals as Pb, Cu, and Zn. A value pH < 8.3 is advisable for limiting zinc wash-out from brass.

By the WHO recommendations for the pH values, an interval 6.5–8.0 is considered to be acceptable; this interval is determined, on one hand, by the considerations of ensuring corrosion resistance of the pipes and, on the other hand, by the conditions of effective disinfection of water.

For ammonia the threshold concentration of smell with alkaline pH constitutes ~ 1.5, while for taste—35 mg/dm^3. Normally, the ammonia content in drinking water is lower than those values which may produce a toxic effect; therefore, the maximum value for it was not set.

In addition, recommendations were given for the content of chlorides, nitrates, and sulfates in order to avoid corrosion of iron and steel. These recommendations are based on Germany's standards (DINEN 12502:2004/2005) [433]:

$$\left(\left[Cl^{+}\right]+\left[NO_3^{+}\right]+2\left[SO_4^{2+}\right]\right)/\left[HCO_3^{-}\right]<1.$$

As to the relative content in water of inorganic microcomponents an analysis of the literature data shows the following.

It is well known that the presence in water of such elements as *arsenic, cadmium, chrome, lead, and mercury* exerts a toxic effect on human health and is strictly limited. As for beryllium, it has been found that the given element and its compounds are carcinogenic for humans [434, 435]. Perhaps this is a reason why beryllium in the last edition of the Guidelines for Drinking Water Quality Control [429] was included as one of a number of impurities whose presence in drinking water is not desirable.

In the second edition of the Guidelines for Drinking Water Quality Control (1993) [81], the recommended value for manganese was 0.1 mg/dm^3, while in the third edition it increased to 0.4 mg/dm^3. However, it has been noted that an excess of the 0.1 mg/dm^3 value was not desirable in terms of beverage taste, and it also causes the emergence of spots on sanitary-engineering porcelain wares during washing. Even at a concentration of 0.2 mg/dm^3 manganese will form deposits on the pipes of distribution systems. The concentration less than 0.1 mg/dm^3 is normally acceptable for the consumer. However, a medically substantiated standard is 4 times higher than the acceptability threshold of 0.1 mg/dm^3. Therefore, it is expedient to stop at the value 0.1 mg/dm^3.

For arsenic the level 10 μg/dm^3 was set in 2001 [429]. Earlier the existing standards determined this value at the level 50 μg/dm^3 [436].

In the Guidelines for Drinking Water Quality Control (1994), the recommended level for copper is 1 mg/dm^3. However, in the third edition (2003) this value which was accepted as safe for humans was increased to 2 mg/dm^3. In the literature, opinions differ with regard to setting the recommended value. Thus, the USEPA (the US Environmental Protection Agency) has set the maximum level for the concentration of contaminants in drinking water as 1,3 mg/dm^3 [437]. The European Commission, based on the analysis of the accumulated experience, proposes a value of 1–2 mg/dm^3 [429, 437].

Concerning *hydrogen sulphide,* the perceptibility threshold of taste and smell were assessed within an interval of 0.05–0.1 mg/dm^3. It is especially perceptible in some underground water. However, hydrogen sulphide quickly oxidizes to sulfate during aeration or chlorination. The level of the negative impact of hydrogen sulfate on human health has not been set. Therefore, the recommended value according to the GOST Standard 2874–82 is absent. As for the underground water sources which quality by all indexes meets the requirements of the GOST Standard 2874–82, the standard value for sulphides is zero. The normal zinc requirement for adult human organisms constitutes 15–20 mg/day. The level of zinc content in surface and underground water normally does not exceed 0.01–0.05 mg/dm^3. However, at the concentration >3 mg/dm^3 water cannot be acceptable for consumers [429] by its organoleptic indexes. Therefore, in the column of maximum possible values one may use a value not in excess of 0.01–0.02 mg/dm^3.

Oil products impart smell to water even at very low concentrations. Formal data for setting standards for summary concentration of oil products are absent, but experience shows that a mixture of oil products drastically reduces this threshold compared with individual substances. Some hydrocarbons, especially alkylbenzenes (e.g., trimethylbenzene) may impart a "diesel-like" smell at concentrations in several μg/dm^3. It is therefore advisable to accept a maximum possible total standard for oil products at the level < 100 μg/dm^3.

Recommended values for concentrations of *pesticides* depend on their chemical nature and their toxic effects on the human organism. A choice of admissible concentrations for pesticides should be determined by the range of pesticides used in Ukraine. The list of pesticides considered by the WHO includes 33 substances. Their permissible recommended concentrations in water vary from 0.03 μg/dm^3 (for aldrin and dildrin) to 100 μg/dm^3 (for chlorpon). For DDT and its metabolites the recommended value constitutes 1 μg/dm^3. In Ukraine SanPin No. 383 for 1996 includes 10 titles of pesticides and herbicides [20].

The general class of *phenol compounds* in the Guidelines for Drinking Water Quality Control has not yet been given. However, there are data on some derivatives of phenols and chlorophenols. 2-phenylphenol is hazardous for human health therefore its presence in drinking water is undesirable. Therefore, the recommended value for it is not set. Pentachlorophenol constitutes 9, for trichlorophenol: 2,4,6–200, which makes it possible to assess maximum admissible levels <10 μg/dm³. The content of phenol in water, which is subjected to chlorination, is standardized at the level 10 μg/dm³.

Let us address in greater detail the consideration of *units of radiological and microbiological indexes* and secondary contamination of water during its decontamination.

Special attention is needed for the unit of radiological indexes of water quality in Ukraine, and this is determined by the radiological situation in Ukraine in connection with the Chernobyl disaster. The origin of radioactive substances in Ukraine may be both natural and anthropogenic. The Guidelines for Drinking Water Quality Control contains recommended levels of activity for 197 radionuclides in drinking water. However, the implementation of control of such indexes does not seem possible. Therefore, in the second and the third editions of the WHO Guidelines for Drinking Water Quality Control, the recommended level of activity dosage is set at 0.1 Sv (sievert) per 1 year of consumption of drinking water, in which case Sv is a reduced effective dose expressed in sieverts. It is a measure of the total effective dose acting within the lifetime of a human after radionuclides has penetrated his organism.

In this case, the radiation dose depends on several chemical and biological factors. They include activity, which is absorbed by tissues and other organs through which radionuclides are transported, and also time, within which radionuclides remain in organs or tissues before excretion. The nature of emission radiation and sensitivity of radiated organs or tissues to radiation should also be taken into account.

The activity concentration in water has been determined as Bq/dm³. This value may be reduced to an effective dose per 1 year (mSv/year) by using a dose coefficient (mSv/Bq) and the average quantity of water consumption per 1 year [429].

For the purposes of routine screening over the quality of drinking water in the radiological aspect, the average summary α and β activity can be used [81, 429]. An increase of summary α and β activity respectively >0.1 and 1 Bq/dm³ indicates a growth in the concentration of individual radionuclides in drinking water.

It should be noted that a contribution to the total level of radiation received by the Earth's population from nuclear weapons in 2000 was assessed as 0.005. Likewise, the level of radiation from the Chernobyl disaster was assessed at 0.0002, and the production of atomic power was assessed at 0.0002 Sv [429]. Despite a relatively small contribution from the Chernobyl disaster, its after-effect should be taken into account for assessing the quality of drinking water in Ukraine. Therefore, it is advisable that the indexes of radiological aspects, along with general indexes, include normalization by individual radionuclides, whose presence in water may be expected due to the Chernobyl disaster, namely: ^{90}Sr; ^{137}Cs; U; ^{226}Ra; ^{3}H.

Apart from this, it is necessary to take into consideration that a small share of natural radiation is related to ^{222}Rn. Therefore, normalizing the content of the given radionuclide in drinking water is expedient if this is determined by the specific feature of the region. It is believed to be impossible to set a suitable concentration of radon in drinking water. However, paper [429] recommends monitoring radon content in a centralized water supply if its level exceeds 100 Bq/dm^3. In this case, measures should be taken to reduce the level to less than 100 Bq/dm^3. If in the geographic locality near large water sources contain a great number of radon-producing minerals, then at large water stations one should carry out an assessment of the radon content at least periodically (e.g., every 5 years).

The list of pathogens of water-borne diseases which present serious hazards for population health is given in Table 4.14. If a number of epidemics of bacterial origin, caused by such bacteria as cholera and salmonella, decreases, the situation with enteropathogenic microorganisms (e.g., *E. coli, Mycobacterium avium, Helicobacter pylori, Campylobacter jejuni*) does not effectively improve. Especially dangerous are enumerated bacteria for humans with weakened immunity or with a syndrome of immunodeficiency. This group of microorganisms includes *Pseudomonas aeruginosa,* species of genera *Flavobacterium, Acinetobacter, Klebsiella, Serratia, Aeromonas,* etc.

Our knowledge about pathogens of diseases transferred via water is constantly being refined. Thus, it has been found that in water, especially near water banks, one encounters *Mycobacterium xenopi*, which may pass through water treatment facilities at water stations and be a cause of pulmonary diseases like pneumonia and tuberculosis. Lately, the cases of detection of such microorganisms as *Norwalk virus, Legionella, Giardia, Cryptosporidium* in water have become frequent. According to the data of [438], a main representative of enteric protozoans is *Cryptosporidium parvum,* which are found in rivers and lakes and may get into drinking water, resulting in the outbreak of epidemics. In the US *Cryptosporidium* is considered a main threat to the water supply due to its high rate of infectiousness, resistance to chlorine and small size, which makes the removal of the given microorganism during water filtration difficult [438]. Even the most advanced technology of water treatment cannot completely guarantee the absence of these protozoans in water. Therefore, for the first time in the US an index of the presence of *Cryptosporidium parvum* has been introduced into drinking water standards.

Cysts of lamlias and oocysts of cryptosporidia possess a higher resistivity to the effect of disinfectants, compared with bacteria and viruses, and may be transferred via drinking water which meets the standards of bacteriological indexes. Microbiologists and parasitologists also identified other related microorganisms.

In recent years, the information about contamination of drinking water with various types of fungi has appeared in the literature [439]. The role of fungi, including both micellar and yeast, in contamination of the water distribution system, however, has not yet been effectively studied. As the investigations conducted in the Institute of Colloid and Water Chemistry, National Academy of Sciences of Ukraine have shown, mesophilic fungi, more often than not of general *Apergillus* and *Penicillium,* are found in drinking water. As for natural water, the spectrum of isolated fungi is wide and depends on the time of year and the sampling place (see Table 4.15).

In the US 50% of the studied samples of drinking water from the distribution system contained genera of fungi: *Alternaria, Apergillus, Penicillium and Cladosporium.* The fungi enter the distribution system as a result of soil contamination.

Table 4.14 Water-borne disease agents

Agent	Disease	Survival rate in water[a]	Resistance to chlorine[b]	Relative infectious dose[c]
Bacteria				
Campylobacter jejuni	Gastro-enteritis, colitises	Average	Low	Average
E coli	Ditto	Ditto	Ditto	Ditto
Salmonella typhi	Salmonolloses	Short	Ditto	Ditto
Shigella spp.	Dysentery	Ditto	Ditto	Average
Vibrio cholerae	Cholera	Ditto	Ditto	High
Yersinia enterocolitica	Coliform yersinioses	Continuous	Ditto	Ditto
Pseudomonas aeruginosa	Purulent disease	Short	Average	Ditto
Aeromonas spp.	Ditto	Ditto	Low	Ditto
Legionelles	Pneumonia	Ditto	Ditto	Ditto
Mycobacterium xenopi	Pneumonia, tuberculosis	Continuous	High	Ditto
Viruses				
Adenovirusis	Keratoconjuctivital fever, adenoviral pneumonia, ARD	Undefined	Average	Low
Enteroviruses	Serous meningitis, poliomyelitis, etc.	Continuous	Ditto	Ditto
Hepatites A	Hepatitis	Undefined	Ditto	Ditto
Vorvalk virusis	Gastroenteritis	Ditto	Undefined	Average
Rotavirusis	Ditto	Ditto	Ditto	Ditto
Fungi				
Aspergillus	Aspergilloses, myco-toxicoses, allergosses	Average	Average	High
Protozoans				
Entamoeba histolilica	Amebic dysentery	Ditto	High	Low
Giardia itestinalis	Giardiases	Ditto	Ditto	Ditto
Cryptosporidium parvum	Cryptosporidioses	Continuous	Ditto	Ditto
Helminthes				
Dranculus medinensis	Helmithiases	Average	Average	Average

[a] survival rate in water at 20 °C: short one—up to a week, medium—from one week to one month, continuous one—over one month;
[b] resistance to chlorine—a high one when chlorine has no effect whatsoever, a medium one—when pathogenic agents may be incompletely inactivated when treating water with traditional doses and durability, a low one—when pathogenic agents are completely inactivated;
[c] a relative infectious dose—a dose necessary for infecting 50 % of healthy volunteers. Maybe very low when the infectious dose of some viruses equals unity

Among the predominant isolators are *A.flavus, A.niger* and *A.fumigatus*, which are producers of toxic agents which are conducive to cause, apart from mytotoxicosis, aspergilloses and allergoses. It was found that the spores of aspergilles may flow into the respiratory track by inhaling an aerosol, for example, from the shower head, causing pulmonary infection.

Table 4.15 Variations of the fungi specific composition in the soil of Kiev's city beach[a]

March	April	May	June	July	August
Aspergillus Niger	*Aspergillus niger*	*Aspergillus Niger*	*Aspergillus niger*	*Aspergillus niger*	*Aspergillus niger*
Penicillium sp.	*Fusarium sp.*	*Asp. versicolor*	*Asp. nidulans*	*Asp. flaws*	*Asp. Flavus*
	Penicillium cyclopium	*Penicillium chrysogenum*	*Asp. fumigatus*	*Asp. oryzae*	*Asp. Terreus*
	P. decumbens	*P. funiculosum*	*Gliocladium sp.*	*Asp. ochraceus*	*Torulopsis glabrata*
	P. funiculosum	*P. cyclopium*	*Paecilomyces varioti*	*Asp. nidulans*	*Candida albicans*
	Trichoderma viride	*P. expansum*	*Penicillium cyclopium*	*Aureobasidium pullulans*	*Paecilomyces Varioti*
		P.sp.	*Penicillium sp.*	*Allernaria sp.*	*Asp. Nidulans*
		P. frequentans	*Trichoderma viride*	*Cladosporium cladosporioides*	*Asp. Candidus*
			Geotrichum	*Penicillium frequentans*	*Penicillium expansum*
			Candida albicans	*Penicillium sp.*	*Penicillium sp.*
			Trichphylon rubrum	*Scopulariopsis brevicaulis*	*Cladosporium sp.*
				Candida albicans	*Trichphylon rubrum*
				Saccharomyces sp.	
				Trichphylon rubrum	
				Microsporium cariis	
				Epidermophyton sp	
				Trichothecium roseum	

[a] The samples were taken at the distance of 5 m from the tide band at the depth of 5 cm

It became clear that the spores of fungi are a main component of biofilms on the surface of pipelines. Fungi may survive after normal water treatment at water stations concurrently causing corrosion of pipelines and adversely affecting organoleptic indexes of drinking water.

Thus, deterioration of water–supply sources and isolation of new pathogens, reducing the population's immunity, add urgency to the problem of water disinfection; therefore, effectively all countries of the world initiate and continue to implement strict monitoring of and set stricter requirements for the drinking water quality by microbiological indexes (see Table 4.16).

Table 4.16 International standards or guidelines in microbiology and parasitology of drinking water [427, 439, 440]

Type of microbiological contamination	Canada	USA	Australia	New Zeeland	Great Britain	EU guidelines	Russian Federation	Ukraine	WHO
Total coliform bacteria	0/100 cm^3 in 90 %	0/100 cm^3 in 95 %	0/100 cm^3	–	0/100 cm^3	–	0/100 cm^3	Not more than 3 in 1 cm^3	–
Thermotolerant bacteria	–	0/100 cm^3	0/100 cm^3 in 98 %	–	–	–	0/100 cm^3	0/100 cm^3	0/100 cm^3
E. Coli	0/100 cm^3	0/100 cm^3	0/100 cm^3	0/100 cm^3	0/100 cm^3	0/100 cm^3	–	–	0/100 cm^3
Enterococci	–	–	–	–	0/100 cm^3	0/100 cm^3	–	–	–
Cryptosporidium parvum	–	99 %-removal or absence	–	–	<1 *cysts* per 10 cm^3	–	–	Absent in 25 cm^3	–
Clostridium perfringens, including spores	–	–	–	–	0/100 cm^3	0/20 cm^3	0/20 cm^3	–	–
Pseudomonas aeruginosa	–	–	–	–	–	0/250 cm^3	–	–	–
Total microbial number at 22 and 37 °C	–	–	–	–	Without anomalous changes	Not more than 20 in 1 cm^3	Not more than 50	Not more than 100 for 55 % of samples	–

Table 4.17 Chemical disinfectants and main by-products, which are formed during disinfection

Disinfectant	Main by-products of disinfection
Gaseous chlorine	THM and other haloorganic matter
Hypochlorite	Ditto, and also bromates, chlorates
Chlorine dioxide	Chlorites, chlorates
Ozone	Oxyorganic compounds, bromates
Chloramine	Nitrites, cyanogen chloride

One of the central issues of modern technologies for drinking water treatment has been determined by the necessity of a reliable water disinfection, which results in the emergence of new disinfection by-products whose qualitative and quantitative composition is determined by a type of a disinfectant. The composition can also be determined by monitoring a content level in the water being disinfected of organic matter. All this in great measure determines drinking water quality.

The most widely used disinfectants for a water supply are chlorine and its derivatives, as well as ultraviolet light. This is related to the fact that chlorine is comparatively cheap, does not cause large difficulties when being used and possesses a broad spectrum of antimicrobial effects. However, the main advantage of chlorine, compared with other disinfectants, lies in the presence of an after effect which causes water to remain disinfected when transported through pipelines over great distances. Nonetheless, there is a tendency to reduce its consumption, especially in the Netherlands, Germany and Denmark. Chlorine dioxide is most intensively used in Germany and Italy, and to a lesser degree in France and the Netherlands [441–444].

However, the presence of a broad spectrum of organic matter both of natural and anthropogenic origin in water results in the chlorination of a number of organic compounds. Recently, numerous data from various biological tests [445] have been accumulated which convincingly indicate that chlorination of drinking water causes high levels of mutagenic activity and toxicity. By-products which possess high genotoxicity, such as trihalomethanes (THM), chloroform, *n*-nitrochlorobenzene and bromoform, have been detected and isolated. Additionally, a new by-product has been isolated and identified, which possesses a powerful mutagenic effect: 3-chloro-4(dichloromethyl)-5 hydroxy-2(SH)-furanon and its geometric isomer. Trihalogenmethanes and carbon tetrachloride belong to basic of 24 halogenorganic compounds that are formed in drinking water after chlorination and are considered as substances presenting a hazard for human health.

With the goal of reducing the amount of by-products, a proposal was made to use other chlorinating reagents such as, for instance, chloramines and recently chlorine dioxide [446–448]. Chloramine possesses a lower bactericidal effect, just as gaseous chlorine prevents the propagation of microorganisms in water system networks and better penetrates biofilms. In this case a, lower amount of organochlorine compounds is formed, but among them highly toxic compounds are encountered, in particular, chloride cyanogen [449].

Chloramine is used for disinfection of distribution systems in Estonia, Finland, Great Britain, Sweden, and Ukraine [442–444]. Table 4.17 gives examples of by-products formed by different methods of water disinfection [444]. As for chlorine dioxide, its activity is much higher than that of other chlorine-containing disinfectants. Among toxic substances, mainly chlorites and chlorates, and in rare cases bromates, are formed.

Thus, chlorine and its preparations are most expedient for use in the transportation of treated (mainly from organic matter) drinking water with a low residual concentration of chlorine (0.3–0.5 mg/dm^3). As a number of countries have demonstrated (Japan, Sweden, Great Britain, Germany, Denmark, to name a few), during treatment and disinfection of water prior to its transportation via pipelines there is a tendency to use chlorine dioxide, ozone and UV-radiation.

As an alternative to chlorination, the method of water ozonation has been introduced. This method is widely used at present all over the world, Ukraine included. Ozonation makes it possible to substantially improve drinking water quality by reducing its color, smell, taste, oil products and SAS, among other characteristics. Also, the efficiency of coagulation and filtration substantially increases with this process. In this case, the safety of drinking water in the sanitary-hygienic respect is secure.

The principal limited use of ozone as a disinfectant is related to its instability since fast decomposition of ozone in water cannot guarantee its residual concentration in the distribution system. Normally, after ozonation, water is treated with chlorine or chlorine dioxide for final disinfection. In addition, under the effects of chlorine, biodegradable organic compounds are formed in water, which results in the appearance of the secondary growth of microorganisms. This occurs especially in the distribution system and requires additional chlorination. Among side toxic organic compounds during ozonation, oxyorganic compounds and bromates have been detected during ozonation.

The use of UV disinfection does not result in the formation of organochlorine products and bromates. A high disinfection effect of UV-radiation has been observed with respect to pathogenic microorganisms (bacteria, viruses, spores, and even to protozoans (*Giardia lamlia cysts, Cryptosporidium parvum oocysts*) and also regarding many ozone-resistant microorganisms [450, 451]. However, it has also been noted that UV-radiation, even at the level of dosage necessary for decontamination (400 J/m^2), may lead to transformation of organic chemical components in water. Particularly, natural organic compounds (NOC) [452] could result from this form of radiation. Paper [453] showed that preoxidation of water with ozone leads to lower THM content after chlorination, while UV-radiation does not affect the formation of THM in water.

Ultraviolet radiation, just as ozone, does not ensure the aftermath effect therefore chlorination is required as a final stage of water conditioning.

Taking into consideration the use of various methods of disinfection of water in Ukraine at water treatment stations of different capacity, it is expedient to include all main by-products of decontamination in a number of maximum admissible controlled indices of drinking water quality after various types of disinfectants.

It is known that the following by-products are volatile compounds, considered potentially carcinogenic and mutagenic [453]: chloroform, bromodichloromethane, dibromochloromethane, bromoform (THM), benzene, trichloroethane (TCE), tetrachloroethane (RCE), and 1,2 dichloroethane (1,2 DCE). These compounds, which present substantial health risks for humans, are priority in the EU and are included into the EC Council Guidelines 98/83/EC [441], which set the maximum level of contaminants in drinking water: 100 μg/dm^3 for THM, 10 μg/dm^3 for the sum of TCE and RCE, 3 μg/dm^3 for 1.2 DCE and 1 μg/dm^3 for benzene. Chloroform is a main component of THM and constitutes ~70 % of its total content [454].

Though the EC Guidelines do not require the detail deciphering of decontamination by-products according to individual chemical compounds, many of the EC countries, except France, Portugal, and Spain, introduced their own set of by-products for national standards [455]. Table 4.18 summarizes standards or guidelines for the content level of these products, whose main group is THM, in different countries. The content of the latter varies from 10 to 100 μg/dm^3 for countries which have standards for individual substances.

In addition to THM, which is considered when assessing the quality of drinking water, it is also essential to take into account the formation of haloidacetic acids (HAA), due to chlorination. These acids form in water with a low value of the pH and a high level of total organic carbon (TAC) [448]. Research into the formation of HAA in water with an increased content of humic substances and high color showed that after chlorination the concentration of HAA may reach 120 μg/dm^3. From this amount, 44 % possess dichloroacetic acid, and 4 % possess trichloroacetic acid. The highest HAA concentration is noted during water's period of maximum temperature when the content of NOC is also the highest [456]. In [105] the value for a sum of five haloidacetic acids is regulated at the level 60 μg/dm^3.

Identified THM and HAA include ~50 % of the total content of haloidorganic products that are formed during chlorination [457]. Other remaining unidentified haloid-containing compounds are also hazardous for human health.

In addition, it is worthwhile to set the maximum admissible content in water after disinfection of chloride cyanogen since many water stations in Ukraine use chloroamination.

The most stringent standards in terms of the content of disinfection products in drinking water have been set in the US and the Netherlands. The total content of disinfection products in Europe varies within the range of 30 to 100 μg/dm^3. The WHO proposes to standardize disinfection by-products by the sum of relative values of chloroform (C_{cf}), bromoform (C_{bf}), dibromochloromethane (C_{dbcm}) and bromodichloromethane (C_{bdcm}), which should not exceed 1:

$$\frac{C_{cf}}{C_{chlor,recomm}} + \frac{C_{bf}}{C_{brom,reccom}} + \frac{C_{dbcm}}{C_{dbcm,recomm}} + \frac{C_{bdcm}}{C_{bdcm,reccom}} \leq 1$$

The list of water quality indexes for the sources of centralized water supply in compliance with the [30] includes only dibrochloromethane and trihalomethanes of three compounds, which are identified as disinfection by-products in standards of the WHO, the EU and the US. In a new law of the Russian Federation on a special technical regulation for the requirements of sanitary-epidemiological safety for water, intended for consumption by human and drinking water supply [458], the list of by-products of disinfection includes 6 indexes: chloroform, formaldehyde, bromodichloromethane, bromoform, dibromochloromethane and 2,4,6-trichlorphenol.

Table 4.18 Maximum admissible concentrations (MAC) of trihalomethanes in drinking water in international and national standards and documents [448]

Country	THM, μg/dm³	Country	THM, μg/dm³
WHO	[a]	Ireland	100
USA, USEPA	80	Luxembourg	50
EU (EU Guidelines 98/83)	100	Norway	100
Italy	30	Sweden	50
Germany	50	Switzerland	25
Spain	100	England and Scotland	100
Austria	30	Greece	100
Belgium	30	Israel	100

[a] In compliance with the WHO recommendations the sum of ratios of each component (chloroform, bromoform, dibromochloromethane, and bromodichloromethane) to the recommended value should not exceed 1

Table 4.19 Maximum admissible concentrations of by-products in drinking water of different countries [20]

Country	By-products	MAC, μg/dm³
Austria	Chlorites	200
Finland	Bromodichlormethanes	60
	Chloroform	200
Germany	Chlorites/chlorine dioxide	200
Netherlands	Bromates	0.5 (without disinfections) 5 (during disinfections)
Great Britain	Chlorites/chlorine dioxide	500

Analyses of Tables 4.18–4.20 show that the list of indexes of the new standard for drinking water in Ukraine includes four main indexes for disinfection by-products [429]. The use of the recommended WHO Guidelines for Drinking Water Quality Control and the rules for the sum of the relative content of four main components of THM <1 make it possible to assess the maximum admissible summary content of THM in drinking water at the level 100 μg/dm³, accepted in most EU countries. These compounds have been studied among other disinfection products and represent the most typical trihalomethanes. They are formed during the reaction of hypochlorous acid (HClO) with NOC in the presence or absence of bromides. The concentration of THM formed depends on such factors as chlorine dosage, concentration of bromides and ammonium in the water treated, as well as the pH, temperature, content, and type of NOC [459–464].

Table 4.19 gives individual indexes of maximum admissible concentrations (MAC) of other by-products for different countries. Table 4.20 shows recommended maximum values for a great number of by-products, which are monitored in the Netherlands, the US, Russia and are recommended by the WHO.

Thus, as can be seen from quoted and analyzed data, the recommended set of quantitative quality indexes of physiologically wholesome drinking water and their maximum admissible values as of today may be supplemented by the data of the four units.

Table 4.20 Recommended maximum admissible concentrations of by-products in drinking water

Products of disinfection	Recommended standard, μg/dm³					
	EU [455]	WHO		Netherlands [455]	USA[a] [448]	Russia [430]
		1993 [81]	2004 [429]			
Bromates	10	25	10	5	–	–
Chlorite	–	200	700	200	800	–
2,4,6-trichlorophenol	–	200	200	–	200	4
Formaldehyde	–	900	900	90	–	50
Bromoform	Sum of concentrations 100	100	100	5	0	100
Bromodichloromethane		60	60	6	0	30
Dibromochloromethane		100	100	5	60	30
Chloroform		200	200	5	0	60
Dichloroacetic acid	–	50	50	50	0	–
Trichloroacetic acid	–	100	200	–	300	–
Chloral hydrate	–	10	10	10	–	–
Dichloroacetonitrile	–	90	20	90	20	–
Dibromoacetonitrile	–	100	70	100	70	–
Trichloroacetonitrile	–	1	[b]	1	–	–
Cyanogen chloride (as CN)	–	70	70	–	–	–

[a] For the USA maximum purposive levels of contaminants in drinking water are given. These indicators are close to the values of maximum admissible concentrations, which are used at present, however with the account of the use in the future of the best technologies of water conditioning and its cost

[b] Accessible data are inadequate in order to set a medically substantiated recommended value

Table 4.21 Residual doses of disinfectants in drinking water after disinfection

Disinfectant	Disinfectant residual doses, mg/dm³						
	USA [447]	Russian Federation [434]	WHO [429]	Ukraine [440]	Germany [26]	Finland [26]	EU [454]
Chlorine							
Residual	4	0.3–0.5	0.5	0.3–0.5	0.3	sum 1.0	–
Free	–	–	–	–	–		–
Chloramine (bound chlorine)	4	0.8–1.2	3	0.8–1.2	–	–	–
Chlorine dioxide	0.8	–	Undefined	–	0.2	–	–
Residual ozone	0.05	0.3	0.5	0.1–0.3	0.01–0.05	–	–

Main residual doses of disinfectant in drinking water are given in Table 4.21.

The first unit represents those impurities, which are most characteristic of Ukraine's natural water and are included in the list of indexes of the new State Standard "Sources of the Centralized Drinking Water Supply". Quantitative values of these indexes for the 1st class of quality of natural water are close to characteristics of the natural state of Ukraine's ecosystems and are in good compliance with the WHO recommendations.

The second unit of indexes is related to the presence of secondary impurities in drinking water due to its treatment and disinfection.

The third unit includes the characteristics of drinking water quality by microbiological, virusological and parasitological indexes. Quantitative values of these indexes should be chosen with the account of available experience in Ukraine and the WHO Guidelines for Drinking Water Quality Control.

Finally, the fourth unit characterizes radiological aspects of water quality indexes.

It should be noted that the number of parameters and their quantitative value is reviewed every several years (in different countries and in different ways) with the account of changes in scientific knowledge, social, and economic policy.

Thus, the EU countries recommended Guidelines 80/778 EU. Based on new scientific data and increased possibilities of the EU countries, these Guidelines were reviewed and the EU Council Guidelines 98/83/EC were recommended. However, at present they have only started to consider new aspects of the Guidelines.

This new perspective follows the latest scientific information on many parameters of water quality. In particular, attention is paid to a link between hardness of drinking water and cardio-vascular diseases [465] and cancer-caused mortality [466]. A generalization of the European Commission data over the period from 1993 to 2005 demonstrated that such chemical parameters as fluorine (of natural origin) and nitrates should be priority problem parameters in Europe. Other characteristic and frequently discussed parameters are arsenic, iron, magnesium, pesticides, trihalomethanes, lead, nickel and nitrites. Attention to the content of magnesium in drinking water was drawn after investigations which showed a protection effect of magnesium consumption from drinking water against cerebrovascular diseases [467]. However, the world literature does not contain any recommendations as to a medically justified standard of magnesium content in drinking water. A special concern has arisen as a result of radioactive contamination of water [465].

In some regions, requirements for drinking water quality are substantially stringent compared with the one recommended by the WHO. Thus, Taiwanese Environmental Protection Administration plans to review standards for drinking water quality for hardness and dry residue in the foreseeable future, having reduced hardness (in terms of $CaCO_3$) from 400 to 150 mg/dm^3 and dry residue from 600 to 250 mg/dm^3 [467]. These standards are more stringent than those of many developed countries (such as the US, Japan, Australia, Canada and France) and the WHO's recommendations.

Some countries guided by the WHO recommendations and EU Guidelines have introduced a relatively large quantity of new parameters to their national standards. In Denmark the amount of added indexes equals 19, in Poland—16, in the Netherlands—15, in Greece—11, in Slovenia—13, in Belgium—12, in the Czech Republic—15. China has increased the number of indexes of the national standard from 35 to 106 [468]. Most often they add or review such parameters as magnesium, total hardness, phenols, zinc, phosphates, potassium and chlorites. More stringent requirements have also been placed on the content of lead, THM and bromates [465]. In 2007, the WHO published a supplement to the third edition of the Guidelines for Drinking Water Quality Control in terms of the content of nickel in drinking water [469].

Special attention in the discussions of recent years has been paid to microbiological aspects. Paper [438], for example, noted that in the US for characterization of water safety in terms of enteroprotozoans they use a definition for the absence of the content of *Cryptosporidium parvum* as a main hazard for water supply. In [470] in was found that prior to 1982 registration of diseases was determined by the presence in water of this type of protozoans. However, as the incidence of AIDS increased, the probability and the number of such diseases increased. If earlier infections which had been caused by animalcular microorganisms had affected humans with a suppressed immune system then, as diagnostic methods were improved, the epidemic outbreaks and other incidents related with contamination via water would have appeared among the healthy population as well [438].

In the European Microbiological Consultative Group, the microbiologists advise the EU Commission to review the EU Guidelines 98/83 with respect to microbiological parameters and analytical methods (such, for example, a coliform method). Much discussion has been caused by parameters such as *Clostridium perfringens* [465]. Therefore, in the future it will also be requisite to review microbiological aspects of drinking water quality in Ukraine, being mindful of the epidemiological situation in the country in terms of incidence of tuberculosis and AIDS.

Special attention when doing further perfection of the National Standard for drinking water should be drawn to the development of the methods of rapid biotesting of water toxicity.

The presence in water of the ascending amount of chemical compounds of anthropogenic origin presenting hazard for human health (including by–products of pharmaceutical industry, household chemistry agents and cosmetology) determines the necessity of a specific search and introduction into practice of modern and cost-effective methods of testing xenobiotics with the use of the relevant animal and plant objects [471].

At present, the testing of various laboratory mammal animals (rodents, dogs and primates) remains a main regulated test. At the same time, there are several reasons which have stimulated the interest of researchers in the last decade to explore alternative methods of biotesting toxicity of new or already known compounds. First of all, as has been pointed out, this is related to an increase of the variety of chemical substances produced and the necessity of determining the whole range of indexes of their effects on the human organism. This, in its turn, results in a substantial increase in costs of toxicological research. Thus, according to calculations, the cost of creating a complete toxicological framework of data only by a group of 12,860 compounds whose production volume exceeds 450,000 tons a year (the US, the data of 1994) constitutes more than $ 10 billion and will require the use of almost 11 million animals [472].

The time for conducting the experiments when revealing acute toxicity varies from 7–14 days and to 6–8 months when determining chronic toxicity [473]. Taking into consideration necessary conditions and possibilities of qualified personnel, the real amount of simultaneously tested toxic substances remains limited [474]. It is difficult to obtain information about the action mechanism of the given substance on a target cell (which constitutes one of the requirements of modern toxicology) [475]. It should also be taken into account that sensitivity of standard tests on laboratory animals may be insufficient for detecting tiny amounts of xenobiotics whose concentration normally is lower by the order of one or two magnitudes compared with registered doses, thus causing a minimal toxic effect.

Moreover, the ever increasing attention in recent years has been drawn to the issue of ethic expediency concerning the large scale use of animals for determination of a toxicological risk of chemical compounds produced by industry. An obvious solution for these issues lies in extending the range of test objects used and, accordingly, in the justification of new biomarkers intrinsic to these objects. Such justifications should include indexes which objectively characterize normal or pathological changes of biological processes and a response of the organism to external effects [476].

The range of similar research encompasses macromolecular complexes, isolated cells, and multicellular aggregates, prokaryotic organisms, protozoans, plants, invertebrate and vertebrate animals. One may underline the following two main areas of research: creation and use of test systems, which are biological "sensors" signaling the human health hazard posed by the presence of a given chemical compound in water; the development of test systems making it possible to obtain reliable information about the avenues for affecting the organism of xenobiotics during its ingress with drinking water.

Testing methods relating to the first group should correspond to the following requirements: to possess high sensitivity to a broad spectrum of substances, to be sufficiently simple and fast in perfection, and also to require minimal costs for their application and adoption. Such methods include tests for luminescent bacteria (natural strains *Vibrio fisheri* and genetically modified *E. coli.*), algae (chlorella), plants (a generally accepted test for toxicity using onion), invertebrate animals (amoeba, hydra, daphnia, ceryodaphnia) and fish (trout, gold fish).

The test results of a group of standard toxic substances (e.g., on onion) may well correlate with the data of epidemiological and clinical observations despite a principle difference in physiology and metabolism of chosen test organisms and human. This may be explained by the effect of a toxicant on universal mechanisms which determine the vitality of cells, such as integrity of a plasmatic membrane, activity of enzymes of mitochondria, etc. In addition, an increase in the number of test organisms (creation of a battery of tests) reduces the probability of emergence of a false-negative result related to a possibility not to register potential toxicity of the given compound [477].

At the same time, it should be taken into consideration that such sensor organisms are not intended for obtaining information about toxic dynamics of the substance studied. The effects of directed action xenobiotics (e.g., medicinal preparations) on human cells and organisms taxonometrically remote from it may differ. To determine the action mechanism of xenobiotics, there are methods based on the use of cells obtained from human and other mammals [478]. Such systems, as a rule, are related to "cultivation", i.e., to adaptation of living cells with artificial creation of conditions outside the organism. The main requirements for the cells is the manifestation of an adequate reaction to an external effect typical of tissues–targets and the possibility of its objective assessment by analytical (first of all, quantitative) methods.

Of special interest are human cells obtained from boundary tissues contacting factors of the external and internal medium and also from organs implementing biotransformation and extraction from the organism of toxic substances. The previously mentioned types include cells of skin, blood vessels, liver and kidneys. Fatty cells may also present interest since a fatty tissue is capable of depositing toxic substances. Also to be considered are the cells of the secretarial epithelium (e.g., endocrinal glands, taking into consideration the issue of detecting in water of compounds violating the function of the endocrinal system).

Interpretation of the data obtained usually does not cause any difficulties. The question lies in correct interpolation of the results obtained using cellular models to the whole organism. This is related not only to the correct choice of the system response indexes (the so-called end–points) but also to the chemical properties of the xenobiotic itself; first of all, its molecular weight and hydrophobicity should be taken into account. With the goal of solving the above mentioned issue, mathematical models have been developed. These models account for possible toxicokinetics of the given compounds in the organism and make it possible to introduce relevant adjustment coefficients when considering its toxic effect [479]. Effective concentrations of xenobiotics causing half of the maximum effect determined for different types of tissue may be used when computing a simple averaged index (e.g., an arithmetic mean of the obtained values).

As can be seen above, the modern methods of biotesting offer a toxicologist working in the field of assessing drinking water quality a wide set of means for the investigation of the effects of xenobiotics on the human organism. Each of the approaches has its own area of optimal application. Thus, finding a toxic effect by means of microorganisms or a relatively simply "battery" of animals hydrobionts may signal a hazard of the presence of the given substance in an aqueous medium. The research carried out on the cells of mammals requires a more complex methodological backing, but their results may be simply extrapolated to the human organism. In turn, the research on subcellular structures may directly point to the mechanism of a possible effect of the given toxic substance on a target tissue. Thus, the integral approach to biotesting seems to be the most promising since it includes all enumerated methods. Together with the data of analytical chemistry and results of traditional toxicological tests, biotesting will secure the obtaining of sufficient information on the presence in drinking water of chemical compounds presenting hazard for human health.

When conducting a preliminary toxicological analysis of xenobiotics present in drinking water, one may recommend the use of a series of biotests with the use (as test objects) of fish and cell culture of animal mammals and humans. Such an approach will secure high sensitivity and a broad spectrum of the potential toxicants that have been found.

So, to summarize, all considered aspects characterizing requirement or sanitary-epidemiological safety of drinking water make it possible to recommend further discussion concerning a list of sanitary-hygienic standards and their quantitative values for physiologically wholesome water and maximum admissible values of drinking water quality indexes in compliance with the modern state of the art in water treatment.

References

1. The Water Code of Ukraine (1995) N 213/95-VR. Off J Verkh Rad of Ukraine 24:190
2. The Law of Ukraine N 2918-III (2002) Drinking water and drinking water supply. Off J Verkh Rad of Ukraine 16:112
3. Principles of Ukraine's and the European community's nature conservation legislation: water resources (1997) Kiev, Ministry of Ecological Safety of Ukraine
4. Ukraine's Ministry of Ecological Safety and the U.S. Center of Policy on Protection of the Atmosphere (1996) Various approaches to ecological management: short course on the practice of assessing risk, establishment of ecological standards and development of programs for reducing pollution of air and water. Kiev, Ministry of Ecological Safety of Ukraine
5. Romanenko VD, Zhukinskii VN, Okiyuk OP (1999) Methodological background for establishment and application of the ecological normative of the quality of surface water. Hydrobiol J 35(3):3–14
6. Amended proposal for a council directive establishing a framework for community action in the field of water policy (1998) Addendum to Document 9265/98 END 258 PRO-COOP 91. EU, Brussels
7. Commission proposal for a Council Directive Establishing a Framework for a Community Water Policy (1996) Consultation draft.
8. Directive 2000/60/EC of the European Parliament and of the Council (2000) Establishing a framework for community action in the field of water policy, EU, Brussels
9. USA Environmental Protection Agency (1995) Final water quality guidance for the Great Lakes system (40 CFR Parts 9, 122, 123, 131 and 132 (FRL-5173-7) RIN 2040-ACO8). Jackson Blvd, Chicago
10. Georgia Department of Natural Resources Environmental Protection Division (1997) Rules and regulations for water quality control Ch. 391-3-6. Atlanta, Georgia
11. A series of publication on water problems No1 (1993) Protection of water resources and ecosystems (ECE/ENVNA/31) prepared by the European Economic Commission in Geneva under the UN aegi, New York
12. Resolution of Ukraine's Cabinet of Ministers (1997) On measure for stage-by-stage introduction in Ukraine of the European Union's guidelines, sanitary Ecological, veterinary, phytosanitary standards, international and European standards No 244, 19 March 1997
13. Zhukinskii VN, Okiyuk OP, Oleinik GN (1981) Principles and experience of constructing the environmental classification of surface water quality. Hydrobiol J 17(2):38–49
14. Zhukinskii VN, Okiyuk OP (1983) Methodological foundations of ecological quality classification of surface water. Hydrobiol J 19(4):59–67
15. Romanenko VD, Zhukinskii VN, Okiyuk OP (1998) Metodyka ekolohichnoï otsinky yakosti poverkhnevykh vod za vidpovidnymy katehoriyamy (Methods of ecological assessment of the quality of surface water according to relevant categories). Sumbol-T, Kiev
16. Romanenko VD, Zhukinskii VN, Okiyuk OP et al (2001) Methods of establishing and use of ecological standards for the quality of land surface water and estuaries of Ukraine. VIPOL, Kiev
17. Romanenko VD, Okiyuk OP, Zhukinskii VN et al (1990) Ekologicheskaya otsenka vozdeystviya gidrotekhnicheskogo stroitel'stva na vodnyye obyekty (Ecological assesment of hydrotechnical construction impact on bodies). Nauk Dumka, Kiev
18. SaNPiN USSR 4630-88 (1988) Sanitary rules and standards for protection of surface water from pollution. Annex 1: hygienic requirements for water composition and properties of water bodies at points of economic-drinking and recreational and household use. Annex 2: sanitary standards of maximum permissible content of harmful substances in water of water bodies of economic-drinking and recreational and household use (MPC and ODU). Annex 3: hygienic classification of water bodies in term of pollution degree. Ministry of Public Health USSR, Moscow

19. State Standard 2874-82 (1982) Voda pit'yevaya. Gigiyenicheskiye trebovaniya i kontrol' za kachestvom (Drinking Water. Hygienic requirements and quality control). State Standard of USSR, Moscow
20. State Sanitary Rules and Standards-383 (1996) Voda pytna. Hihiyenichni vymohy do yakosti vody tsentralizovanoho hospodars'ko-pytnoho vodopostachannya (Drinking water. Hygienic requirements for water quality of centralized household and drinking water supply). Ministry of Health, Ukraine
21. SaNPiN 2.1.4.559-96 (1996) Sanitarnyye pravila i normy. Pit'yevaya voda i vodosnabzheniye naselennykh mest. Pit'yevaya voda. Gigiyenicheskiye trebovaniya k kachestvu vody tsentralizovannykh sistem pit'yevogo vodosnabzheniya. Kontrol' kachestva (Sanitary Rules and Standards. Drinking water and water supply of residential areas. Drinking water. Hygienic requirements for water quality of centralized system of drinking water supply). State Comm for Sani and Epidem Surveill of the Russ Federation, Moscow
22. Russian Federation Committee on Fishery (1995) Perechen' predel'no-dopustimykh kontsentratsiy i oriyentirovochno bezopasnykh urovney vozdeystviya vrednykh veshchestv dlya vody rybokhozyaystvennykh vodoyemov (List of maximum permissible concentrations and roughly safe level of exposure to harmful substances in water of fishery ponds). Medinor, Moscow
23. National primary drinking water regulations (2012) http://water.epa.gov/drink/contaminants/index.cfm. Accessed 5 June 2012
24. Directive of the EU 98/83/EU (2001) Drinking water. Regulatory documents. Reference book. Leonorm, L'viv
25. Fomin GS (2000) Voda. Kontrol' khimicheskoy, bakteriologicheskoy i radiatsionnoy bezopasnosti po mezhdunarodnym standartam: Entsiklopedicheskiy spravochnik (Water. Control of chemical, bacterial and radiation safety by international standards: encecl refer book). Protector, Moscow
26. Council Directive 75/440/EC (1988) Concerning the quality of surface water intended for the abstraction of drinking water in the member states. European community environmental legislation (1967–1987), vol 4. Water. Document N XI-987/87 Commission of European Communities Directorate-General for Environment, Consumer Protection and Nuclear Safety, Brussels
27. State Standard GOST 2761-84 (1984) Istochniki tsentralizovannogo khozyaystvenno-pit'yevogo vodosnabzheniya. Gigiyenicheskiye, tekhnicheskiye trebovaniya i pravila vybora (Sources of centralized household drinking water supply. Hygienic specification and rules of choice). State Standard USSR, Moscow
28. Unified methods of water quality research (1983) Unifitsirovannyye metody issledovaniya kachestva vod. Metody biologicheskogo analiza vod (Methods of biological analysis of water). SEV Publ, Moscow
29. Oleksiv IT (1992) Pokazateli kachestva prirodnykh vod s ekologicheskikh pozitsiy (Quality indices of natural water in terms of ecology). Svit, L'viv
30. DSTU 4808:2007 (2007) Dzherela tsentralizovanoho pytnoho vodopostachannya. Hihiyenichni ta ekolohichni vymohy shchodo yakosti vody i pravyla vybyrannya (Sources of centralized drinking water supply. Hygienic and ecological requirements to water quality and rules of selection). State Comm of Ukraine for Techn Regul and Consum Policy, Kiev
31. Ute S, Rainer B, Pudenz S (2004) Aspects of decision support in water management example Berlin and Potsdam (Germany) I-spatially differentiated evaluation. Water Res 38(2):1809–1816
32. Goncharuk VV, Klimenko NA, Savchina LA et al (2000) Minimizing the genetic risk through development of modern drinking water technologies. J Water Chem Technol 22(5):487–503
33. Goncharuk VV, Zhukinskii VN, Chernyavskaya AP et al (2003) Development of ecological and hygienic classification of Ukraine's surface water quality—sources of centralized water supply. J Water Chem Technol 25(2):3–36
34. Goncharuk VV, Klimenko NA, Savchina LA et al (2008) Current issue in the technology of drinking water conditioning. J Water Chem Technol Spec Issue, Part I:2–98

35. Garnier G, Mouner S, Benaim JY (2004) Influence of dissolved organic carbon content on modeling natural organic matter acid-base properties. Water Res 38(3):3685–3692
36. Buffle J (1988) Analytical chemistry. Harwood, Chichester
37. Byrne RH (1996) Specific problems in the measurement and interpretation of complexation phenomena in seawater. Pure Appl Chem 68(8):1639–1656
38. Swietlik J, Dabrowska A, Raczyk-Stanislawik U et al (2004) Reactivity of natural organic matter fractions with chlorine dioxide and ozone. Water Res 38(1):547–558
39. Maurice PA, Pullin MJ, Cabanies SE et al (2002) Comparison of surface water natural organic matter in raw water samples, XAD, and reverse osmosis isolates. Water Res 36(9):2355–2371
40. Leenheer JA (1981) Comprehensive approach to preparative isolation and fractionation of dissolved organic carbon from natural water and wastewater. Environ Sci Technol 15(5):578–587
41. Peurovuori J, Pihlaja K (1998) Multimethod characterization of lake aquatic humic matter isolated with two different sorbing solids. Anal Chem Acta 363(2):235–247
42. Peurovuori J, Ingman P, Pihlaja K et al (2001) Comparison of sorption of aquatic humic matter by DAX-8 and XAD-8 resins from solid-state ^{13}C NMR spectroscopy's point of view. Talanta 55:733–742
43. Perminova IV, Frimmel FH, Kudryavtsev AV et al (2003) Molecular weight characteristics of humic substances from different environments as determined by size exclusion chromatography and their statistical evaluation. Environ Sci Technol 37(1):2477–2485
44. Aiken GR, McKnight DM, Thorn KA et al (1992) Isolation of hydrophilic organic acids from water using non-ionic macroporous resins. Org Geochem 18(4):567–573
45. Korshin GV, Li C-W, Benjamin MM (1997) Monitoring the properties of natural organic matter through UV spectroscopy: a consistent theory. Water Res 31(7):1787–1795
46. Erk M, Raspor B (2001) Anodic stripping voltammetry in the complexation study of the peptide lys-Cys-Thr-Cys-Cys-Ala MT I and cadmium: application in determination of complexing capacity and stability constant. J Electroanal Chem 502:174–179
47. Gonzalez-Davila M, Santana-Casiano IM, Laglera LM (2000) Cooper adsorption in diatom cultures. Marine Chem 70(1/3):161–170
48. Ramos MA, Fiol S, Loper R et al (2002) Analysis of the effect of pH on Cu^{2+} fulvic acid complexation using a simple electrostatic model. Environ Sci Technol 36:3109–3113
49. Linnik PN (2003) Causes of water quality deterioration in the Kiev and Kaniv reservoirs. J Water Chem Technol 25(4):56–69
50. Linnik PN, Nabivanets YuB, Vasilchuk TA et al (1995) Role of organic substances in iron migration in the Kiev reservoirs. Hydrobiol J 31(3):106–112
51. Chernysheva NN, Svintsova FD, Gindullina TM (1995) Humic substances of natural water—possible source of toxic substances under water treatment. J Water Chem Technol 17(6):601–608
52. Rook II (1974) Formation of haloforms during chlorination of natural water. J Water Treat Exam 23(2):234–243
53. Bieber TI, Trehy ML (1985) Dihaloacetonitriles in chlorinated natural water. In: Jolley RL (ed) Water chlorination: environmental impact and health effects, vol 4. Ann Arbor Science, Ann Arbor, p 85
54. Servais P (1992) Studies of BDOC and bacterial dynamics in the drinking water distribution system of the Northern Parisian Suburbs. Rev Sci Eau 5:68–69
55. Escobar IC, Randall AA (2001) Assimilable organic carbon and biodegradable dissolved organic carbon: complementary measurements. Water Res 35(18):4444–4454
56. Myllykangas T, Nissinen TK, Rantakokko P et al (2002) Molecular size fractions of treated aquatic humus. Water Res 36:3045–3053
57. Garnier C, Piñeta I, Mounier S et al (2004) Influence of the type of titration and of data treatment methods on metal complexing parameters determination of single and multiligand systems measured by stripping voltammetry. Anal Chim Acta 505:263–275

58. Peurovuori J, Pihlaja K, Valimaki N (1997) Isolation and characterization of natural organic matter from lake water: two different adsorption chromatographic methods. Environ Int 23(4):453–464
59. Marhaba TF, Van D, Lippincott RL (2000) Changes in NOM fractions through treatment: a comparison of ozonation and chlorination. Ozone Sci Eng 22:249–266
60. Barber LB, Leenheer JA, Noyes TI et al (2001) Nature and transformation of dissolved organic matter in treatment wetlands. Environ Sci Technol 35:4805–4816
61. Swietlik J, Sikorska E (2004) Application of fluorescence spectroscopy in the studies of natural organic matter fractions reactivity with chlorine dioxide and ozone. Water Res 38:3791–3799
62. Egeberg PK, Christy AA, Morten E (2002) The molecular size of natural organic matter (NOM) determined by diffusionmetry and seven other methods. Water Res 36(4):925–932
63. Joret I, Levi Y, Dupin T et al (1998) Rapid method for estimating bioeliminable organic carbon in water. Paper presented at the AWWA Annual Conference, Orlando, Florida, 21–25 June 1998
64. Cipparone LA, Diehl AC, Speitel GE (1997) Ozonation and BDOC removal: effect on water quality. J Am Water Works Assoc 89(2):84–97
65. Raczyk-Stanislawiak U, Swietlik J, Dabrowska A et al (2004) Biodegradability of organic by-products after natural organic matter oxidation with ClO_2-case study. Water Res 38:1044–1054
66. Escobar IC, Randall AA (2001) Ozonation and distribution system biostability. J Am Water Works Assoc 93(10):77–89
67. Andrews SA, Huck PM (1996) Disinfection by-products in water treatment. The chemistry of their formation and control. Lewis Publ, New York
68. Volk CI, Le Chevallier M (2000) Assessing biodegradable organic matter. J Am Water Works Assoc 92(5):64–76
69. Weiss WJ, Bouwer EJ, Ball WP et al (2004) Riverbank filtration effect of ground passage on NOM character. J Water Suppl Res Technol 53(2):61–83
70. Richardson SD, Simmons IE, Rice G (2002) Disinfection by-products: the next generation. Environ Sci Technol 36:199A–205A
71. Weinberg H (1997) Disinfection by-products in drinking water: the analytical challenge. Anal Chem 71(10):801A–808A
72. Nawrocki J, Swictlik J, Raczyk-Stanislawiak U et al (2003) Influence of the ozonation's conditions on the aldehyde and carboxylic acid formation. Ozone Sci and Eng 25(1):53–62
73. Elrashidi MA, O'Connor GA (1982) Boron sorption and desorption in soils. Soil Sci Soc Am J 46:27–31
74. Gupta UC (1968) Relationship of total and hot water soluble boron, and fixation of added boron, to properties of podzol soils. Soil Sci Soc Am J 32:45–48
75. Chauveheid E, Denic M (2000) The boron-organic carbon correlation in water. Water Res 38:1663–1668
76. Brinkmann Th, Hörsch Ph, Sarforius D et al (2003) Photoformation of low-molecular weight organic acids from brown water dissolved organic matter. Environ Sci Technol 37(18):4190–4198
77. Southworth B, Voelker BM (2003) Hydroxyl radical production via the photo-Fenton reaction in the presence of fulvic acid. Environ Sci Technol 37(6):1130–1136
78. Tseeba YaYa, Maystrenko YuG (1972) Kiyevskoye vodokhranilishche: Gidrokhimiya, biologiya, produktivnost' (Kiev reservoir: hydrochemistry, biology, productivity). Nauk Dumka, Kiev
79. Linnik PN, Nabivanets BI (1986) Formy migratsii metallov v presnykh poverkhnostnykh vodakh (Forms of migration of metals in fresh surface water). Gidrometeoizdat, Leningrad
80. Linnik PN (2001) Role of humic substances in complexation and detoxification processes (taking the Dnieper reservoirs as example). Hydrobiol J 37(5):98–112
81. World Health Organization (1994) Guidelines on drinking water quality control. Recommendation, vol 1. WHO, Geneva

82. Hjort JO, Ericsson PA (1998) Artificial groundwater recharge in Stocholm I. In: Peters JH (ed) The project and its general aim artificial recharge of groundwater. Balkema, Rotterdam, pp 383–385
83. Hozalski RM, Goel S, Bouwer E (1995) TOC removal in biological filters. J Am Water Works Assoc 87(12):40–45
84. Kozyatnik IP, Klimenko NA, Savchina LA (2010) The effect of water ozonation on the efficiency of extracting natural organic matter during filtration through biologically activated carbon. J Water Chem Technol 32(3):139–144
85. Tuiveringsstrategieen VH (1998) OEDI-concept: cen natuulijke keuze. Tijdschr Watervoorz En Waterbeheer 31(9):15–21
86. Graun GF, Hauchman FS, Robinson DE (2001) Microbial pathogens and disinfection byproducts in drinking water: health effects and management of risks. International Life Science Institute, Washington
87. Rittmann BE, Snoeyink VL (1984) Achieving biologically stable drinking water. J Am Water Works Assoc 76(10):106–107
88. Van derKD, Visser A, Hijnen WAM (1982) Determing the concentration of easily assimilable organic carbon in drinking water. J Am Water Works Assoc 72(10):540–544
89. Storey MV, Ashbolt NJ (2002) A comparison of methods and models for the analysis of distribution of pipe biofilms. Water Sci Technol Water Supply 2(4):73–80
90. Graham NJD (1999) Removal of humic substances by oxidation/biofiltration processes—a review. Water Sci Technol 40(9):141–148
91. Huck PM (1999) Development of a framework for quantifying the removal of humic substances by biological filtration. Water Sci Technol 40:149–156
92. Goel S, Hozalski RM, Bouwer E (1995) Biodegradation of NOM: effect of NOM source and ozone doze. J Am Water Works Assoc 84(1):90–105
93. Hozalski RM, Bouwer EI, Goel S (1999) Removal of natural organic matter (NOM) from drinking water supplies by ozone-biofiltration. Water Sci Technol 40(9):157–163
94. Melin ES, Odegard H (1999) Biofiltration of ozonated humic water in expanded clay aggregate films. Water Sci Technol 40(9):165–172
95. Siddiqui MS, Amy GL, Murphy BD (1997) Ozone enhanced removal of natural organic matter from drinking water sources. Water Res 31(12):3098–3106
96. Huck PM, Fedorak PM, Anderson WB (1991) Formation and removal of assimiable organic carbon during biological treatment. J Am Water Works Assoc 83(12):69–80
97. Costerton JW, Lewandowski ZI, Caldwell DE et al (1995) Microbial biofilms (review). Annu Rev Microbiol 49:711–745
98. Stoodley P, Yang S, Lappin-Scott H et al (1997) Relationships between mass transfer coefficient and liquid flow velocity in heterogeneous biofilm using microelectrodes and confocal microscopy. Biotechnol Bioeng 56(6):681–688
99. Langmark JV, Storey MV, Ashbolt NI et al (2004) Artificial groundwater treatment: biofilm activity and organic carbon removal performance. Water Res 38:740–748
100. Brauch H, Sacher F, Denecker E et al (2000) The effectiveness of bank filtration for the removal of polar organic trace elements. Wasser Abwasser 141(4):226–234
101. Heberer T, Schmidt-Bumler K, Stan H-J (1998) Occurrence and distribution of organic contaminants in the aquatic system in Berlin. Acta Hydrochim et Hydrobiol 26(5):272–278
102. Emtiazi F, Schwartz T, Marten SM et al (2004) Investigation of natural biofilm formed during the production of drinking water from surface water embankment filtration. Water Res 38:1197–1206
103. Van der Hock GP, Graveland AR (1994) Water supply syst.: new technol. Paper present at the proc. NATO adv. study jnst. “New technol large water supply proj”, Varna, 24 October–4 November 1994
104. Helmisaari HS, Lindroos A-G (2000) Intkimustieto tekopohjaveden muodostamisesta. Kemia-Kemi 27(5):359–362
105. Wang LA, Chin YP, Traina SJ (1997) Adsorption of (poly)maleic acid and aquatic fulvic acid by goethite. Geochim et Cosmochim Acta 61:5313–5324

106. Kaiser K, Zech W (1997) Competitive sorption of dissolved organic matter fractions to soils and related mineral phases. Soil Sci Soc Am J 61:64–69
107. Meier M, Chin YP, Maurice P (2004) Variations in the composition and adsorption behavior of dissolved organic matter at a small, forested watershed. Biogeochemistry 67:39–56
108. McKnight DM, Smith RL, Harnish RA et al (1993) Seasonal relationships between planktonic microorganisms and dissolved organic material in an alpine stream. Biogeochemistry 21:39–53
109. Hornberger GM, Bencala KE, McKnight DM et al (1994) Hydrological controls on dissolved organic carbon during snowmelt in the Snake River near Montezuma, Colorado. Biogeochemistry 25:147–165
110. Murphy EM, Zachara JM, Smith SC et al (1994) Interaction of hydrophobic organic compounds with mineral-bound humic substances. Environ Sci Technol 28:1291–1299
111. Stott DE, Martin JP (1989) Organic matter decomposition and retention in arid soils. Arid Soil Res Rehabil 3:115–148
112. Almendros G, Dorado J (1999) Molecular characteristics related to the biodegradability of humic acid preparations. Eur J Soil Sci 50:227–236
113. Oades JM (1988) The retention of organic matter in soils. Biogeochemistry 5:35–70
114. Grunheid S, Amy G, Jekell M (2005) Removal of bulk dissolved organic carbon (DOC) and trace organic compounds by bank filtration and artificial recharge. Water Res 3(14):3219–3228
115. U.S. Environmental Protection Agency (1998) National primary drinking water regulations. Disinfectants and disinfection byproducts. Fed Regist 63(241):69389–69476
116. Chiristman RF (1983) Identity and yield of major halogenated products of aquatic fulvic acid chlorination. Environ Sci Technol 17(10):625–629
117. Reckhow DA, Singer PC (1984) Removal of organic halide precursors by preozonation and alum coagulation. J Am Water Works Assoc 76(4):151–158
118. Singer PC, Change SD (1989) Correlations between trihalomethanes and total organic halides formed during water treatment. J Am Water Works Assoc 81(8):61–69
119. Singer PC, Arlotta Ch, Snider-Sajdak N et al. Effectiveness of pre-and intermediate ozonation on the enhanced coagulation of disinfection by-product precursors in drinking water. J Am Water Works Assoc 95:453–471
120. White MC (1997) Evaluating criteria for enhanced coagulation compliance. J Am Water Works Assoc 89(5):64–70
121. Singer PC, Reckhow DA, Letterman RD (eds) (1999) Chemical oxidation. AWWA water quality and treatment: handbook of community water supplies. McGraw-Hill, New York
122. Chaiket T, Singer P, Miles A et al (2002) Effectiveness of coagulation, ozonation, and biofiltration in controlling DBPs. J Am Water Works Assoc 94(12):81–95
123. Goncharuk VV, Klimenko NA, Vakulenko VF et al (1999) Investigation of the efficiency of process of ozonation and sorption on activated carbon in the treatment of the Dnieper water. J Water Chem Technol 21(4):23–31
124. Edwards MN, Benjamin MM (1992) Transformation of NOM by ozone and its effect on iron and aluminum solubility. J Am Water Works Assoc 84(6):56–61
125. Edwards MN, Benjamin MM (1992) Effect of preozonation on coagulant—NOM interactions. J Am Water Works Assoc 84(8):63–69
126. Paralkar AA, Edzwald JK (1996) Effect of ozone on NOM and coagulation. J Am Water Works Assoc 88(4):143–151
127. Yavich AA, Lee K-H Chen K-C et al (2004) Evaluation of biodegradability of NOM after ozonation. Water Res 38:2839–2846
128. Klevens CM, Collins MR, Negm RA et al (1996) Natural organic matter characterization and treatability by biological activated carbon filtration. In: Minnear RA, Amy GL (eds) Water disinfection and natural organic matter: characterization and control, ACS symposium series 649. American Chemical Society, Washington, pp 211–246
129. Klimenko NA, Nevinnaya LV, Sidorenko YuV et al (2007) The impact of preoxidation of SAS on biosorption efficiency on activated carbon. J Water Chem Technol 29(1):15–22

130. Klimenko N, Winter-Nielsen M, Smolin S et al (2002) Role of physical-chemical factors in the purification process of water from surface-active matter by biosorption. Water Res 36:5132–5140
131. Cleasby J, Logsdon G (1999) Granular bed and per coat filtration. Water quality and treatment, 5th edn. McGraw-Hill, New York
132. Dugan NR, Williams DJ (2004) Removal of cryptosporidium by in-line filtration-effects of coagulant type, filter loading rate and temperature. J Water Supply Res Technol 53(1):1–15
133. US EPA (2001a) Draft long term 2 enhanced surface water treatment rule preamble language
134. US EPA (2001b) Draft long term 2 enhanced surface water treatment rule regulatory language
135. Ongerth J, Pecoraro J (1995) Removing cryptosporidium using multi-media filters. J Am Water Works Assoc 87(12):83–89
136. Nieminski E, Ongerth J (1995) Removing *Giardia* and *Cryptosporidium* by conventional treatment and direct filtration. J Am Water Works Assoc 87(9):96–106
137. Edzwald J, Kelley M, Dunn H et al (1996) Control of *Cryptosporidium* by coagulation, flotation and filtration. Paper presented at the AWWA water quality technology Denver AWWA, Boston, 17–21 Nov1996
138. Edzwald J, Kelley M (1998) Control of *Cryptosporidium*: from reservoirs to clarifiers to filters. Water Sci Technol 37(2):1–8
139. Swaim P, Heath M, Patania N et al (1996) High-rate direct filtration for *Giardia* and *Cryptosporidium* removal. Paper presented at the AWWA Annual Conference, Toronto, Canada, 23–27June 1996
140. Goncharuk VV, Klimenko NA, Solomentseva IM et al (2002) Through treatment of high-color natural water. J Water Chem Technol 24(1):35–43
141. Jiu-Hiu Q, Hui-Juan L, Suo-Xiang L (2003) Reduction of fulvic acid in drinking water by ferrate. J Environ Eng 129(1):17–24
142. Schreyer JM, Ockerman LT (1951) Stability of the ferrate (VI). Anal Chem 23(3):1313–1318
143. De Luca SI, Chao AC, Smallwood IC (1983) Ames test of ferrate treated water. J Environ Eng 109(5):1159–1167
144. De Luca SI, Chao AC, Smallwood IC (1983) Removal of organic priority pollutants by oxidation-coagulation. J Environ Eng 109(1):36–46
145. De Luca SI, Idle CN, Chao AC (1996) Quality improvement of biosolids by ferrate(VI) oxidation of offensive odour compounds. Water Sci Technol 33(3):119–130
146. Waite TD, Gilbert M (1978) Oxidative destruction of phenol and other organic water residuals by iron (VI) ferrate. J Water Pollut Control Fed 50(3):543–551
147. Waite TD (1979) Feasibility of wastewater treatment with ferrate. J Environ Eng Div. Am Soc Civ Eng 105(6):1023–1034
148. Murman FK, Robinson PR (1974) Experiments utilizing $FeO^{2-}{}_4$ for purifying water. Water Res 47(8): 543–547.
149. Tian BZ, Qu IH (1999) A study on the producing of potassium ferrate by chemical oxidation from a recycled solution. Environ Chem 18(2):173–177
150. Camel V, Bermond A (1998) The use of ozone and associated oxidation process in drinking water treatment. Water Res 32(1):3208–3222
151. Milyukin MV, Goncharuk VV, Vakulenko VF (1999) Changes in concentrations of carbonyl compounds at technological stages of the water conditioning process. J Water Chem Technol 21(6):1–7
152. Goncharuk VV, Vakulenko VF, Sova AN et al (2004) Removal of organic impurities from natural water by ozonation combined with UV-radiation. J Water Chem Technol 26(1):14–26
153. Wataru N, Mukaidaini T, Mitsumasa O (2003) DOC removal by multi-stage ozonation-biological treatment. Water Res 37:150–154

154. Yavich AA, Masten SJ (2003) Use of ozonation and FBT to control THM precursors. J Am Water Works Assoc 95(4):159–171
155. Carlson K, Amy G (1987) The formation of filter-removable biodegradable organic matter during ozonation. Ozone Sci Eng 19(2):179–199
156. Urfer DA, Huck PM, Booth SD et al (1997) Biological filtration for BOM and particle removal: a critical review. J Am Water Works Assoc 89(12):83–98
157. Gul S, Abbt-Braun G, Frimmel F (2003) Use of gel permeation chromatography to characterize the changes in natural organic matter through oxidative treatments. Int J Environ Anal Chem 83(9):761–768
158. Frimmel FH, Jahnel JJ, Hesse SJ (1998) Characterization of biogenic organic matter (BOM). Water Sci Technol 37:97–103
159. Hesse SJ, Kleiser GJ, Frimmel FH (1999) Characterization of refractory organic substances (ROS) in water treatment. Water Sci Technol 9:1–7
160. Gul S, Abbt-Braun G, Frimmel F (1999) Fractionation and characterization of ozonated and post chlorinated aquatic fulvic acid using gel chromatography. Int J Environ Anal Chem 75:275–284
161. Werner P (1984) Investigation on the substrate character of organic substances in connection with drinking water treatment. Zbl Bakt Hyg 108(1):46–61
162. Hambsch B, Schmiedel U, Werner P et al (1993) Investigations on the biodegradability of chlorinated fulvic acids. Acta Hydrochim Hydrobiol 21:167–173
163. Anderson LJ, Johnson JD, Christman FG (1986) Extent of ozone's reaction with isolated aquatic fulvic acid. Environ Sci Technol 20:739–742
164. Legube B, Croué JP, Dore MS et al (1989) Ozonation of an extracted aquatic fulvic acid: theoretical and practical aspects. Ozone Sci Eng 11:69–92
165. Coleman WE, Munch JW, Ringhand HP et al (1992) Ozonation/post-chlorination of humic acid: a model for predicting drinking water disinfection by-products. Ozone Sci Eng 14:51–69
166. Goncharuk VV, Vakulenko VF, Sova AN et al (2003) Effect of UV-radiation mode on the kinetics and degree of degradation of humic and fulvic acids by ozone. J Water Chem Technol 25(5):1–21
167. Chang EE, Chung-Huei L, Ya-Wen K (2002) Effect of ozone dosage for removal of model compounds by ozone. GAC treatment. Ozone Sci Eng 24:357–367
168. Dojlido Q, Zbiec E, Swietlik J (1999) Formation of the haloacetic acids during ozonation and chlorination of water in Warsaw waterworks. Water Res 33(14):3111–3118
169. Galapate RP, Bacs AU, Okada M (2001) Transformation of dissolved organic matter during ozonation: effects on trihalomethane formation potential. Water Res 35(9):2201–2206
170. Nishijima W, Kim WH, Shoto E et al (1998) Performance of an ozonation-biological activated carbon process under long term operation. Water Sci Technol 38(6):163–169
171. Volk C, Renner P, Paillard H, Joret J (1993) Effect of ozone on the production of biodegradable dissolved organic carbon (BDOC) during water treatment. Ozone Sci Eng 15(5):389–404
172. Wrickle B, Petzoldt H, Heiser H et al (1996) NOM-removal by biofiltration after ozonation—results of a pilot plant test. In: Graham N, Collins R (eds) Advances in slow sand and alternative biological filtration. Wiley, New York
173. Shu S, Zhang J (2008) Ozonation-BAC process for the control of disinfection by-products and biostability in drinking water. J Biotechnol 136(1):S668–S669
174. Hongwei W, Wenjun L, Zhansheng W et al (2002) Determination of biodegradable organic matter in drinking water and study of its biostability in drinking water in Beijing city. Chin J Environ Sci 21(4):29–33
175. Gorenflo A, Guping L, Frimmel FH (2002) Ozonung am bespiel von ligninsulfonsäure und den Kläranlagenablant einer papierfabrik. Chem Ind Technol 74(4):512–517
176. Goncharuk VV, Vakulenko VF, Samsoni-Todorov AO et al (2010) Substatiation of technological modes for effective operation of oxidative-sorption units of water treatment. J Water Chem Technol 32(1):39–49

177. Krasner SW, Selimenti MJ, Coffey BM (1993) Testing biologically active filters for removing adehydes formed during ozonation. J Am Water Works Assoc 85(5):62–71
178. Herzberg MJ, Dosoretz CG, Sheldon JJ et al (2003) Patchy biofilm coverage can explain the potential advantage of BGAC reactors. Environ Sci Technol 37(18):4274–4280
179. Amy GL, Tan LK, Davic MK (1991) The effects of ozonation and activated carbon adsorption on trihalomethane speciation. Water Res 25(2):191–202
180. Owen DM, Amy GL, Chowdhury ZK et al (1995) NOM characterization and treatability. J Am Water Works Assoc 87(1):43–46
181. Gracia RJ, Aragues JL, Ovelleiro JL (1996) Study of the catalytic ozonation of humic substances in water and their ozonation byproducts. Ozone Sci Eng 18(3):195–208
182. Weinberg HS, Glaze WH (1997) Unified approach to the analysis of polar organic byproducts of oxidation in aqueous matrices. Water Res 37(7):1555–1572
183. Yamada H, Somiya I (1989) The determination of carbonyl compounds in ozonated water by the PFBOA method. Ozone Sci Eng 17(1):53–69
184. Marhaba TF (2000) A new look at disinfection by-products in drinking water. Water Eng Manag 1:30–34
185. Ljubas D, Ruñinski N, Dobrovic S (1999) Utjecaj redoslijeda primjene koagulanta i adsorbenta na smanjenje udjela prirodnih organskih tvari u jezerskoj void. Stojarstoo 41(5/6):191–200
186. Klimenko NA, Koganovskii AM, Smolin SK (1997) Adsorption purification of river and drinking water and role of biodegradation of adsorbed substances in this process. J Water Chem Technol 19(8):13–26
187. Masayuki S (1991) Purification of water using aactivated carbon. Ind Water 393:45–65
188. Takash K, Tatsuya H, Kohei U (1997) Changes of adsorption capacity and pore distribution of biological activated carbon on advanced water treatment. Water Sci Technol 35(7):155–162
189. Clements M, Haarhoff Q (2004) Practical experience with granular activated carbon (GAC) at the water treatment plant. Water SA 30(1):89–95
190. Shmidt QL, Pimenof AV, Lieberman AI (2001) Composite adsorbent element US patent 6299771. 9 Oct 2001
191. Shvets DI, Openko NM, Mametyeva YePetal (2001) Sorption-filtration material for water treatment Ukr patent 41513. 17 Sept 2001
192. Slavinskaya GV (2003) Purification of natural water from the humic acid by a combination anion exchange and active carbons. Sorp Chromat Proc 3(3):286–291
193. Slavinskaya GV, Kuznetsova NS (2003) Sorption purification of water from fulvic acids by a combination of various types anion exchanges. Sorp Chromat Proc 3(4):455–459
194. Kilduff JE, Karanfie TN, Weber IrWJ (1998) TCE adsorption by GAC preloaded with humic substances. J Am Water Works Assoc 90(5):76–89
195. Gillogly TET, Snoeyink VL, Elarde JR et al (1998) ^{14}C-MIB adsorption on PAC in natural water. J Am Water Works Assoc 90(1):98–108
196. Bernazeau F, Mandra V, Charles P et al (1996) Pesticides removal of activated carbon: competitive adsorption with natural organic matter. Water Supply 14(2):43–48
197. Quinlivan PA, Lei L, Knappe DRU (2005) Effects of activated carbon characteristics on the simultaneous adsorption of aqueous organic micropollutants and natural organic matter. Water Res 39(1):1663–1673
198. Newcombe G, Morrison J, Hepplewhite C et al (2002) Simultaneous adsorption of MIB and NOM onto activated carbon. II. Competitive effects. Carbon 40:2147–2156
199. Li Q, Snoeyink VL, Marinas BJ et al (2003) Elucidating competitive adsorption mechanisms of atrazine and NOM using model compounds. Water Res 37:773–784
200. Li Q, Snoeyink VL, Marinas BJ et al (2003) Pore blockage effect of NOM on atrazine adsorption kinetics of PAC: the roles of PAC pore size distribution and NOM molecular weight. Water Res 37:4863–4872
201. Pelekani C, Snoeyink VL (1999) Competitive adsorption in natural water: role of activated carbon pore size. Water Res 33(5):1209–1219

202. Pelekani C, Snoeyink VL (2000) Competitive adsorption between atrazine and methylene blue on activated carbon: the importance of pore size distribution. Carbon 38:1423–1436
203. Pelekani C, Snoeyink VL (2001) A kinetic and equilibrium study of competitive adsorption between atrazine and Congo red dye on activated carbon: the importance of pore size distribution. Carbon 39:25–37
204. Ebie K, Li F, Azuma Y et al (2001) Pore distribution effect of activated carbon in adsorbing organic micropollutants from natural water. Water Res 35(1):167–179
205. Kilduff JE, Srivastava R, Karanfil T (2002) Preloading of GAC by natural organic matter: effect of surface chemistry or TCE uptake. Stud Surf Sci Catal 144:553–560
206. Roche P, De Traversay C (2002) Gestion de l'etape de filtration sur charbon act if en grains pour l'elimination de l'atrazine et ses sous-produits. Technol Sci Muth 3:69–79
207. Gia RB, Zhang XH, Zhang WH et al (2003) Fluctuation of microcystins in water plant. J Environ Sci Health A 38(12):2867–2875
208. Goncharuk VV, Vakulenko VF, Klymenko NA et al (2002) Oxidative pretreatment of solutions of synthetic surfactants prior to biosorption. J Water Chem Technol 24(3):1–11
209. Klymenko NA, Marutovsky RM, Pidlisnyuk VV et al (2002) Biosorption processes for natural and wastewater treatment. Part 1: literature review. Eng Life Sci 10(2):317–324
210. Qian Y, Yuahgming Z, Yuangming B et al (2003) Use of biofiltration method under drinking water treatment. Ind Water Treat 23(6):8–13
211. McGuire MJ, Suffet IH (eds) (1983) Treatment water granular activated carbon. American Chemical Society, Washington
212. Kunio E, Gal-Ho L (2001) Formation and transformation of haloacetic acids in drinking water in Beijing city. J Jpn Water Works Assoc 70(9):2–13
213. Polyakova TV, Klimenko NA, Savchina LA (2011) The impact of active carbon oxidation on adsorption of fulvic acids from aqueous solutions. J Water Chem Technol 33(1):20–25
214. Shuahg L, Xiaojian Z, Wenjun L et al (2001) Water treatment using ozone and activated carbon. J Environ Sci Health A 36(4):475–481
215. Masten SQ, Javich AA (2002) Method for treatment of organic matter contaminated drinking water. US patent 6365048, 2 April 2002
216. Le Chevaller MW, Becker WC, Schorr P et al (1992) Evaluating the performance of biologically active rapid filters. J Am Water Works Assoc 84(4):136–141
217. Miltner RJ, Rice EW, Summers RS (1992) Pilot-scale study of biological treatment process. In abstract of the AWWA annual conference, Vancouver, British Columbia, 19–22 June 1992
218. Yasui H, Miyaji Y (1992) Novels approach to removing refractory organic compounds in drinking water. Water Sci Technol 26(7–8):1503–1512
219. Dietrich AM, Orr MP, Gallagher DL et al (1992) Tastes and odors associated with chlorine dioxide. J Am Water Work Assoc 92(6):82–87
220. Knocke WR, Van Benschoten JE, Kearney MJ et al (1991) Kinetics of manganese and iron oxidation by potassium permanganate and chlorine dioxide. J Am Water Work Assoc 83(6):80–85
221. Long BW, Husley RA, Hochn RC (1990) Complementary use of chlorine dioxide and ozone for drinking water treatment. Ozone Sci Eng 21:465–471
222. Swietlik J, Dabrowska A, Raczyk-Stanislawiak U et al (2004) Reactivity of natural organic matter fractions with chlorine dioxide and ozone. Water Res 38(2):547–558
223. Swietlik J, Raczyk-Stanislawiak U, Bilozor S et al (2002) Effect of oxidation with chlorine dioxide on the adsorption of natural organic matter on granular activated carbon. Pol J Environ Stud 11(4):435–439
224. Karpel von Leitner N, DeLaat J, Dore M et al (1996) Disinfection by-products in water treatment. In: Minear RA, Amy GL (eds) The chemistry of their formation and control. chap 19. Lewis Publ., Boca Raton, p 393
225. Slunjski M, Nguyen H, Ballard M et al (2000) MIEX—good research commercialized. Water 29(2):42–47

226. Singer PC, Bilyk KJ (2001) Enhanced coagulation using a magnetic ion exchange resin. Water Res 36(16):4009–4022
227. Lee N, Sinha S, Amy G (2003) Removal of polar natural organic matter (NOM) with a magnetically impregnated ion exchange (MIEX) media. In: Abstract of the American water works association annual conference and exhibition, Anaheim, San Francisco, 13–15 June 2003
228. Johnson CJ, Singer PC (2003) Impact of a magnetic ion exchange on ozone demand and bromate formation during drinking water treatment. Water Res 38(17):3738–3750
229. Drikas M, Chow CWK, Cook D (2001) The impact of recalcitrant organic character on disinfection stability, trihalomethane formation and bacterial regrowth: an evaluation of MIEX and alum coagulation. J Water Supply Res Technol 52(7):475–481
230. Boyer TH, Singer PC (2005) Bench-scale testing of a magnetic ion exchange resin for removal of disinfection by-products precursors. Water Res 39:1265–1276
231. Jones KL, O'Melia CR (2000) Protein and humic acid adsorption onto hydrophilic membrane surfaces: effects of pH and ionic strength. J Membr Sci 165(1):31–46
232. Lin CF, Lin TY, Hao OJ (2000) Effects of humic substance characteristics on UF performance. Water Res 34(4):1097–1106
233. Speth TF, Gusses AM, Summers RS (2000) Evaluation of nanofiltration pretreatments for flux loss control. Desalination 130(1):31–44
234. Berube PK, Mavinic DS, Hall ER et al (2002) Evaluation of adsorption and coagulation as membrane pretreatment steps for the removal of organic material and disinfection-by-product precursors. J Environ Eng Sci 1:465–476
235. Adham SS, Jacangelo JG, Lain JM (1996) Characteristics and costs of MF and UF plants. J Am Water Works Assoc 88(5):22–31
236. Freeman SDN, Logsdon GS, Harris AT et al (1996) Evaluation of microfiltration performance with bacillus spore, particle count and particle index measurements. Paper presented at the American Water Works Association Annual Conference, AWWA, Toronto, 23–27 June 1996
237. Ponomarev MI, Dedechek VL, Ovcharov LF et al (1999) Ultra- and nanofiltration of water that contains microorganisms. J Water Chem Technol 21(11):30–35
238. Siddiqui M, Amy G, Joseph R et al (2000) Membranes for the control of natural organic material for surface water. Water Res 34(13):3355–3370
239. Maatens A, Swart P, Jacobs EP (1999) Feed-water pretreatment: methods to reduce membrane fouling by natural organic material. J Membr Sci 163(1):51–62
240. Amy GL, Cho JJ (1999) Interactions between natural organic material and membranes; rejection and fouling. Water Sci Technol 40(9):131–139
241. American Water Works Association Research Foundation (1996) Lyonnaise des Eaux, and water research commission of South Africa. Water treatment processes. McGraw Hill Publ. Co, New York
242. Lebeau T, Lelievre C, Buisson H et al (1998) Immersed membrane filtration for the production of drinking water: combination with PAC for NOM and SOCs removal. Desalination 117:219–231
243. Lin CF, Huang YJ, Hao OJ (1999) Ultrafiltration processes for removing humic substances; effect of molecular weight fraction on PAC treatment. Water Res 33(5):1252–1264
244. Schäfer AI, Fane AG, Waite TD (2001) Cost factors and chemical pre-treatment effects in the membrane filtration of water containing natural organic material. Water Res 35(6):1509–1517
245. Lee JD, Lee SH, Jo MH et al (2000) Effect of coagulation conditions on membrane filtration characteristics in coagulation-microfiltration process of water treatment. Environ Sci Technol 34(17):3780–3788
246. Eikebrokk B, Saltest T (2001) Removal of natural organic matter (NOM) using different coagulants and light weight expanded clay aggregates filters. Water Sci Technol 1(2):131–140

247. Snodgrass WJ, Clark MM, O'Melia CK (1984) Particle formation and growth in dilute Aluminium (III) solutions. Water Res 18(4):479–488
248. Lebeau T, Lelievre C, Wolbert D et al (1999) Effect of natural organic matter loading on the atrazine adsorption capacity of an aging powdered activated carbon slurry. Water Res 33(7):1695–1705
249. Best G, Mourato D, Singh M et al (2001) Application of immersed ultrafiltration membranes on high turbidity and high TOC surface water. Paper presented at the American Water Works Association Membrane Technology Conference, AWWA, San Antonio, 4–7 March 2001
250. Tsapyuk EA (1999) Colloid-chemical aspects of pre-treatment of solutions subjected to pressure-driven membrane separation. J Water Chem Technol 21(6):16–28
251. Pirbazari M, Badriyha B, Ravindran V (1992) MF-PAC for treating water contaminated with natural and synthetic organics. J Am Water Works Assoc 84(12):95–103
252. Crozes Q, Anselme C, Mallevialle J (1993) Effect of adsorption of organic matter on fouling of ultrafiltration membranes. J Membr Sci 84(2):61–77
253. Song W, Ravindran V, Koel BE, Pirbazari M (2004) Nanofiltration of natural organic matter with H_2O_2/UV pretreatment: fouling mitigation and membrane surface characterization. J Membr Sci 241:143–160
254. Zhu H, Nystrum M (1998) Cleaning results characterized by flux, streaming potential and FTIR measurements. Coll Surf A Physicochem Eng Aspects 138(2/3):309–321
255. Lee H, Amy JG, Cho JW et al (2001) Cleaning strategies for flux recovery of an ultrafiltration membrane fouled by natural organic matter. Water Res 35(14):3301–3308
256. Koganovskii AM, Klimenko NA, Levchenko TM et al (1990) Adsorbtsyya orhanycheskykh veshchestv iz vody (Adsorption of organic substances from water). Khimiya, Leningrad.
257. Tu SC, Ravindran V, Den W et al (2001) Predictive membrane transport for nanofiltration process in water treatment. AIChE 47(6):1346–1362
258. Backlund O (1992) Degradation of aquatic humic material by ultraviolet light. Chemosphere 25(2):1869–1878
259. Schefer AI, Fane AG, Waite TD (1998) Nanofiltration of natural organic matter: removal, fouling and influence of multivalent ions. Desalination 118(13):109–122
260. Fan LH, Harris JL, Roddick FA (2001) Influence of the characteristics of natural organic matter on the fouling of microfiltration membranes. Water Res 35(18):4455–4463
261. Edwards M, Benjamin M (1992) Influence of the characteristics of natural organic matter on the fouling of microfiltration membranes. J Am Water Works Assoc 84(6):56–66
262. Tran-Ha MH, Wiley DE (1998) The relationship between membrane cleaning efficiency and water quality. J Membr Sci 145(1):99–110
263. Braghetta AH, DiGiano FA, Ball WP (1998) NOM accumulation at NF membrane surface: impact of chemistry and shear. J Environ Eng 124(1):1087–1098
264. Hong S, Elimelech M (1997) Chemical and physical aspects of natural organic matter (NOM) fouling of nanofiltration membranes. J Membr Sci 132(2):159–181
265. Flemming HC (1997) Reverse osmosis membrane biofouling. Exp Therm Fluid Sci 14:382–391
266. Basu OD, Huck PM (2004) Integrated biofilter-immersed membrane system for the treatment of humic water. Water Res 38:655–662
267. Fukada S, Tsuji T, Minegishi T et al (2000) Fouling performance in the filtration of water containing humic acids and/or kaolin with microporous membrane. Water Sci Technol 41(10/11):317–325
268. Gensen KE, Knudsen QH (2000) Removal of humic substances from water using nanofiltration system. In: Abstract of the international conference EUROMEMBRANE-2000, Jerusalem, Israel, 24–27 September 2000
269. Beros M, Ventresque C, Bablon G (2000) Angewandte nanofiltration in einen wasserkwork in Frankreich. Wasser Boden 52(2):25–32
270. Neparidze GT, Neparidze RSh (2000) Water supply of settlements and fish factories in coastal areas of Kamchatka. In: Abstract of the 4th international congress "water: ecology and technology" EKVATEK-2000, Moscow, 30 May–2 June 2000

271. Bodzek NL, Bodzek DL, Luks-Betleg KJ et al (2000) Pressure-driven membrane techniques for the removal of organic micropollutants from water. In: Abstract of the international conference EUROMEMBRANE-2000, Jerusalem, Israel, 24–27 September 2000
272. Schellekens R (2003) Kristallklares resin wasser. WWT: Wasserwirt Wassertechn 6:42–43
273. Agua E, Voutchkov N, Sedlak DL (2009) Disinfection by products and their potential impacts on the quality of water produced by desalination systems: a literature review. Desalination 237(1/3):214–237
274. Meyer M (1996) How common is methemoglobinemia from nitrate contaminated well. Water Cond Purif 1:78–81
275. Weyer PJ, Cerhan JP, Kross BC et al (2001) Municipal drinking water nitrate level and cancer risk in older women: the Iowa Women's Health Study. J Epidemiol 11(3):327–338
276. Khan JA, Spalding RF (2004) Enhanced is situ denitrification for a municipal well. Water Res 38:3382–3388
277. Shrimali MJ, Singh KP (2001) New methods of nitrate removal from water. Environ Pollut 112(3):351–359
278. Ivleva GA, Rodina IS, Kozina AK et al (2000) Purification of underground water from nitrate and boron compounds for drinking water supply. In: Abstract of the 4th international congress "Water: ecology and technology" EKVATEK-2000, Moscow, 30 May–2 June 2000
279. Matuju V, Cizinska S, Krejci J et al (1992) Biological water denitrification—a review. Enzyme Microbial Technol 14(3):170–183
280. Ivanov VN, Ulanov MN, Stabnikova YeV (2001) Denitrification of drinking water by cells of *Paracoccus denitrificans* in natural and artificially formed biofilms. J Water Chem Technol 23(2):64–70
281. Ergas SJ, Reuss AF (2001) Hydrogenotrophic denitrification of drinking water using a hollow fiber membrane bioreactor. J Water Supply Res Technol Aqua 50(3):161–171
282. Schipper L, Vojvodic-Vokovic M (1998) Nitrate removal from ground water using a denitrification wall amended with sawdust: field trial. J Environ Qual 27:664–668
283. Interstate technology and regulatory cooperation work group (2000) Technology overview-emerging technologies for enhanced in situ biodenitrification (EISBD) of nitrate-contaminated ground water. IIRC, Washington.
284. Khan IA, Spalding RF (2003) Development of a procedure for sustainable in situ aquifer denitrification: remediation. J Environ Clean Technol 13(2):53–69
285. Peyton BM (1996) Improved biomass distribution using pulsed injection of electron donor and acceptor. Water Res 30(3):756–758
286. Constantin H, Fick M (1997) Influence of C-source on the denitrification rate of high-nitrate concentrated industrial wastewater. Water Res 31(3):583–589
287. Ergas SJ, Rheinheimer DE (2003) Drinking water denitrification using a membrane bioreactor. Water Res 38:3225–3232
288. Cho J, Amy G, Pellegrino J (2000) Membrane filtration of natural organic matter: factors and mechanisms affecting and flux decline with charged ultrafiltration (UF) membrane. J Membr Sci 164(1/2):89–110
289. McCleaf PR, Schroeder ED (1995) Denitrification using a membrane-immobilized biofilm. J Am Water Works Assoc 87(3):77–86
290. Fuchs W, Schatzmayr G, Braun R (1997) Nitrate removal from drinking water using a membrane-fixed biofilm reactor. Appl Microbiol Biotechnol 48(2):267–274
291. Mansell BO, Schroeder ED (1998) Biological denitrification in a continuous flow membrane reactor. Water Sci Technol 38(1):9–14
292. Mansell BO, Schroeder ED (1999) Biological denitrification in a continuous flow membrane reactor. Water Res 33(8):1845–1850
293. Kim J-HM, Kishida N et al (2004) Integrated real-time control strategy for nitrogen removal in swine wastewater treatment using sequencing batch reactors. Water Res 38(14/15):3340–3348
294. Fanning JC (2000) The chemical reduction of nitrate in aqueous solution. Coord Chem 199:159–179

295. Ottley CJ, Davison WP, Edmunds WM (1997) Chemical catalysis of nitrate reduction by iron (II). Geochim et Cosmochim Acta 61(9):1819–1828
296. Daum JM, Vorlop KD (1999) Kinetic investigation of the catalytic nitrate reduction: construction of the test reactor system. Chem Eng Technol 55:159–179
297. Siantar DP, Schreier CG, Reinhard MN (1996) Treatment of 1,2-dibromo-3-chlorpropane and nitrate-contaminated water with zero-valent iron or hydrogen/palladium catalysts. Water Res 30(10):2315–2322
298. Hu HY, Goto N, Fujie K et al (2001) Reductive treatment characteristics of nitrate by metallic iron in aquatic solution. J Chem Eng Jpn 34(9):1097–1102
299. Chi J, Shu-Ting Z, Lu Y et al (2004) Chemical reduction of nitrate by metallic iron. J Water Supply Resear Technol 53(1):37–41
300. World Health Organization (1996) Guidelines for drinking water-quality, 2nd edn, recommendations, vol 1. WHO, Geneva
301. Lipunov IN, Sanakoyev VN (1999) Preparation of underground water for drinking water supply. In: Abstract of the international scientific conference "Socioecon and ecolog problems of the forest complex", Ural State Forest Academy, Yekaterinburg, 1 Feb 1999
302. Sleptsov GV, Orlov VO (2001) Opportunities to increase productivity of deferrization station. Water Supply San Technol 2:26–28
303. Kröning H (2003) News wasserwerk für Hamburg. Abwassertechnik 9:10–16
304. Vandenabeele J, De Beer D, Germonpru R et al (1992) Manganese oxidation by microbial consortia from sand filters. Microbiol Ecol 24:91–108
305. Dmitrenko GN, Shum OA (2002) Dissimilative reduction of Mn(IV) by collection strainsoft aerobic bacteria. J Water Chem Techn 24(1):56–60
306. Bourgine FP, Gennery MK, Chapman JJ et al (1994) Biological process at Saints Hill Water treatment plant, Kent. J Inst Water Environ Manag 8:379–392
307. Seppánen HT (1992) Experiences of biological iron and manganese removal in Finland. J Inst Water Environ Manag 6:333–341
308. Mouchet P (1992) From conventional to biological removal of iron and manganese in France. J Am Water Works Assoc 8(4):158–167
309. Hope CK, Bott TR (2004) Modeling of manganese biofiltration using biofilms of *Leptothrix discophora*. Water Res 38:1853–1861
310. Vandenabeele J, De Beer D, Germonpre R et al (1995) Influence of nitrate on manganese consortia from sand filters. Water Res 29:579–587
311. Zhang J, Lion LW, Nelson YM et al (2002) Kinetics of Mn(II) oxidation by *Leptothrix discophora SSI*. Geochim et Cosmochim Acta 65(5):773–781
312. Pontius FW, Brown KG, Chen CJ (1994) Health implications of arsenic in drinking water. J Am Water Works Assoc 86(9):52–63
313. Huges MF (2002) Arsenic toxicity and potential mechanisms of action. Toxicol Lett 133(1):1–16
314. Katsoyiannis IA (2004) Application of biological process for the removal of arsenic from groundwater's. Water Res 38(1):17–26
315. World Health Organization (1996) Guidelines for drinking water quality, 2nd edn. Health criteria and other supporting information, WHO, Geneva
316. Tallman DE, Shaikh AU (1980) Red ox stability of inorganic arsenic(III) and arsenic(V) in aqueous solution. Anal Chem 52: 199–201.
317. Cullen WR, Reimer KJ (1989) Arsenic speciation in the environment. Chem Rev 89:713–764
318. Zouboulis AI, Katsoyiannis IA (2002) Removal of arsenates from contaminated water by coagulation-direct filtration. Sep Sci Technol 37(12):2859–2873
319. Kartinen EO, Martin CJ (1995) An overview of arsenic removal processes. Desalination 103:79–88
320. Driehaus W, Seith R, Jekel M (1995) Oxidation of As(III) with manganese oxides in water treatment. Water Res 29(1):297–305

321. Kim MJ, Nriangu JA (2000) Oxidation of arsenite in groundwater using ozone and oxygen. Sci Total Environ 247:71–79
322. Vinokurov SV, Kantor LI, Tsypysheva AG (1999) Problems, ways and facilities to protect the environment from pollution by oil and petroleum products. In: Abstract of the 3rd science and technology conference, VIMI, Moscow, 6–9 April 1999
323. Cristal-Campos C, Baudin J (2002) L'elimination des pesticides par le procéde. Technol Sci Méth 3:49–53
324. Piet C, Peltier S, Gauber J-Y et al (2002) Elimination par le CAP de l'atrazine et de la DEA. Technol Sci Méth 3:43–48
325. Sutherland J, Adams C, Kekobad J (2004) Treatment of MTBE by air stripping, carbon adsorption and advanced oxidation: technical and economic comparison for five groundwater. Water Res 38(1):193–205
326. Squillace P, Pankow J, Kortes N et al (1997) Review of the environmental behavior and fate of methyl-tert-butylether. Environ Toxicol Chem 16:1836–1844
327. Cater SR, Stefan MI, Bolton JR et al (2000) UV/H_2O_2 treatment of methyl tert-butyl ether in contaminated water. Environ Sci Technol 34(4):659–662
328. Shon HK, Vigneswaran S, Ngo HH (2005) Chemical coupling of photocatalysis with flocculation and adsorption in the removal of organic matter. Water Res 39(12):2549–2558
329. Sanchez-Polo M, VonGunten U, Rivera-Utrila J (2005) Efficiency of activated carbon to transform ozone into OH· radicals: influence of operational parameters. Water Res 39(14):3189–3198
330. Kimiaki I, Youichi S, Nobuo J et al (2001) Investigation process of water treatment using biofilters. J Jpn Water Works Assoc 70(8):2–12
331. Nicolella C, Zolezzi M, Rabino M et al (2005) Development of particle-based biofilms for degradation of xenobiotic organic compounds. Water Res 39(12):2459–2780
332. Dallas L, Meyer CL, Ja-Chin SB et al (2002) Method and apparatus for biodegradation of alkyl ethers and tertiary butyl alcohol US Patent 6458276, 28 Oct 2002
333. Tuxen N, Tüchsen PL, Rügge K et al (2000) Fate of seven pesticides in an aerobic aquifer studied in column experiments. Chemosphere 41(9):1485–1494
334. Council directive 98/83 EC (1998) On the quality of water intended for human consumption. Off J Eur Commun 330:32–54
335. Block JC (1992) Biofilm in drinking water distribution systems. In: Melo LF et al (eds) Biofilms scince and technology. Klüwer Acad Publ, Netherlands, pp 469–485
336. Niquette P, Servais P, Savoir R (2000) Impact of pipe materials on densities of fixed bacteria biomass in drinking water distribution system. Water Res 34(6):1952–1956
337. Zacheus OM, Jivanainen EK, Nissinen TK et al (2000) Bacterial biofilm formation on polyvinyl chloride, polyethylene and stainless steel exposed to ozonated water. Water Res 34(1):63–70
338. LeChevallier MW (1990) Coliform regrowth in distribution water: a review. J Am Water Works Assoc 82(1):74–86
339. Schmitt J-P (1974) La qualete de l'eau dans le rèseau de distribution. le programme de recherches "Biofilm". Courants 26:29–33
340. Haas CN (1999) Benefits of using disinfectant residual. J Am Water Works Assoc 91(1):65–69
341. Zacheus OM, Lehtola MI, Korhonen LK (2001) Soft deposits, the key site for microbial growth in drinking water distribution networks. Water Res 35(7):1757–1765
342. Szewzyk U, Szewzyk R, Manz W (2000) Microbiological safety of drinking water. Annu Rev Microbiol 54:81–127
343. Rogers J, Dowsett A, Dennis P et al (1994) Influence of plumbing materials on biofilm formation and growth of *Legionella pneumophila* in potable water systems. Appl Environ Microbiol 60(6):1842–1851
344. Schwartz T, Hoffman S, Obst U (1992) Formation and bacterial composition of young, natural biofilms obtained from public bank-filtered drinking water systems. Water Res 32(9):2787–2797

345. Schwartz T, Kalmbach S, Hoffman S et al (1998) PCR-based detection of mycobacteria in biofilms from a dinking water distribution systems. J Microbiol Method 34:113–123
346. Camper A, McFeters G, Characklis W et al (1991) Growth kinetics of coliform bacteria under conditions relevant to drinking water distribution systems. Appl Environ Microbiol 57(8):2233–2239
347. Mackay W, Grobbon L, Barer M (1999) Biofilms in drinking water systems: a possible reservoir for *Helicobacter pylori*. J Appl Microbiol 85(Suppl 1):52S–59S
348. van Loosdrecht MCM, Eikelboom D, Gjaltema A et al (1995) Biofilm structures. Water Sci Technol 32(8):35–43
349. Picioreanu C, van Loosdrecht MCM, Heijenen JJ (1998) Mathematical modeling of biofilm structure with hybrid differential discrete cellular automaton approach. Biotechnol Bioeng 58(1):101–116
350. Picioreanu C, van Loosdrecht MCM, Heijenen JJ (2001) Two-dimensional model of biofilm detachment caused by internal stress from liquid flow. Biotechnol Bioeng 72(2):205–218
351. Davies DG, Parsek MR, Pearson JP et al (1998) The involvement of cell-to-cell signals in the development of bacterial biofilm. Science AAAS Weekly Pap Edit 280:295–297
352. Kjelleberg S, Molin S (2002) Is there a role for quorum sensing signals in bacterial biofilms. Curr Opin Microbiol 5(3):254–258
353. Sauer KA, Camper AK, Ehrlich GD et al (2002) *Pseudomonas aeruginosa* displays multiple phenotypes during development as a biofilm. J Bacteriol 184(4):1140–1154
354. Telgmann U, Horn H, Morgenroth E (2004) Influence of growth history on sloughing and erosion from biofilms. Water Res 38:3671–3684
355. Horn H, Reiff H, Morgenroth E (2003) Simulation of growth and detachment in biofilm systems under defined hydrodynamic conditions. Biotechnol Bioeng 81(5):607–617
356. Lehtoba MJ, Nissinen TK, Miettinen IT et al (2004) Removal of soft deposits from the distribution system improves the drinking water quality. Water Res 38:601–610
357. Sarin P, Snoeyink VL, Bebce J et al (2001) Physico-chemical characteristics of corrosion scales in old iron pipes. Water Res 35:2961–2969
358. Sarin P, Snoeyink VL, Bebce J et al (2004) Iron release from corroded iron pipes in drinking water distribution systems: effect of dissolved oxygen. Water Res 38(5):1259–1269
359. Benjamin MM, Sontheimer H, Leroy P (1996) Corrosion of iron and steel: internal corrosion of water distribution systems. AWWA Research Foundation, Denver, pp 29–70
360. Pontius FW (1992) A current look at the federal drinking water regulations. J Am Water Works Assoc 3:36–50
361. Frateur I, Deslouis C, Kiene L et al (1999) Free chlorine consumption induced by cast iron corrosion in drinking water distribution systems. Water Res 33(8):1781–1790
362. Miettinen IT, Vartiainen T, Martikainen PI (1997) Phosphorus and bacterial growth in drinking water. Appl Environ Microbiol 63(8):3242–3245
363. Sathasivan A, Ohgaki S, Yamamoto K (1997) Role of inorganic phosphorus in controlling regrowth in water distribution system. Water Sci Technol 35(8):37–44
364. Kanakoudis VK (2004) A Troubleshooting manual for handling operational problems in water pipe networks. J Water Supply Res Technol 53(2):109–124
365. Kanakoudis VK, Tolikas DI (2001) The role of leaks and breaks in water networks-technical and economical methodic solutions. J Water Supply Res Technol 50(5):301–311
366. Van den Boomen M, van Mazijk A, Beuken RHS (2004) First evaluation of new design concepts for self-cleaning distribution networks. J Water Supply Res Technol 53(1):43–50
367. Camper A, Buis I, Goodrum L (2001) Effects of UV-disinfection on humic substance and biofilms. In: Abstract of the annual AWWA water quality technology conference, Nashville, 11–14 November 2001
368. Pozos N, Scow K, Wuertz S et al (2004) UV disinfection in a model distribution system: biofilm growth and microbial community. Water Res 38(10):3083–3091
369. Momba M, Cloete T, Venter S et al (1998) Evaluation of the impact of disinfection processes on the formation of biofilms in potable surface water distribution systems. Water Sci Technol 38(8/9):283–289

370. Lund V, Ormerod K (1995) The influence of disinfection processes on biofilm formation in water distribution systems. Water Res 29(4):1013–1021
371. Goeres DM, Palys T, Sandel BB et al (2004) Evaluation of disinfectant efficacy against biofilm and suspended bacteria in a laboratory swimming pool model. Water Res 38:3103–3109
372. Critchley MM, Fallowfield HI (2001) The effect of distribution system bacterial biofilms on copper concentrations in drinking water. Water Sci Technol Water Supply 1(4):247–252
373. Lehtola MJ, Miettinen IT, Keinánen MM et al (2004) Microbiology, chemistry and biofilm development in a pilot drinking water distribution system with copper and plastic pipes. Water Res 38:3769–3779
374. Rossman LA, Clark RM, Grayman WM (1994) Modeling chlorine residuals in drinking water distribution systems. J Environ Eng (ASCE) 120(4):803–820
375. Shah M, Sinai G (1988) Steady state model for dilution in water networks. J Hydraul Eng (ASCE) 114(2):192–206
376. Wood DJ, Ormsbee LE (1989) Supply identification for water distribution systems. J Am Water Works Assoc 81(7):74–80
377. Boulos PF, Altman T, Sadhal K (1992) Computer modeling of water quality in large networks. J Appl Math Modeling 16(8):439–445
378. Boulos PF, Altman T, Jarrige PA (1995) Discrete simulation spproach for network water quality models. J Water Resour Plan Manag (ASCE) 121(1):49–60
379. Rossman LA, Boulos PF, Altman T (1993) Discrete volume element method for network water-quality models. J Water Resour Plan Manag (ASCE) 119(5):505–517
380. Boulos PF, Altman T, Jarrige PA (1994) An event-driven method for modeling contaminant propagation in water networks. J Appl Math Model 18(2):84–92
381. Islam MR, Chaudhary MH (1998) Modeling of constituent transport in unsteady flows in pipe networks. J Hydraul Eng (ASCE) 124(11):1115–1124
382. Rossman LA, Boulos PF (1996) Numerical methods for water quality in distribution systems: a comparison. J Water Resour Plan Manag (ASCE) 122(2):137–146
383. Munavalli GR, Mohan Kumar MS (2004) Dynamic simulation of multicomponent reaction transport in water distribution systems. Water Res 38:1971–1988
384. Munavalli GR, Mohan Kumar MS (2004) Modified Langrangian method for modeling water quality in distribution systems. Water Res 38:2973–2988
385. Graham NJ, Collins RD (1996) Advances in slow sand alternative biological filtrations. Wiley, Chichester
386. Weber-Shirk ML, Dick RI (1997) Biological mechanisms in slow sand filters. J Am Water Works Assoc 89:73–83
387. Weber-Shirk ML, Dick RI (1997) Physical-chemical mechanisms in slow sand filters. J Am Water Works Assoc 89:87–100
388. Mauclaire L, Schůrmann A, Thullner M et al (2004) Sand filtration in water treatment plant: biological parameters responsible for clogging. J Water Supply Res Technol 53(2):93–108
389. Hynes JP (2000) Water treatment apparatus. Brit patent 2350357, 29 Nov 2000
390. Goreangab project (2003) Windhoek, namibia: a unique technology—municipal wastewater is turned into drinking water. In: Abstract of the 27 international exhibition-congress on chemical engineering, environmental protection and biotechnology, ACHEMA, Frankfurt a. M., 19–24 May 2003
391. Sheng-JL, Guin-J L, Guin Je L et al (2001) Water treating apparatus. British patent 2354515, 28 March 2001
392. Ripperger S (2000) Wasseranfbereitung in Namibia. Filtr Sep 14(3):155–156
393. Montiel A, Welte B, Maucolin P et al (1997) Etude du relargage des nitrites des filter a carbon actif en grains. Technol Sci Meth 7/8:11–15
394. Korabelshchikov V (2003) The complex methods of water treatment. Industry 2:36–37
395. Modern waterwork"Vozdvizhenye" (1999) New technoloies and equipment in water supply and water management 2:185–186
396. Ceronio AD, Basson ND, Kruger M et al (2002) The in-depth evaluation of three filtration facilities. Water SA Spec. Iss:6–9

397. Lukas A (2000) Rohwasser aus der warnow. Wasserwirt—Wassertechn 1:17–22
398. Jguel N (2001) L'apport des nouvelles technologies pour la gestion du cycle de l'eau. Eau, ind, nuisances 245:53–54
399. Hays G, Haaland J (2002) Australian patent 747194, 9 May 2002
400. Urban F, Buchsteiner D (2000) Flockungzanlage zur aufbereitung von talsperrenwasser. Teil 1. Planung und bau. BBR: Brunnenbau, Ban Wasserwork, Rohrleitunggfau 51(9):16–22
401. Tomoyoshi M (2004) Water supply in Tokyo. Water 21(6):25–26
402. Monarca S, Zani C, Richardson SD et al (2004) A new approach to evaluating the toxicity and genotoxicity of disinfected drinking water. Water Res 38:3809–3819
403. Hwang BF, Jaakkola JJ (2003) Water chlorination and birth defects: a systematic review and meta-analysis. Arch Environ Health 58:83–91
404. Monarca S, Zanardini A, Feretti D et al (1998) Mutagenicity of lake drinking water treated with different disinfectants in bacterial and plant tests. Water Res 32:2689–2695
405. Goncharuk VV, Potapchenko NG, Savuk OS et al (2004) Combined antimicrobial effect of ozone and ultraviolet radiation generated by various sources. J Water Chem Technol 26(2):60–67
406. Goncharuk VV, Potapchenko NG, Savuk OS et al (2001) Disinfection of water by ozone: effect of inorganic impurities on kinetics of water disinfection. J Water Chem Technol 23(2):55–63
407. Goncharuk VV, Potapchenko NG, Savuk OS et al (2001) The effect of humic compounds on the kinetic of dying off of *Echerichia coli* in disinfection of water by ozone. J Water Chem Technol 23(3):54–61
408. Richardson SD, Thruston AD, Caughran TV et al (2000) Identification of new drinking water disinfection by-products from ozone, chlorine dioxide, chloramines and chlorine. Water Air Soil Pollut 123(1/4):95–102
409. Milov VR, Boldyrev VV, Baranov SV (2001) Experience in water disinfection by membrane electrolysis at the city waterworks Monchegorsk. Water Ecol Probl Decis 3:17–20
410. Khramenkov SV, Malyshev BV (2000) Test results of Moscow river water disinfection by chlorine dioxide. Water Supply San Technol 10:2–4
411. Niquette P, Pre'vost M, Merlet N et al (1999) Influence de facteurs contrólant l'enlevemnt de la demande en chlore et de precurseurs de sous-produits de chloration dans des filters biologiques. Water Res 33(10):2329–2344
412. Goncharuk VV, Potapchenko NG, Savuk OS et al (2003) A study of the various procedures of water disinfection by O_3/UV. J Water Chem Technol 25(5):65–72
413. Stockholm setzt tukünftig auf UV-desinfektionstechnik (2001) BBR: Brunnenbau, Ban Wasserwerk, Rohzleitungsbau 52(1):41–44
414. Anlagen V (2001) Stockholm erteilt WEDECO zweiten grobauftrag. Galvanotechnik 92(12):3408–3410
415. Hoger O (2000) Desinfektion mit UV-anlagen nach DVGW-arbeitsblatt W 294. BBR: Brunnenbau, Ban Wasserwerk, Rohzleitungsbau 51(10):76–83
416. Auckenthaler A (2003) Pathogene mikroorganismen im grund- und trinkwasser: transport, nachweismethoden, wassermanagement, basel etc. Birkhduser 1:95–130
417. An experience of practical application of drinking wastewater disinfection by ultraviolet radiation (2001) New technologies and equipment in water supply and water management 3:127–129
418. A Drinking Water Production (2011) http://www.cryobzkm.ru/water. Accessed 01 Jan 2011
419. Zohra B, Driss O, Godfrey W (2004) EVAPRO: economic and financial evaluation of water supply projects. J Hydroinform 6(2):109–121
420. Goncharuk VV (ed) (2003) Byuvety Kyyeva. Yakist' artezians'koï vody (Kiev's artesian wells. Quality of artesian water). Geoprint, Kiev
421. Goncharuk VV (2008) The concept of choosing a list of indices and their regulatory values for determination of hygienic requirements and control over the drinking water quality in Ukraine. J Water Chem Technol Spec Iss(Part II):52–111

422. Wadha SG, Khaled GH, Edberg SC (2002) Comparative microbial character of consumed food and drinking water. Crit Rev Microbiol 28(3):249–279
423. Cotran RS, Kumar VI, Robbins SL (eds) (1994) Pathological basis of disease, 5th edn. Saunders, Philadelphia
424. Mandell GL, Douglas JE, Benett RI (eds) (2000) Principles and practice of infectious diseases, 5th edn. Churchill Livingston, Philadelphia
425. Ducan HE, Edberg SC (1995) Host-microbe interaction in the gastrointestinal tract. Crit Rev Microbiol 21(2):85–100
426. Edberg SC, Gallo PI, Kontnick CI (1996) Analysis of the virulence characteristics of bacteria isolated from bottled, water cooler, and tap water. Micro Ecol Health Dis 9:67–72
427. Edberg SC, Kops S, Kontnick C (1997) Analysis of cytotoxicity and invasiveness of heterotrophic plate count bacteria (HPC) isolated from drinking water on blood media. J Appl Microbiol 82:455–481
428. Reasoner DJ (1990) Drinking water microbiology. Springer-Verlag, New York
429. World Health Organization (2004) Guidelines for drinking water quality. Recommendations, vol 1. WHO, Geneva
430. Water FG. (2003) Joint strategy of introduction. Working group 2A. Ecological state (ECO-STAT) in general approach to classification of ecological potential, Rome.
431. Sinclair M, Schlosser O (2007) Water softening: epidemiological evidence on calcium and magnesium in drinking water and cardiovascular disease. Water 21(2):18–19
432. Rosenlund M, Berglind N (2005) Daily intake of magnesium and calcium from drinking water in relation to myocardial infarction. Epidemiology 16(4):570–576
433. Mons M, VanDuk H, Gatel D et al (2007) Softening, conditioning and the optimal composition of drinking water. Water 21(2):19–20
434. International Agency for Research on Cancer (1993) Beryllium, cadmium, mercury, and exposures in the glass manufacturing industry: IARC monographs on the evaluation of carcinogenic risks to humans. IARC, Lyon
435. Vaessent Hubert AMG, Szteke B (2000) Beryllium in food and drinking water—a summary of available knowledge. Food Addit Contam 17(2):149–159
436. Anderson A (2004) Clean water report. http://www.pmmag.com/Articles/Feature_Article/7652f6d1030d7010VgnVCM100000f932a8c0. Accessed 1 June 2004
437. Fewtrell L, Kay D (2001) A review of the science behind drinking water standards for copper. Int J Environ Health Res 11:161–167
438. American Public Health Association (2001) Drinking water quality and public health (Position paper). J Publ Health 91(3):499–500
439. Goncharuk VV, Rudenko AV, Koval EZ et al (2004) The problem of water infection with pathogens of mycoses and prospects of resolving this problem. J Water Chem Technol 26(2):1–18
440. Gang D, Clevenger TE, Banerji SK (2003) Relationship of chlorine decay and THMs formation to NOM size. J Hazard Mater 96(1):1–12
441. The guidelines of the European Union Council 98/83/EC (1998) Water quality intended for consumption by man. Off J Eur Commun 1:2–5
442. Environmental Institute (1997) Standards and strategies in the European Union to control trihalomethanes (THMs) in drinking water, vol. 2, ECSC-EC-EAEC, Brussels, Luxembourg
443. Environmental Institute (1997) Exposure of the European population to trihalomethanes (THM_S) in drinking water, vol. 2, ECSC-EC-EAEC, Brussels, Luxembourg
444. Meeting strict DBP standards—the view of the European water industry (2006) Dr. R.A. Breach, head of quality and environmental services, severn trent water, chairman. EUREAY Commission 1:140–145
445. Charrois JWA, Graham D, Hrudey SE (2004) Disinfection by-products in small Alberta community drinking water supplies. J Toxicol Environ Health A 67:1797–1803
446. Swietlik J, Raczyk-Stanislawik U, Bilozor S et al (2002) Adsorption of natural organic matter oxidized with ClO_2 on granular activated carbon. Water Res 36(9):2328–2336

447. Swietlik J, Raczyk-Stanislawik U, Bilozor S et al (2003) Reduction of ClO_2 demand by ClO_2 oxidation and subsequent GAC filtration. Water Res 37(19):4693–4702
448. Collivignarelli C, Sorlini S, Belluati M (2002) Rimozione del clorito con carbone attivo granulare e effetti della riattivazione chimica. Ing Ambient 21(5):265–271
449. Roccaro P, Mancini G, Vagliasindi FGA (2005) Water intended for human consumption. Part 1: compliance with European water quality standards. Desalination 176:1–11
450. Linden KG, Shin GA, Faubert G et al (2002) UV-disinfection of *Giardia* lamblia cysts in water. Environ Sci Technol 36(11):2519–2522
451. Shin G, Linden K, Arrowood M et al (2001) Low-pressure UV-inactivation and DNA repair potential of *Cryptosporidium parvum* oocysts. Appl Environ Microbiol 67(7):3029–3032
452. Legrini O, Oliveros E, Braun A (1993) Photochemical processes for water treatment. Chem Rev 93(2):671–698
453. Brass HJ (2000) Status of the drinking water standards program in the United States. Water Air Soil Pollut 123(1/4):1–9
454. Caro J, Serrano A, Gallego M (2007) Direct screening and confirmation of priority volatile organic pollutants in drinking water. J Chromatogr A 11(38):244–250
455. Fielding M, Farrimond M (1999) Disinfection by-products in drinking water: current issues. Royal Society of Chemistry, UK
456. Dojlido Q, Zbiec E, Swietlik J (1999) Formation of the haloacetic acids during ozonation and chlorination of water in Warsaw waterworks (Poland). Water Res 33:3111–3118
457. Shannon MA, Bohn PW, Elimelech M et al (2008) Science and technology for water purification in the coming decades. Nature 452:301–310
458. Russian F (2004) The federal law "On special technical regulation for requirements for sanitary-epidemiological safety to water intended for human consumption and drinking water supply" (draft). Res Inform Collect 19(1):27
459. El-Dib MA, Ali RKA (1995) THMs formation during chlorination of raw Nile river water. Water Res 29(9):375–378
460. Lin T-F, Hoang SW (2000) Inhalation exposure to THMs from drinking water in South Taiwan. Sci Total Environ 246:41–49
461. Boccellia DL, Trybyb ME, Uberg JG et al (2003) A reactive species model for chlorine decay and THM formation under rechlorination conditions. Water Res 37(11):2654–2666
462. Gang D, Clevenger TE, Banerji SK (2003) Relationship of chlorine decay and THMs formation to NOM size. J Hazard Mater A96: 1–12.
463. Abdulahha MP, Yewa CH, Ramlib MS (2003) Formation, modeling and validation of trihalomethanes (THM) in Malaysian drinking water: a case study in the districts of Tampin, Negeri Sembilan and Sabak Bernam, Selangor, Malaysia. Water Res 37(19):4637–4644
464. Abd El-Shafy M, Grunwald A (2000) THM formation in water supply in South Bohemia, Czech Republic. Water Res 43(13):3453–3459
465. Cortvriend J, Hulsmann A (2006) Europe paves the way for the revision of the drinking water directive. Water 21(1):17–19
466. Sauvant M-P, Pepin D (2002) Drinking water and cardiovascular disease. Food Chem Toxicol 40:1311–1325
467. Yang CY (1998) Calcium and magnesium in drinking water and risk of death from cerebrovascular disease. Stroke 29:411–414
468. Water global news digests (2007) http://www.iwapublishing.com/ Accessed 23 Jan 2007
469. Water global news digests (2007) http://www.iwapublishing.com. Accessed 6 Feb 2007
470. Guerrant RL (1997) Cryptosporidiosis an emerging, highly infectious treat. Emerg Infect Dis 3(1):51–57
471. Goncharuk VV (ed) (2005) Ekologicheskiye aspekty sovremennykh tekhnologiy okhrany vodnoy sred (Ecological aspects of modern technologies of water medium protection). Nauk Dumka, Kiev
472. Balls M (2004) Progressing toward the reduction, refinement and replacement of laboratory animal procedures: thoughts on some encounters with Dr. Iain Purchase. Toxicol In vitro 18:165–170

473. Pleteneva TV, Solomatin EM, Syroezhkin AV et al (2005) Toksikologicheskaya khimiya (Toxiocological chemistry). GEOTAR-Media, Moscow
474. Sundstrom L, Morrison B, Bradley M et al (2005) Organotypic cultures as tools for functional screening in the CNS. Drug Discov Today 10(14):993–1000
475. Ferro M, Doyle A (2001) Standardization for *In Vitro* testing. Cell Biol Toxicol 17:205–212
476. Ilyin S, Belkowski S, Plata-Salaman C (2004) Biomarker discovery and validation: technologies and integrative approaches. Trends Biotechnol 22(8):411–416
477. Arkhipchuk VV, Goncharuk VV (2004) Assessment of the quality of bottled drinking water by biotesting methods. J Water Chem Technol 26(5):38–60
478. Freshney I (2001) Application of cell cultures to toxicology. Cell Biol Toxicol 17:213–230
479. Gulden M, Dierickx P, Seibert H (2006) Validation of a prediction model for estimating serum concentrations of chemicals which are equivalent to toxic concentrations *In Vitro*. Toxicol In Vitro 20(7):1114–1124

Chapter 5
Peculiarities and Problems of Water Decontamination

Abstract The data on methods applied worldwide for drinking water decontamination has been systemized and generalized. The negative aspects of traditional decontamination technologies in terms of modern level of bacillary and chemical pollution of water sources, which are used for drinking water treatment, have been described. Special attention has been paid to the problem of water infection contamination by activators of fungal infection. The advantage of decontamination of drinking water by ozone treatment and ultraviolet irrigation has been theoretically and experimentally justified.

Keywords Decontamination of drinking water · Chlorination · Ozonation · Bactericidal action of ultrasonic · Bactericidal action of ultraviolet rays · Combination of ozone treatment and ultraviolet irradiation · Fungal infection activators

5.1 Chlorination

5.1.1 The Mechanism of Intensifying the Antimicrobial Action of Activated Chloride with Copper

The significance of the water medium in the propagation of epidemic diseases may be explained by the following circumstances: a noticeable increase of microbial contamination of water; variation of the pathogens' properties—their virility and resistance to the effects of the outside medium increases; total nonspecific resistance of microorganisms' decreases; number of disease incidences increased due to man's frequent contact with water (bathing, conditioning systems, etc.). Water has become a medium for transferring not only the pathogens of known diseases, but also those whose role in the infection pathology of human is considered hypothetical [1, 2].

The most widespread method of water disinfection until now is chlorination. Despite the fact that chlorine and its compounds with a series of organic compounds form toxic substances of the type of trigalogenmethanes, it is the only accessible disinfectant which possesses a prolongation effect [3]. Thanks to this quality, chlorine is widely used when transporting water along pipelines. Alternatives to chlorine

V. V. Goncharuk, *Drinking Water*, DOI 10.1007/978-3-319-04334-0_5,

in the visible future are not expected since the increased degree of water purification of organic matter results in the introduction of 0.3 mg/dm^3 of activated chlorine to water, which does not cause any toxic phenomena [4, 5].

However, considering the increasingly unfavorable infectious situation involving the water factor, the necessity of intensifying the antimicrobial effect of chlorine occurs at the concentration of 0.3 mg/dm^3 by way of creating complex methods of water disinfection. This, on the one hand, will make it possible to create a spectrum of the antimicrobial effect of disinfectants and, on the other hand, will be conducive to obtaining synergetic effects.

There is a special interest in the use of copper in the ionic form and in the form of powders in various degrees of dispersity. This is related to the fact that copper possesses an alhicidic effect (unlike chlorine and other disinfectants) and makes for penetration of antimicrobial agents inside the cells [6].

The most effective condition for the use of copper for intensifying the antimicrobial effect of disinfectants, in particular, chlorine, is shown in paper [7]. These experiments were carried out on the culture *E.coli* 1257, obtained from the collection of the A. Tarasevich Institute for Control of Medical and Biological Preparations of Russia's Ministry of Public Health. The techniques of preparing bacterial cultures for experiments and their accounting are described in paper [8]. Results of water disinfection in various experiments were represented as a logarithm of the ratio of bacteria number, which survived after treatment with disinfectants (N_t) to their initial number (N_0). The effect of a joint action of antimicrobial agents was assessed by the ratio *T/E*, where *T* is the theoretically calculated fraction of the bacteria that survived, while *E* is the experimentally defined one. If $T/E < 1$, then one can observe an antagonistic effect, given the joint effect of disinfectants; at $T/E = 1$ it is additive while at $T/E > 1$ it is synergic.

The data obtained were subjected to statistical processing [9] with the allowance for a mean arithmetic error of the fraction of the survived cells (N_t/N_0); the Student coefficient and the integrity of the *T/E* difference were calculated.

The working solution of sodium hypochlorite was prepared from the commercial chemical Merk (the content of activated chlorine (AC) > 13 %), which then was metered to the water treated to the set concentrations (0.3 and 2 mg/dm^2). The concentration of activated chlorine in the solution was determined by the calorimetric method [10].

In experiments on investigating permeability of membranes of the cells *E.coli* 1257 for potassium ions in the presence of different bacterial agents and their combinations used the method of radioisotope recording of ionic flows [11].

The buildup of potassium radioactive isotope and determination of its yield from the cells were carried out using the modified methods [11]. For this goal 2 μM/cm^3 of Rb^{86} was added to the liquid synthetic medium M 9 containing 4 mM KCI and it was grown to a stationary growth phase at 37 °C. When treating the cells with chemical agents the culture in the stationary phase was washed twice with a 10 mM solution of tris-Cl (pH 7.5) centrifuging at 5,000 rpm for 5–10 min. The sediment was suspended in 10 mM tris-Cl until it reached the final concentration of 10^9 indiv/cm^3, and then it was diluted 10-fold and incubated at room temperature at a set time in the

Table 5.1 Impact of the joint action Cu^{2+} (0.5 mg/dm^3) with activated chlorine (0.3 mg/dm^3) on survivability of the cells *E.coli* in the artesian water

Contact time, min	Quantity of survived bacteria (%) under action of			*T/E*
	Cu^{2+}	AC	Cu^{2+} and AC	
1	97.3	39.4	31.5	1.22
10	74.2	19.4	3.5	4.17
20	50.0	8.0	0.2	18.93
30	40.9	3.8	0.1	25.98

presence of antimicrobial agents. Upon the end of the incubation with disinfectants, an aliquot was applied to a filter with the pore diameter of 0.1–0.2 μm, and the filter was washed with a 10 cm^3 of a nonradioactive solution. The membrane filter was submerged in a 5 cm^3 of the scintillating liquid ZHS-1. Radioactivity was determined on a liquid scintillating counter 1217/1218 Rackbeta.

The violation of the properties of a cytoplasm membrane (CPM) of bacterial cells under the impact of Cu^{2+} is accompanied by the variation of the magnitude of the transmembrane potential ($\Delta\varphi$), which was determined by means of a selective electrode by measuring the distribution of penetrating lipophilic tetraphenilphosphonium chitins between the cells and the medium [12].

Investigation [13, 14] carried out on tap water demonstrated that copper in the presence of such antioxidants as chlorine and hydrogen peroxide cause the synergetic effects. With a detailed investigation of the water disinfection process with these disinfectants, it turned out that the synergetic effect of copper and chlorine depend on the physicochemical composition of the treated water.

Table 5.1 provides data on the impact of chlorine at concentration 0.3, copper—0.5 mg/dm^3 and their joint action with respect to *E.coli* 1257 in the artesian water, pH 7.2–7.4. As can be seen, the number of survived microorganisms with the joint action of disinfectants—activated chlorine and copper ions—is smaller compared with the results obtained when treating with the given agents individually. We may judge it also by the value of the ratio *T/E* whose magnitude is less than 1 within the whole time of exposure; one can observe a gradual growth of this value as the contact time increases, which indicates the synergetic action of the previously mentioned disinfecting substances.

Similar data were obtained for a broader spectrum of microorganisms: *Bac. Subtilis, Str. Faecalis, E. coli.* Figure 5.1 shows kinetic curves for dying the enumerated microorganisms with the joint action of chlorine (at the concentration of 0.3 mg/dm^3 for *Str. Faecalis* and *E. coli* and 4 mg/dm^3—for *Bac. Subtilis)* and copper (at the concentration 0.5 mg/dm^3—for *Str. Faecali* and *E.coli* and 4 mg/dm—for *Bac. Subtilis).* An increase of the dosage of chlorine and copper for *Bac. Subtilis* is related to the fact that the given type of organisms is spore-forming and that for its inactivation it needs higher concentrations of disinfectants. Sensitivity of the microorganisms to the antimicrobial agent (or to their complex) was assessed by the constants of the death rate (K, min^{-1}), which were determined graphically as a slope angle tangent of straight sections of kinetic curves of death of test-microorganisms $lg(N_t/N_0)=f(t)$ and constituted for *Bac. Subtilis:* 0.03, *E.coli* 0.25 and for *Str. Faecalis:* 0.76 min^{-1}.

Fig. 5.1 Kinetics of the death of the microorganisms in artesian water in complex treatment with activated chlorine and ions of Cu^{2+}: *1—Str. Faecalis* ($C_{Cu^{2+}}$=0.5 and C_{AC}=0.3 mg/dm³); *2—E. coli* ($C_{Cu^{2+}}$=0.5 and C_{AC}=0.3 mg/dm³); *3—Bac. Subtilis* ($C_{Cu^{2+}}$ and C_{AC}—by 4 mg/dm³)

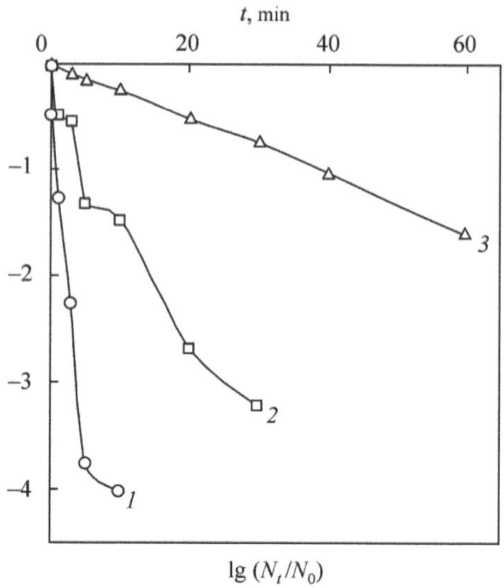

However, when using distilled or bidistilled water as a dispersion medium, the increase of the chlorine antimicrobial effect with copper was not observed (Table 5.2). As follows from the data shown, independent of the contact time and experimental conditions ($C_{Cu^{2+}}$ = 1, C_{AC} = 1, $C_{Cu^{2+}}$ and C_{AC}—by 1 mg/dm³), the value of the ratio *T/E* equals either less than unity, which indicates an additive, or an antagonistic nature of the copper or chlorine effect.

Figure 5.2 shows the relationship between the antimicrobial action of disinfectants and the water composition. The given series of experiments are carried out using three types of water: bidistilled, tap and artesian. The water under research was treated with activated chlorine and copper ions at the concentration of 0.3 and 0.5 mg/dm³, respectively. *E.coli* 1257 was used as a test-microorganism. It became clear that the tap water and the water from an artesian well are conducive to the emergence of the synergetic effect, unlike bidistilled water whereby one can observe an additive effect, and sometimes an antagonistic one.

Such results seem illogical when taking into account that in the distilled and, furthermore, the bidistilled water, the cells are weakened and greatly subjected to the effect of microbial agents, owing to the disruption of osmotic equilibrium between the cell and the medium.

Intensification of the antimicrobial effect of the activated chlorine we found (in this case—sodium hypochlorite) may occur either as a result of the catalytic action of copper ions in the treated water or directly, by way of copper's action within the cell. This, in fact, results in the intensification of chlorine's antimicrobial action. Paper [15] shows that copper ions intensify oxidation with sodium hypochlorite of humic acids. The achieved increase in the rate of oxidizing organic compounds involving copper, which is present in the natural water, may indirectly make for intensification of the chlorine's antimicrobial effect. This is related to the fact that

Table 5.2 Impact of joint action of Cu^{2+} (1 mg/dm^3) with activated chlorine (1 mg/dm^3) on survivability of the cells *E.coli* in distilled water

Contact time, min	Number of survived bacteria (%) under the action of			*T/E*
	Cu^{2+}	AC	Cu^{2+} and AC	
10	76.2	41.2	50.3	0.624
30	77.9	32.4	33.2	0.760
45	61.5	27.2	22.6	0.740
60	62.8	14.1	15.4	0.571

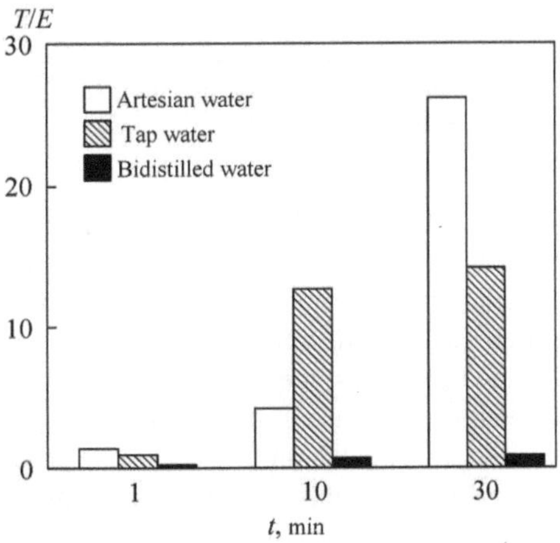

Fig. 5.2 Variation of the *T/E* parameter in the disinfection process of the water of different physicochemical composition. The test-microorganism—*E.coli* 1257 (initial contamination 10^4 CFU/dm^3). We used complex treatment with activated chlorine and Cu^{2+} ($C_{Cu^{2+}} = 0.5$ and $C_{AC} = 0.3$ mg/dm^3)

the organic matter present in the water "protects" microorganisms from the effects of disinfectants. As a result of oxidation of organic compounds, an access of copper ions to the cell's interior is substantially simplified. On the other hand, it is not an exception that copper intensifies the oxidation of organic matter on the surface of the cell membrane. Consequently, main mechanisms of raising the chlorine antimicrobial effect with copper, perhaps, are related to the processes occurring at the level of the cell itself.

The research into the mechanism of copper's effect on the microbial cell has demonstrated that in order to ensure the copper antimicrobial effect, it is necessary to fulfill two conditions: the interaction of Cu^{2+} with the targets of the type of thiol groups, which are located on the outer side of the cellular membrane; that the ionic exchange between the cell and the medium is at the level of CPM [16].

If the first condition is mandatory for tap, artesian and bidistilled water, then for the second condition the cell should be in a physiologically active state (bidistilled water is classically used for bringing the cells to the state of hunger, and in this case the physiological activity of the cells is suppressed).

The normal functioning of the cell, first of all, is related to the active transport between the cell and the surrounding environment, which is brought about through special channels and is characterized by the CPM permeability of microbial cells. Paper [17] found that permeability of the cellular membrane, as a result of the

Table 5.3 Impact of Cu^{2+} (1 mg/dm^3) and activated chlorine (1 mg/dm^3) in distilled water[a] on the content of Rb^{86} in the cells *E.coli*

Medium content	Incubation time, min	Amount of Rb^{86} in cells[b], puls/min	Amount of Rb^{86} in cells, % of reference
Reference	1	9,000	100
	45	8,800	97.8
Cu^{2+}	1	8,400	93.3
	45	8,300	92.2
AC	1	8,800	97.8
	45	9,000	100
Cu^{2+} and AC	1	8,100	90.0
	45	8,400	93.3

[a] pH was established by means of a tris-Cl-buffer
[b] average of three experiments

leakage of the ions K^+ from the cell, depends on the composition of the medium treated, the initial value of the transmembrane potential between the cell and the medium, and the intracellular pH. The most effective removal from the cells of the ions K^+ and the decrease of membrane permeability under the impact of Cu^{2+} was detected from tris-acetate and 50 mM tris-NCl. Dilution of these media leads to a substantial decrease of $\Delta\varphi$ and respectively the leave of K^+ from the cell. This fact leads to the conclusion that Cu^{2+} ions penetrate the cells the same way that ions do, and this is important for the cell in terms of energy.

The special experiments were carried out for clearing up the dependence mechanism of the antimicrobial effect during the joint use of copper and hypochlorite on the physicochemical composition of the treated water. These experiments investigated the permeability of the CPM of microbial cells. The state of CPM was assessed by the leakage of potassium from the *E.coli* 1257 cells. Radioactive Rb^{86} was used as an analogue. The experiments demonstrated that the treatment of the cells for 20 min in distilled water (adjustment up to pH 7.2 by means of a tris-Cl-buffer) does not change the amount of the radioactive marker inside the cell. Copper (1 mg/dm^3) and also together copper with chlorine (1 mg/dm^3) insignificantly decrease the amount of the radioactive marker inside the cells. When the contact time of the disinfectants was prolonged to 45 min, the impact of copper and chlorine on the amount of the radioactive marker inside the cell was insignificant (Table 5.3). Table 5.4 provides data on the impact of copper and chlorine in a concentration of 1 mg/dm^3, together and separately, on the amount of the radioactive rubidium, which remained in the bacteria cells after their incubation for 45 min in the physiological solution. From this data, one can see that under these conditions copper and chlorine (especially chlorine together with copper) cause complete leakage of Rb^{86} from the cells.

It is necessary to note that copper and chlorine, apart and together, substantially affect the permeability of the membrane of *E.coli* 1257 for ions of rubidium—the analogue of the potassium ion in the cell. In this case the effects obtained depend on the medium composition in which the cells are suspended.

Table 5.4 Impact of Cu^{2+} (1 mg/dm^3) and activated chlorine (1 mg/dm^3) in the physiological solution[a] on the content of Rb^{86} in the cells *E.coli*

Medium content	Incubation time, min	Amount of Rb^{86} in cells[b], puls/min	Amount of Rb^{86} in cells, % of reference
Reference	1	11,200	100
	45	10,000	89.3
Cu^{2+}	1	10,100	90.2
	45	250	2.2
AC	1	–	–
	45	280	2.8
Cu^{2+} and AC	1	3,100	27.7
	45	150	1.3

[a] pH was established by means of a tris-Cl-buffer
[b] average of three experiments

The probability of penetrating Cu^{2+} ions inside the CPM cell as a result of the active transport followed by the transformation of Cu^{2+} to a single-valence form is noted in [16, 17]. In distilled water, where the cells are under conditions of hunger and their physiological activity is substantially retarded, copper ions cannot penetrate inside the cell, and therefore the antimicrobial activity of the latter was reduced to a minimum. On the other hand, these data indicate the intensification of the antimicrobial effect of the disinfectants by the copper ions of a nonspecific nature.

Thus, the mechanism of enhancing the antimicrobial effect of chlorine with copper ions is of a multifold nature. On the one hand, copper enhances oxidation of organic matter both in solution and on the surface of the cells, and on the other hand, the intensification mechanism is related to the CPM permeability of microbial cells. With the aid of labeled atoms of Rb^{86}—an analogue of cell potassium—we established that membrane permeability depends on the composition of the water treated. Thus, if in the physiological solution introduction of copper and chlorine results in complete leakage of potassium then in the distilled water potassium leakage under the effect of disinfectants was not detected. These results are confirmed by the fact that copper may penetrate the cell wall in the course of the exchange of ions between the cell and the medium surrounding it. Investigation of the relationship between the transmembrane potential, the copper concentration and organic matter (in our experiments tri-acetate) also showed that the copper antimicrobial effect had substantially decreased in the diluted solutions.

Based on the obtained data, one can conclude that the absence of the synergetic action of copper and chlorine in the bidistilled water is related to the decrease of the CPM permeability of the cell for copper ions while, as in tap and artesian water, a pronounced synergetic effect is obvious.

5.2 Ozonation

5.2.1 Electron Structure of the Ozone Molecule and Reactivity of Ozone

The interest in the chemistry of the ozone is determined by its role in the protection of our planet's surface from the harmful effect of solar ultraviolet radiation, as well as its chemical properties. These properties account for a wide range of practical uses, such as a powerful oxidant of organic impurities in industrial wastewater and a sterilizing reagent in potable water production processes.

The ozone molecule is characterized by a significant electron affinity (according to different data, it ranges between 1.9 and 2.7 eV [18]) which correlates with the high oxidative activity of this compound. Unlike the triangular carbon framework of the cyclopropane molecule with an equidistant arrangement of carbon atoms, the oxygen atoms in ozone form an isosceles triangle with the O–O–O angle close to 120. This is indicative of sp^2-hybridization of the central atom's valency orbitals [19]. The length of the O–O (Roo) bond in the ozone molecule is intermediary between the sum of two covalence radii of the oxygen atom in the mono- (O–n) and bivalent (O=) states [20]. This suggests that its order is close to 1.5 (as in the case of the C–C bond in the benzene molecule) making the structure of an isolated molecule rather stable. This is at variance with the common notions about the ease of detachment of the atomic oxygen with which the oxidative properties of ozone are often associated [21]. Since this compound has a relatively low dipole moment (0.52 D [19]), there may be (because of the obtuse angle at the central oxygen atom and, consequently, a short dipole arm) a significant asymmetry of the electron distribution, with localization of appreciable negative charges on the end atoms and a positive charge on the central oxygen atom. This might lead one to expect a predominantly heterolytic pattern of reactions involving the ozone molecule.

While the properties of ozone have been studied quite extensively, some aspects of its chemical behavior are still far from being fully elucidated. In particular, should the ozone molecule's tendency to dissociate under the effect of light excitation be viewed as a capacity for decomposition under soft conditions and the principal cause of chemical activity? It is not clear why this polar compound, which exhibits a high electron affinity, is relatively poorly soluble in water [18] (less soluble by approximately one order of magnitude than chlorine, whose molecule has a similar electron affinity [22]). Our understanding of the details of the main paths of transformation of ozone dissolved in water also leaves much to be desired. Regarding the better studied reactions involving ozone, one should mention its interaction with olefins, the mechanism which has not yet been definitively understood [23, 24].

The natural approach to solving a number of problems concerning the chemistry of the ozone would be to examine them at the molecular level using reliable quantum-chemical methods. Quantum-chemical calculations using the software complex GAMESS [24] within the framework of the Restricted Hartree-Fock method (RHF) and the multiconfigurational theory of self-consistent field (MC SCF) [25] with the Danning-Hay [26, 27] basis sets have been employed. In cases where used MC SCF we took into consideration, unless otherwise specified, all the spin states, S_0, S_0+1, S_0+2 ($2S_0+1$—multiplicity of the ground state of the molecule or radical under study).

Electron Structure of the Ozone Molecule, its Possible Spin States, and Features of Chemical Transformations First of all it will be appropriate to consider the highly reactive nature of ozone. While the ozone molecule is diamagnetic, liquid ozone is slightly paramagnetic [18]. This is why it may be assumed that this compound exists in the form of a biradical. Normally biradical forms are realized in the case of compounds characterized by a relatively small singlet-triplet *(S—T)* splitting of the ground state [28]. In the case of the equilibrium structure of the ozone molecule, it may be inferred from MC SCF calculations (including all states with $S=0$, 1, 2, 3 and the Danning-Hay basis set with polarization *d*-functions) that the energy of the above-mentioned *S—T* splitting is actually quite high, although this is in obvious contradiction with the previous assumption. However, our calculations indicate that this characteristic closely depends on the geometric parameters of the ozone molecule. In particular, the lengthening of one O–O bond by 0.05 nm (39.6 %) followed by optimization of the other parameters reduces the singlet-triplet splitting energy 4.7 times. Further growth of the ROO length of this bond (with a relatively low absorption of energy) results in a rapid decrease of the ΔE *(S—T)* value. By the $R_{OO}>0.18$ nm the triple state becomes the ground. According to paper [28], the contributions of the biradical ($O^{\bullet}$–O–$O^{\bullet}$) and the ionic (O^{-}–O^{+}=O) structures to the ground state of the ozone molecule are 52 and 48 %, respectively. Therefore, it is believed that ozone participates in both hetero- and hemolytic reactions.

As is demonstrated in paper [29], ozone is characterized by the typical properties of nonrigid molecules (several equilibrium configurations of nuclei with low interconfiguration transition barriers [< 100 kJ/mol], and small changes of the molecule's total energy after noticeable changes of the valence angles). In particular, in addition to the symmetric equilibrium conformation ($R_{100}=R_{200}=0.1264$ nm; O–O–O angle 117.97°), an asymmetrical arrangement ($R_{100}=0.1229$, $R_{200}=0.1328$ nm; O–O–O angle 116.19°) that differs in energy from the symmetrical one by just 18.2 kJ/mol has been discovered in ozone molecules. The latter conformation corresponds to the ionic structure of ozone. Another symmetrical structure ($R_{100}=R_{200}=0.1475$ nm) with an acute angle of 59.77° at the central atom has also been found. The total energy of this structure is higher than that of the first conformation by 117.5 kJ/mol.

In the case of ozone, as follows from above-mentioned calculations, a characteristic feature of the hypersurface of potential energy in the neighborhood of the minimum is its flat pattern, which is confirmed by small changes of the molecule's energy and significant deviations of the bond lengths and valence angles from the equilibrium values. Thus, for example, the lengthening of one O–O bond by 38.9 % and a simultaneous shortening of the other one by 5.5 % results in an energy increase of only 32.3 kJ/mol (0.005 %). Likewise, an appreciable widening or narrowing of the valence angle O–O–O, e.g. by 10°, involves a rather small consumption of energy (17.6 and 20.9 kJ/mol, respectively). Note that variation of the molecular

Table 5.5 Charge distribution on oxygen atoms in ozone molecule and energy of lower molecular orbital (λ_{LMO}) depending on molecular structure parameters as calculated by method of multiconfigurational theory of self-consistent field

Structure	Bond length, nm		O–O–O angle, degrees	Charges on atoms, atomic units			$\lambda_{LMO}{}^{a}$, eV
	O_1–O_2	O_2–O_3					
1	0.1264	0.1264	117.93	−0.099	0.198	−0.099	1.09
2	0.1277	0.1277	107.93	−0.095	0.190	−0.095	1.33
3	0.1336	0.1336	97.93	−0.071	0.142	−0.071	1.82
4	0.1261	0.1261	127.93	−0.107	0.214	−0.107	0.95
5	0.1266	0.1266	137.93	−0.103	0.206	−0.103	0.91
6	0.1305	0.1305	116.96	−0.086	0.172	−0.086	1.60
7	0.1405	0.1405	114.47	−0.046	0.092	−0.046	2.66
8	0.1940	0.1940	114.85	0.014	0.060	−0.074	0.58
9	0.1464	0.1464	112.88	−0.027	0.054	−0.027	3.12
10	0.1476	0.1476	58.77	0.0015	−0.003	0.0015	−2.66

[a] λ_{LMO} values were calculated in the framework of the limited Hartree-Fock-Rotan method (positive—λ_{LMO} value corresponds to increased electron-acceptor ability)

structure parameters tangibly affects the charge distribution on the oxygen atoms as well as the energy of the lower vacant molecular orbital, whose value is related to the electron-acceptor property of the molecule, i.e., its oxidizing ability (see Table 5.5). In addition, as has already been noted, the ozone molecule may respond to a significant deformation of its equilibrium structure by transitioning to the biradical state.

It is known that ozone in concentrated gas mixtures and its condensed state is dangerously explosive [18, 19, 21]. On the other hand, without exposure to catalysts or ultraviolet gases, ozone degrades quite slowly, at just 250 °C [21]. Since the causes of ozone decomposition are as yet imperfectly understood, a theoretical discussion of the elementary processes of its transformations may be of interest here.

It has been demonstrated above that an isolated molecule of ozone does not undergo decomposition despite quite considerable deviations of the parameters of its molecular structure from the equilibrium values. Considering the tendency for ozone's dependence on the stability of its concentration, it may be assumed that its decomposition is of a biomolecular nature. From the viewpoint of electrostatics, the most efficient interaction of two ozone molecules is ensured by their parallel orientation (see Fig. 5.3).

Calculations with the Hartree-Fock method indicate that the process of interacting molecules approaching each other involves a relatively low consumption of energy. Even with simultaneous fixation of all the three distances between the oxygen atoms of ozone molecules, the maximum value of the complex's total energy (E_t) on the path of intersection exceeds E_t for the equilibrium structure by only 206 kJ/mol. Although the transition state was not exactly localized by us, it is reasonable to suppose that the activation barrier of the reaction $2\ O_3 \rightarrow 3\ O_2$ will be appreciably lower than this value. This is why the most probable path of degradation of ozone in the gaseous

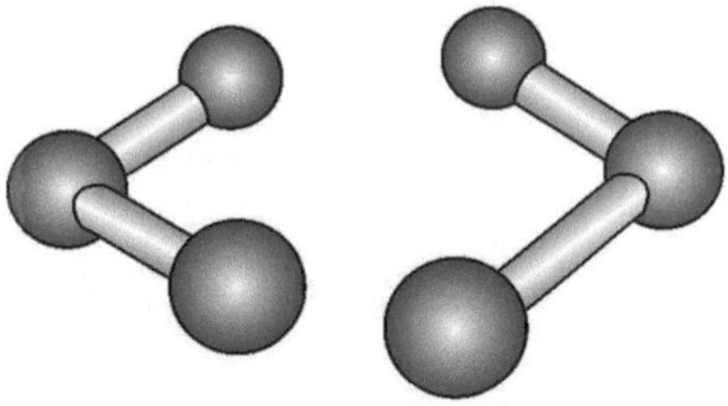

Fig. 5.3 Mutual orientation of ozone molecules in O_3–O_3 complex

phase, in the absence of reducing agents, catalysts and sources of UV-radiation, is apparently associated with its bimolecular transformation into molecular oxygen.

Mechanism of the Initial Stage of Oxidation of Olefins by Ozone One typical reaction of ozone is its reaction with olefins, which proceeds under soft conditions and produces ozonides:

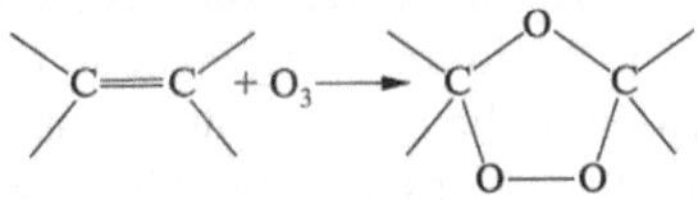

The generally accepted scheme for the course of this reaction, proposed by Kriege [21], includes three principle stages (cycloaddition of ozone via the double bond with formation of an unstable mole ozonide [30] and formation of an intermediate in the form of a=C=O aldehyde or ketone and the zwitterion=C=O^+–O^- whose condensation yields the end product). We have to admit, though, that many of the details of this mechanism are still far from being conclusively established. For example, according to Kriege, the first stage consists of a dipolar 1,3-addition of ozone via the double bond:

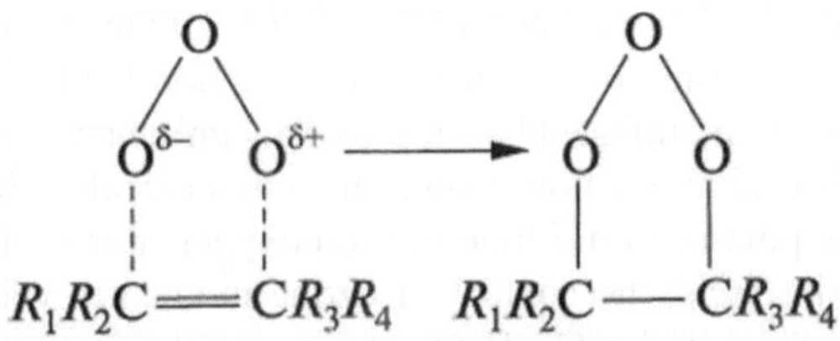

However, the assumed formation of mole ozonide ignores the fact that the end atoms of oxygen in the ozone molecules are negatively charged. Paper [31] considered another version of the first stage of ozonization of olefins that corresponds to 1,2-addition of ozone via a quasidouble inter-oxygen bond:

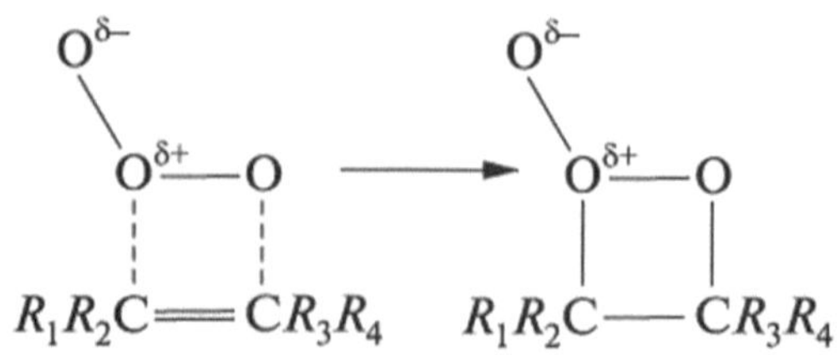

On the other hand, the simultaneous addition of the ozone molecule is a forbidden process. Therefore, a four-member cycle should be able to form only via intermediate products with corresponding activation barriers [24], which is at variance with the easy course of the overall reaction. Logically, it is more probable that the first stage of ozonation will be realized, where the ozone molecule is added as a biradical.

To reveal the mechanism of the initial stage of the reaction of ozone with olefins, we studied its interaction with the ethylene molecule in terms of the RHF. We found that no potential barrier had to be overcome as the interacting molecules approached each other along the minimum energy reaction path (MERP). This would suggest a nonbarrier formation of the intermediate compound mole ozonide. In order to support our assumption, we carried out the MC SCF calculate, which showed that the reactive system assumes a triplet (biradical) state at the O–C distances $R_{10c}=R_{20c}=0.2$ nm. On the other hand, the ozone and C_2H_4 molecules approach each other via the MERP, which corresponds to the interaction of the reagents, and involves, as expected, significant energy consumption (the respective section of the reagent molecules' potential energy lies appreciably higher than product molecules). It may therefore be assumed that the interactive features of ozone and olefins are determined by ozone's ability to exist in the form of a biradical.

Energy of the Reactions of Transformation of Ozone in Water and Assessment of its Oxidation Potential Considering the ozone's high electron affinity, the water molecule's electron-donor properties and the polarity of these compounds, it would be logical to expect an effective electrostatic attraction between their molecules. This attraction could be manifested by means of a significant solubility of ozone in water and their chemical interaction. However, actual calculations (RHF, Danning-Hay basis including polarization *d*-functions on oxygen atoms) indicate that, in the absence of external effects, the contact of ozone molecules with water produces only the Van der Waals complex $O_3 \cdot H_2O$, whose equilibrium structure is characterized by a rather low stability. Table 5.6 shows the energies of ozone and water interaction, their components for different interatomic distances and also a comparison of the respective data for the water dimer.

A comparison of the data listed in this table indicates that the main stabilizing component of the interaction energy E_{int}(electrostatic energy E_{es}) in the case of the equilibrium structures of ozone hydrate and the dimer $(H_2O)_2$ exceeds in absolute value exchange repulsion energy (E_{ex}) of the interacting molecules' electron shells by a factor of ~1.6–2. With closer contact between ozone and water molecules, all the stabilizing contributions to E_{int} grow rapidly, but an even quicker significant

Table 5.6 Energy of interaction between ozone and water molecules and its components (kJ/mol) in complexes with different interatomic distances

Complex[a]	$-E_{int}$	$-E_{es}$	E_{ex}	$-E_p^{4x}$	$-E_{CT}^{5x}$
A[b]	14.8	22.6	11.5	1.4	4.2
B[c]	16.8	28.9	59.2	2.6	11.9
C[d]	−125.9	130.2	313.8	22.2	56.9
$(H_2O)_2$[e]	32.2	53.2	32.3	3.9	7.7

[a] in O_3H_2O complexes the difference between the central oxygen atom of the ozone molecules, and the oxygen atom of the water molecule is designated as R_{OO}, and that between the end atoms of oxygen and the hydrogen atoms as—R_{OH}

[b] equilibrium structure of Van der Waals complex: R_{OO}=0.2757 nm, R_{HH}=0.3317 nm

[c] R_{OO}=0.3029 nm, R_{OH}=0.2 nm

[d] R_{OO}=0.1989 nm, R_{OH}=0.2 nm

[e] R_{OO}=0.2839 nm

increase of E_{ex} makes the formation of stable ozone hydrates rather improbable and explains the relatively poor solubility of ozone in water.

In conclusion, we will consider the features of chemical transformations of ozone in pure water [31]. It has been established that hydrogen peroxide accumulates with the passage of time in water containing ozone. It may be assumed that one of the paths of H_2O_2 formation is associated with the direct interaction of the ozone and water molecules: $O_3+H_2O \rightarrow H_2O_2+O_2$. Although this is an exothermic reaction with a thermal effect of 38.9 kJ/mol (hereafter we refer to thermal effect figures as calculated by the MC SCF method with allowance made for all the spin states S_0, S_0+1, S_0+2, and $S_0=0$ or 1 determined by the Danning-Hay basis), it can be realized only by means of a vigorous external action. This is because of its considerable activation barrier (386.4 kJ/mol, as estimated by the RHF with the Danning-Hay basis and polarization *d*-functions on oxygen atoms). Transformations of ozone with the participation of hydroxyl anions are much more probable, as evidenced by the rapid degradation of ozone in an alkaline medium [32]. The interaction of ozone with OH^- and transformations of the products formed [32] can be described by the following chain of reactions:

$$
\begin{aligned}
&O_3 + OH^- \rightarrow O_3^- + OH^\cdot && (-\Delta H = 273.2);\\
&O_3^- + H^+ \rightarrow HO_3^\cdot && (-\Delta H = 1695.1);\\
&HO_3^\cdot \rightarrow HO^\cdot + O_2 && (-\Delta H = 536.5);\\
&HO^\cdot + O_3 \rightarrow HOO^\cdot + O_2 && (-\Delta H = 291.9);\\
&HO^\cdot + OH^- \rightarrow H_2O + O^- && (-\Delta H = 48.9);\\
&O^- + O_3 \rightarrow O_2^- + O_2^- && (-\Delta H = 444.9).
\end{aligned}
$$

where ΔH, kJ/mol is the reaction thermal effect.

The exothermic nature of these elementary processes is a weighty argument in terms of the validity of a number of elements within the general scheme of ozone transformations in water (in accordance with the Bell-Evans-Polany principle [33]). This can be supplemented with the following reactions:

$$O_3 + HO^- \rightarrow HOO^- + O_2 \quad (-\Delta H = 288.6);$$
$$HOO^- + O_3 \rightarrow HO^- + 2O_2 \quad (-\Delta H = 442.1);$$
$$HOO^{\cdot} + O_3 \rightarrow HO^{\cdot} + 2O_2 \quad (-\Delta H = 420.6);$$
$$O_2^- + O_3 \rightarrow 2O_2 + O^- \quad (-\Delta H = 267.6).$$

It should be noted that water, when playing the role of an active reaction medium, is not directly involved in the processes of ozone degradation, due to the considerable exothermicity of its molecules' possible reactions with the products of these processes. The above scheme of transformations of ozone in water would be incomplete if we failed to take into account the previously mentioned possibility of degradation of ozone as a result of interaction of ozone molecules with each other. Consequently, ozone dissolved in water gradually degrades via a ramified chain of ionic and biradical reactions initiated by hydroxyl anions, and also through interaction between ozone molecules.

A look at the reactions involved in the oxidation of methane can give an idea of how the oxidizing potential of ozone differs from that of the particles forming as a result of its decomposition:

$$CH_4 + 4/3O_3 \rightarrow 2H_2O + CO_2 \quad (-\Delta H = 837.3);$$
$$CH_4 + 2O_2 \rightarrow 2H_2O + CO_2 \quad (-\Delta H = 337.0);$$
$$CH_4 + 8\,HO^{\cdot} \rightarrow 6H_2O + CO_2 \quad (-\Delta H = 974.5);$$
$$CH_4 + 8/5\,HO_3^{\cdot} \rightarrow 14/5H_2O + CO_2 \quad (-\Delta H = 132.2);$$
$$CH_4 + 8/3\,HOO^{\cdot} \rightarrow 10/3H_2O + CO_2 \quad (-\Delta H = 699.2);$$
$$CH_4 + 4HOO^- \rightarrow 2H_2O + 4OH^- + CO_2 \quad (-\Delta H = 644.1).$$

It may be inferred from the comparison of the thermal effects of these reactions that the oxidizing ability of ozone and the products of its transformations decrease in the series: $HO_3\cdot > HO\cdot > O_3 > HOO\cdot > HOO^- > O_2$. Consequently, the initial stage of the degradation of dissolved ozone is accompanied by the formation of a number of active intermediate products with high oxidizing potentials that can effectively oxidize organic impurities in drinking water.

Thus, it may be concluded that the chemical activity and transformations of ozone are due to its high electron affinity, a considerable plasticity of its structure, an organic combination of polarity and an ability to change to the biradical state.

5.2.2 *Theoretical Analysis of Disinfection Processes of Water in Ozonization*

In the continuing investigations [34, 35] dealing with the theoretical analysis of physicochemical processes occurring in water ozonation, and their impact on efficiency of water conditioning in paper [36], represent the theoretical analysis of disinfection processes of water in ozonation.

One of the most important moments in developing a model for destruction of organic matter and water disinfection is considering the place where a reaction occurs: on the surface or near the surface of emerging bubbles with an ozone-air mixture (OAM) or in the bulk of a solution.

A theoretical model of the destruction of organic matter when passing OAM through water, developed in [34], implies that a reaction between ozone and organic matter occurs neither on the surface of emerging bubbles not in a convection-diffusion layer formed by ozone molecules diffusing from the bubbles, but this reaction does occur in the bulk of the liquid. Such an approach is based on theoretical assumptions about the strong difference between the diffusion velocities of ozone and the larger molecules of organic compounds, which ensures an encounter of reactive molecules outside the convection-diffusion layer. An additional argument for such a supposition was also premised on experimental data we obtained about the absence of a noticeable floatation entrapping of humic acids by air bubbles.

A similar approach may be used for modeling disinfection. At first sight, the direct application of bacteria with air bubbles is an efficient way of disinfection. The ozone concentration in the bubbles is higher than in the water (at least, when the bubbles begin to float up and sparge, and when the ozone concentration in the water is very low). In addition, diffusion mobility of ozone in a gas phase exceeds its mobility in the liquid phase, i.e., ozone molecules could be rapidly brought to the contact point of the bubble and the bacteria.

However, such a variant of deactivating bacteria could hardly be realized. First, as a result of much research into floatation, one may come to the conclusion that the floatation entrapment of bacteria is low. Indeed, with floatation of the particles of the same nature but different size, one can observe the curve of the entrapping efficiency of the particles by the bubbles normally to a minimum corresponding to the size of the particles from fractions of the micron to several microns [37].

On the one hand, given the size of the particles of inertia, movement to the surface of a bubble, which is characteristic of larger particles, is not that obvious. On the other hand, the Brownian mobility of the particles is still not high enough to ensure the diffusion mechanism for the particles to approach a bubble, a mechanism of which is characteristic of submicron particles. The presence of the minimum hardly depends on the surface charge of the bubble and floating particles and therefore is universal. Since the size of bacteria corresponds to the minimum of the floatation degree, one may come to a conclusion about the impossibility or scant likelihood of bacteria's attachment to the surface of the bubble.

Second, the lifetime of the bacteria on the surface of the bubble cannot be long. From the instant of the presumable entrapment of the bacterium, it elapses sometimes, and after the bubble reaches the water surface in the reactor (in the cells—it takes 2.5–3 s from the time the bubble appears) it is destroyed and the bacterium again enters the water. In addition, ozone concentration in the bubble during the time of its surfacing noticeably decreases [34]. Moreover, the decrease of the ozone concentration in the bubbles is expressed mostly in the first seconds after the start of the sparging, when the water is still not saturated with ozone, and the concentration gradient is at a maximum point. However, it is within this time interval that disinfection is most effective.

Taking into account the above considerations, we will assume the bacteria deactivation takes part in the liquid bulk. In this case, an important component in the model of bacteria deactivation provisions is developed for the exchange reaction between ozone and organic matter. Thus, paper [34] showed that the velocity of ozone dissolution in water and, accordingly, the destruction degree of organic impurities depend not only on the ozone diffusion coefficient, the size of the bubbles and the OAM pumping-through speed, but also on the mobility of the surface of the bubbles, features of the diffusion and hydrodynamic ozone transport through the convection-diffusion layer and the rate of ozone decomposition. It is natural that the rate of deactivation of the bacteria in the liquid bulk should depend on the concentration of dissolved ozone, i.e., on the physicochemical processes enumerated above. In addition, since ozone concentration in the liquid form decreases, due to its interaction with organic impurities, bacteria deactivation also should depend on the concentration of organic impurities. Therefore, our work deals with the theoretical modeling of these mutually related processes.

General Notions about the Deactivation Process The interaction of ozone and bacteria is mainly similar to the interaction of ozone and organic compounds since it deals with the destruction of organic molecules on the surface of a bacterium. At the same time, these processes are conspicuously different. Upon the reaction of ozone with an organic molecule, destruction of this or another bond occurs, which, in the long run, leads to the destruction of the molecule as a whole. Furthermore, although the constants of organic matter destruction in [34] have not been calculated but were obtained from the comparison of theoretical models with experimental ones, it is clear that impossibility of calculating the constants is only the result of insufficient information on the fraction make-up and conformation of organic molecules, the probability of the ozone molecule approach to a specific bond, and the rate of the reaction taking place, etc.

The process of bacteria deactivation is rather complicated. The existing models of bacteria death [38–41] are very simplified (normally only the slope of the curve is analyzed) and, unfortunately, these models are not based on concrete processes. The only fact, which is confirmed in virtually every publication, is the relationship between the beginning of bacteria death and the achievement of some critical value of the parameter Ct (the product of the ozone concentration C by the time t from the start of its feeding).

According to [41], the presence of the latent period and the relationship between the speed of disinfection and the parameter *Ct* may be determined with various methods. Among them one can distinguish the relationship between the complexity of achieving the surface of the bacteria, disinfectant diffusion time through the bacterium membrane and their rates of response with vitally important centers of the bacteria [40], which would lead to the cell vitality [43].

To clarify the nature of the latent period of existence, one may also assume that dying-off the accumulation on a multitude of defects on their surface is necessary for bacteria. These defects may be caused by destruction of some or other molecules. It is natural, then, that the greater the ozone concentration, the faster the damage of the bacteria surface. However, it should be remembered that bacteria has the ability to regenerate small damaged sections, but for a substantial damage and loss of vitality it is necessary to have a certain amount of defects emerging through brief time interval. Therefore, it is likely that given a low concentration of ozone, the amount of *Ct* necessary for the start of the disinfection process may increase. It is natural that this, also depend on the type of bacteria.

On the other hand, it has not been ruled out that the latent period and, respectively, the value *Ct* in many cases is linked with not only the properties of the bacteria themselves, but also with the presence of various inorganic and organic impurities in the water, including products of bacteria vital activity. These could also be organic macromolecules present in any natural water disinfectants adsorb and react with these organic macromolecules. Similar macromolecules may cover the surface of the bacteria, impeding the access of disinfectants.

It should also be noted, that as a result of the experiments carried out, the ozone concentration in water highly depends on the presence of bacteria. Therefore, it is possible to analyze only bacteria deactivation and ignore its impact on the ozone concentration. At the same time, the ozone concentration in OAM is comparatively low, and therefore saturation of the water with ozone takes a rather long time. As a result, due to interaction with the bacteria and organic impurities, ozone concentration in the water is not enough to be reduced. Therefore, under present conditions, instead of the *Ct* a more complex value taking into account the relationship between the concentration of dissolved ozone and time appears.

However, the most important issue is not the quantitative description of the processes taking place setting and resolving the issues concerning their reasons and nature. Therefore, instead of approximation equations used in the literature [38–41], it is necessary to obtain equations which take into consideration simultaneous variation in the numbers of living bacteria and the concentration of dissolved ozone and organic impurities.

Disinfection of Distilled Water First of all, let us consider the disinfection of bacteria in distilled water. In this case, one can ignore the residual amount of inorganic and organic impurities, their impact on the ozone concentration and the behavior of the bacteria.

The results of the experimental research of disinfection of the solution containing the bacteria *E.coli* are shown in Fig. 5.4.

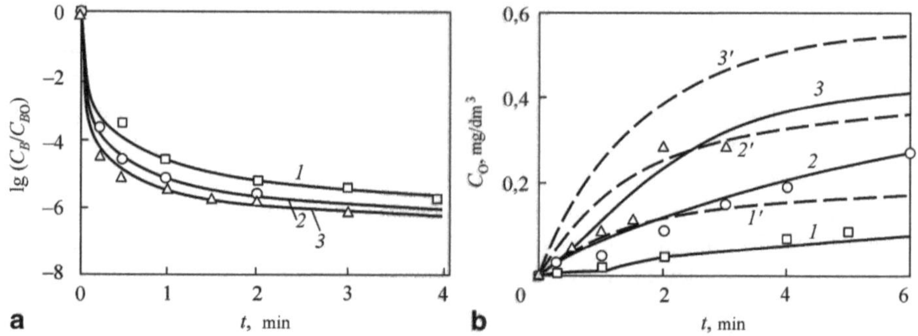

Fig. 5.4 Variation of living *E.coli* (**a**) and the concentration of ozone dissolved in distilled water in the presence of *E.coli* (**b**). The ozone concentration in the ozone-air mixture mg/dm³: *1*—0.5; *2*—1.0; 3—1.5; *1*—*3*—theoretical curves of ozone dissolution in water with bacteria; *1'*—*3'*—the same without bacteria

As can be seen from Fig. 5.4a, in all considered cases, disinfection starts virtually instantaneously so the *Ct* for the beginning of the process tends to be zero. Hence, one can draw the conclusion that bacteria deactivation is related to the destruction of the cell surface. Even so, it has not been ruled out that the defects may spread inside bacteria and disrupt important centers after bacteria sampling for analysis.

The modeling of the deactivation process is carried out [36], taking into account the continuous convection-diffusion dissolution ozone and its consumption for deactivation. Since the characteristic sizes of bacteria are in the order of one micron, they represent Brownian particles, which make it possible to develop such an approach to their description as in the case of macromolecules.

Differential equations for concentration of active bacteria and concentration of dissolved ozone will be represented in the form similar to the one used earlier for describing the interaction between ozone and organic compounds [34]:

$$\frac{dC_B(t)}{dt} = -k_B C^n_{O,L}(t) C^m B(t) \tag{5.1}$$

$$\frac{dC_{O,L}(t)}{dt} = k_1^* \left(\alpha C_{O,G} - C_{O,L}\right) - k_{B2} C^k_{O,L}(t) C^1_B(t). \tag{5.2}$$

where C_B—bacteria concentration; $C_{O,G}$; $C_{O,L}$—ozone concentration in OAM and the liquid; *n*, *m*, *k*, *l*—the parameters characterize the reaction degree; coefficients k_{B1}, k_{B2} describe the rate in the decrease of the concentration of active bacteria and the rate in the concentration of ozone dissolved in water corresponding to this process; *α*—Henry constant; the coefficient k_1^* describes the ozone dissolution rate with delayed surface of the bubbles [34]:

$$k_1^* = \frac{3ShD_O v_G t_{cr}}{V_L} \left[\frac{1 - \exp\left(-3\alpha ShD_O t_{cr} / a^2\right)}{3\alpha ShD_O t_{cr}} \right] \tag{5.3}$$

where D_O—coefficient of ozone diffusion in water; v_G—volumetric velocity of OAM feed; V_L—liquid volume in the reactor; $t_{cr}=h/u$—time at bubble surfacing (h— water height in the reactor; u—bubble surfacing rate); Sherwood number

$$\mathrm{Sh}_{\mathrm{r}} = \left(0.625\sqrt[3]{\mathrm{Pe}} + 0.461\right)\left(1 + \mathrm{Re}^{1/3} / 8\right) \tag{5.4}$$

where $\mathrm{Pe}=au/D_O$—Peclet number; $\mathrm{Re}=au/\nu$—Reynolds number (ν—liquid kinematic viscosity, a—bubble radius).

The solution of the equation should be carried out with the initial conditions, namely:

- At the instant of OAM feed to the reactor of the concentration of living bacteria is equal to the set initial value C_{BO}

$$C_B(t=0) = C_{BO} \tag{5.5}$$

- The initial concentration of the dissolved ozone is equal to zero

$$C_O(t=0) = 0 \tag{5.6}$$

Use of Eq. (5.1) and (5.2) implies that disinfection occurs in the bulk of the solution rather than at the surface of the bubbles; the process which limits bacteria deactivation is mutual diffusion of ozone and the bacteria (i.e., there are no additional processes whose diffusion rate is lower than that of the rate of diffusion ensuring the number of collisions of the ozone molecules with the bacteria).

The coefficient k_1^* in (5.2) is responsible for physicochemical processes at the surface of the bubble. This is obtained according to formulas (5.3) and (5.4), based on the theoretical model developed in [34], and is equal to 0.72 min^{-1}. Unlike [34], where the coefficient k_2 in the differential equations for ozone destruction and destruction of organic impurities was one and the same, two different coefficients (k_{B1}, k_{B2}) were introduced to (5.1), (5.2). This is related to the fact that bacteria deactivation has its specific features, which do not allow us to describe it by analogy with organic matter. Let us assume that the coefficients are identical, while all the powers equal to unity, i.e., the reactions of the first order are run according to the concentration of each interacting substances. Several variants of the calculation for this case are presented in Fig. 5.5 (curves *1*–*4*). The calculations for these curves are fulfilled with the following parameters: $n, m, k, l=1$; $k_{B1}=10^{-9}$ (curve *1*); $k_{B1}=10^{-8}$, $k_{B2}=10^{-8}$ (curve *2*); $k_{B1}=10^{-6}$, $k_{B2}=10^{-6}$ (curve *3*); $k_{B1}=10^{-6}$ dm³/min·mg, $k_{B2}=3\times10^{-7}$ dm³/min CFU (curve *4*),where CFU—are colony-forming units.

Neither of the theoretical curves presented here, according to the assumption of linear relationship between the reaction rate and concentration, qualitatively approaches the experimental curves. Moreover, an experimental curve for disinfection is concave, while theoretical ones are convex. Similar convex curves were obtained during calculations with other values of the constants. Such curves correspond to disinfection with a large latent period and *Ct* parameters. However, in the

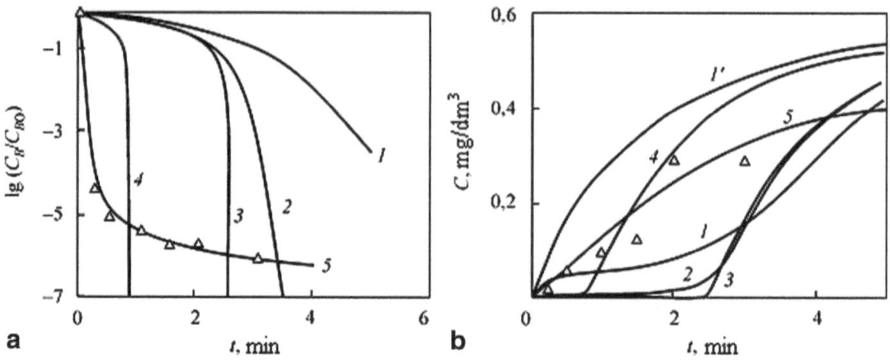

Fig. 5.5 Variation of the number of living *E. coli* (**a**) and concentration of ozone dissolved in bidistilled water (**b**) at $C_{O,G}$=1.5 mg/dm³ (*b*) (Δ—experimental data): *1*–*5*—theoretical relationships (coefficients used when designing these curves are given in the text); *1′*—a theoretical curve of ozone dissolution in the water without bacteria

experiment the latent period is virtually absent and, consequently, *Ct* for the initial period of disinfection tends to be zero. Thus, all or part of the powers should differ from unity.

Since the experimental curves have a pronounced nonlinear nature, we will discuss what mechanism may be responsible for the deviation from linearity.

Emerging differences between the theoretical and experimental data with long periods of water treatment may be related to various factors, for instance, to the presence of the zones of stagnation in which the ozone of concentration is decreased, while disinfection is delayed. It cannot be excluded that when the bacteria are introduced into the water, they do not completely decompose into a separate bacterium but remain in the form of aggregates of bacteria. In this case, the access of ozone to the central bacteria is slowed down. The special stability of individual bacteria, whose surface appear to be protected, is possible, perhaps by a layer of adsorbed organic impurities introduced to the bidistilled water, together with the bacteria.

After the decrease of the concentration of living bacteria by many digital orders, even the small amount of bacteria that survived is enough for deterioration of the general process of disinfection. However, such a slow-down of disinfection can be observed not only in our research but also in other investigations [41]. Therefore, it should be assumed that there are several processes slowing down disinfection.

As noted earlier, when a certain concentration of ozone is achieved, sufficient defects leading to the death of bacteria appear. Since survivability of different cells is different than for the death of a specific bacterium, a different number of defects are needed. Therefore, the degree of disinfection has a certain statistic nature for the given aggregate of bacteria. However, this does not mean that after the death of absolutely all cells the ozone consumption should be terminated. It may continue to form new defects on the surface of already dead bacteria or be spent on interaction (adsorption or reaction) with the products of cell lysis. Hence, it follows that the decrease in the concentration of living bacteria and ozone take place at a substantially different rate, and therefore the values *n, k, l, m* may differ from unity. In addition, it may turn out that $n \neq k$, $l \neq m$. Usually, in the equations on chemical kinetics these powers are identical. However, their inequality may be used when approximating the resultant relationships; it may also be an indication of the fact that the system is where some processes are taking place which are not expressed in an explicit form in the equations. The difference between the powers reflects on the coefficients, which acquire not only different values but also unequaled dimension.

Optimization of the compliance of theoretical relationships obtained during numerical solution of Eq. (5.1), (5.2) with the experimental data on deactivation of *E.coli* has led to the following values of the constants:

$$k_{B1} = 10^{-6}\,\text{dm}^3/(\text{min mg})\left(\text{dm}^3/\text{CFU}\right)^{1,3}; k_{B2} = 3\times10^{-5}\,\text{dm}^3/\text{min CFU};$$
$$n,k,l = 1;\ m = 2.3.$$

Kinetic curves obtained in this case are shown in Fig. 5.4 (curves *1–3*) and in Fig. 5.5 (curves *5*). It is necessary to note that the same parameters provide good correspondence between the number of survived bacteria and the concentration of dissolved ozone at different values of ozone concentration in OAM (see Fig. 5.5).

The value of the parameter $m=2.3$ in Eq. (5.1) indicates that the greater the number of initial living cells, the slower their death. This can be explained by the following: during lysis of dying cells their internal contents enter the solution, thereby creating a medium making it difficult for ozone to have an access to the surface of a dying bacterium or other living cells surrounding it. The fewer the number of living cells, the greater the number of organic substances (determined by lysis) which negatively affect the deactivation of the cell that remains living. This is not necessarily absorption or decomposition of ozone by organic matter entering the solution, but, most likely, it is a simple creation of conditions for lowering the rate of ozone diffusion in the bulk of the liquid or for aggravating the direct contact of ozone with the surface of the bacteria. At least, an attempt of introducing Eq. (5.2) the component responsible for absorption or decomposition of ozone to lysis (i.e., the component proportional to the number of dead cells) did not result in better conformity with the experiment. Additionally, it did not only improve, but even became worse. In this case the decrease of ozone concentration, due to its reaction with organic matter, determined by lysis, leads to even more convex kinetic curves than those shown in Fig. 5.5.

Since it is very complicated to model this process, it is very likely that the model does not take into account the whole variety of bacteria's behavior during lysis. Thus, the above theoretical model implies that during lysis, organic matter enters the solution and spreads it uniformly in the course of destruction or right after bacteria destruction, and continues to emanate for several seconds. However, for the arising or uniform distribution of these organic substances some time is necessary. Even if it is only several seconds or several tenths of seconds, it is enough for the first, i.e., fast stage of disinfection. Perhaps a formally chosen parameter $m=2.3$ reflects the kinetics of the given process: many bacteria, i.e., a few organic substances enter the

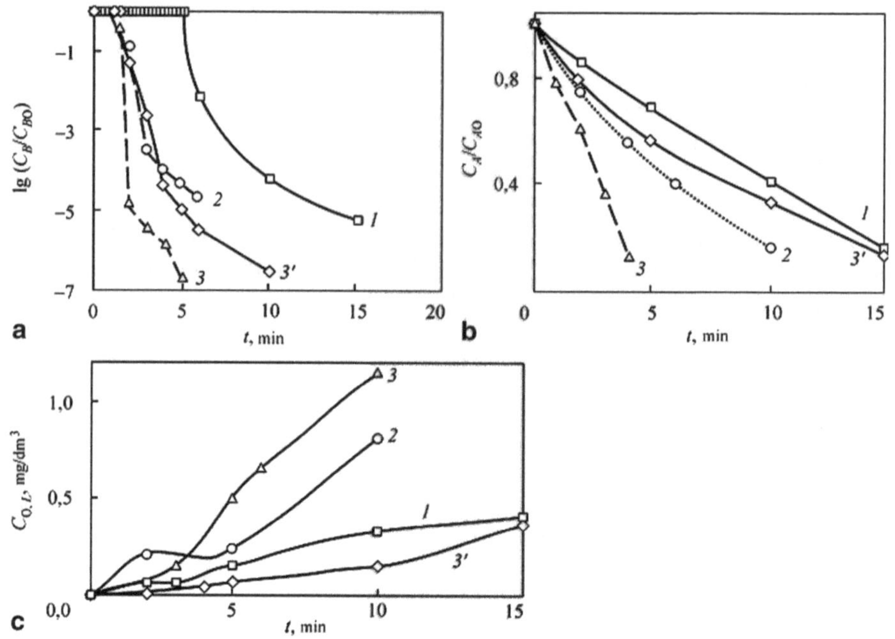

Fig. 5.6 Experimental relationships between the concentrations of living bacteria (**a**), humic acid (**b**), dissolved ozone (**c**) and time. $C_{O,G}$ (mg/dm^3): 5(*1–3*) and 2 (*3′*); C_{AO}, (mg/dm^3): 20 (*1*); 10 (*2*); 5 (*3, 3′*)

solution because of the lysis, disinfection proceeds quickly; there are fewer living bacteria, i.e., the higher concentration of organic matter entering the solution, and disinfection proceeds very slowly.

Since the decrease in the concentration of living bacteria after the start of the sparging proceeds very fast, the *Ct* value for the start of the disinfection process constitutes not more than several seconds and therefore cannot be determined experimentally. At the same time, the fact that the latent period for distilled water with rather low concentrations of ozone used in the experiment tends to be zero means that the reasons discussed in [40–43] for the emergence of the latent period are not real, at least for the *E.coli* bacteria studied in the paper [36].

We come to the conclusion that when using natural or model water with the inclusion of organic impurities, the latent period and the *Ct* parameter are linked not so much by the interaction between ozone and bacteria as by the presence of additional factors determined by the destruction of organic matter. This process, in fact, will be analyzed in the following section.

Disinfection of Water Containing Organic Impurities Experimental data for several initial concentrations of organic impurities and two of ozone in OAM are shown in Fig. 5.6. As one can see in the aforementioned Fig. 5.6, the presence of humic acid results in the increase of the *Ct* parameter, and its value increases along with the concentration of organic impurities.

To explain the above in paper [36], we will make a few assumptions. First of all, we will assume that the processes occur in distilled water in the presence of bacteria and in water containing organic impurities, but not containing bacteria in a new more complex system, are independent. This means that all constants used in previous calculations in paper [36] and in [34, 35] remain intact. However, the equations for the above mentioned processes should be generalized.

Supplement Eq. (5.1) and (5.2) with an equation for destruction of organic compounds [34, 35] and introduction of the corresponding terms to Eq. (5.2) for the concentration of dissolved ozone:

$$\frac{dC_A(t)}{dt} = -k_2 C_{O,L}(t) C_A(t) \tag{5.7}$$

$$\frac{dC_B(t)}{dt} = -k_1 C^n_{O,L}(t) C^m B(t) \tag{5.8}$$

$$\frac{dC_{O,L}(t)}{dt} = k_1(aC_{O,G} - C_{O,L}) - k_2 C_{O,L}(t) C_A(t) - k_{B2} C^k_{O,L}(t) C^l_B(t) \tag{5.9}$$

satisfying the initial condition for the concentration of organic matter:

$$C_A\ (t = t_1) = C_{AO} \tag{5.10}$$

and the initial conditions for bacteria and dissolved ozone (5.5), (5.6).

Let us give several theoretical curves (Fig. 5.7a–c, curve *1*), obtained during numerical calculations by these equations which take into account the constants k_1=0.72; k_2=0.25 min^{-1} when modeling destruction of humic acids [34, 35] and constants $k_{B1}=4\times10^{-6}$ dm^3/min mg (dm^3/CFU)^{1,3}; $k_{B2}=3\times10^{-5}$ dm^3/min CFU; n, k, l=1; m=2, 3, used when modeling disinfection of bidistilled water.

Kinetic curves obtained concurrently with these parameters for disinfection do not correspond to the experimental curves. Although much of the ozone's energy is spent on the destruction of organic impurities, the theoretical parameter Ct for bacteria tends to zero, as in distilled water. Thus, ozone consumption for decomposition of humic acids is not responsible for the emergence of the latent phase in the course of disinfection. In addition, the constant k_2=0.25 min^{-1} very accurately describes the data presented in [34, 35] and leads to the theoretical curve, characterizing a weaker destruction than obtained experimentally. Although the investigations were carried out in the same cell to explain the resultant data, we still may infer that the experimental conditions were not absolutely identical, which led to a somewhat faster destruction of humic acid.

Based on these considerations, we carried out calculations for other constants: k_2=0.5 min^{-1} (see Fig. 5.7a, curve *2*), k_2=1 min^{-1} (curve *3*) and even k_2=5 min^{-1} (curve *4*). Despite such unusually large constants, the theoretical curves for destruction of humic acid approached the experimental curves. However, the theoretical

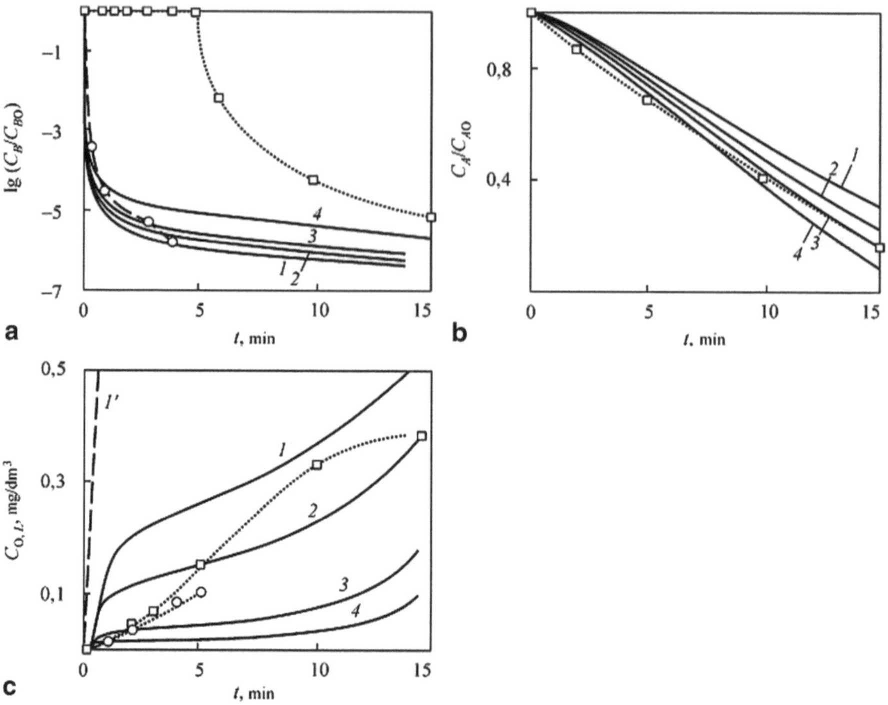

Fig. 5.7 Experimental and theoretical relationships between the concentrations of living bacteria (**a**), humic acid (**b**), dissolved ozone (**c**) and time: $C_{O,G}=5$; $C_{AO}=20$ mg/dm^3; ○, □—experimental data respectively on disinfection in bidistilled water and in the presence of organic impurities; curve *1′*—theoretical relationship for ozone dissolution in distilled water. k_1: 0.25 (*1*), 0.5 (*2*), 1 (*3*) and 5 min^{-1} (*4*) (other coefficients used during calculations are given in the text)

concept of ozone in water turned out to be much underrated, while the latent period of the solution, despite large ozone consumption for the destruction of organic compounds, disinfection failed to arise. In addition, all initial sections of theoretical curves were close to the disinfection curve in bidistilled water.

Let us quote several calculations for other models. We can assume that a certain mechanism which inhibits the negative effects of ozone on bacteria appears although ozone diffuses freely in a solution, and when it enters the surface of the bacteria its concentration decreases. This case may be described by Eq. (5.7)–(5.9) and in (5.9) we also introduce C_{BO} instead of $C_B'(t)$, i.e., we believe that concentration of living bacteria does not depend on time, which corresponds to the experimental data at small times of water treatment. In this case Eq. (5.8) takes the form:

$$\frac{dC_B(t)}{dt} = 0 \qquad (5.11)$$

Using the coefficient from the previous calculations ($k_1=0.72$; $k_2=0.25$; 0.5; 1; 2 min^{-1} and $k_{B2}=3\times10^{-5}$ dm^3/min CFU) we find that in this case ozone is absorbed so completely by bacteria that its presence in water is virtually equal to zero, although this is not in agreement with the experimental data.

It is possible choose also another approach. We will infer that in the latent period, ozone interacts not only with organic impurities but also with bacteria. In making this connection we divided the task into two time intervals: before and after the beginning of bacteria death. In the first time interval [0, *t*] was used the equations for dissolution of ozone, for the destruction of organic compounds and also for the absorption of bacteria with ozone, without lowering the concentration of Irving bacteria:

$$\frac{dC_A(t)}{dt} = -k_A C_{O,L}(t) C_A(t) \tag{5.12}$$

$$\frac{dC_{O,L}(t)}{dt} = k_1(\alpha C_{O,G} - C_{O,L}) - k_A C_{O,L}(t) C_A(t) \tag{5.13}$$

where k_A—constant characterization of the ozone's interaction with organic compounds; C_A (*t*)—concentration of organic compounds satisfying initial conditions (5.10).

In the second interval [*t*, ∞], these two equations are supplemented with the equations describing bacteria death, i.e., in place of (5.1), (5.2), (5.12) and (5.13) we will obtain the following equations:

$$\frac{dC_A(t)}{dt} = -k_A C_{O,L}(t) C_A(t) \tag{5.14}$$

$$\frac{dC_A(t)}{dt} = -k_A C_{O,L}(t) C_A(t) \tag{5.15}$$

$$\frac{dC_{O,L}(t)}{dt} = k_1(\alpha C_{O,G} - C_{O,L}) - k_2 C_{O,L}(t) C_A(t) - k_{B2} C_{O,L}^k(t) C_B^l(t) \tag{5.16}$$

which do not meet the initial conditions, namely:

- The concentration of organic matter equals the value C_A^* obtained when solving Eq. (5.12), (5.13) at time instant t_1

$$C_A(t = t_1) = C_A^* \tag{5.17}$$

- The concentration of living bacteria equals the initial one

$$C_B(t = 0) = C_{BO} \tag{5.18}$$

- The concentration of dissolved ozone equals the value C_A^* obtained when solving Eq. (5.12), (5.13) at time instant t_1

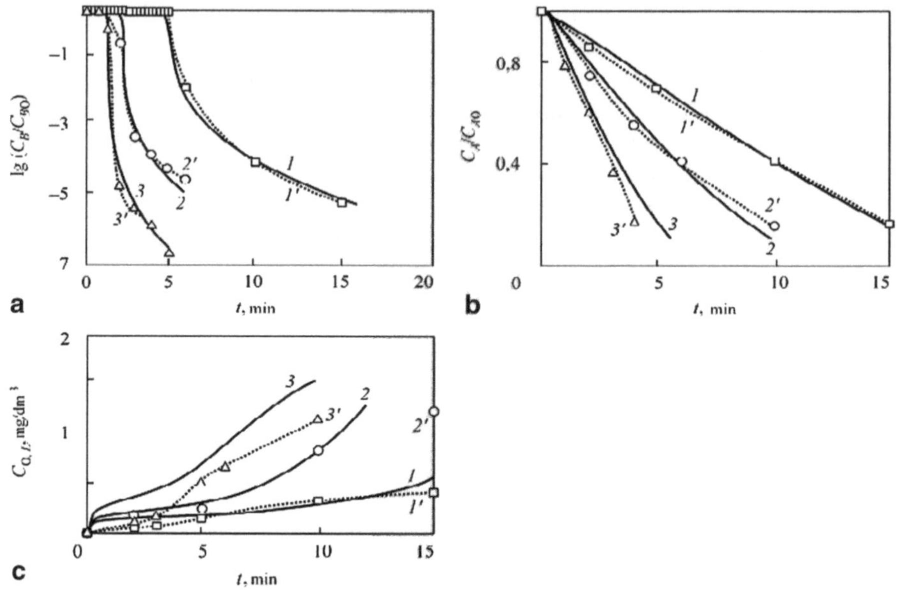

Fig. 5.8 Theoretical (*1*—*3*) and experimental (*1′*—*3′*) relationships between the concentrations of living bacteria (**a**), humic acids (**b**), dissolve ozone (**c**) and time: $C_{O,G}=5$ mg/dm³; $C_{AO}=20$ (*1*, *1′*); 10 (*2*, *2′*) and 5 mg/dm³ (*3*, *3′*)

$$C_O(t = t_1) = C_O^* \tag{5.19}$$

Numerical calculations done using these equations (Fig. 5.8) have demonstrated that the maximum possible approximation for experimental relationships is achieved when the initial concentration of organic compounds and ozone concentration in OAM is changed. This occurs when the coefficients change in (5.13), (5.16).

The only coefficient which retains its value for all the curves is $k_1=0.72$. The coefficient k_2 varies from 0.5 at $C_{AO}=20$ mg/dm³ to 0.75 at $C_{AO}=5$ mg/dm³. Unfortunately, papers [34, 35] did not investigate the relationship between the destruction rate of organic compounds and their concentration. However, despite the sufficiently low concentration of impurities, the efficiency of ozone- and macro-molecular collisions decreases when their own concentration and the concentration of the products of their destruction increases. In addition, the latter may also decompose molecular ozone themselves, which indicates the kinetic coefficient k_2.

As for the coefficients k_{B1} and k_{B2}, it turns out that the maximum agreement between the theoretical curves and the experimental ones is achieved when we assume that *n*, *k*, *l*=1 and *m*=1.4. Since the reaction order also varies naturally, not only the value but also the dimension of the constant k_{B1} changes. The optimal variants of the constant k_{B1} for $C_{O,G}=5$ mg/dm³ and $C_{AO}=20$ (curve *1*), 10 (curve *2*) and 5 mg/dm³ (curve *3*) are respectively: 0.01; 0.02 and 0.05 dm³/min mg (dm³/CFU)$^{0.4}$. The constant k_{B2} in the all cases is equal 10^{-9} dm³/min CFU.

If we analyze the relationship of constants k_{B1} for different concentrations of impurities, then we can see that their values are inversely proportional to the value of computed effective values Ct_{eff} based on experimental values in the formula:

$$\int_0^T C_{O,L}(t)dt = CT_{eff} \tag{5.20}$$

The greater Ct_{eff}, the smaller the disinfection rate will be. This fact and the rules governing the variations of the constants k_{B1} and k_{B2} make it possible to put forward a hypothesis about the change of the disinfection mechanism.

The decrease from the value k_{B2} to values at which the variation of these values by several times virtually does not affect the calculated values means that direct interaction of ozone and bacteria compared with disinfection in bidistilled water is substantially lowered. This is determined by deterioration in ozone access to the surface of bacteria due to faster collisions of ozone molecules with organic impurities. This cause of deterioration may be proved by comparing the efficiency of ozone molecule collisions with bacteria and organic molecules.

Actually, the number of collisions Z between two molecules depends on their radius (a_1, a_2), the molecular weight (m_1, m_2) and the number of molecules in the unit volume (n_1, n_2) [44]:

$$Z_{1,2} = \pi(a_1 + a_2)^2 n_1 n_2 \left(\frac{4kT}{\pi m^*}\right) \tag{5.21}$$

where kT is the energy of heat movement;

$$m^* = \frac{m_1 m_2}{m_1 + m_2} \tag{5.22}$$

Taking into consideration that ozone molecules are substantially smaller than the molecules of organic impurities (for instance, $a_1 >> a_2$ and $m_1 >> m_2$ or $m^* \approx m_2$), the expression (5.21) may be simplified:

$$Z_{1,2} = \pi a_1^2 n_1 n_2 \left(\frac{4kT}{\pi m_2}\right) \tag{5.23}$$

Additionally, bacteria are Brownian particle expressions (5.21)–(5.23) which may also be used for the analysis of the number of collisions of bacteria with ozone molecules and humic acid. Taking into account (5.23) the number of organic impurities—ozone Z_{AO} and bacteria—ozone Z_{BO} and bacteria—organic impurities Z_{BA} collisions will be represented in the form:

$$Z_{AO} = \pi a_A^2 n_A n_O \left(\frac{4kT}{\pi m_O}\right); \; Z_{BO} = \pi a_B^2 n_B n_O \left(\frac{4kT}{\pi m_O}\right);$$
$$Z_{BA} = \pi a_B^2 n_B n_A \left(\frac{4kT}{\pi m_A}\right) \tag{5.24}$$

where a_A and a_B are the effective radii of the molecules of humic acid and bacteria; n_A, n_B and n_O are the numerical concentration of interacting molecules and bacteria; a_B and $n_B = C_{BO}$—a radius and numerical concentration of bacteria.

Compare the number of collisions

$$\frac{Z_{AO}}{Z_{BO}} = \frac{a_A^2 n_A}{a_B^2 n_B} \text{ and } \frac{Z_{AO}}{Z_{BO}} = \frac{a_A^2 n_O m_A}{a_B^2 n_B m_O} \tag{5.25}$$

Taking into account the sizes of bacteria and humic acids, and inaccuracies in determining the effective radius of macromolecules and bacteria, we obtain: $Z_{AO}/Z_{BO} \sim 10 \div 50$ and $Z_{AO}/Z_{BA} \sim 10^{10}$.

Thus, the number of collisions between organic impurities and ozone Z_{AO} compared with the number of collisions between bacteria and ozone is at least 10 times lower. This means that the reaction is diffusion-limited and it is quite likely that ozone is spent on decomposition of organic impurities and then on disinfection, meaning that it arrives on the surface of the bacteria later and in small amounts. The higher the concentration of impurities, the greater the ozone consumption for their decomposition, which leads to more destructive products and more complicated ozone diffusion, both through the bulk of the liquid to the bacteria and through the adsorption layer on its surface.

The relationship between Z_{AO} and Z_{BA} shows that the probability of collisions of organic impurities with ozone is substantially higher than with the bacteria. Hence, one may come to the conclusion that there are finer and more mobile products of organic matter destruction rather than substances that adsorb the surface of bacteria. However, the adsorption layer of both the initial organic substances and products of their destruction is sufficiently filled in a way that may prevent ozone penetration to the surface of the bacteria and thereby slow down disinfection.

Since the reaction constants during diffusion-limited reactions reflect the exact number of collisions between interacting substances, the relationship between the constants k_{B1} and the concentration of the impurities also reflects the same process. With regard to the relationships Z_{AO} and Z_{BA}, disinfection at low values of the ozone absorption coefficient k_{B2} suppresses the vital activities of the bacteria. This is not only due to ozone, but also to the negative effect of the destruction products of organic compounds.

Thus, the theoretical model of water disinfection we developed takes into consideration convection-diffusion processes at the surface of the floating bubbles, containing OAM, and the interaction of dissolved ozone with bacteria. Ozone with organic impurities made it possible to describe the resultant experimental kinetic curves for dissolved ozone, the deactivation of bacteria and the destruction of humic acids. We have analyzed possible mechanisms of altering the disinfection kinetics during the changes of water chemical composition.

5.3 Bactericide Effect of Ultrasound

5.3.1 *The Mechanism of the Effect of Ultrasound on Aqueous Systems*

The effect of ultrasound on aqueous systems has traditionally been explained by the formation of areas with high concentrations of mechanical energy that produce changes in the sample. Ultrasound exerts a disinfecting action on drinking and waste water [45–47]. The destruction of organic matter in water with ultrasound has been attributed to free-radical processes [48–51]. In particular, it was demonstrated [48] that ultrasonic treatment was accompanied by the formation of products of sonochemical reactions, with a possible attendant formation of radicals enhancing the oxidizing action of hydrogen peroxide [49]. It was established [50] that certain oxidants could be completely removed from water by ultrasonic treatment. Formation of free radicals in the process of ultrasonic treatment was confirmed in [51] by the EPR method with the use of spin scavengers.

It is customary to assume that cavitation is a phenomenon that lies at the heart of the above-mentioned and other manifestations of the chemical action of ultrasound. This, however, is only qualitatively related to the treatment results achieved. This is explained by the complexity of the phenomenon and the dependence of the development of cavitation on many conditions (presence of impurities, temperatures, characteristics of the ultrasonic field, etc.). On the other hand, the role of other factors that determine the effect of ultrasound has not been adequately explored [52]. The variety of effects produced by ultrasonic treatment of aqueous systems suggests that in addition to cavitation, other mechanisms may operate. In light of this investigation, paper [53] studied the conditions of initiation of electrochemical processes in aqueous systems exposed to the action of ultrasound.

Charges on the Surface of Phase Interfaces of Aqueous Systems According to the present-day notions, at the interfaces of phases (liquid and solid or liquid and gaseous) surface charges that form a double electric layer (DEL) exist [54]. Water and solutions may be viewed as disperse systems because they always contain gases that constitute a dispersal phase in the form of microbubbles or bubbles. In addition, water contains molecular or ionic associates, i.e., formations that differ in local density from the bulk water. Filled clusters [55], whose concentration in water reaches 32 % [56], may also be regarded as a disperse phase with respect to the rest of the water. Apparently, despite the clusters' short lives, they have an interface with properties characteristic of disperse systems.

Relative Phase Displacement in Ultrasonic Treatment Ultrasonic treatment causes the phases of a disperse system to move relative to each other, which was experimentally confirmed in [57]. The vibratory motion of the acoustic transmitter induces displacement of the disperse system components, which is not uniform for the disperse phase and the disperse medium [58]. The acceleration can acquire under the constantly applied force F depends, according to

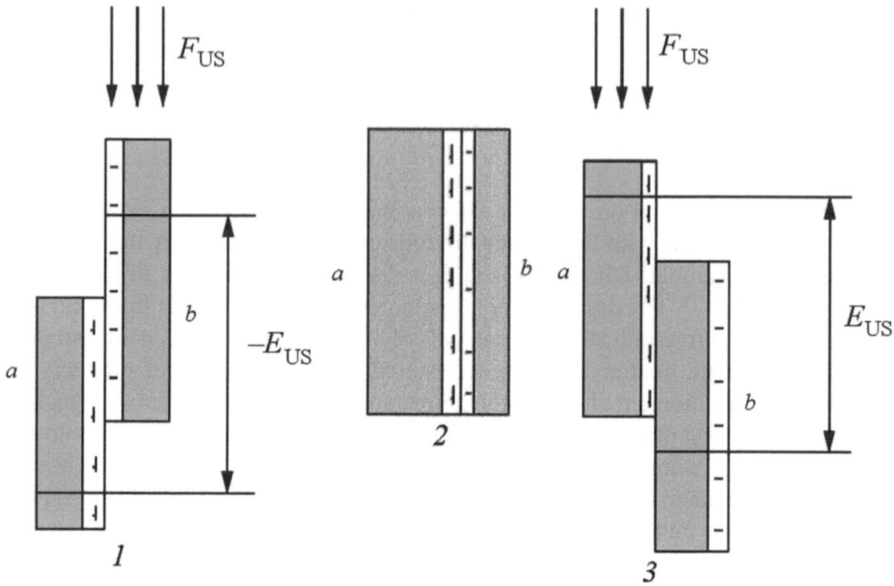

Fig. 5.9 Schematic of formation of polarization emf, E_{us}, at phase interface under the effect of ultrasound: *a, b*—disperse phase and disperse medium; *1*—density of disperse phase higher than that of disperse medium, *2*—phase boundary in the absence of ultrasonic action, *3*—density of disperse phase lower than that of disperse medium (F_{us}—force operating during time $\tau_0 = 0.5/f$)

$$a = F / m \tag{5.26}$$

on the particle mass *m*. This means that in the vicinity of the transmitter generating the force *F*, the speed of a bubble or a solid disperse particle differs from the speed of the adjacent liquid because of their different densities and, accordingly, masses. If the density of a solid particle is higher than that of the disperse medium (e.g., for water and quartz particles the densities are 1.0 and 3.3 g/cm^3, respectively), then, according to Eq. (5.26), the acceleration acquired by a quartz particle is 3.3 times lower than that of the adjacent water layer of the same volume. This is why a moving quartz particle exposed to the mechanical force of ultrasound lags behind the motion of the adjacent water layer. For a gas bubble, whose density is much lower than that of water, the picture is different: bubbles move faster than the adjacent water layers. It should be noted that in an unconfined space containing a moderately viscous liquid (such as water) displacement of a gas bubble under the effect of ultrasound occurs constantly, whereas the subsequent reduction of its size through cavitation may or may not take place.

Thus, because of different densities of the disperse medium and particles of the disperse phase, a tangential displacement of the liquid at the phase interface occurs (see Fig. 5.9). The magnitude of this displacement is related to the duration of the ultrasonic pulse, which depends on the frequency *f* and the amplitude of the ultrasonic field. The frequency of the ultrasonic field determines the duration of the mutual phase displacement process, τ_0:

$$\tau_0 = 1/2f \tag{5.27}$$

For example, for ultrasound of $f=22$ kHz, in the case of square pulses, the time of possible tangential displacement in one direction is: $\tau_0 = 1/2 \times 22000 = 2.3 \times 10^{-5}$ s.

The velocity of displacement grows with the ultrasonic frequency. However, even in low-frequency fields (20–50 kHz) it exceeds the mobility of charged particles in the solution. This creates conditions under which lingering relaxation processes result in a local accumulation of charges of the same sign, thereby producing polarization of the DEL.

The amplitude of oscillations of the transmitter determines the magnitude of the tangential phase displacement, S_0. This value, and the time during which the displacement takes place, can be used to calculate the velocity of relative phase displacement, v:

$$v = S_0 / \tau_0 \tag{5.28}$$

Thus, for instance, with a vibrational amplitude (A) of 1 μm (corresponding to $S_0=1$ μm), $v=1\times10^{-6}/2.3\times10^{-5}=4.3\times10^{-2}$ m/s. The vibrational amplitude of the transmitter depends on the ultrasound power and, with high powers, may reach 1 mm. In this case the phase displacement velocity at a frequency of 22 kHz will be 43 m/s.

Mechanical Polarization of the Double Electric Layer in Ultrasonic Treatment The effects produced by the displacement of phases relative to each other are associated with changes in the state of the DEL at the interface between the disperse medium and the disperse phase, and are described by equivalent formulae for electrophoresis, electro-osmosis, the flow potential, and the sedimentation potential [59]. The classical Smolukhovsky formula relates the velocity of the motion of phases relative to each other to the potential gradient in the tangential plane along the direction of motion θ and the potential of the DEL diffuse part, ξ:

$$\zeta = \eta v / \varepsilon_0 \varepsilon \theta$$

where η and ε are the viscosity and the permittivity of the disperse medium, and ε_0 is the electric constant.

This formula makes it possible to assess the potential gradient θ arising as a result of displacement of phases under the effect of ultrasound:

$$\theta = \eta v / \varepsilon_0 \varepsilon \zeta \tag{5.29}$$

For example, for the case of displacement of an air bubble that has a ζ value in water of 20 mV with $\eta=1$ mPa, $\varepsilon=81$, $\varepsilon_0=8.85\times10^{-12}$ F/m and $v=43$ m/s (i.e. a vibration amplitude of 1 mm), $\theta=30\times10^6$ V/cm. At amplitude of 1 μm the θ value is 30×10^3 V/cm.

Combining the expressions (5.27), (5.28), and (5.29), and in view of $\theta=EIS_0$, we obtain a formula relating the polarization potential difference, E, to the characteristics of the disperse medium (η, ε), the phase interface (ζ), and the ultrasound parameters (amplitude and frequency):

$$\mathrm{E} = 2\,\eta\, S_0^2 f\,/\,\varepsilon_0 \varepsilon\, \zeta = \eta\, S_0^2 f\,/\,\varepsilon_0\, \varepsilon \zeta \tau_0 \tag{5.30}$$

One direct confirmation of the existence of polarization potentials is luminescence observed in ultrasonic treatment of liquid samples, which was believed to be associated with ionization of gas in a bubble in the process of cavitation [57]. Since there must be a potential difference for ionization of a gas, the chief cause of luminescence may be the rising emf, according to Eq. (5.30), as a result of the slow running-off of accumulated charges; this is because of the low conductivity or dielectric rigidity of the liquid.

Electrolysis Caused by Polarization Potential Difference Thus, phase displacement, as a result of mechanical polarization in ultrasonic treatment, produces high local electric tensions in the liquid phase. The electric discharge caused by the potential difference triggers chemical processes. Since the disperse medium is conductive, an electric current traverses the solution initiating electrochemical reactions. The current magnitude (I)

$$I = E\chi \tag{5.31}$$

with the specific conductivity of water $\chi_0 = 1 \times 10^{-5}$ S/cm and the cell constant $C_{cl} = \chi_0 / \chi$, may reach, for the above E values, 0.3 to 300 A, causing, in addition to chemical reactions, a rapid heating of the sample. A peculiar feature of the ultrasonic treatment is the initiation of electrochemical processes in the absence of electrodes in the ordinary sense (i.e., without use of metal electrodes). The role of electrodes is performed by areas of the solution with increased concentrations of charges of the same sign, formed in the process of polarization of DEL. Charges travel from these areas through the conductive solution, causing electrolytic processes upon attainment of the decomposition potential.

Knowing the ultrasonic treatment time (τ_{us}), we can calculate the quantity of electricity (Q) that passes through the solution

$$Q = I\tau_{us} \tag{5.32}$$

and the concentration of the substance taking part in the electrolysis

$$\mathrm{C} = Q\,/\,FV, \tag{5.33}$$

where F is the Faraday number and V is the volume of the liquid.

Combining (5.31), (5.32), and (5.33) we obtain an expression relating the change in the substance concentration under the effect of the current of the polarization potential difference to the duration of exposure to ultrasound:

$$\mathrm{C} = E_{\chi o}\tau_{us}\,/\,\mathrm{C_{cl}}FV \tag{5.34}$$

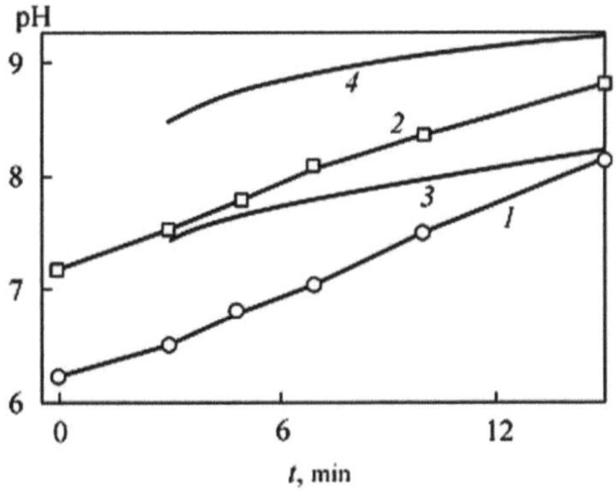

Fig. 5.10 Effect of duration of ultrasonic treatment on water pH: *1, 3*—distilled water; *2, 4*—tap water; *1, 2* experimental curves; *3, 4*—curves calculated from Eqs. (5.30) and (5.34)

Effect of the Disperse Phase Charge Sign on Direction of Electrolysis in Ultrasonic Treatment Since in the ultrasonic polarization charges of one sign are located at the surface of the disperse phase and are therefore more bonded than charges of the other sign (which are in the solution), it is the latter charges that can have a greater effect on the end products of the electrolysis. In other words, depending on whether the surface of the disperse phase is charged positively or negatively (as judged from the ζ-potential sign) electrolysis during ultrasonic treatment will proceed, respectively, by the cathode or the anode mechanism. Thus, for example, it is known that electrolysis of water in the cathode region proceeds according to the reaction

$$H_2O + e^- \rightarrow \tfrac{1}{2}H_2 + OH^- \tag{5.35}$$

and in the anode region according to

$$H_2O \rightarrow SO_2 + H^+ + 2e^- \tag{5.36}$$

The experiments on the ultrasonic treatment of distilled and tap water, using a UZDN-2 generator operating at 22 kHz (with a transmitter diameter of 1.5 cm and a sample volume of 80 cm^3), showed pH increases (Fig. 5.10) indicating that the ultrasound-induced electrolysis of water proceeded primarily by the cathode mechanism described by reaction (5.35). Consequently, as a result of polarization of the bubble surface/water interface, the negative charges were concentrated in the water and the positive ones at the bubble surface. After adding a small quantity of an anionic surfactant (of the sodium dodecyl sulfate type) to the water, the bubble surface was recharged and the ultrasonic treatment reduced the pH value.

It is known [59] that the effect of ultrasound on an aqueous system can be changed by changing the nature of the gas dissolved in the liquid. In light of the latest notions of the electrochemical effect of ultrasound, it is necessary to take into account not only the oxidation-reduction properties of a gas but also the polarization characteristics of its bubbles' surfaces.

Figure 5.10 shows theoretical curves for the relationship between the water pH and the ultrasonic treatment time calculated from Eqs. (5.30) and (5.34). It can be

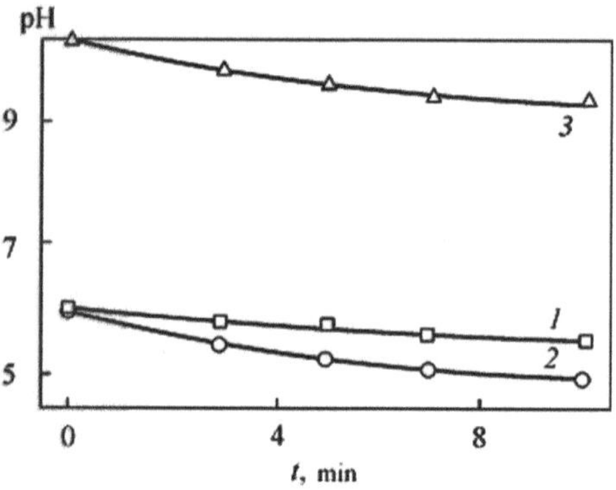

Fig. 5.11 Effect of ultrasonic treatment time on pH of aqueous suspensions: *1*—0.0063 % $Fe(OH)_3$; *2*—0.013 % $Fe(OH)_3$; *3*—2.5 % anthracite

seen that the calculated and the experimental data are in agreement in the order of magnitude of pH changes. The higher pH values for the theoretical curves, as compared to the experimental ones, are probably due to unaccounted losses of the electrolytic current in the real systems.

If the predominant role in a suspension is played by a dispersal phase, the negatively charged surface, ultrasonic treatment reduces pH. This was confirmed by experiments in which suspensions exposed to ultrasound had initial pH values both below and above the neutral point (Fig. 5.11). As can be seen from Fig. 5.11, increases in the concentration of iron hydroxide were made for pH decreases in the process of exposure to ultrasound, which is explained by a rise in the current magnitude because of increasing numbers of particles involved in the polarization. The pH decrease in the alkaline region for the anthracite suspension indicates that the coal surface was charged negatively, so that the phase interface polarization was accompanied by a buildup of positive charges in the solution, with the result that electrolysis of water preceded the anode mechanism (5.36).

Comparison of Energies of the Thermal and the Electrochemical Effects of Ultrasound It is known that ultrasonic treatment of aqueous systems is accompanied by heating [60]. In certain cases, the degree of heating is used to determine the power of the ultrasonic transmitter [59]. Our experiments showed that the heating of water under the effect of ultrasonic treatment did not follow a uniform pattern. The temperature of a water sample rose rapidly with increasing ultrasonic treatment time in the range of 25 to 60 °C, after which the rate of growth of the temperature declined (see Fig. 5.12). The specific temperature increment, $\Delta T/\tau_{us}$ for distilled and tap water reached the maximum (5–5.5 °/min) at 50 °C and then decreased to 0.5 °/min at 70 °C.

Using the $\Delta T/\tau_{us}$ value we can calculate the ultrasound thermal power (N_Q):

$$N_Q = c_p \mathrm{r} V \Delta T / \tau_{us} \tag{5.37}$$

where c_p is the specific heat, and r is the density of a liquid sample of the volume V.

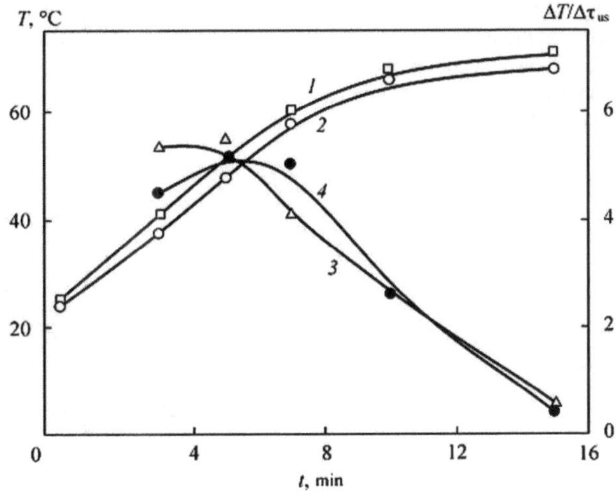

Fig. 5.12 Effect of ultrasonic treatment time on water temperature (*1*), (*2*) and $\Delta T/\tau_{us}$ coefficient (*3*, *4*) (*1*, *3*—distilled water; *2*, *4*—tap water)

It was only at the beginning of the ultrasonic treatment that N_Q maintained its maximum value (Fig. 5.13). After 5–7 min of treatment time, the efficiency of the thermal action decreased despite the fact that the power of the transmitter remained unchanged.

Knowing the change produced by the ultrasonic treatment in pH we can calculate the quantity of electricity spent on the water electrolysis (Q_e):

$$Q_e = \Delta C_{pH} V F \tag{5.38}$$

and then determine the current magnitude:

$$I = Q_e / 60\tau_{us} \tag{5.39}$$

Using (5.38) and (5.39), one can find the ultrasonic power spent to affect the pH change (N_{pH}):

$$N_{pH} = I^2/\chi = (\Delta C_{pH} V F / 60\tau_{us})^2 C_{cl} / \chi_0. \tag{5.40}$$

As is obvious from Fig. 5.13, for the distilled and tap water the N_{pH} value, as calculated by Eq. (5.40), grows with the ultrasonic treatment time.

Role of Cavitation in Polarization Processes. Considering the dispersal properties of aqueous systems (water layers of different density, phase interfaces, clusters, and other structural anomalies), it may be assumed that ultrasound will always exert an effect with the polarization mechanism. Even in relict water, where there are no cavitation centers created by high-energy particles [57], conditions persist under which a relative displacement of layers under the effect of ultrasound will produce polarization charges and emf. Growth of the sizes and concentration of vapor bubbles in a liquid as a result of a developing cavitation enhances the operation of

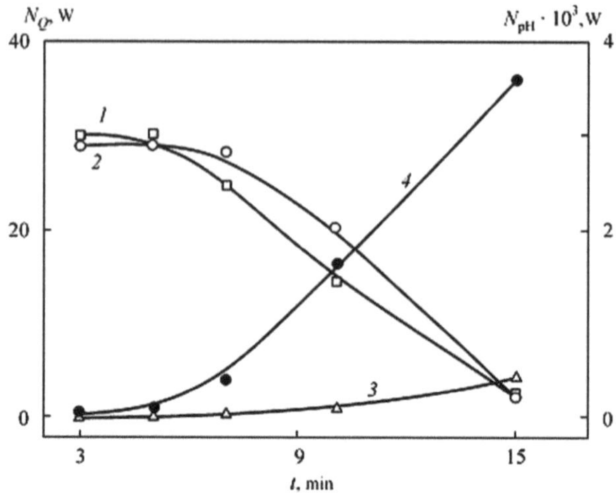

Fig. 5.13 Relationship between thermal power (*1*, *2*), power spent on pH changes (*3*, *4*) and ultrasonic treatment time (*1*, 3—distilled water; *2*, *4*—tap water)

the polarization mechanism, but in this case it will involve not only the anomalous water structures but also a stable disperse phase with boundary layers having longer relaxation times.

According to Eq. (5.27), the polarization potential pulse duration increases with decreasing frequency of ultrasound. This is why at low frequencies ultrasound is particularly efficient in cases where treatment is associated with an electrochemical action. For example, stronger bactericidal actions of a low-frequency ultrasound, as compared to a high-frequency one, have been reported in [46, 47, 57]. Owing to this effect, ultrasonic disinfection can probably be carried out using low-power sources (0.2–0.4 W/cm^3) with exposure times of 1 to 2 h [47]. Although the mechanism of ultrasonic disinfection calls for further investigation, it can already be assumed that the main role in the destruction of live cells is played by high polarization potentials not found in the conventional electrochemical treatment.

5.4 The Bactericide Effect of Ultraviolet Rays

It is known that the main mechanism of the antimicrobial action of ultraviolet radiation involves the destruction of bacterial DNA as a result of absorption of the ultraviolet quantum. The absorption band of DNA lies in the region between 200 and 300 nm with a peak at 254 nm [61, 62]. However, it is only in the case of low-pressure mercury lamps that almost the whole radiation falls within this region. The predominant portion of light generated by other sources has longer wavelengths. Therefore, spectral distribution of the radiation must be taken into account in comparing the bactericidal effect of the various lamps. Similarly, the very notion of dose must be specifically defined for every concrete case.

The principal purposes of investigation represented in paper [63] was to determine the average radiation dosages applied to the reactors from the various sources with different illumination geometry, and also to study the relationship between the dying-off of a test organism and the radiation dosage under experimental conditions. In addition, the present investigation examines the possibility of inactivating bacteria by heating them with a powerful pulse radiation.

The water disinfection processes were studied on a laboratory unit, described in paper [61], in two cylindrical reactors: one was of glass ($V=625$ cm^3), in which a source of UV-radiation was immersed; the other was of quartz ($V=300$ cm^3) exposed to external lateral illumination. In the glass reactor the height of the liquid column was 394 mm with a layer thickness of 15 mm, and in the quartz one 290 and 40 mm, respectively. To test the radiation source we used lamps of three types: an 8-W low-pressure mercury lamp DRB-8, a 100-W medium-pressure mercury Tungsram-100, and a pulse xenon lamp IFP-800 of 800 W. The pumping energy of IFP-800 was 60 J/pulse with pulse duration of 100 μs. The discharge measured 5.2 mm in diameter and 80 mm in length. The radiation intensity of the low-pressure mercury lamp was measured by varying the distance from the quartz reactor and the medium-pressure lamp was measured by means of an aperture diaphragm. The procedure used for preparing a culture of *Escherichia coli* 1257 is also described in paper [61].

For model water we used an aqueous solution containing, mg/dm^3: Na^+—23, HCO_3^-—61, SO_4^{2-}—12, Cl^-—35.5; Ca^{2+}—20, Mg^{2+}—2.45; K^+—1.95, humid acids—5. The solution alkalinity was 1 mg-eq/dm^3, and pH 8.8.

The optical density of the *E. coli* suspension (10^7 cfu/cm^3) in a 1 cm thick cuvette at $\lambda=254$ nm varied between 0.19 and 0.23.

Direct determination of the UV-radiation dosage under our conditions involved considerable difficulties. This is why the basic method chosen for its determination was actinometry [64]. Measured by this method (einsteins) the dosage is equal to the number of photons absorbed by a solution (suspension) during the exposure time divided by Avogadro's number. The light intensity in the 200–480 nm wavebands was measured for the given experimental conditions in the same way.

In establishing the relationship between the measured radiation dosage and the dosage in the predetermined spectral region, it was necessary to know the spectral distribution of the lamps. This was determined by means of a calorimetric meter of light energy, power IMO-2N and a set of wide-band filters with different shortwave transmission cutoffs. The longwave transmission cutoffs of all filters fell at the wavelength of ~3 μm. The whole spectral range was divided into sections between the shortwave transmission cutoffs of the filters; the light intensity in every section was determined by the difference in readings of the device following a change of filters. These measurements enabled us to determine only the relative intensity of radiation over the spectrum. Also, as the filters' transmission cutoffs were not clearly defined, the tables below the radiation energies of lines located in the vicinity of the cutoffs may be divided among two contiguous regions.

Table 5.7 Distribution of radiation intensity (I) of DRB-8 lamp by wavelengths

λ, nm	$C_\lambda = I_\lambda{}^a/I$	I^b, W
254	0.89	1.13
407 + 436	0.075	0.10
546	0.035	0.04
Total	1.0	1.27

[a] Intensity by wavelength λ
[b] Full intensity of the lamp

Low-Pressure Mercury Lamp As measured by the actinometric method, the radiation intensity in the glass reactor with an immersed DRB-8 lamp was 2.72×10^7 einsteins/s.

Relative and absolute distribution of the radiation intensity is shown in Table 5.7. Absolute intensity was determined on the basis of the actinometrically measured dose in the immersion reactor, in which the entire range of the lamp's radiation was absorbed by the solution.
Energy at wavelength X (E_K) was calculated by the formula

$$E_\lambda = C_\lambda E, \tag{5.41}$$

where C_λ is the coefficient determined from calorimetric measurements; E—is full radiation energy of the lamp which can be calculated as follows:

$$N = E \sum_\lambda \frac{C_\lambda}{h\nu_\lambda} F_\lambda \tag{5.42}$$

where N is the number of Fe^{2+} ions formed in actinometric solution; $h\nu_\lambda$ is the quantum energy at wavelength λ; F_λ is the quantum yield of the reaction of formation of Fe^{2+} at this wavelength.

Medium-Pressure Mercury Lamp The radiation intensity in the reactor measured by the actinometric method was 3.39×10^{-7} einsteins/s. The configuration of the reactor is described below.

Distribution of radiation over the spectrum of this lamp is shown in Table 5.8. The reactor's longitudinal section area is $S = 103$ cm^3. The lamp was located at a distance of 62 cm from the reactor, and its glowing column was projected onto the reactor plane by means of a short-focus quartz lens. For calorimetric measurements, the head of the energy meter was placed in the reactor plane. As in the case of the DRB lamp, absolute intensity values for Tungsram-100 were determined from results of actinometric measurements by Eq. (5.41) and (5.42).

Pluse Xenon Lamp IFP-800 The radiation dosage of this lamp, measured by the actinometric method, was 1.2×10^{-5} einsteins per pulse. Its spectral and photometric characteristics are listed in Table 5.9. Against the background of a continuous spectrum, the lamp radiated several lines in the 230–260 nm, but they were quite narrow and their combined energy was small. The whole energy was ~20% of the pumping energy, and the 200–290 nm bands accounted for 10% of total or 2% of

Table 5.8 Spectral distribution of radiation intensity for Tungsram-100 lamp

Range λ, nm	$C_\lambda = I_\lambda^{a}/I^{b}$	$I_{s\lambda}^{c}$, mW/cm²	I_λ, mW
200–260	0.17	0.34	35.0
260–290	0.10	0.20	20.6
290–370	0.25	0.50	51.5
370–410	0.06	0.12	12.3
410–450	0.06	0.12	12.3
450–510	0	0	0
510–550	0.05	0.10	10.3
550–600	0.14	0.28	28.8
600–640	0	0	0
640–700	0	0	0
700–3,000	0.17	0.34	35.0
Total:	1.0	2.0	205.8

[a] Full radiation power in the given wavelength range in the reactor plane
[b] Total spectral radiation power in reactor at the solution inlet
[c] Average density of radiation power in given wavelength range in the reactor plane

Table 5.9 Spectral distribution of energy radiated by IFP-800 lamp

Range λ, nm	$C_\lambda = E_\lambda/E$	E_λ, J/pulse
200–260	0.053	0.60
260–290	0.048	0.55
290–370	0.105	1.19
370–410	0.055	0.62
410–450	0.041	0.47
450–510	0.107	1.21
510–550	0.038	0.43
550–600	0.060	0.68
600–640	0.068	0.77
640–700	0.057	0.65
700–3,000	0.368	4.18
Total:	1.0	11.35

the pumping energy. The absolute value of the energy radiated by the IFP-800 lamp was determined in two ways. In one case, actinometric measurements were taken after complete immersion of the lamp into the solution, and energy was calculated by Eq. (5.41) and (5.42). In the other case we compared the IFP and DRB energies as described above. The measurements were taken on the similarity principle. We took account of the fact that sensitivity of a calorimetric sensor did not depend on the wavelength and assumed that the radiation intensity of the lamps was uniform along the discharge.

The DRB lamp was shielded in its upper and lower parts, leaving open only its middle section 80 mm in length, which equaled the length of discharge of the IFP lamp. The radiation intensity of the two lamps was measured with the sensor located equidistant from the center of the exposed part of the discharge. If the sensor registered pulse energy E_1 of the IFP lamp, the fall radiation energy was

$$E = kE_1 \tag{5.43}$$

$$k = Ih / I_1 H \tag{5.44}$$

where for DRB lamp I_1 is the registered intensity; I is the full intensity; H is the full length of discharge, and h is length of exposed part of discharge.

Determining the radiation energy with the two methods described above we, obtained results differing by not more than 10 %.

As noted above, determination of the average bacterium irradiation dosage under conditions of nonuniform illumination represented a complex problem which was examined by us in two cases with different configurations of the photoreactors employed in the present investigation. We assumed that the exposure time was much longer than the full solution mixing time (in our reactors the mixing time was ~5 s, and the mixing was done by blowing air at a flow rate of 200 cm^3/min). By convention, the solutions were assumed to be transparent to UV-light so that its attenuation or scatter as a result of passage across solution could be disregarded.

We studied two versions of positioning UV-light sources relative to reactor.

The First Version A tubular lamp of radius r with discharge length h was completely immersed in the reactor of the solution, the reactor radius $R<h$. Then the average energy dosage will be

$$D = E(R - r) / V_0 \tag{5.45}$$

where E is the full lamp radiation energy during exposure time and V_0 is the solution volume.

The Second Version A cylindrical reactor with solution was uniformly illuminated from outside by a tubular lamp. As a result of focusing action on the reactor, zones appeared in the solution that is not reached by the light, and, on the other hand, the concentration of light at the reactor back wall was increased. Calculations showed that the effects of these factors on the average irradiation dosage were approximately equal in magnitude but opposite in sign. We also allowed for an increase of Fresnel reflection on the periphery of the reactor. This was found to have little effect on the average dosage. As a result, we can consider that with this illumination of geometry, the average energy dosage is described by the formula

$$D = I_s t \tag{5.46}$$

where I_s is the radiation power density at reactor position (W/cm^2) and t is the exposure time.

If the lamp is positioned close to the reactor, and the discharge is shorter than the reactor length, the right-hand side in Eq. (5.46) must be multiplied by the ratio of illuminated volume V to full solution volume V_0:

$$D = (V / V_0) I_s t \tag{5.47}$$

Equations (5.45)–(5.47) obtained without allowance for absorption and dispersion of light are applicable for calculation of average dosages at layer optical density <0.2. At higher optical densities, real average dosages will be smaller than those calculated from these formulae. However, it should be brought to mind that at short wavelengths attenuation of the direct passage of light through a bacterial suspension is due to dispersion rather than absorption. The cross-section of dispersion of *E. coli* bacteria at 250 nm is 4×10^{-8} cm^2 [65], which correspond to the optical density of a 1-cm layer with a bacterium concentration of 10^7 cells/cm^3 (0.17). In current experiments, this value differed slightly from the optical density of the suspensions (0.19–0.23). Scattered quanta did not disappear but only changed the direction of propagation, which is why they contributed to the average irradiation dosage both before and after the moment of scatter. These data lead one to believe that real average dosages differed little from those calculated by the above formulae.

Heating of Bacteria by Pulse Radiation A supposition is made in paper [66] that a powerful pulse UV-irradiation of microorganisms in the air and aqueous solutions produces an additional bactericidal effect through thermal destruction of bacteria as a result of their local heating to several tens of degrees. It is believed that this is made possible by a difference in absorption between bacteria and the medium. To verify this hypothesis, we did the necessary calculations, taking inactivation of *E. coli* as an example. It is known that most bacteria are practically transparent at wavelengths >300 nm. In the radiation region between 200 and 300 nm ultraviolet light is absorbed by DNA of bacteria. It is this absorption that is the only reason for possible heating of the bacterial cell. Contributions of light absorption at other wavelengths can be ignored. In our further reasoning, we will visualize this cell as a ball 1 μm in diameter with the same density, thermal capacity and heat conductivity as those of water.

The cross-section of *E. coli* absorption at 250 nm is of the order of 10^{-9} cm^2 [65]. Then, without allowing for heat exchange of the cell with the medium, the UV-radiation dosage necessary to heat it by 50 °C, for example, will be ~ 100 mJ/cm^2. With powerful pulse lamps such dosages can be achieved, if only in small volumes. However, most bacterial species die at much lower dosages as a result of the well-known DNA photodestruction mechanism [62]. Thus, the thermal inactivation mechanism should be considered only for certain species of microorganisms that are especially resistant to UV-radiation.

On the other hand, a microscopic object has a high ratio of surface area to volume and, consequently, a high rate of heat exchange with the environment. Let us assess the maximum temperature of a cell in water, allowing for heat exchange, on a point radiator model with typical lamp pulse duration of 100 μs.

Assume that a point source of radiation (in our case this is an absorbing DNA molecule) is situated at the center of a cell. After time t and its actuation the temperature at distance r from it will be [67]

$$U(r,t) = \frac{q}{4\pi k}\frac{1}{r}\frac{2}{\sqrt{\pi}}\int_b^\infty e^{-\alpha^2}\,d\alpha \tag{5.48}$$

where q is the source of power; k is the heat conductivity coefficient; $b = \frac{r}{2\sqrt{a^2 t}}; a^2 = k / c\rho$ is temperature conductivity coefficient; c is the specific heat capacity, and ρ is density.

The maximum value of the integral in Eq. (5.48) is $\sqrt{\pi}/2$, which is why to determine the maximum cell temperature we must average the expression

$$U(r) = \frac{q}{4\pi k}\frac{1}{r} \tag{5.49}$$

over the volume of the cell, i.e. a ball with the radius $r=0.5$ μm.

If in a pulse of duration $\tau=100$ μs the cell receives a UV-dosage (D = 100 mJ/cm²), it will be heated by the temperature (°C)

$$U_m = \frac{3q}{8\pi k r_c} = 0.4 \tag{5.50}$$

where $q=D_\sigma/\tau$, $\sigma=10^{-9}$ cm² is cross-section of *E. coli* absorption at the wavelength of 250 nm.

Because of the small size of the object, a thermal wave travels from the center to the surface in fractions of a microsecond before releasing heat into the environment. The heat removal rate will be determined by the heat conductivity coefficient of the medium. Air, for example, it is only one-twentieth that of water. Our calculations showed under the same conditions a cell will be heated in the air by a few degrees. Thus, in studying the mechanism of thermal destruction of bacteria by a light pulse, it will be necessary to develop new sources of UV-radiation of high intensity and pulse duration considerably shorter than 100 μs. The data given below confirm that the mechanism of the antimicrobial action of a typical pulse lamp is the same as for a low-pressure mercury lamp.

Results of the determination of the antimicrobial action of radiation generated by low- and medium-pressure mercury lamps are shown in Fig. 5.14. The tests were made in a quartz reactor with external illumination of varying intensity. Plotted on the abscissa is the radiation energy dosage falling within the 200–290 nm range.

It was only the UV-radiation in the range from 200 to 290 nm that determined the antimicrobial effect of a source, irrespective of its type and portion of radiation in other regions of the spectrum.

Experiments with the IFP-800 pulse lamp were conducted in a glass reactor with the lamp immersed in solution. For comparison the antimicrobial action of the low-pressure mercury lamp DRB-8 was studied under the same conditions (Fig. 5.15). In this case, too, the antimicrobial effect of radiation generated by the pulse xenon lamp IFP-800 was analogous to that of the low-pressure mercury lamp, assuming

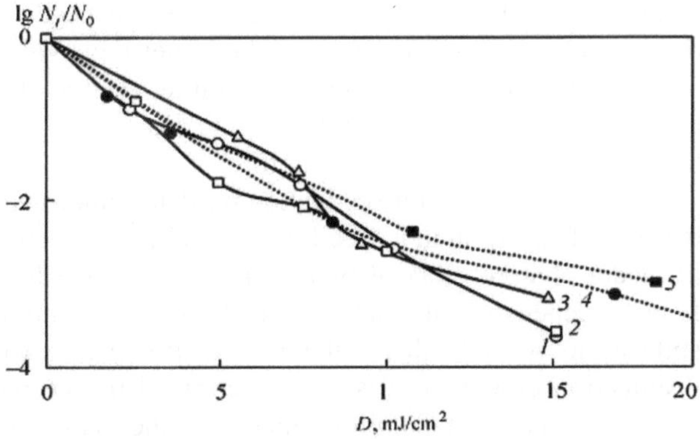

Fig. 5.14 Antimicrobial action of UV-radiation generated by mercury lamps of low (DRB-8, curves *1*–*3*) and medium pressure (Tungsram-100, curves *4*, *5*) with intensity density in UV-region mW/cm²: 0.025 (*1*); 0.2 (*2*); 1.85 (*3*); 0.56 (*4*), and 0.06 (*3*) in quartz reactor

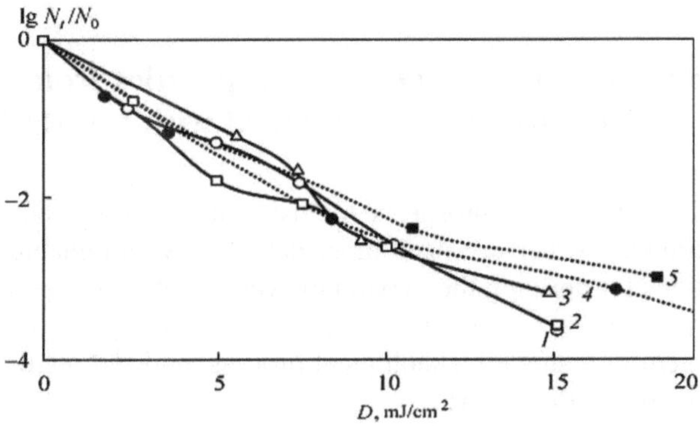

Fig. 5.15 Antimicrobial action of UV-radiation from pulse lamp IFP-800 (*1*) and low-pressure mercury lamp DRB-8 in quartz (*2*) and glass (*3*) reactors

that the UV-radiation in the 200–290 nm range was the sole operating factor. But the antimicrobial action of the DRB-8 lamp in the quartz reactor (curve *2*) was higher than in immersion tests in the glass reactor (curve *3*). In the initial section (up to 10 mJ/cm²), the curves coincided; as the UV-radiation dosage was increased, the killing rate constant for the quartz reactor remained practically unchanged whereas the glass reactor there was a decrease by a factor of almost five.

This difference was not associated with attenuation of light as a result of its passage through the suspension layer because in the quartz reactor the layer was, on the average, twice as thick as in the glass one. The most probable reason for slower inactivation in the glass reactor is as follows.

The useable volume of the reactor was contained within a hollow cylinder with the vertical circular wall of small thickness, as compared to height (see above for the dimensions). As the air was bubbled through it, visually observable constant turbulent pockets whose liquid mixed very slowly with the rest of the suspension were formed. The DRB-8 lamp illuminated only 55 % of the suspension volume. The maximum exposure time in this experiment was 35 s. Part of the cells caught in a turbulent zone in the unlit area of the reactor may be retained there for a considerable period or the entire exposure time, which is why the real dosage for them will be much lower than the calculated average. The negative effect of such turbulences on the end result will be greater; the larger their volume, the longer bacteria are retained in them, and the higher the required degree of suspension treatment. In the quartz reactor the layer width-to-height ratio was several times higher, and no stable turbulent zones were observed in it.

Thus, in the present investigation we determined average combined dosages and dosages of radiation in a specified interval of wavelengths for three types of lamps in model photoreactors of two configurations. We found that the disinfection efficiency depended only on the intensity of radiation in the range from 200–290 nm and was not related either to the full radiation intensity or to the type of source. The disinfection efficiency was also determined by the reactor design, which had to provide for a rapid and uniform mixing of the suspension. We assessed the possible thermal heating of bacteria by pulse light with high dosages of UV-radiation.

5.5 Removal of Organic Impurities from Natural Water by Ozonation Combined with UV-Radiation

To evaluate the prospects for using the O_3/UV technology in the production of drinking water, we need more data that would enable us to determine the conditions for deriving the maximum synergic effects from the action of ozone and UV-radiation on organic impurities occurring in natural water, methods for combining them in a single technological process, and the relative position of O_3/UV in the treatment flow chart.

It is known that the bulk of natural organic impurities consists of humic and fulvic acids. In river water fulvic acids predominate with concentrations of approximately 20 to 40 times as high as those of humic acids. In particular, in Kiev water storage, the concentration of fulvic acids varies between 14 and 54 mg/dm^3 and 0.3–2.0 mg/dm^3 for humic acids [68]. Also, the removal of trace quantities of anthropogenic impurities (pesticides, surfactants, petroleum derivatives, polynuclear aromatic hydrocarbons, etc.) represents a serious problem in drinking water production. The rates of reactions of molecular ozone are low, with most substances occurring in water in trace quantities, which is why these impurities can pass into the drinking water. These and other vulnerabilities of ozonation can be reduced or eliminated by combining with UV-radiation [69].

Destruction of natural organic impurities typically occurring in surface water supply sources by combined use of ozone and ultraviolet has not yet been adequately explored. Experiments were made on real river or lake water only in a few studies. It was demonstrated that the O_3/UV technology could be used for removing volatile organochloric compounds from natural water [70, 71] and for destroying substances responsible for the unpleasant tastes and offensive smells of surface water [72]. The efficiency of this technology in reducing the combined total content of organic matter in natural water was assessed in [73, 74]. Previously [75] explored the effect of the various modes of UV-irradiation upon the kinetics and the degree of decomposition of humic and fulvic acids by ozone in model solutions. In paper [76], we studied the separate and simultaneous action of ozone and UV-radiation on natural water of the Dnieper and Desna rivers in the spring-summer season.

The purpose of [76] was to yield the most efficient processes for combining O_3 with UV-radiation in removing organic matter from natural water.

Treatment of water samples with ozone and ultraviolet was performed in a laboratory-type direct-contact reactor unit with a submersible UV-radiation source in a quartz case [75]. Continuous irradiation was produced by a DRB-8 low-pressure mercury quartz lamp, and for pulse radiation we used a xenon pulse lamp IFP-800. The pumping energy of the pulse lamp was 60 J/pulse with pulses of 100 us and a recurrence frequency of 2.6 s. The UV-radiation intensity, as measured by the actinometric method, was 4.36×10^{-6} einsteins/(dm^3 s) for the DRB-8 lamp and 6.8×10^{-6} einsteins/(dm^3 s) or 1.76×10^{-5} einsteins/(dm^3 pulse) for IFP-800. Experiments were made at a UV-irradiation density in the 200–290 nm range, calculated from the spectral radiation distribution of the sources used: 2.72 mJ/(cm^2 s) for DRB-8 and 1.06 mJ/(cm^2 s) or 2.76 mJ/(cm^2 pulse) for IFP-800.

The procedures used to control the technological parameters of the oxidation process, the treated water quality and the methods of kinetic studies and calculation of the organic impurity decomposition rate constants are described in paper [75]. The kinetics of the destruction of organic matter contained in the natural water in the process of ozonation, photolysis, and prontoozonation were studied by the rate of discoloration (A_{364}), the rate of decomposition of the impurities' aromatic structure (A_{254}) and the rate of oxidation from changes in COD values and total organic carbon (TOC) content. We also assessed the specific light absorption of natural water at 254 nm per 1 mg/dm^3 or 1 mol/dm^3 TOC (SUVA). It was demonstrated in [77] that natural water absorbed ozone more vigorously the higher the value of specific light absorption in the 254–280 nm range. Electronic spectra of the test water were registered by means of a SPECORD UV VIS spectrophotometer. Natural water COD values were determined by the standard method (ISO 8467) and TOC by the method of liquid-phase oxidation in combination with reaction gas chromatography [78].

Carbonate alkalinity of the natural water (concentration of HCO_3^-) was determined by potentiometric titration. In our test water, the concentration of bicarbonate ions ranged from 2 to 4 mg-eq/dm^3. Bicarbonate ions serve as traps for OH-radicals and can significantly influence the efficiency of O_3/UV treatment of surface water. The average value of constants of the interaction rate of OH-radicals with natural organic impurities is estimated at $(2.3 \ldots 3.6) \times 10^8$ (mol C/dm^3)$^{-1}$ s^{-1} [76, 79], and with bicarbonate ions at $(0.85 \ldots 1.5) \times 10^7$ M^{-1} s^{-1} [76].

The river water was prefiltered across a paper filter. The characteristics of the Dnieper and Desna water are shown in Table 5.10.

The effect of the ozone concentration in the ozone-air mixture (OAM) on the destruction kinetics of organic impurities in the water of the Dnieper river was studied in the range 2.5 to 8.0 mg/dm^3 with an OAM feed rate of 0.2 dm^3/min.

Oxidization of natural organic matter in the various O_3 and O_3/UV processes proceeded in two stages (slow and rapid), which was consistent with data obtained for model humic and fulvic acids in [75, 80]. The first stage was characterized by a

Table 5.10 Characteristics of Dnieper and Desna water

Water source	Color, Degr.	A_{364}, (5 cm)	COD, mg O/dm³	TOC, mg/dm³	A_{254}, (1 cm)	SUVA, cm⁻¹ (mol C/dm³)⁻¹	pH	Alkalinity, mg-eq/dm³
Desna (July)	30	0.246	26	14.7	0.27	220	8.6	3.7
Dnieper (May–July)	37–54	0.316–0.443	31–42	17.6–21.2	0.375–0.425	229–320	8.0–8.7	2.5–2.8
Ditto (March)	30	0.245	27	–	0.28	–	8.1	2.3

vigorous absorption of ozone which then slowed down at the second stage. Depending on the rate of feed of O_3 into the reactor (0.8–4.2/(min dm³)), the ozone dosage absorbed at the first stage by the Dnieper water was 0.45–0.54 of the amount introduced. This was 0.26–0.47 at the second stage. However, unlike the model, humic and fulvic acid solutions in [75], in the case of the Dnieper water the boundary between the stages is not very distinct (Fig. 5.16a, b) and becomes quite blurred at a C_{oz} in OAM of 2.5 mg/dm³.

A comparison of the ozonograms and the ozone absorption kinetics with changes in the spectrophotometric characteristics of the water studied (Fig. 5.16c) shows that the first stage coincides in time with thorough discoloration, and a significant reduction of A_{254} is characteristic of the interaction of ozone with the chromophore chains responsible for the color of humic and fulvic acids.

At the first stage of ozonization of the Dnieper water the concentration of dissolved ozone changed from 0.1 to 0.3 mg/dm³ in the range of C_{oz} in OAM from 2.5 to 7.8 mg/dm³ (Fig. 5.16a). At the second ozonation stage it grew gradually at C_{oz} in OAM > 5 mg/dm³.

Ozonogram patterns for O_3/UV treatment of the Dnieper river water are shown in Fig. 5.19. In the process of this treatment, the equilibrium concentration of dissolved ozone was ≤ 0.2 mg/dm³, irrespective of C_{oz} in OAM (see Fig. 5.17) and decreased with increasing treatment time.

The kinetic curves describing the discoloration of the Dnieper water and the breakup of the aromatic structure of its impurities in the process of continuous O_3/UV treatment are similar to those obtained for ozonation alone (see Fig. 5.18).

The interrelationship between changes in the spectrophotometric characteristics of the Dnieper water and the absorbed ozone dosage is shown in Fig. 5.19. In addition to humic and fulvic acids, natural water contains a broad spectrum of other organic compounds [81]. The composition of organic matter in river water is subject to substantial fluctuations, which complicates the comparison of the quality of treated water samples taken at different times. Nevertheless, we believe that in this case, too, the correlation between the spectrophotometric characteristics of treated water and the absorbed ozone dosages is satisfactory. Therefore, our conclusion in regard to model solutions of humic and fulvic acids [75] are also valid for the natural water: the degree of water purification is mainly determined by the dosage of absorbed ozone.

The absorbed ozone dosage required to discolor the Dnieper water to the standard varied between 4.3 and 5.8 mg/dm³ in ozonation and from 4.8 to 6.5 mg/dm³ in

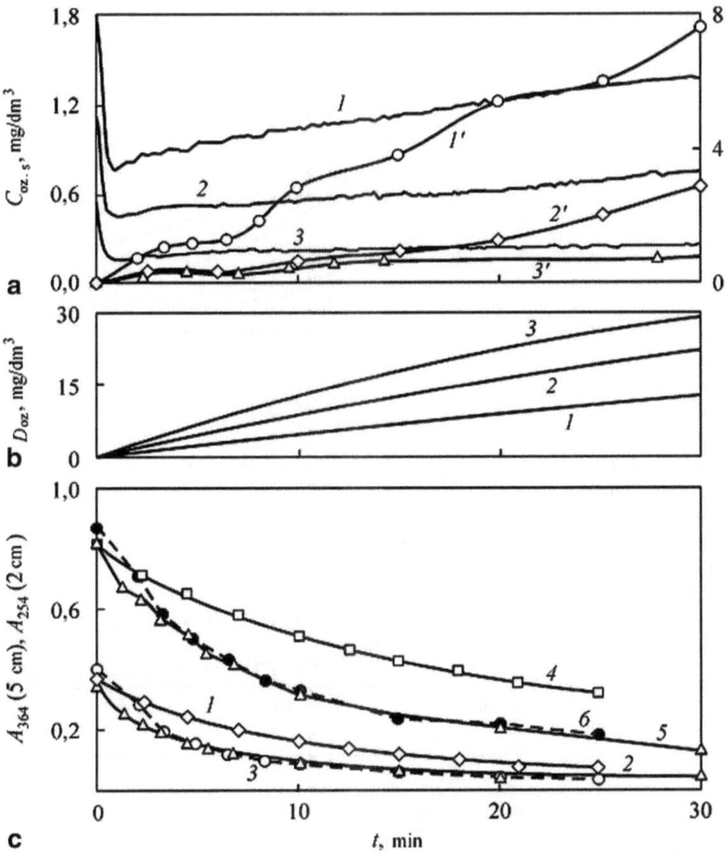

Fig. 5.16 Changes of ozone concentration in OAM (*1–3*) and solution (*1′–3′*) (**a**), dosage of absorbed ozone (**b**) and optical density (**c**) at 364 (*1–3*) and 254 nm (*4–6*) in the process of ozonation of Dnieper water with C_{OZ} in OAM: 2.3 (*1, 4*), 4.9 (*2, 5*), and 7.5 mg/dm³ (*3, 6*)

Fig. 5.17 Changes of ozone concentrations in OAM (*1, 2*) and in solution (*1′, 2′*) in the process of O_3/UV treatment of Dnieper river water with initial C_{OZ}: 8.1 (*1*) and 4.9 mg/dm³ (*2*)

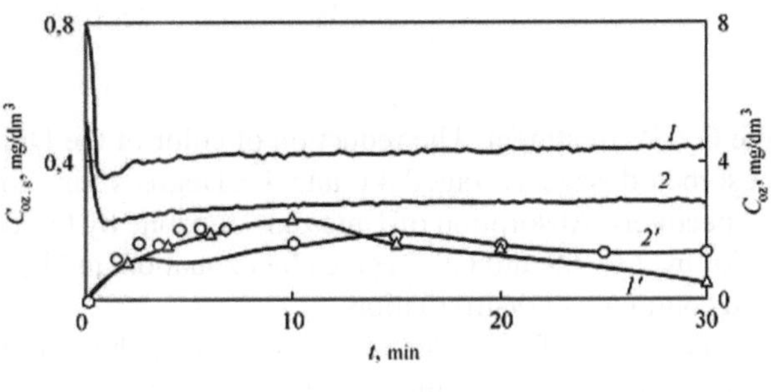

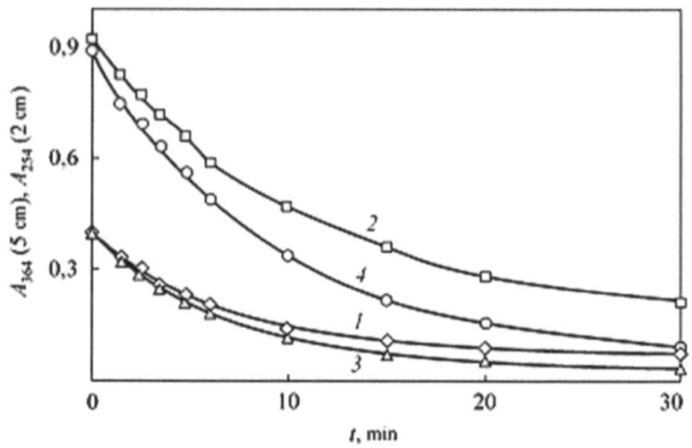

Fig. 5.18 Changes in A_{364} (*1*, *3*) and A_{254} (*2*, *4*) of Dnieper water in the process of treatment with ozone (*1*, *2*) and O_3/UV (*3*, *4*)

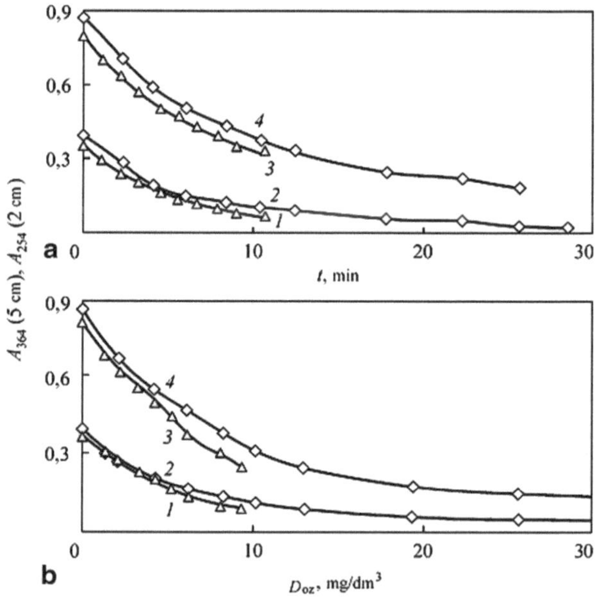

Fig. 5.19 Changes of A_{364} (*1*, *2*) and A_{254} (*3*, *4*) of Dnieper water in ozonation (**a**) and O_3/UV-treatment (**b**) on absorbed ozone dosage with C_{OZ} in OAM: 2.3 (*1*, *3*) and 7.4 mg/dm^3 (*2*, *4*)

the O_3/UV treatment. The reduction of color of the Dnieper water per 1 mg/dm^3 of absorbed dosage averaged 4.6 and 4.5 Degrees for ozonation and photoozonation, respectively. Absorption of 1 mg/dm^3 of ozone by the Desna river water reduced its color by 7.6, 7.9 and 6.9 Degrees by ozonation and by O_3/UV treatment with pulse and continuous UV-irradiation.

The pseudo first order rate constants of the variations in spectrophotometric characteristics of the Dnieper water in the process of ozonation and O_3/UV treatment are listed in Table 5.11.

It follows from the above data that the rates of destruction of natural organic impurities at the first and the second stages of the oxidative treatment differ by factors of

Table 5.11 Pseudo first-order rate constants of decomposition of organic impurities in Dnieper river water in the process of O_3 and O_3/UV-treatment

C_{oz} in OAM, mg/dm^3	Stage	K, min^{-1}			
		A_{364}		A_{254}	
		O_3	O_3/UV	O_3	O_3/UV
2.0–2.6	1	0.088	0.076	0.047	0.060
	2	0.072	0.075	0.040	0.056
4.6–4.9	1	0.097	0.121	0.066	0.095
	2	0.029	0.050	0.035	0.059
7.5–7.8	1	0.167	0.174	0.103	0.133
	1	0.238	0.212	0.128	0.143
	2	0.061	0.051	0.038	0.053
	2	0.141	0.126	0.074	0.093

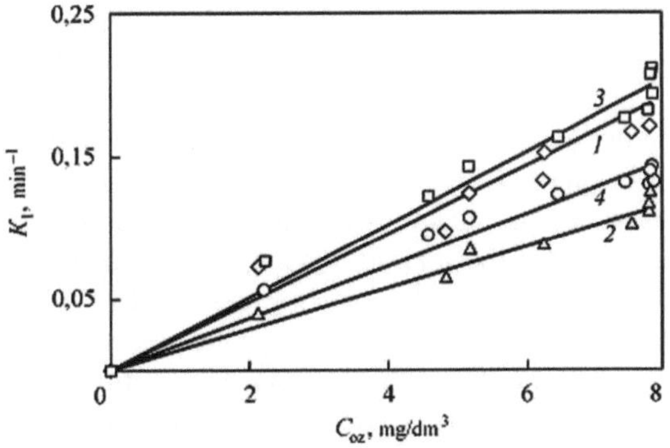

Fig. 5.20 Relationship between pseudofirst-order constants of the rates of change of A_{364} (*1, 2*) and A_{254} (*3, 4*) of Dnieper river water and ozone concentrations in OAM, in the process of O_3 (*1, 2*) and O_3/UV (*3, 4*)

2 to 3. Qualitative changes in the composition of the water from May through July result in a significant increase in the decomposition rate of organic matter at the same as C_{oz} in OAM. Water discoloration at the first stage proceeded faster than the breakup of the impurities' aromatic structure; the K_{A254}/K_{A364} ratio is 0.5–0.7 and 0.7–0.8 for O_3 and O_3/UV, respectively.

At the first stage, the variations rate constants A_{364} and A_{254} grow in direct proportion to the ozone concentration in OAM (see Fig. 5.20) are dependent on the rate of ozone feed into the reactor. Consequently, the processes of discoloration of the natural water and decomposition of the aromatic structure of organic matter by ozone or O_3/UV treatment are controlled by the ozone mass transfer coefficient.

The constant of the rate of discoloration (A_{364}) of the river water under the effect of UV-irradiation was 0.002–0.005 min^{-1}, and that of the decomposition of the natural impurities' aromatic structure was (A_{254}) 0.005–0.009 min^{-1}. Thus, the rate of photolytic destruction of organic matter in the Dnieper and Desna water was lower than the rate of their destruction by ozone or O_3/UV by 1–2 orders of magnitude (see Table 5.11).

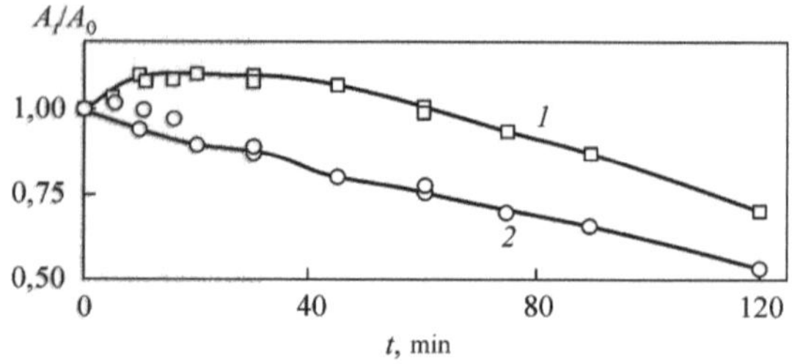

Fig. 5.21 Changes of A_{364} (*1*) and A_{254} (*2*) of Dnieper water in the process of UV-irradiation (DRB-8)

These data indicate that at the first stage the processes of destruction by ozone of the natural organic impurities were mainly reactions of molecular ozone with conjugate double bonds and the aromatic structures of humic and fulvic acids responsible for the color of the natural water. The water discoloration proceeded in a diffuse mode or a fast reaction mode. The interaction between ozone and impurities dissolved in the water took place directly in the thin film at the interface, which is why the oxidant concentration in the body of the solution was low.

Simultaneous use of ozonation and UV-radiation at the first stage was not particularly effective since the gain in the rate of destruction of natural organic impurities in comparison to ozonation alone was only slight (see Fig. 5.20). Moreover, we registered an antagonistic effect from the combined use of O_3 and O_3/UV on our test water in the summer season, because over the first 0.5 h of photolysis the water color actually rose by 2–5 Degrees (see Fig. 5.21).

Using the method of automatic registration of A_{364} we studied the kinetics of discoloration of the Dnieper water in two successive stages (O_3 followed by O_3/UV). Variation of the O_3 and O_3/UV treatment stage times from 2.5 to 10 min had practically no effect on the water discoloration rate. The rate constant was $0.0068 \pm 0.002\ min^{-1}$.

Treatment of the natural water by pulse UV-radiation (IFP-800) in our reactor was carried out under conditions where the solution column height (395 mm) significantly exceeded that of the illuminated layer since the discharge length of IFP-800 was 80 mm. This is why in this series of tests water samples in the process of ozonation, UV-irradiation and O_3/UV treatment were taken from a special sampling device fitted in the exposed area.

Results of our comparative studies of the pulse and continuous modes of UV-irradiation in O_3/UV treatment of the Dnieper and Desna water are given in Table 5.12. The discoloration rate of the Dnieper water by photooxidization did not depend on the UV-irradiation mode and was close to that produced by ozonation. The effect of pulse UV-irradiation was more pronounced in the case of the Desna water aromatic structure discoloration. The destruction rate of the organic impurities contained in our test water in the process of O_3/UV treatment was somewhat lower for the pulse lamp than in the case of DRB-8 (K_{pls}/K_{cont} = 0.8–0.85). The constants of the decomposition rate of organic impurities in the Desna water were significantly higher at the first stages of the ozonation and O_3/UV treatment processes than in the case of the Dnieper water.

Table 5.12 Comparison of pseudofirst-order constants of the rates of decomposition of natural water organic impurities in O_3 and O_3/UV-treatment with pulse (IFP-800) and continuous (DRB-8) UV-irradiation

Water source	Stage	*t*, min	K, min^{-1}						K_{pls}/K_{cont}	
			O_3		O_3/UV (DRB-8)		O_3/UV (IFP-800)			
			A_{364}	A_{254}	A_{364}	A_{254}	A_{364}	A_{254}	A_{364}	A_{254}
Dnieper river	1	0–4.75	0.16	0.09	0.15	0.10	0.15	0.08	0.98	0.81
	2	4.75–13	0.08	0.07	0.12	0.09	0.10	0.07	0.84	0.79
Desna river	1	0–4.75	0.21	0.12	0.21	0.14	0.25	0.12	1.16	0.85
	2	4.75–13	0.08	0.07	0.12	0.09	0.10	0.08	0.83	0.81

Table 5.13 Changes in COD and TOC of Dnieper water in the process of ozonation and O_3/UV-treatment

pH	C_{oz} in OAM, mg/dm^3	*T*, min	O_3			O_3/UV (continuous)		
			D_{oz} mg/dm^3	COD, mg O/dm^3	TOC, mg/dm^3	D_{oz}, mg/dm^3	COD, mg O/dm^3	TOC, mg/dm^3
8.6	0	0	0	38	17.6	0	38	17.6
	5.1	4.5	4.0	32	16.4	4.1	31	15.3
	5.0	30	19.5	28	13.1	25.1	24	12.4
8.0	0	0	0	35	21.2	0	35	21.2
	7.8	7.25	9.2	29	17.1	9.7	25	15.1
	7.9	15.75	18.2	26	14.2	21.6	22	13.1
	7.8	29.75	27.0	23	13.8	39.3	17	10.9

At the second stage of the ozonation process, there was a gradual buildup of dissolved ozone in the water (see Fig. 5.16a). Simultaneous use of UV-radiation at this stage accelerated the decomposition of organic matter in the test water 1.3–2.2 times.

Monitoring of the oxidization of the river water impurities by COD and TOC (see Table 5.13) indicates that purification of the Dnieper water by O_3/UV treatment was somewhat better than by ozonation over the entire D_{oz} range.

The O_3/UV treatment produced better purification of the test water by COD and TOC than ozonation alone, irrespective of the UV-irradiation mode (see Table 5.14). Although the aggregate of our data for the water of the Dnieper and Desna rivers is indicative of a greater efficiency of continuous UV-irradiation, the difference between the absolute COD and TOC values for the samples compared is within the limits of the error of analytical determination of these indices. Consequently, the effects of pulse and continuous irradiation in the O_3/UV treatment of our test water on the destruction of organic impurities were identical, despite different UV-irradiation densities.

With absorbed ozone dosages of 10–12 mg/dm^3, which were acceptable for the drinking water production technology, ozonation of the Dnieper river water reduced COD by 17–19 % and TOC by 19–31 %. Purification of the Desna water by COD

Table 5.14 Changes of COD and TOC of Dnieper and Desna river water in the process of ozonation and O_3/UV-treatment with pulse (IFP-800) and continuous (DRB-8) UV-irradiation

Water source	pH	HCO_3^-, mg-eq/dm^3	*t*, min	D_{oz} mg/dm^3	O_3		O_3/UV (DRB-8)		O_3/UV (IFP-800)	
					COD, mg O/dm^3	TOC, mg/dm^3	COD, mg O/dm^3	TOC, mg/dm^3	COD, mg O/dm^3	TOC, mg/dm^3
Desna river	8.6	3.7	0	0	26	14.7	26	14.7	26	14.7
			2.5	2.0–2.2	24	12.9	22	12.0	23	12.4
			4.8	4.1–4.4	22	10.7	18	9.4	20	10.2
			7.0	6.1–6.4	18	9.5	16	8.9	17	9.2
			13	10–12.1	17	8.8	15	8.3	15	8.5
Dnieper river	8,3	2.75	0	0	31	17.9	31	17.9	31	17.9
			4.8	4.1–4.3	30	13.6	28	12.2	30	12.6
			7.0	6.2–6.5	27	–	23	–	25	–
			13	9.9–11.3	25	12.4	22	11.6	24	11.7

and TOC in this D_{oz} range was 35 and 40 % respectively. O_3/UV treatment of the same duration increased the degree of purification of the Dnieper water by COD by 10–12 % and by TOC by 4–10 %, and of the Desna water by 2–3 % respectively.

Bubbling of ozone-free air with simultaneous UV-irradiation during a time corresponding to the duration of the O_3/UV treatment reduced COD of our water samples by 3–11 % and TOC by 14–20 %. These results suggest that oxygen participates in the process of photooxidative decomposition of organic matter contained in the river water. Aeration may also remove part of volatile impurities,

With ozone dosages corresponding to the ozone absorption of the Dnieper (22–27 mg/dm^3) and the Desna (~17 mg/dm^3) water the maximum removal of organic matter by ozone reached 41–49 %. O_3/UV treatment of the same duration increased the purification by a further 4–6 %.

To reduce the color of the Dnieper and the Desna water by one degree, it took an average of 0.22 and 0.14 mg/dm^3 of ozone respectively. The specific rates of ozone consumption for the reduction of COD and TOC were significantly higher. They depended on the degree of oxidation, the treatment mode, and the water quality (including alkalinity). As the alkalinity of natural water rises, the effect of the O_3/UV treatment decreases [73, 75, 82].

As the ratio of absorbed ozone to initial TOC (D_{oz}/TOC$_{ini}$) rose in the ozonation of the Dnieper water from 0.3 to 1.7, the specific rate of ozone consumption for the reduction of COD increased from 1.0 to 2.6 mg/mg and for the reduction of TOC from 1.0 to 3.6 mg/mg. In the O_3/UV treatment the consumption of ozone for the reduction of COD and TOC was 0.8–2.2 mg/mg and 0.8–3.8 mg/mg respectively. In the treatment of the Desna water (D_{oz}/TOC$_{ini}$ = 0.1…1.7) by ozone and O_3/UV the specific rate of ozone consumption for the reduction of COD was 1.0–1.5 and 0.7–1.1 mg/mg and for the mineralization of 1 mg/dm^3 of TOC 1.0–2.3 and 0.8–3.2 mg/mg respectively. At the rather low D_{oz}/TOC$_{ini}$ ratios (0.3–0.6 for the Dnieper and 0.1–0.4 for the Desna water) characteristic of drinking water conditioning, the specific rate of oxidant consumption for reducing COD and TOC values by 1 mg/dm^3 was lower in the case of the O_3/UV treatment.

It was demonstrated previously [83, 84] that the combined action of ozone and UV-radiation substantially accelerated the destruction of the various classes of pesticides and sodium alkylbenzene sulfonate (ABS) in model solutions. It was also found that the advisability of employing destructive methods involving the use of hydroxyl radicals for removing trace quantities of pesticides depended on the total content of organic matter in the water. While O_3/UV treatment improved purification by the integral indexes (COD and TOC), its rate of destruction of certain pesticides was lower than that of ozonation due to the nonselectivity of radical oxidative processes and considerable differences in the concentrations of natural organic impurities and pesticides. It was, therefore, of interest to determine the most efficient procedure combining O_3 and O_3/UV for purification of water containing impurities of both natural and anthropogenic origin.

The above information on the ozonation and O_3/UV treatment of natural water indicates that no significant improvement in the removal of hardly oxidizable impurity traces can be achieved by combined O_3 and UV treatment at the first stage (in the first contact areas) since the concentration of dissolved ozone necessary for the formation of hydroxyl radicals begins to increase only after the water discoloration. A two-stage process—ozonation followed by O_3/UV treatment—is more efficient for the treatment of natural water than either of these processes used separately. This method takes into account the qualitative heterogeneity of the water under treatment and changes in its quantitative composition in the various contact areas. For the discoloration of natural water, destruction of readily oxidizable impurities and in cases with high initial concentrations of oxidizable matter there is no need to enhance the ozonation process by combining it with ultraviolet. The combination of ozonation with simultaneous UV-irradiation at the second stage is designed for effective decomposition of hardly oxidizable compounds and improvement of the overall water purification degree.

The advantages of this two-stage process were demonstrated in the treatment of a model solution consisting of the natural Dnieper water (see Table 5.10, March) and 6 pesticide compounds in concentrations of 0.1–0.29 mg/dm^3 plus 0.2 mg/dm^3 of ABS. The treatment was carried out on a continuous-flow unit comprising two reactors of ~4 dm^3 each connected in tandem. One was for ozonation and the other for O_3/UV treatment. The photoozonization reactor contained a submersible source of UV-radiation (lamp DB-15). The water and OAM were fed from the bottom upwards into the first reactor; spent OAM was then fed at the bottom into the second reactor, and the water from the upper part of the first reactor passed into the upper part of the second one and was discharged from the bottom. The total volume of solution undergoing treatment was 6.5 dm^3 and the treatment time was 22 min. The initial ozone concentration in the OAM at the inlet was 11.5 mg/dm^3 and the OAM feed rate was 0.55 dm^3/min. We compared the purification of the model solution by ozonation in both reactors (the first method) and by ozonation in the first reactor and O_3/UV in the second one (the second method). The water ozonation time in the first reactor was 9 min, and the O_3/UV treatment time in the second one 13 min

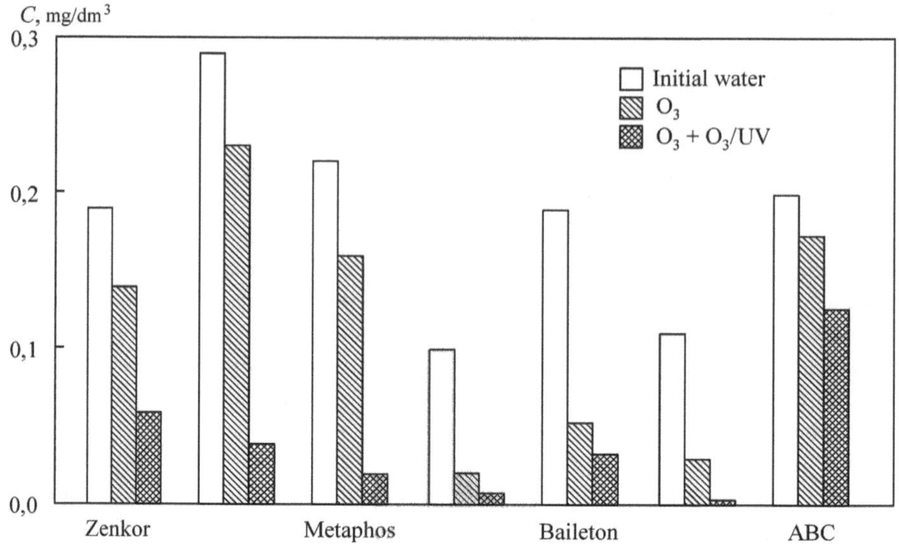

Fig. 5.22 Comparison of the purification of the Dnieper river water polluted with pesticides and ABS by ozonation and two-stage O_3 and O_3/UV treatment

(see Fig. 5.21). The introduced ozone dosage was 21.2 in both procedures, and the absorbed ozone dosage was 18.05 and 18.82 mg/dm^3 respectively for single-stage and two-stage processes.

It follows from the data in Fig. 5.22 that the two-stage treatment significantly improved removals of all added organic traces, with the overall degree of purification by COD rising from 30 to 52 %.

Thus, to derive the maximum benefit from the synergic effect produced by the action of O_3, ultraviolet natural water should be treated in two stages: by ozonation in the first stage and by O_3/UV in the second. This two-stage treatment significantly reduces the residual concentrations of various anthropogenic traces and improves the overall water purification in terms of COD.

5.5.1 A Study of the Various Procedures of Water Disinfection by O_3/UV

In connection with the detection of toxic byproducts of disinfection in drinking water, such as chlorination (chlororganic compounds) and some disinfection-resistant microorganisms (cysts of Cryptosporidia, lambliae, etc.) much greater interest has been shown in such alternative water treatment methods as ozonation and ultraviolet irradiation. This is especially true of the technology providing for simultaneous use of these techniques, which belongs to a group of particularly promising methods referred to as advanced oxidation processes (AOP). Basically, AOP lies in a generation of hydroxyl radicals which are more effective in destroying organic substances that are difficult to oxidize by ozone [85, 86]. As regards the disinfection process, use of the O_3/UV-method may, as demonstrated in [87], result in a variety of effects. Thus, for example, it was found that the combination of different ozone dosages with high dosages of UV-radiation in O_3/UV-treatment produced a lower degree of disinfection than the same dosage of UV-radiation without ozone, which suggested an antagonistic effect. An additive effect was observed when two conditions were met: first, when water was exposed to UV-radiation in the initial period corresponding to the latent phase on the kinetic curve describing the killing of *Escherichia coli* by ozone alone; and, second, when treatment combined low dosages of the two disinfectants, which could be achieved, for example, by reducing the ozone concentration in the ozone-air mixture and the UV-radiation intensity.

The purpose of the investigation [88] was to make a more comprehensive study of these two treatment procedures.

A detailed schematic of the experimental unit appears in [8]; the radiation spectrum of the DRB-8 lamp, the procedure for determining intensity, and irradiation dosages are given in [63]; the preparation technique, the method of counting the test microorganism *E. coli 1257,* and the experimental procedure are covered in [87].

The combined effect was assessed by the ratio *T/E*, where *T* is theoretically calculated portion of surviving bacteria, and *E* is the same value obtained experimentally. In calculating the theoretical value the separate effects of the individual agents were assumed to be independent of each other [89]. At $T/E<1$ the interaction of the disinfectants is antagonistic, at $T/E=1$ it is additive, and at $T/E>1$ it is synergistic.

UV-Irradiation in the Initial Period of the Ozonation Process The duration of the latent phase of the dying-off of *E. coli* under the effect of ozone varied in the range 120–180 s and 60–30 s, with the ozone concentrations C_{0Z} in OAM of 2, 5 and 10 mg/dm^3 respectively [87]. During this period the suspension undergoing ozonation was exposed to UV-light of different intensity, also using different dosages of ultraviolet radiation. For comparison, we treated bacterial suspensions under identical conditions by O_3 and UV-radiation separately and simultaneously.

Figure 5.23 shows kinetic curves describing the dying-off of *E. coli* test microorganisms at different ozone concentrations (2, 5, and 10 mg/dm^3) in the OAM feed and with different dosages of UV-radiation (81.7, 12, and 4.5 mJ/cm^2). The exposure time varied from 30 to 180 s, depending on the UV-radiation intensity and ozone concentration in OAM. The higher the ozone concentration in OAM (i.e., the shorter the latent phase of the dying-off of microorganisms in ozonation), the higher the intensity of UV-radiation used. It can be seen from Fig. 5.23 that disinfection proceeded most rapidly during the first 30 s with a rather high O_3 concentration in OAM (10 mg/dm^3) and an elevated UV-radiation intensity (2.72 mJ/cm^2 s) (see Fig. 5.24, curve *1*). Disinfection increased with increased UV-radiation dosage in the initial period of O_3-treatment (curves *2* and *3*) or with increasing C_{oz} in OAM at a constant UV-radiation dosage (curves *3* and *5*).

Based on the data in Fig. 5.23, we calculated test microorganism killing rate constants for our model water (see Table 5.15).

The rate constants for the first 30–180 s of exposure of suspensions undergoing ozonation to UV-radiation were close to those obtained in UV-irradiation (without

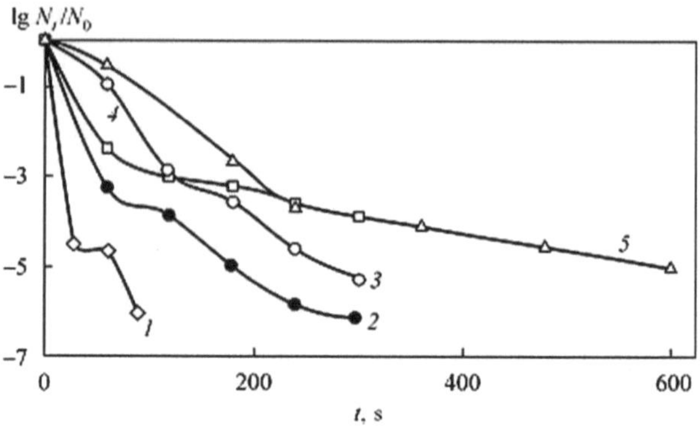

Fig. 5.23 Kinetics of *E. coli* dying-off during UV-irradiation of model water in the initial period of ozonation: *1*—C_{oz} in OAM—10 mg/dm³, D_{uv}—81.7 mJ/cm², I_{uv}—2.72 mJ/(cm² s), *t*—30 s; *2*—C_{oz} in OAM—5, D_{uv}—12, I_{uv}—0.2, *t*—60; *3*—C_{oz} in OAM—2, D_{uv}—12, I_{uv}—0.2, *t*—60; *4*—C_{oz} in OAM—2, D_{uv}—4.5, I_w—0.025, *t*—180 s

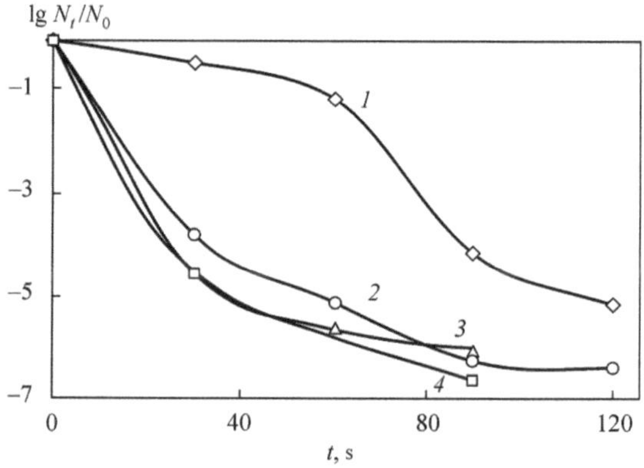

Fig. 5.24 Kinetics of the dying-off of *E. coli* in model water as a result of the following treatment: *1*—by ozone; *2*—by UV-radiation (submersible lamp DRB-8); *3*—by O_3/UV during the first 30 s followed by O_3 alone; *4*—by O_3/UV. C_{oz} in OAM—10 mg/dm³, I_{uv}—2.72 mJ/(cm² s)

ozonation) of the same intensity: 0.128, 0.054, and 0.010 lg unit s^{-1} at I_{uv} values of 2.72, 0.2, and 0.025 mJ/(cm² s) respectively. The highest disinfection rate in both phases (rapid and slow) was achieved by combining a UV-radiation dosage of 81.7 mJ/cm² during the first 30 s of ozonation with the highest ozone concentration in OAM—10 mg/dm³.

Figure 5.23 compares data on the efficiency of disinfection of model bacterial suspensions by ozone (C_{oz} in OAM 10 mg/dm³), UV-radiation at the maximum intensity of 2.72 mJ/(cm² s), and jointly by ozone and UV-radiation depending on the mode of UV-radiation: during the first 30 s of ozonation or throughout the ozonation process. As can be seen from this figure, UV-irradiation in the initial period of treatment (30 s), when ozone reacted with the readily oxidizing portion of humic acids and made practically no contribution toward disinfection, inactivated a considerable part of bacteria contained in the water (4–5 orders of magnitude) at a rate adequate to its effect. Longer exposure to UV-radiation was inexpedient as it offered no further gain in the rate of disinfection. The kinetic curves describing the dying-off of *E. coli* under the effect of O_3/UV-treatment and UV-irradiation during the first 30 s of ozonation practically coincide (see Fig. 5.24, curves *3* and *4*). Part of the UV-radiation was absorbed by dissolved ozone, which reduced its equilib-

Table 5.15 *E. coli* killing rate constants *(k)* for UV-treatment in the initial period of ozonation

C_{oz}, mg/dm^3	I_{uv}, mJ/(cm^2 s)	D_{uv}, mJ/cm^2	UV exposure time, s	k, lg unit s^{-1} in bacteria dying-off phases	
				Rapid	Slow
10	2.72	81,7	30	−0.151	−0.025
5	0.2	12	60	−0.054	−0.013
2	0.2	12	60	−0.040	−0.006
2	0.025	4.5	180	−0.015	−0.004

Table 5.16 Technological parameters necessary for disinfection of model water to required levels by exposing suspensions to UV-radiation in the initial period of ozonation

Treatment procedure	Degree of disinfection, %	Duration of treatment, s	Dosage of ozone, mg/dm^3	
			fed in	Absorbed
1	99.9	20	1.1	0.2
	99.99	24	1.3	0.2
	99.999	72	3.8	1.3
2	99.9	60	3.6	0.6
	99.99	190	11.6	2.7
	99.999	288	17.4	4.1
3	99.9	120	2.8	0.8
	99.99	312	7.2	1.8
4	99.9	205	4.8	1.2
	99.99	340	8.2	2
	99.999	600	14.4	3.2

rium concentration from 0.383 mg/dm^3 in ozonation (C_{oz} in OAM 10 mg/dm^3) to 0.096 mg/dm^3 in O_3/UV-treatment resulting in a lower degree of disinfection.

From the data in Fig. 5.23, we calculated treatment times and dosages of ozone fed and absorbed necessary to achieve the various degrees of disinfection: by 3, 4, and 5 orders of magnitude, corresponding to 99.9, 99.99, and 99.999 % respectively (see Table 5.16).

The data in Table 5.16 corroborates the above conclusion that the most efficient treatment procedure (the highest degree of disinfection in a short time with a lower ozone consumption rate) is combining a short (30 s) exposure of a bacterial suspension to UV-radiation in the initial period of its ozonation with a high ozone concentration in OAM (10 mg/dm^3). It reduced the treatment time required to achieve a disinfection of 99.99 % by a factor of 8–12, and the needed ozone feed dosages by a factor of 5.5–9.

UV-Treatment During the Entire Ozonation Process Water undergoing ozonation can be exposed to a certain dosage of UV-radiation by either continuous or intermittent irradiation varying the UV-radiation intensity. Results of the former procedure were presented in [87]. The latter consisted in UV-irradiation of a bacterial suspension in the process of ozonation (C_{oz} in OAM—2 and 5 mg/dm^3) with intermittent pulses of 4 to 36 s once a minute depending on the intensity of UV-radiation (D_{uv} 0.45–22.2 mJ/cm^2 per pulse). As a rule, in 5 min of ozonation we

Table 5.17 *Escherichia coli* dying rate constants for model water in the process of O_3/UV-treatment (intermittent exposure to UV-irradiation, lamp DRB-8)

Experimental conditions						In rapid phase under the effect of		
C_{oz}, mg/dm³	I_{uv}, mJ/(cm² s)	D_{uv} in pulse, mJ/cm²	Pulse time, s	Number of pulse	Total UV dosage, mJ/cm²	UV	O_3	O_3/UV
2	0.025	0.45	18	10	4.5	−0.002	−0.013	−0.016
2	0.2	2.4	12	5	12	−0.011	−0.013	−0.016
2	1.85	7.4	4	5	37	−0.006	−0.013	−0.043
2	1.85	22.2	12	5	111	−0.055	−0.013	−0.061
5	0.025	0.45	18	5	2.22	−0.002	−0.022	−0.047
5	0.025	0.9	36	5	4.5	−0.002	−0.022	−0.033
5	0.2	2.4	12	5	12	−0.011	−0.022	−0.042
5	1.85	7.4	4	5	37	−0.019	−0.022	−0.050
5	1.85	22.2	12	5	111	−0.055	−0.022	−0.070

applied 5 UV-radiation pulses of different duration using a dosage range from 2.22 to 111 mJ/cm². This enabled us to uniformly reduce the UV-radiation dosage and simulate the conditions of a continuous-flow unit with several O_3/UV-treatment chambers, and also to use pulse sources of UV-radiation. With such a procedure of UV-irradiation in the above-mentioned dosage range we observed no appreciable differences in dissolved ozone concentrations between ozonation and O_3/UV-treatment. At C_{OZ} in OAM of 5 mg/dm³ the concentration of dissolved ozone varied between 0.085 and 0.123 mg/dm³, and at C_{oz} = 2 mg/dm³ varied from 0.033 to 0.045 and from 0.087 to 0.092 mg/dm³ within 5 and 10 min respectively. Based on kinetic curves describing the dying-off of *E. coli* in model water we calculated the killing rate constants listed in Table 5.17.

It follows from the data in this table that the bacteria killing rate constants in the various O_3/UV-procedures were close to, or higher than, the sum of constants obtained for the separate actions of ozone and UV-radiation in their respective dosages. However, at I_{uv} = 1.85 mJ/(cm² s) and a UV pulse time of 12 s with a rate of one pulse per minute the disinfection rate constants in O_3/UV-treatment (C_{oz} in OAM—2 and 5 mg/dm³) were lower than the sum of constants for disinfection by ozone and UV-radiation separately—which is an indication of antagonistic interactions between the disinfectants. Consequently, exposure of a suspension undergoing ozonation to a high combined dosage (111 mJ/cm²) of intermittent UV-radiation produced an antagonistic effect, just as in the case of continuous irradiation [87]. Note also that we did not observe any appreciable decomposition of ozone dissolved in the water under the effect of such a dosage of UV-radiation. On the other hand, in continuous O_3/UV-treatment with the same UV-radiation intensity C_{oz} was just 0.022 mg/dm³ after 5 min, whereas ozonation alone reduced it only to 0.116 mg/dm³. The combined total dosage of UV-radiation, after which ozone decomposed, was 555 mJ/cm².

This data confirms our supposition [87] that UV-radiation decomposed ozone in the thin film at the water-gas interface (ozone-air bubble), which was also where the

Table 5.18 Technological parameters of disinfection of *E. coli* in model water by O_3/UV-treatment at positive *T/E* values

Test	Treatment time, min	C_{oz}, mg/dm^3	UV-radiation pulse parameters *t*,s	I_{uv}, mJ/(cm^2 s)	D_{uv}, mJ/cm^2	Combined total D_{uv}, mJ/cm^2	D_{fd} O_3,mg	*T/E*
1	2	5	4	1.85	7.4	14.8	7.08	0.80
2	2	5	12	0.2	2.4	4.8	7.43	0.55
3	2	5	18	0.025	0.45	4.5	7.24	1.00
4	2	5	36	0.025	0.9	4.5	7.24	0.93
5	3	5	18	0.025	0.45	1.4	10.85	0.87
6	4	2	18	0.025	0.45	1.8	5.82	0.64

first, rapid stage of disinfection took place. This is why an antagonistic effect was also observed in cases where there was no difference between the concentrations of dissolved ozone in solutions treated only with O_3 and with O_3/UV.

For every experimental sampling point (every minute) we calculated the *T/E* values. A value close to unity was obtained for treatment times of 1 min (C_{oz} in OAM 5 mg/dm^3) and 1–3 min (C_{oz} in OAM 2 mg/dm^3), which corresponded to the latent period of the separate ozonation process. Here an additive effect arose, as in the case described above, where bacteria were killed only by the UV-radiation while ozone was expended on the oxidation of organic matter. A true additive effect was registered with treatment times of up to 2–3 min (C_{oz} in OAM 5 mg/dm^3) and 4 min ($C_{oz}=2$ mg/dm^3). Since the T/E values were somewhat lower than unity with unreliable differences between them *($P<90$ %)*, one may therefore assume that the interaction between the two disinfectants was additive in nature (see Table 5.18).

Further increase of the treatment time produced an antagonistic effect. The variant of Test 6 is of no interest, because the bacteria killing rate constant obtained under these conditions was small—−0.016 lg un s^{-1}. In tests 3, 4, and 5 ultraviolet radiation with an intensity of 0.025 mJ/(cm^2 s) produced a very weak antimicrobial action reducing the number of bacteria by 0.4 orders in 4 min of treatment, with the result that the overall effect of the O_3/UV-treatment approached that of ozone alone. The most pertinent and of major interest are tests 1 and 2, in which bacterial suspensions in the process of ozonation (C_{oz} in OAM 5 mg/dm^3) were irradiated for 12 s every minute ($I_{uv}=0.2$ mJ/(cm^2 s) or for 4 s ($I_{uv}=1.85$ mJ/(cm^2 s). The bacteria killing kinetics for these two cases are shown in Fig. 5.25.

Comparing the curves in Fig. 5.25a and b it can be seen that the most efficient treatment of a bacterial suspension in model water by O_3/UV was achieved by using the following procedure: C_{oz} in OAM—5 mg/dm^3, I_{uv}—1.85 mJ/(cm^2 s), irradiation with 1 pulse of 4 s per every minute of ozonation (see Fig. 5.25a). With this treatment a high, six-order degree of disinfection (99.9999 %) was achieved within 3 min. The total exposure time in this process was 12 s and the UV-dose 22.2 mJ/cm^2. With ozonation alone it took more than 5 min to obtain the same effect, and without ozonation the same dose of UV-radiation reduced the numbers of bacteria by less than four orders of magnitude. With Procedure 2, the same disin-

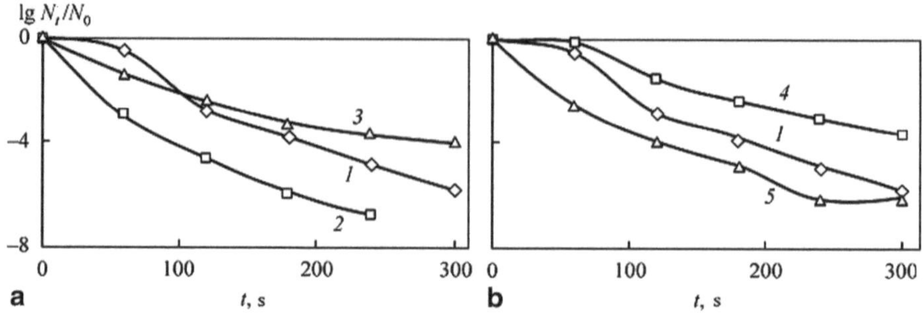

Fig. 5.25 Kinetics of the *E. coli* dying-off in model water under the effect of ozone (*1*), UV-radiation in pulses of 4 (*2*) and 12 s recurring once a minute (*4*), and their combined action (*3* and *5* respectively). I_{uv}—1.85 (**a**) and 0.2 mJ/(cm² s) (**b**)

Table 5.19 Comparison of various O_3/UV-procedures in treatment of model water containing test microorganisms

Mode of exposure to UV during ozonization	C_{oz} mg/dm³	I_{uv}, mJ/(cm² s)	Total *Duv*, mJ/cm²	Exposure time, s	*k*, lg unit s⁻¹	Parameters required for disinfection by 99.99%	
						t, s	D_{fd}, mg
Intermittent	2	0.025	4.5	180	−0.016	>360	9
Initial	2	0.025	4.5	180	−0.015	340	8.2
Intermittent	2	0.2	12	60	−0.016	300	7.15
Initial	2	0.2	12	60	−0.040	312	7.2
Intermittent	5	0.2	12	60	−0.042	220	9
Initial	5	0.2	12	60	−0.054	190	11.6
Intermittent	2	1.85	37	20	−0.043	210	5.6
Intermittent	5	1.85	37	20	−0.050	180	6
Initial	10	2.72	81.7	30	−0.151	24	1.3
Continuous	2	0.025	6	240	−0.032	>300	5.8

fection by O_3/UV-treatment could be achieved in 4 min, but the total UV-dose was lower—9.6 mJ/cm².

The various modes of UV-irradiation of bacterial suspensions in the process of ozonation which showed an additive interaction of the disinfectants (exposure at the initial stage of ozonation, intermittent irradiation, and continuous exposure throughout the ozonation process) are compared in Table 5.19. In the last of the above-mentioned procedures studied a suspension undergoing ozonation with the minimal ozone feed (C_{oz} in OAM—2 mg/dm³) could be continuously exposed to UV-radiation of low intensity (0.025 mJ/(cm² s)) only during 4 min.

Further treatment by this procedure produced an antagonistic effect. As follows from the data listed in Table 5.19, in cases where suspensions ozonized were exposed to equal low dosages of UV-radiation (4.5 or 12 mJ/cm^2) the required degree of disinfection was achieved in about the same time and with the same ozone consumption rate irrespective of the mode of irradiation—whether at the initial stage or intermittently throughout the process of water treatment. In these cases increase of the ozone concentration in the OAM fed into the bacterial suspension with the same dosage of UV-radiation, reduced the treatment time and slightly increased the ozone feed dosage required to obtain a disinfection of 99.99 %.

Exposure of a bacterial suspension undergoing treatment by an ozone-rich (C_{oz} = 10 mg/dm^3) OAM feed to a UV-dosage of 81.7 mJ/cm^2 (I_{uv} = 2.72 mJ/(cm^2 s)) resulted in a significant reduction of the required treatment time. It should be noted, however, that UV-irradiation of a bacterial suspension in the process of ozonation with high dosages is possible only in the initial period, because exposure to the same or higher dosages in the intermittent mode or during the entire process results in an antagonistic effect.

It can thus be concluded from the foregoing that continuous or intermittent irradiation of bacterial suspensions undergoing ozonation with high UV-dosages (D_{uv} > 14.8 mJ/cm^2) produces antagonistic effects in the process of disinfection. In the intermittent procedure it is advisable to expose a bacterial suspension ozonized to short UV-pulses of high radiation intensity. The most efficient disinfection can be achieved by irradiating a bacterial suspension with high UV-dosages in the initial period of ozonation, equal in duration to the latent phase showing on the kinetic bacteria killing rate curve for ozonation alone. In this case the degree of disinfection will be sufficient to the ultraviolet radiation dosage applied.

5.5.2 Combined Antimicrobial Effect of Ozone and Ultraviolet Radiation Generated by Various Sources

It was established that the combination of high dosages of UV-radiation and ozone leads to the emergence of antagonistic effects and lower efficiency of water disinfection process by the O_3/UV-method rather than by UV-radiation or O_3 [87, 90] processes only. However, the above pattern was established when using mercury lamps of low pressure, where the short-wave region of 254 nm accounts for − 90 % of their radiation.

As is known, bacteria dying-off under the effect of short-wave UV-radiation (λ = 240…280 nm) occurs, mainly, at the expense of irreversible lesions in DNA. However, alike as short-wave UV-radiation, the long-wave UV-radiation (λ = 315…400 nm) is also capable of inducing photo-destructive reactions in DNA. In addition, apart from DNA lesions it causes other damage the majority of which is disturbance of the permeability barrier resulting in bacteria dying-off [91, 92]. This region of radiation is generated to a greater degree by mercury lamps of medium pressure. Thus, for instance, a Tungsram-100 lamp's short wave range accounts for 17, while a long-wave one accounts for 41 % of the overall radiation spectrum [63].

All processes of photo-oxidation (including traditional AOP) using mercury-quartz lamps of high and low pressure belong to the processes of the first generation. Active radicals, which are formed from H_2O_2 or O_3 when water is radiated with UV-light ($\lambda=254$ nm), being powerful oxidants are the basis of these processes. New generations of UV-lamps are capable of implementing direct photodissociation of water impurities by splitting chemical bonds in organic compounds. Pulse xenon lamps, which operate at current density 5 kA/cm^2<j>1 kA/cm^2 and pulse duration 150 μs<τ>1 μs are very promising in this respect. In this case the radiation in the region of 300 nm may account for 5–8% of the overall radiation [93]. Photo-disintegration products may additionally be oxidized by oxygen, hydrogen peroxide, or ozone [93, 94]. The data presented in paper [93] indicated that the bactericidal effect of the pulse xenon lamp with the pulse energy of 29 J was higher than that of a traditional mercury lamp of medium pressure with the identical power consumption (40 and 60 W per 1 cm of the lamps' length) and one pulse reduced the number of bacteria in water by a factor of 10^2. We believe that the mechanism of disinfection by pulse UV-radiation with the UV-flux of 10 J/cm^2, whereby the inactivation level reached 9–12 orders of magnitude per one pulse, contains two components: the first one—the bactericidal effect of lamp radiation in the region of 200–280 nm and the second one obtained by calculation—the heating of bacteria by the entire spectrum of a pulse lamp. Paper [94] implies that the advantage of high-intensity pulse sources of UV-radiation of a continuous spectrum lies in a greater depth of penetration and smaller dosages necessary for disinfection than in the case of UV-radiation with a linear spectrum.

The study of the processes of water disinfection when carrying out O_3/UV-treatment was conducted in a laboratory unit [95] in three cylindrical reactors: a glass one (V=625 cm^3) and a plastic one (V=200 cm^3) into which a source of UV-radiation was immersed and a quartz one (V=300 cm^2) with external lighting. In the glass reactor the height of the liquid column was 394 mm and the layer thickness 15 mm, the plastic column was 80 and 23 mm, while in the quartz it was 290 and 40 mm respectively. For radiation source we used lamps of two types: a mercury lamp of medium pressure (Tungsram-100), power 100 W; and a pulse xenon lamp (IFP-800), power 800 W. Specifications of these lamps and their spectral distribution are given in [63]. UV-radiation intensity (I_{UV}) of the mercury lamp was changed by varying its aperture diaphragm. A detailed calculation technique for UV-dosages in the reactors is given in the same source. Paper [8] provides an account of a technique for preparing the *E. coli* 1257 culture and a summary of the bacteria that survived.

Results of disinfecting bacteria in different experiments were presented as the logarithm of the ratio of different bacteria, which survived (N_t), to their original number (N_0).

All the experiments were conducted on a model solution whose composition is shown in [63]. The dosages of the absorbed (S_{ab}) and fed (DFD) ozone were calculated in terms of 1 dm^3 of the water studied. Dissolved ozone (C_d) was measured according to [96] using indigo.

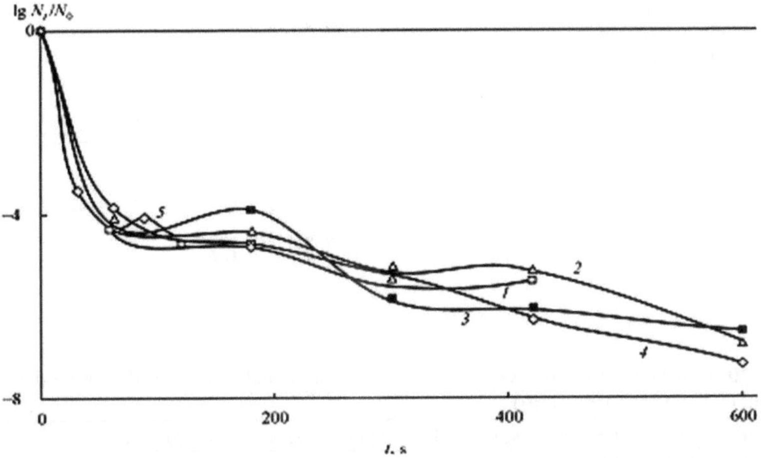

Fig. 5.26 Combined effect of UV-radiation (*1*) (a Tungsram-100 lamp, I_{uv}—0.56 mJ/(cm² s) and ozone at C_{oz} in OAM: 1 (*2*), 2 (*3*), 5 (*4*), and 10 mg/dm³ (*5*) for survival of *E. coli* in the model water

The effect of combined action was assessed by the ratio *T/E*. When assessing the theoretically obtained value we assume the independent effect of every agent separately [89].

The combined action of ozone with UV-radiation was investigated with at least triple experiment repetitions while in the course of the experiments we always studied the antimicrobial effect of each disinfectant individually under the same conditions as their combined effect.

The data obtained was subjected to statistical processing [9] with the account of a mean square error of the surviving cells (N_t/N_0) within a relevant time length for each disinfecting agent (O_3, UV-radiation, O_3/UV). With the obtained data we computed Student's coefficient and reliability of differences in *T* and *E* values *(P)* to determine the effect of interaction of ozone with ultraviolet radiation at every experimental sampling point.

The experiments using the mercury lamp Tungsram-100 were conducted only in the quartz reactor with external lighting. In the course of an O_3/UV-treatment of a model solution containing *E. coli* we investigated the dynamics of their dying-off when feeding ozone with C_{oz} in the OAM—1, 2, 5, and 10 mg/dm³ and I_{uv}—0.56 and 0.06 mJ/(cm² s). The curves of *E. coli* dying-off with an unvarying I_{UV} 0.56 mJ/(cm² s) and varying C_{oz} in OAM are given in Fig. 5.26, while in the case of a constant C_{oz} in OAM 5 mg/dm³ (the ozone feed rate—3.8–3.9 mg/min^{-1}) and the varying I_{uv} in Fig. 5.27.

The change of ozone concentration in OAM (C_{oz}—(1–10) mg/dm³) given the invariable intensity of UV-radiation 0.56 mJ/(cm² s) in the process of O_3/UV-treatment did not affect the kinetics of dying-off of test-microorganisms in model water (see Fig 5.27). Constants of the *E. coli* dying-off rate in the fast phase with such a treatment by ozone in corresponding concentrations are effectively identical. They

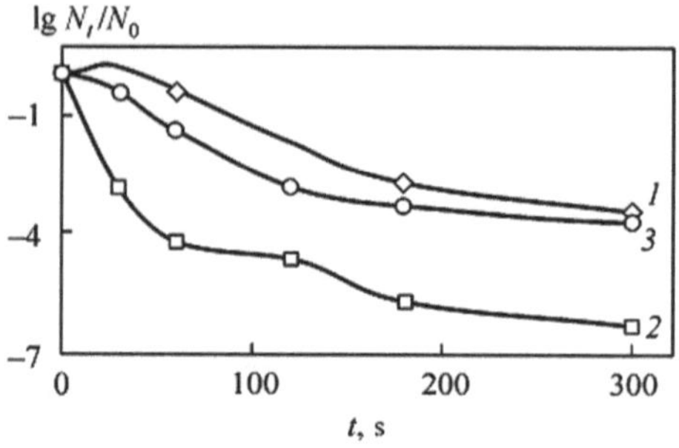

Fig. 5.27 *E. coli* dying-off kinetics in the model water when treated with O_3/UV (a Tungsram-100 lamp) with UV-radiation intensity: 0(*1*); 0.56 (*2*); 0.06 mJ/(cm² s) (*3*) and invariable C_{oz} in OAM—5 mg/dm³

Table 5.20 Constants of *E. coli (k)* dying-off rate in the model water at different modes of (UV-treatment (a Tungsram-100 lamp))

Phases of dying-off kinetic curve	I_{uv}, mJ/(sm² s)	CC_{oz} in OAM, mg/dm³	k, s^{-1} with time period (t) when treated with					
			O_3		UV		O_3/UV	
			t, s	*k*, s^{-1}	*t, s*	*K*, s^{-1}	*T, c*	*k*, s^{-1}
Latent	0.56	1	0–180	0	–	–	–	–
	0.56	2	0–120	0	–	–	–	–
	0.56	5	0–60	−0.008	–	–	–	–
	0.56	5	0–60	−0.008	–	–	–	–
	0.56	1	180–240	−0.012	0–60	−0.076	0–60	−0.076
Fast	0.56	2	120–180	−0.024	0–60	−0.076	0–60	−0.073
	0.56	5	60–180	−0.036	0–60	−0.076	0–60	−0.081
	0.56	10	0–60	−0.079	0–60	−0.076	0–60	−0.081
	0.56	5	60–180	−0.036	0–60	−0.021	0–120	−0.038
	0.56	1	420–600	−0.004	60–600	−0.007	60–600	−0.009
Slow	0.56	2	180–600	−0.006	60–600	−0.007	60–600	−0.007
	0.56	5	180–240	−0.007	60–600	−0.007	60–600	−0.009
	0.56	10	60–180	−0.024	60–600	−0.007		
	0.56	5	180–600	−0.007	60–300	−0.008	60–300	−0.012

were higher than when treated by ozone in corresponding concentrations and are equal to the constants obtained when treating the model water by UV-radiation only (see Table 5.20).

The decrease of I_{uv} with the invariable feed of ozone (C_{oz} in OAM—5 mg/dm³) at O_3/UV-treatment of a model solution leads to reduction of disinfection intensity (see Fig. 5.27). In this case the constant of the test-microorganism dying-off in the "dying-off fast phase" decreased from 0,081 to 0,038 s^{-1} with the lowering of I_{uv} of the Tungsram-100 lamps from 0.56 to 0.06 mJ/(cm² s) and effectively did not change in the slow phase (see Table 5.20).

From the data given in Table 5.20 it can be observed that the constants of the *E. coli* dying-off rate in the fast phase in the model water in the course of O_3/UV-treatment (I_{uv}—0.56 mJ/(cm² s)) were equal to the constants of dying-off rate under the effect of UV-radiation within the first 60 s of treatment if C_{oz} in OAM did not exceed 5 mg/dm³. With the increase of C_{oz} in OAM to 10 mg/dm³ the constant of

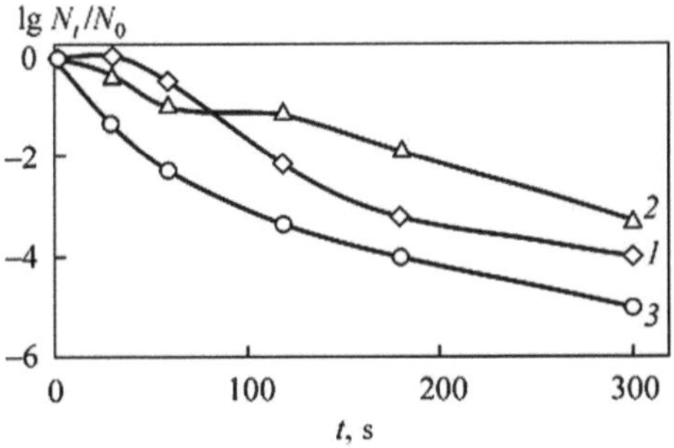

Fig. 5.28 *E. coli* dying-off kinetics in the model water when ozonized (C_{oz} in OAM—5 mg/dm³) (*1*), treated with UV-radiation (I_{uv}—0.06 mJ/(cm² s)) (*2*) and O_3/UV-treatment (*3*)

the *E.coli* dying-off rate with O_3/UV-treatment (−0.081 s^{-1}) was much smaller than the sums of the constants when treating only with O_3 (−0.079 s^{-1}) and with UV-radiation only (−0.076 s^{-1}) in the corresponding dosages, which strongly indicates a pronounced effect of antagonism.

The analysis of the *T/E* values calculated in the experiments using the lamp of medium pressure in each time point, when sampling was made, showed that the combination of high dosages of ozone and UV-radiation leads to an antagonistic effect; the higher the dosage of each reagent in the mixture—the greater this effect. The additive effect was noted at the initial period of treatment (0.5–5 min) during photoozonation of the model solution depending on C_{oz} in the OAM (1–5 mg/dm³). The higher C_{oz} in the OAM—the shorter the period within which the additive effect was observed. With an O_3 concentration of 10 mg/dm³ in the course of the O_3/UV-treatment of the model solution containing *E. coli* the additive effect was not observed. The duration of the O_3/UV-treatment of the model solution within which one noted the additive effect corresponds to the duration of the latent phase on the kinetic curve of the dying-off of test-microorganisms under the effect of ozone alone.

From the data given in Table 5.20 it can be seen that during this period the bacteria dying-off under the effect of the O_3/UV-treatment occurs at a rate commensurate with the dying-off rate under the effect of UV-radiation only. Consequently, the additive effect obtained is the action of UV-radiation only while ozone is spent only on oxidation of organic matter. Further ozone feed into the water treated together with its simultaneous radiation leads to the appearance of antagonistic effects. A true additive effect in the fast phase of bacteria dying-off during O_3/UV-treatment occurs at I_{uv} 0.06 mJ/(cm² s) and C_{oz} in OAM 5 mg/dm³ only within the first 2 min when the *T/E* values are close to unity and reliability of difference between the values *T* and *E* are low (16.2–43 %) (see Fig 5.28). When the exposure is continued the *T/E* values sharply drop (after 3 and 5 min of treatment they constitute respectively 0.026 and 0.00019 with the reliability level of 72.9 and 62.9 %), which indicates the antagonistic interaction between the disinfectants being investigated. The results obtained are similar to such when using the O_3/UV-treatment by a DRB-8 mercury lamp of low pressure [90].

The other source of UV-radiation under study in the O_3/UV-treatment was an IFP-800 pulse lamp having an automatic pulse feeding system at a frequency of one pulse per 2.6 s and a manual pulse feeding system at a selected frequency. The

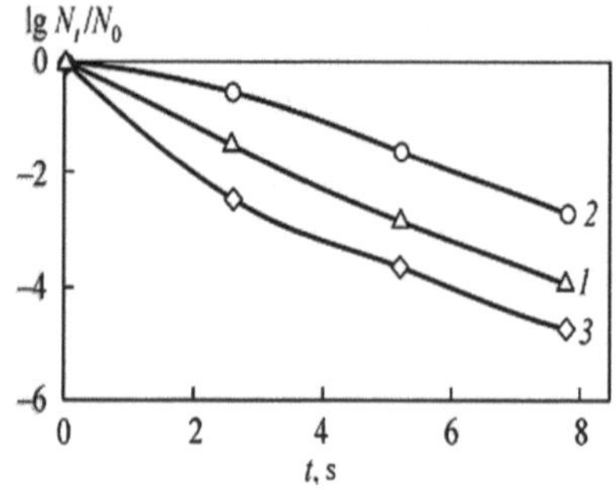

Fig. 5.29 *E. coli* dying-off kinetics in the bidistilled water under the effect of ozone (*Coz*—0.38 mg/dm^3) (*1*), pulse UV-radiation (*2*) and combined O_3/UV-treatment (*3*)

assessment of the O_3/UV-treatment disinfection method using the IFP-800 lamp in the plastic reactor (V=200 cm^3) was carried out in the following way: bidistilled water saturated with ozone in a sparging reactor (the concentration of dissolved ozone—0.38 mg/dm^3) was poured into the plastic reactor, where 2 cm^3 of the *E. coli* suspension with a density of 10^9 CFU/cm^3 was introduced and a source of radiation was immersed. After irradiation with 1, 2, and 3 pulses the solution was drained and the number of bacteria that survived was determined as well as the concentration of dissolved oxygen. For comparison under the same conditions the suspension of bacteria was treated only with the UV-radiation or ozone. When ozonation of the bacteria suspension was carried out, the sampling was conducted at an interval of 2.6, 5.2, and 7.8 s, which corresponds to the duration of the UV-treatment with 1, 2, and 3 pulses. The obtained results are given in Fig. 5.29, from which show that the O_3/UV-treatment of the *E. coli* suspension using a pulse radiation source under the same conditions of the experiment led to a higher dying-off rate ($k=-0.651$ s^{-1}) than under the action of ozone only ($k=-0.514$ s^{-1}) or UV-radiation ($k=-0.314$ s^{-1}). The computation of the *T/E* values showed that only when the ozone solution containing test-microorganisms was irradiated with one pulse of UV-radiation (D_{uv}—14 mJ/cm^2) could observation of an additive effect of disinfectants (*T/E*—3.16; *P*—23 %).

Further irradiation of suspension with 2 and 3 pulses (D_{uv} respectively 28 and 42 mJ/cm^2) led to the emergence of the antagonistic effect (*T/E* respectively 0.19 and 0.014). When treating the ozone solution with bacteria with one pulse of UV-radiation a moderate reduction of the dissolved ozone concentration compared with the treatment by ozone alone (respectively 0.047 and 0.068 mg/dm^3) is observed. However, irradiation with 2 and 3 pulses led to the concentration drop of dissolved ozone to 0.004 mg/dm^3. Thus, the antimicrobial antagonistic effect, occurring due to the use of two disinfectants, is determined by the competitive absorption of UV-radiation with dissolved ozone.

Although the efficiency of using pulse UV-radiation (an IFP-800 lamp) in the plastic reactor is much higher than in the glass one (the density of UV-radiation of one pulse in the first case constituted 5.4, while in the second one it was 1.06 mJ/cm^2) for further investigation we had to use the latter reactor, which was adapted for continuous feed of OAM to the water treated. The dying-off kinetics of a test-microorganism in the model water in this reactor during its different treatment is shown in Fig. 5.30.

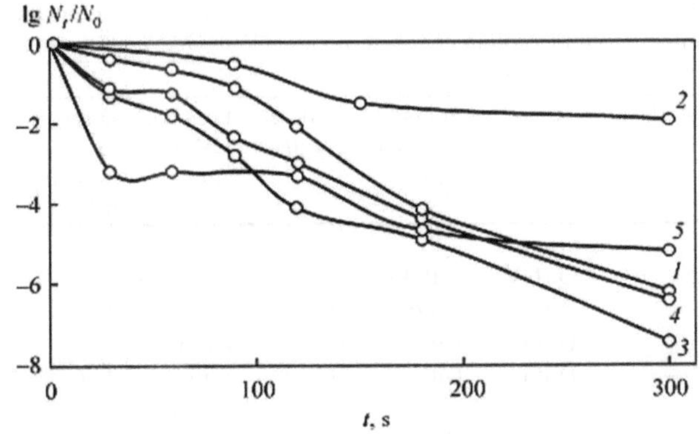

Fig. 5.30 *E. coli* dying-off kinetics in the model water when treated with ozone (C_{oz} in OAM—5 mg/dm³) (*1*); pulse UV-radiation in the mode of a single pulse every 30 s (Σ—10 pis) (*2*) and their combined effect (*3*); at O_3/UV-treatment within the first 30 s of ozonation with 10 pulses of an IFP-800 lamp (*4*) and UV-radiation with a DRB-8 lamp (*5*)

We investigated two conditions of introducing UV-pulses: (1) within the whole process of ozonation (C_{oz} in OAM—5 mg/dm³, treatment time—5 min) at the frequency of one pulse per each 30 s (2.76 mJ/cm² per one pulse) (2) feeding of pulses within the first 30 s of ozonation (8, 10, and 12 pulses). The summary dosage of UV-radiation in this case was respectively 22.1, 27.6 and 33.1 mJ/cm². For comparison, we used the feeding mode of UV-radiation generated by a DRB-8 lamp of low pressure within the first 30 s of ozonation (D_{uv}—81/6 mJ/cm²). From kinetic curves of the bacteria dying-off we calculated the constants of dying-off rate, the dying-off duration necessary for disinfection by 99.99 % ($t_{99,99}$) and the *T/E* values for the first 30 s of treatment. This data is shown in Table 5.21.

The constant of the *E. coli* dying-off rate in the model water during the O_3/UV-treatment in both modes is larger than under the effect of ozone or UV-radiation only with the same feeding modes. Thus, in the case of uniform irradiation with 10 pulses of UV-radiation within the whole ozonation process, the constant of the bacteria dying-off rate obtained was—0.027 s^{-1}, which is somewhat higher when treating respectively with O_3 and UV-radiation (−0.015 and—0.007 s^{-1}).

At UV-irradiation with 8–12 pulses within the first 30 s of ozonation the constants of the *E. coli* dying-off rate were somewhat lower or equal to the sum of constants obtained under the effect of each of the disinfectants individually. A much higher constant of *E. coli* dying-off rate was found when subjecting the ozonized suspension to UV-radiation by a DRB-8 lamp within the first 30 s (−0.108 s^{-1}). However, in the given case the dosage of UV-radiation was much higher—81.6 mJ/cm² and consequently, the constant of the bacteria dying-off rate under the effect of UV-radiation only (−0.080 s^{-1}) was higher. The result obtained agrees with the calculated *T/E* values, which indicate that the effect of interaction between the disinfectants is close to the additive effect. Judging by the treatment time necessary for reducing the number of bacteria by four orders of magnitude (99.99 %), the disinfection rate, and the *T/E* value, the most acceptable are treatment modes 5 and 6 (see Table 5.21).

Therefore, as a result of the work done, it was established that a combination of high dosages of disinfectants in the O_3/UV-treatment leads to the emergence of antagonistic effects, which is caused by competitive absorption of light quanta by

Table 5.21 Some technological parameters during ozonation and photoozonation of model water infected with *E. coli* in different modes (C_{oz} in QAM—5 mg/dm³)

Treatment mode						k, s^{-1}	$T_{99,99}$, s	T/E value in the first 30 s
Mode no	Treatment mode	Sourse of UV-radiation	Number of pulses	Perion of UV-radiation	Dosage of UV-radiation mJ/sm²			
1	O_3	–	–	–	–	−0.015	260	–
2	O_3/UV	IFP-800	8	First 30 s	22.1	−0.018	180	0.63
3	Ditto	Ditto	10	Ditto	27.6	−0.023	160	0.67
4	Ditto	Ditto	12	Ditto	33.1	−0.018	160	0.63
5	Ditto	Ditto	10	Within 5 min	27.6	−0.027	140	8.3
6	Ditto	DRB-8		First 30 s	81.6	−0.018[a]	150	2.6

[a] Within the first 30 s of treatment, for other modes—within the whole experiment (5 min)

bacteria and ozone dissolved in water. The efficiency of the O_3/UV-treatment in such case was higher than O_3, but lower than UV-radiation. The combination of low rates of ozone feed and intensities of UV-radiation for the O_3/UV-treatment, the pulse feed of ultraviolet radiation in the process of ozonation, and UV-irradiation of water ozonized at the initial period equal to the latent phase when treating with ozone only leads to the additive effects. The efficiency of the O_3/UV-treatment in this case was higher than each of the disinfectants separately. The optimal conditions of the combination of O_3 and UV-radiation in the case of the O_3/UV-treatment mainly depend on the correctly chosen dosages of UV-radiation. The efficiency of the O_3/UV-treatment does not depend on the source of UV-radiation used, but only on its dosage—in the region of 200–290 nm.

5.6 Disinfecting Action of a Space Spark Discharge in Water

Current water treatment practices have become increasingly oriented to a search for new treatments and disinfection techniques as well as development of combined procedures. The combined use of chemical and physical factors facilitiates the enhancement of antimicrobial action in order to produce a synergic effect that would overcome the limitations inherent in using individual methods [97, 98].

From this standpoint, one such new method that appears to hold considerable promise is space discharge treatment, which combines a variety of physical, chemical, physicochemical, and electrochemical effects in a single technological process. It has been known that combining more than one process results in synergic action. A major role is played by a close-range factor, owing to spatial distribution of sources of electric-spark radiation in the treated water. Additionally, depending on the medium, space spark treatment can directly produce fresh aluminum and iron

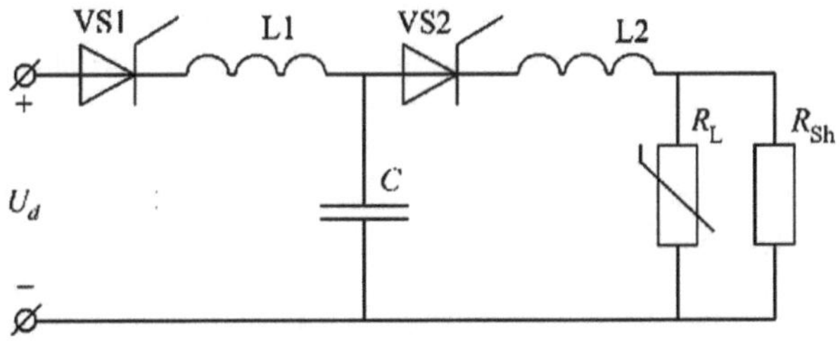

Fig. 5.31 The diagram of the shaper power unit

hydroxides that are effective for removing various microorganisms, including those with the highest resistance to disinfectants [99, 100].

The proposed method consists of the creation of multichannel fast-migrating space spark discharges between particles of a coagulant-forming metal immersed in water. This produces a fine-dispersed sol with particle sizes of 3 to 70 μm, interacting with water to form hydroxide of a metal (in this case aluminum). The particle sizes and their concentration depend on the quantity of electric power applied to the Technological Apparatus (TA), the slope, amplitude, and duration of the discharge pulse, the rate of heat dissipation from the sparking contact areas, and also the physical properties of the current-carrying particle material, thermal capacity and temperature of the water undergoing treatment, etc. [101].

Due to small particle sizes and an extensive active surface (ratio of surface area to particle volume is of the order of $1.2–3 \times 10^5\ m^{-1}$), the sol obtained in this way has in itself good supportive properties. By selecting the appropriate medium material, e.g. aluminum, it is possible to obtain a highly active $Al(OH)_3$ coagulant.

In addition to coagulation and physical effects, the process of spark treatment of a liquid includes such factors as oxidation by ozone, hydrogen peroxide, free radicals, high-temperature plasma, high-intensity electricity, magnetic fields and currents with amplitudes of up to several thousand amperes [98, 102].

A laboratory unit for space spark treatment of water consists of a Discharge Pulse Shaper (DPS) based on an open-input natural-commutation sequential resonance inverter without reverse diodes [103], a constant-voltage regulator which is a regulated rectifier furnishing voltage U_d, a technological apparatus, and a Control Unit (CU) that generates the sequence of charge and discharge control pulses with regulated frequency *f*. The electric part of DPS is shown in Fig. 5.31.

Thyristor VS1, choke L1, and capacitor C form a charge circuit. The discharge circuit consists of thyristor VS2, connecting wire inductance coil L2, load resistor R_L, and capacitor C. The discharge operation is mainly aperiodic, although in certain periods where resistance of R_L decreases, the discharge may be of an oscillatory nature with a slight recharge of capacitor C.

To eliminate abnormal operating conditions that arise as a result of sharp changes of the resistance of R_L, this resistor was shunted with a resistive shunt R_{Sh} with a resistance of 0.75 Ω.

Without the shunt a considerable increase of the R_L load would result in a longer discharge pulse and at the point of triggering the VS1 charge thyristor discharge thyristor, VS2 could open. However, at a very low R_L value capacitor C recharges without losses, the initial charge conditions for the next period are changed, and the capacitor voltage and choke currents increase due to resonance. Voltage U_C on the discharge capacitor on completion of the charge process is determined by the equation

$$U_c = (U_d + \left(U_d - U_0)\exp\left(\frac{-\pi}{2Q_1}\right)\right)\exp\left(\frac{-\pi}{\sqrt{4Q_2^2 - 1}}\right) \tag{5.51}$$

where U_0 is the voltage on capacitor C at the moment of the beginning of charge; Q_1 is the charging circuit Q-factor; Q_2 is the discharge circuit Q-factor equal to

$$Q_2 = \frac{(R_H - R_w)}{R_H \cdot R_w}\sqrt{\frac{L_2}{C}} \tag{5.52}$$

In its turn, U_0 is the described by equation

$$U_c = \frac{U_d \exp\left(-\pi/\sqrt{4Q_2^2}\right)}{1 - \exp\left[-\pi\left(\frac{1}{2Q_2} + \frac{1}{\sqrt{4Q_2^2 - 1}}\right)\right]} + \frac{U_d \exp\left[-\pi\left(\frac{1}{2Q_1}\right) + \frac{1}{\sqrt{4Q_2^2 - 1}}\right]}{1 - \exp\left[-\pi\left(\frac{1}{2Q_1} + \frac{1}{\sqrt{4Q_2^2 - 1}}\right)\right]} \tag{5.53}$$

The technological apparatus consisted of a cylinder of organic glass measuring 60 mm in inner diameter and 130 mm in height with two aluminum electrodes sized $130 \times 15 \times 2$ mm, 50 mm apart and mounted opposite each other. It was filled with pieces of aluminum chips $15–30 \times 4 \times 0.5$ mm to load height $h_1$and water to be treated of volume V height h_2.

The equivalent electrical resistance of the TA designated R_L and defined as a ratio of full power integral to the Joule integral is of a nonlinear stochastic nature and varies in range from 0.2 to 1.5 Ω. A stable process in the TA with the design described above has h_1=40 mm, pulse voltage amplitude U_m=300…600 V and discharge pulse frequency range f=5… 150 Hz.

$$\mathrm{R_H} = \left[\int_0^{\tau_1} \mathrm{u(t)i(t)dt}\right] / \left[\int_0^{\tau_1} \mathrm{i^2(t)dt}\right] \tag{5.54}$$

where τ_p is the duration of discharge pulse; $u(t)$ is the time function of voltage under load; i(t) is the time function of current under load.

The dynamics of resistance R_{CH} is affected by an individual low-voltage spark channel over the sparking time, and its relationship to current and parameters of the discharge gap are given in [104]:

$$R_K = \delta\sqrt{2p/\left[\sigma_{max}\int_0^{\tau_1} k \cdot i^2(t)dt\right]} \qquad (5.55)$$

where δ is the length of the discharge channel, p is the pressure in the discharge channel, k is a coefficient characterizing the proportion of power per unit of discharge gap that is spent for channel expansion, i is discharge gap current, σ_{max} is conductivity of plasma in the channel center defined [105] for discharges in the air as $\sigma_{max} \approx 3 \times 10^4$ S/m.

Resistance of the TA active area, just as that of a single contact, is of a stochastic nature [101]. This can be explained by different mobility of conductive particles, uneven gas-dynamic action of vaporized and molten metal in the sparking area, electrochemical phenomena in the active area, and a large number of probable electric contacts [102].

Controlled parametric feedback loops can be applied to stabilize electrical parameters of discharge pulses, e.g. voltage [106]. This can best be done by implementing an effective algorithm for the stabilizing voltage, pulse power or discharge current on the discharge capacitor. Further studies are needed to determine priorities to be followed in stabilizing the above parameters. The simplest way to decrease the influence of changes of the load resistance over a wide range is to shunt the load resistor with another resistor—as was done in laboratory unit.

Disinfection tests were made under static conditions, i.e., without a water flow [107]. Water of fixed volume V contaminated with the strain *Escherichia coli* 1257. With initial concentration, N_0 was treated in the TA active area over time t_0. The TA contents were poured into a sterile retort and left in the active medium for aftereffect time t_a and then filtered across filter paper with pores of 2 μm. In all tests the water had temperatures of 21 to 35 °C and a pH value of 5.

Numbers of microbes that survived this treatment were determined by the direct plating in the Endo medium followed by incubation during 24 h at 37 °C and count of number N_t of colony-forming units. The antimicrobial effect was assessed by the logarithm of the N_t/N_0 ratio.

The electrical parameters of discharges were measured by means of Cl-73 and Cl-96 oscilloscopes equipped with standard 1:10 dividers with a resistance of 10 MH and a capacitance of up to 20 pF and a noninductive current shunt with a resistance of 6 MΩ.

Figure 5.32 presents the relationship between the bactericidal effects of spark discharges, aftereffect time t_a, and treatment time t_{tr}. The graph shows results of two experiments. The parameters of the first experiment (curves *1* and *1'*) were: h_1=25 mm, h_2=55 mm, V=230 cm³, C=100 μF, U_m=600 V, and f=25 Hz. Those of the second experiment (curves *2* and *2'*) were: h_1=h_2=40 mm, V=100 cm³, C=200 μF, U_m=300 V, and f=10 Hz.

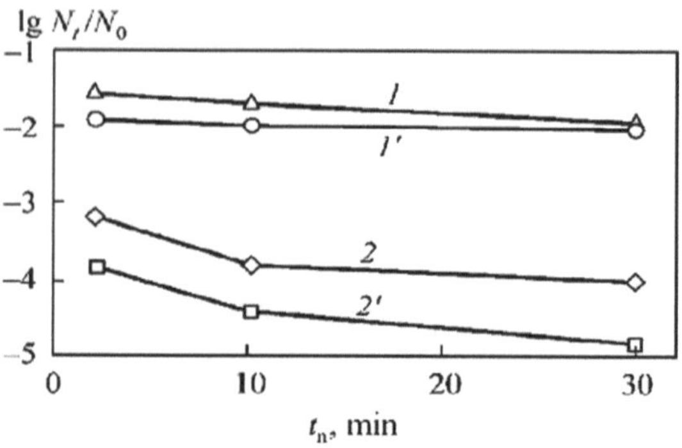

Fig. 5.32 The relationship between antimicrobial effect and the after-effect time t_n and treatment t_{tr}: *1*, *2*—30, *1'*, *2'*—60 s

As can be seen from the figure, the greatest influence on the inactivation of microorganisms is exerted by the distance from the centers of action (sparking) to the most remote point of the liquid being treated. Even though specific operative energy was higher in the first experiment than in the second, the antimicrobial effect of discharges was found to be lower by an order of magnitude.

It may be expected that the effects of most physical factors arise as a result of a discharge (ultraviolet radiation, water hammer, and ultrasound) and sharply decrease with increasing distance to an individual sparking source. It is known that in media without absorption of energy the intensity of ultraviolet radiation is inversely proportional to the square of the distance to source, while in media with absorption of energy, of which the medium under consideration most probably is one, the intensity is inversely related to the exponent of distance to source.

It can be noticed, in analyzing the relationship between the antimicrobial effect of the method in question and time of active medium after effect (see Fig. 5.32), that this curve consists of two sections. The first of these (t_a=2…10 min) apparently reflects the effect of the freshly obtained coagulant and oxidants, as well as that of free radicals produced by exposure of the water suspension of bacteria to the sparking treatment, whereas the second section (t_a= 10…30 min) shows only the effect of the $Al(OH)_3$ coagulant [100]. This is why the first section of the relationship curve is steeper.

Figure 5.33 shows the relationship between the inactivating effect of this method, current amplitude I_m, voltage U_m of discharge pulses, and their frequency *f*. Conditions of the experiment were: C=200 μF, h_1=h_2=40 mm, V=100 cm^3, t_0=30 s, and t_a=2 min.

The relationship between the antimicrobial action and discharge pulse voltage amplitude in the above frequency range is close to parabolic, which is especially pronounced at low (5–10 Hz) frequencies. Considering the nature of the current relationship between the resistances of an individual spark discharge channel and flowing current, it can be expected that with equal pulse power the antimicrobial effect relationship will be described by a power function of exponent 2.5, allowing for the proportionality factor.

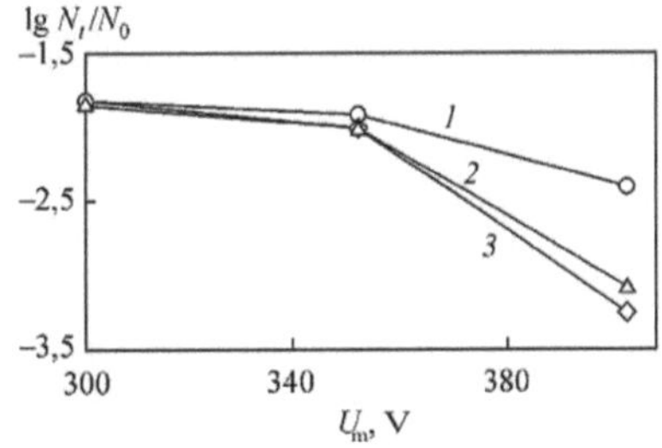

Fig. 5.33 The relationship between antimicrobial effect on voltage U_m and the frequency of pulse sequence *f*, Hz: 150 (*1*), 10 (*2*), 5 (*3*)

Regarding the pulse frequency function of the bactericidal effect of the method under study, an increase of frequency was accompanied by a certain decrease of the antimicrobial effect, despite higher combined pulse energy per unit of water volume.

We will now discuss the mechanism whereby the electric resistance of the TA interelectrode space increases with the pulse frequency. At the moment of sparking the contacting aluminum, particles receive a mechanical pulse, most probably of gas-dynamic and electromagnetic forces that arise under similar conditions [101]. Aluminum particles, which have a certain mobility depending on their mass, effective section area, degree of caking, hydrodynamic resistance of the water, etc., move under the action of the mechanical pulse, impairing or breaking the electric contact. As the pulse frequency increases, the number of electric contacts restored after the passage of the previous pulse is reduced, which means that the equivalent electric resistance of the TA interelectrode space rises. Depending on the magnitude of the mechanical impulse and the particle mobility, it takes a certain time to restore a contact's resistance. If the ratio of the linear, dimensions of the TA active area to the particle sizes is, as it was in our case, the number of probable sparking contacts is insignificant.

It is the permanent, rather than periodic, factors that have the greatest effect on microorganisms [108, 109]. In our experiment, the relative pulse duration was quite high: 27–2,000. Under such conditions there is probably, for cases where limited parcels are treated, a certain threshold frequency that determines the operating energy per unit of water volume. It is assumed that further increase of frequency in the treatment of fixed parcels will fail to enhance method efficiency.

Figure 5.34 shows the function of the method disinfecting effect of the discharge pulse frequency at constant pulse power $P_p = 22.8$ kW and aftereffect time $t_a = 2$ and 20 min. The experiment conditions were: $C = 200$ μF, $h_1 = h_2 = 40$ mm, $V = 100$ cm^3 and $t_0 = 30$ s.

As follows from Fig. 5.34, in the given frequency range and at a constant pulse power the antimicrobial effect of the method does not depend on the pulse frequency, which corroborates the above conclusions. Certain differences can be attributed to errors arising from measurement and data processing.

Pulse power P_p was calculated as the product of amplitude values of voltage U_m and current I_m of the technological apparatus with allowance for pulse shape coefficients. Here the voltage pulse shape with a short front and a gently sloping back was approximated by a linear function diminishing from the U value to zero over pulse

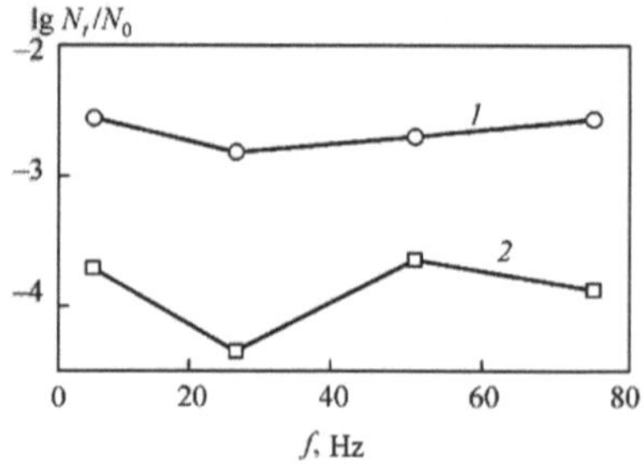

Fig. 5.34 The relationship between antimicrobial effect and the frequency of pulse sequence at P_p=22.8 kW. Treatment time: 2 (*1*) and 20 min (*2*)

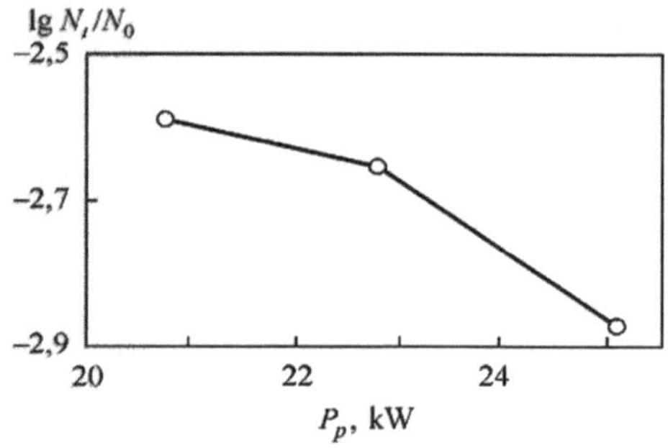

Fig. 5.35 The relationship between antimicrobial effect and the value of the charge pulse power

duration time τ_p. On the other hand, the current pulse shape was close to sinusoidal and was described by a harmonic function. Pulse power was calculated by formula:

$$P \approx \frac{U_m I_m}{\tau} \int_0^{\tau} \left(1 - \frac{1}{\tau}\right) \sin\frac{\pi t}{\tau} dt = \frac{U_m I_m}{\sqrt{6}} \tag{5.56}$$

Figure 5.35 presents the relationship between the experimental pulse power of the inactivating effect of water disinfection through space-spark dispersion and coagulant-forming metals. The conditions of this experiment were the same as for the previous one. As the pulse power increases, so does the antimicrobial effect of the method under consideration. This function is close to linear.

Based on the results of our investigation, basic principles were formulated for designing experimental technological facilities for water disinfection and treatment by the electric-spark method:

- to enhance the efficiency of water treatment and to provide for maximum use of the effect of close-range action, measures must be taken to eliminate the shaded zone in the passage of treated water through the active area of the discharge chamber;
- in disinfecting water by the method under consideration, the power consumption rate can be reduced more effectively by raising the voltage of discharge pulses than by increasing their frequency;

- increased stability of the method's disinfecting effect can be achieved by stabilizing the pulse power or the discharge pulse current rather than the pulse voltage since the antimicrobial effect function of the pulse voltage is significantly nonlinear;
- to lessen the pulse frequency dependence of the TA electric resistance on the frequency and expand control range of the method in question, it is necessary to make sure that the ratio of the TA active area dimensions to the linear sizes of current-conducting particles is within the range of 20 to 100. Also, it is desirable to ensure the required particle mobility;
- for higher efficiency of water disinfection, the aftereffect factor should be used. This can be done by incorporating in the facility a settling tank in which water must remain for at least 10 min after treatment in the TA.

5.7 The Problem of Water Infection with Pathogens of Mycoses and Prospects of Resolving this Problem

The modern world has faced the problem of ecological systems degradation and gradual depletion of natural resources, such as the water resources on which the existence of man on Earth depends. The Ukraine has lived through an especially painful experience since the Chernobyl disaster occurred. This event arguably marked the lowest point in the decline of traditional industrial civilization concomitant with anti-ecological development.

The Earth's biogeosphere is an open system whose functioning is based on the balanced interaction of biological and abiotic components. The state of the biological component (biocenosis) in great measure is determined by the stable equilibrium in the host-pathogen or parasite system. In the course of evolution, a population of parasitic microorganisms has formed: bacteria, viruses, fungi, and protozoa. These microorganisms limit the number of the populations of the main types of the biocenosis. The effect of the above microorganisms on human is distinguished in great variety: some are causes of epidemics, while others constantly produce a weak toxic effect leading to latent pathological processes, very often with a lethal outcome. The power of this effect is determined by a whole complex of factors among which unfavorable ecological conditions play an important role; these conditions lower the immunity of the organism and considerably enhance the activity of the pathogenic factors.

Against the backdrop of a vast number of papers dealing with the effect of unfavorable factors on the state of the human immune system, the origin of epidemics, epizootics, and epiphytotics remains almost unheeded. Conventionally, pathogenic microorganisms are capable of realizing their pathogenic properties by intensifying the expression of the genes responsible for the production of enzymes and toxins, especially when a human or animals with a lowered functional activity of immune-competent cells are infected. Water is one of the leading courses in propagation of such microorganisms and, in the first place, of microscopic fungi (producers of biologically active and toxic substances). Moreover, the disruption of the ecological equilibrium in the existing air-water-earth system leads to a change of the biological properties of the micro world which feature resistance to an aggressive medium and adaptive properties to extreme factors; this is how new pathogens emerge [110].

The variation of the ecological situation under the influence of the anthropogenic load has led to the change of the microbial background of the environment. At present, one can observe throughout the world the replacement of a pathogenic bacterial component with a more aggressive fungal one, which is customarily, is considered as conventionally pathogenic without taking account of or assuming its potential aggressive qualities. A sharp increase in the number of patients suffering from systematic and local mycoses has forced us to focus maximum attention on this problem and deal more seriously with the detection of individual genera of micromycetes when assessing the infectious danger of the environment. In this case, the criteria of qualitative and taxonometric characteristics of detected microorganisms have been offered [111].

Thus, a majority of the most widespread domestic fungi are referred to the 1^a and 2nd degrees of the risk. The Atlas of Pathogenic Fungi provides information on more than 800 genera present in various regions of the world, which reflects the situation on a global scale [112]. In addition, not only the generic composition of the pathogens of different taxonometric classes but also the nature of their functioning in the human organism varies substantially. Fungi affect not only skin, but cause lesions of effectively all organs and systems of human, animals, birds, fish and insects. The most wide-spread mycoses are: onychomycoses, otomycoses, keratomycoses (especially in patients suffering from diabetes) and fungemia in oncological and hematological patients, in the case of tuberculosis, in patients having bums, in recipients after transplantation of organs and in HIV-infected persons [113–116].

The activity and malignancy of the mycotic infection depends, on the one hand, on the state of the immune defense of each individual and, on the other, on the set of aggressive factors of a pathogen. Taking into account the development of the secondary immune deficiency in the Earth's population due to a multi-factor effect (stress, unsound diet, wide and uncontrolled consumption, including animal husbandry and plant growing, hormones, cytostatics, antibiotics, a redundant amount of vitamins and biologically active substances of varidirectional action), mycoses become a serious threat to human life. In terms of spreading, they follow step by step after viral infections especially determined by the viruses of hepatitis, herpes and HIV. The international community of medical mycologists characterizes mycoses metaphorically as "a waking giant" [117].

The factors conducive to the greater activity of mycotic infection include:

1. an extremely high adaptation ability of fungi [112, 113];
2. ubiquity, i.e. world-wide prevalence in nature and ability of one and the same species to affect plants, birds, fish, animals, and human.

Mycotic diseases include not only mycoses, but also intoxication with toxic matter of the fungi—mycotoxicoses, mycetism, and mycoallergoses [118]. Fungus metabolism products entering blood and lymphatic vessels produce a sensitization effect causing the development of the conditions mentioned. Various scientific data

Table 5.22 Diseases caused by micromycetes

Fungus pathogen	Diseases	Toxins produced
Zigomycetes *Mucor racemosus,* *Mucor pusillus,* *Rhizopus arrhizus*	Dermatomycoses of animals and human, otomycoses, mucormycoses	Mucortoxins, corticoids
Deuteromycetes		
Aspergillus flavus, *Aspergillus fumigatus,* *Aspergillus glaucus,*	Aspergilloses, aspergillomas, mycetomas; hepatotoxicoses, bronchitises, allergies	Fumigacylline, gliotoxin, sterigmatocystine, ochratoxin, cytrinine, patuline
Aspergillus terreus, *Aspergillus versicolor,* *Aspergillus niger,* *Aspergillus nidulans* *Aspergillus alternate*	Otomycoses onychomycoses, dermatomycoses, necroses of skin and various tissues, allergies	Sterigmatocystine, mycotoxins
Cladosporium cladosporioides Cladosporium herbarum	Dermatomycoses	Sterigmatocystine
Paecilomyces varioti, *Penicillium citrinum,* *Penicillium cyclopium,* *Penicillium fellutanum,* *Penicillium notatum,* *Scopulariopsis brevicaulis*	Pycelomycoses, ephrotoxicoses, allergies of children, adults, especially dangerous for post-operational patients, onychomycoses	Vanotine, citrinin, palitantine, tremorgen, aflatoxins, trikventine, notatin, xanthociline

have provided us with the material to compile consolidated tables [114], making it possible to assess the spectrum of pathogenicity of the fungi most often encountered in nature and which are causative factors of human diseases of different nosology (see Table 5.22).

The main mechanisms for implementing a pathological process with fungi or, more specifically, with micromycetes include their ability to produce toxic substances, enzymes, and hormones. A fused reaction results in the formation of biopolymers characteristic only of vital activity of fungi, which take part in biodestruction of biopolymers of the macroorganism cell [114]. Toxic substances, which are produced by fungi, may affect practically all organs and systems of humans and animals (Table 5.23). The pathogenic effect of fungi is realized by the functioning of many systems, making it possible for propagules to implement adhesion, invasion, and produce a toxic effect; this toxic effect manifests itself in the disruption of metabolic and energy processes in the host cells (see Fig. 5.36). When fungi enter a macroorganism in any way (by water, air) there is always the danger of propagation followed by a subsequent development of a local or system mycosis, especially in weakened patients. In addition, it is known that some mycotoxins (especially aflatoxins) which are produced by *Aspergillae,* manifest a carcinogenic effect causing the development of primary cancer of the liver, adenomas, and adenomas-carcinomas in lungs [119], stomach, and kidneys. Table 5.23 shows their tropism to embryonic tissues.

In the light of what has been set out, the problem of sanitary-epidemiological danger, related to the microbial pollution of water, including not only natural water

Table 5.23 Producers of mycotoxins causing lesions of human organs

Toxin	Producer	Effect on organs
Ochratoxins	*P. cyclopium, A. terreus*	Cause lesions of kidneys, which are manifested in the formation of necroses of tubular cells, hyaline degeneration is found in the liver and fatty degeneration of parenchyma cells. Ochratoxin A is most toxic, whose nephrotoxic effect is manifested in necrosis of kidney tubules. Causes a carcinoma of kidneys, lesion of liver and necrosis of lymphoid tissue
Islandiotoxins	*P. islandicum*	Cause degeneration of liver in the form of centerlobular necroses, fatty degeneration, cirrhoses, formation of malignant tumors (sarcoma and carcinoma)
Viridicaptopenicillino-Toxins	*Fungi of genus Penicillium*	Cause chronic lesion of kidneys expressed as a degeneration of tubular cells, formation of connective tissue and cysts. Clinically it is manifested in enhanced consumption of water and discharge of urine
Aflatoxins (B_1, B_2, G_1, G_2, M_1, M_2)	*A. flavus, A. niger, P. cyclopium*	Cause lesions of liver in the form of necroses, proliferation of epithelium cells of biliary ducts, fibroses and development of malignant tumors (carcinoma), inhibition of growth and food assimilation
		Effect of toxins on blood cells leads to reduction of functional activity of all three types of cells of the monocytic-macrophagous system: promonocytes of bone marrow, mature monocytes of peripheral blood and tissue monophages. In addition, they inhibit activation of the complement system and reduce levels of Ig A and Ig G
		Causes reduction of quantitative content and functional activity of K-cells, which may be one of the reasons for raising frequency of malignant tumors
		The most dangerous are toxins of the B_1 fraction. Toxicosis is characterized by rapid course and high mortality rate. It is noted by loss in coordination of movements, convulsions, paralysis, and lesion of digestive tract. Hepatocarcinogenicity is characteristic of chronic toxicosis
		Role of aflatoxicoses is admitted in diseases characterized by fatty degeneration of internal organs, brain edema, kidney deficiency (Reye's syndrome)
Stachybotryotoxins	*Stachybotrys atra*	Cause leukopenia, agranulocytosis, trombopenia and in this connection inhibition or absence of the retraction of a blood clot, emergence of hemorrhage and necroses of mucoid tissues of the alimentary tract, other tissues and parenchimatous organs. Produce a dermanecrotic effect, cause hemorrhage diathesis
Dendrodochines	*Dendrochium sp.*	Cause disruption of blood circulation, hemorrhage of tissues, parenchimatous organs and tissues. Lesion of the cardiovascular system

Table 5.23 (continued)

Toxin	Producer	Effect on organs
Sporotrichiellofusario-toxic	*Fusarium sporotrichiella*	Cause alimentary-toxic aleukia. Clinical cause is characterized by severe degenerative changes in the bone marrow cells followed by aplasia
Citreoviridine	*P. citreovivide*	Causes lesions of motor centers of the spinal cord and *medulla oblongata* with progressive paralysis of movements and respiration. Heart lesions. Acts as an etiological factor of a toxicosis well-known from ancient times in Japan: "cardiac beri-beri"
Zearalenons (more than 50 are known)	*F. oxysporum, F. roseum, F. graminearum*	Cause an extrogenic syndrome, which manifests itself in extension of the uterus, swelling of the vulva, increase of mammary glands and termination of pregnancy. Embryotoxic.
		Cause ergotism, which is characterized by a severe course and high mortality rate.
		They are an etiological factor for the emergence of Kashin-Bell disease during which tubular bones shorten, joints thicken and muscles atrophy ("Urov disease")
Ergotoxins	*ClavicepsPurpurea*	Cause cyanosis. A convulsive form is accompanied by cramps and a gangrene form by necrosis of limbs. Causes sharp uterus contraction especially during pregnancy, hyaline degeneration of brain vascular walls, formation of hyaline thrombi
Sporodesmines	*Pithomyces Chartarum*	Cause lesions of the liver in the form of acute inflammation of biliary ducts, which terminates in their fibrous degeneration and inflammation of the skin of face, lips and other organs not protected against sun light
Rubrotoxins A и B	*P. rubrum, A. Flavus*	Possess pronounced hepatotoxic effect. Affect central nervous system. Embryotoxic
Lmeoscirine	*P.islandicum*	Causes severe alimentary mycotoxicosis. Possesses a mutagenic neurotoxic effect
Patulin	*A. terreus, P. cyclopium, P. expansum, P.urticae*	Possesses carcinogenic and teratogenic effect
T-2 toxin	*F. oxysporum*	Embyotoxic
Chrinine, erythroscirine, xanthomegnine	*Fungi of genus Penicillium*	Possess nephrotoxic effect
Penitremas A–D	*A. Flavus, P. cyclopium*	Possess teratogenic effect

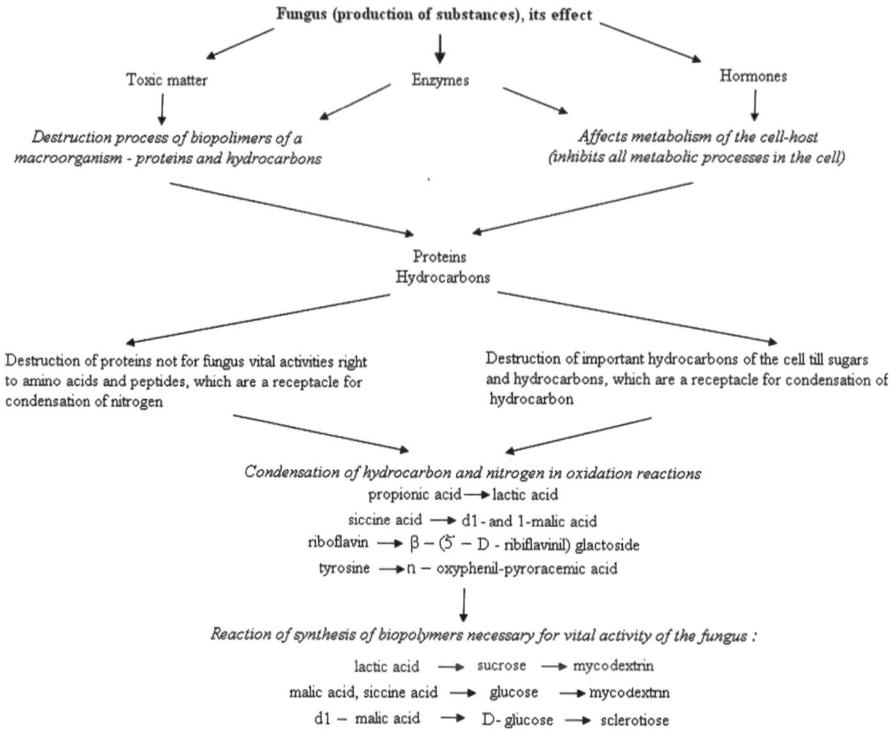

Fig. 5.36 The mechanism of the pathogenetic action of fungi micromycetes by the principle of biodestruction

bodies but also a centralized water supply system, acquires ever greater importance. With the development of industrial construction, greater population density in cities, a need for adequate water supply appears. Increasing deficiency of potable water determines the necessity to consider this problem as one of the most urgent, not only on technogenic terms, but in sanitary terms as well.

Water reservoirs, canals, and fresh-water lakes, formed as a result of building man-made dams across natural waterways, more often than not become the main sources of centralized water supply. Due to these transformations, not only streamways are changed but hydrologic characteristics and the regime of the self-cleaning system are also disrupted. This is related to the fundamentally new destruction processes of bottom deposits, which are formed on the sites where areas, having heavy anthropogenic load, are inundated. Such new conditions of functioning in natural hydrocenoses became the media for the formation of the microbial composition capable of adapting to the anaerobic type of metabolism when trophic sources were colonized with a high content of different organic matter. Bottom deposits contained soils inhabited by common soil fungi possessing the ability to develop new substrates of various origin(s), including animal, owing to which they survive under extreme conditions. Ordinary aqueous successions of microorganisms prevailed with new aggressive components, including micromycetes with a high degree of potential danger for the environment and, in the first turn, for humans [120].

Representatives of the genera *Aspergillus* and *Penicillium* known as potentially pathogenic, cause mycoses, and opportunistic infections are dominant micromycetes.

According to [121] it is possible to realize various mycoses only in cases when natural human resistance is reduced, which is conducive to the penetration of a fungus into an organism from any source of the external medium. The reduction of barrier function of the organism is possible also due to the virulent aggression of fungi.

The authors of paper [122] studied ecological aspects of human mycotic lesions and calculated the correlation coefficients between the pollution of the atmospheric air, water, and soil with fungi and mycotic infection of the stomach, large intestine, and other organs in inhabitants of Vologda and the Vologda region. There is a correlation between the intensity of pollutions of sources of water supply in terms of microbiological indices and mycotic lesions of the stomach, the large intestine, gallbladder and biliation tracts (probability > 95 %). The specific weight of water samples which are unfavorable in terms of bacterial indices has increased (over 5 years only!) from 10.3 to 14.5 %. The authors of paper [122] believe that the main causes for pollution of water bodies with bacteria, fungi, viruses, and protozoa are the unstable and ineffective operation of water treatment facilities, which purify weakly turbid water and lack a barrier function against viruses and dissolved chemical compounds. Unfortunately, traditional water chlorination does not lead to complete obliteration of fungi, viruses, and even bacteria.

The ecological group of "aqueous" fungi may be broken down into a subgroup of obligate-aqueous fungi, which, in order to cover the whole life cycle, need a water medium and a subgroup of facultative-aqueous fungi, which are better known in the traditional ecological classification as ground or soil fungi. The representatives of this group do not have any special devices to ensure their adaptation to an aqueous existence; however, they are constantly discharged from water and bottom layers [123].

General pollution of water bodies and the presence of various chemical compounds promote the efficiency reduction of the barrier function of water-treatment systems with respect to microbial contamination. More often, the barrier function of water treatment facilities is disrupted in the high water period when their load increases [124, 125]. Different authors emphasize that the efficiency of drinking water purification depends on the level of microbial contamination of water sources. As for bacteria and viruses, the latter may be found at all stages of purification and in the main water system. It is shown that many microorganisms isolated from the tap water possess low sensitivity to chlorine. At the same time, the prolonged effect on microorganisms of sub-lethal concentrations of chlorine may lead to changes of biological (growth) properties of pathogens right to the formation of L-forms. Such bacteria cannot be separated via conventional culture media. This also concerns the species composition of the mycobiota whose representatives do not grow on traditionally bacteriological media.

Longitudinal research observations over the mycobiota of the Kuibyshev reservoir made it possible for the authors of paper [123] to classify the fungi in terms of their occurrence and quantitative characteristics into three main groups: constant, inconstant (periodically isolated), and rare species. The group of constant fungi, the residents of this reservoir, includes those species which actively develop in all seasons in different water layers. In the first place, these were representatives of the genus *Trichoderma (T. harzianum, T. hamatum, T. lignorum, T. koningii)* and species of the genera *Penicillium (P. chrysogenum, P. nigricans, P. expansum), Acremonium (A. harticola, A. strictum), Cladosporium (C. cladosporoides,C. herbarum), Mycelia sterilia.*

The group of inconstant species includes fungi, which periodically develop comprehensively or locally. Among the representatives of this group, there are species of the genera *Phoma (P. glomerata, P. herbarum, P. exigua, P. medicaginis var.pinodella, P. pomorum), Aspergillus (A. fumigatus, A. ochraceus, A. niger, A. wentii), Altemaria alternate,* some species of *Penicillium (P. roseopurpurum, P. frequentans, P. funiculosum)* and Mucor fungi. The latter may be a marker for a constant ingress of organic matter.

The following micromycetes *Beauveria, Paecilomyces cyaneum, Phoma fimeti, Phoma cava, Geotrichum candidum* etc. were referred by the authors of [123] to rarely encountered micromycetes.

Micromycetes possess a number of advantages in the competitive struggle for a substrate and, primarily, at the expense of producing biologically active substances, abilities to keep vitality in extreme conditions and adapt to functioning in new substrates.

Reports on fungal contamination of drinking water from a water-distribution system can be found in foreign recourses more often than in Ukrainian [126, 127]. In some papers, authors even attempted to provide results which confirm the development of human diseases related to the presence of fungi in water used [128]. These authors emphasized that the role of fungi, including both mycelial and yeast, in the contamination of the water-distribution system were not effectively discussed, and the suitability of drinking water in terms of the microbiological control was judged by the presence of fecal bacteria. Paper [129] showed the presence in tap water of mesophilic fungi often from the genus *Aspergillus.* Fungi, but within a wider species spectrum and quantitative ratio, were isolated in the samples of the sources of raw water from rivers and lakes.

The information on the presence of molds in water from the distribution system in California is given in [130]. The authors of this paper isolated fungi from sources of water supply and ground water. Most isolates belonged to the genus *Penicillium, Paecilomyces,* and *Acremonium.* In Great Britain, an isolation of fungi from the genus *Cephalosporium, Verticillium, Trichoderma, Phoma* [128]. Paper [126] confirms the presence of different species of fungi in drinking water of the distribution system in 50 % of samples studied. Most often, the representatives of the genera *Alternaria, Aspergillus, Penicillium,* and *Cladosporium* were detected.

We believe that fungi penetrate the distribution system as a result of soil pollution; they justify their conclusion with the fact that most isolates belong to the class of *Deuteromyceles,* which are typical representatives of soil. Among isolates prevailing *A. flavus, A. niger, A.fumigatus*—are producers of toxins capable of causing, apart from mycotoxicosis, aspergilloses, and allergoses. Researchers believe that the spores of aspergilluses may penetrate the respiratory tract by inhaling aerosols, for instance from the spraying head of a shower, and cause pulmonary infection; the spores may penetrate into the mouth and ears while swimming or taking a bath and cause a local infection [126, 131–133].

An interesting report presented in [127] to a certain extent discloses possible mechanisms of tap water contamination by fungi. Samples of tap water were investigated together with the study of the samples of the biofilm from the internal surface of the municipal water-distribution system in the city of Springfield, using both the microbiological method and a scanning electron microscope. The most wide-spread species of fungi in terms of quantity and occurrence were representatives of the genera *of Aspergillus* and *Penicillium* and, rarely, *Mucor racemosus* and *Stysanus stemonites,* and many isolates were included under the heading *Mycelium sterile.* Among yeast fungi, the following were identified: *Aureobasidium pullulans, Candida spp., Cryptococcus sp.,* and *Rhodotorula spp.* This research gives complete data of the specific identification of fungi.

Specific Composition of Fungi Isolated from the Water Distribution System [126]	
Aspergillus sp.	*Nigrospora oryzae*
Aspergillus flavus	*Penicillium oxalicum*
A. fumigates	*Penicillium sp.*
A. janus	*Peyronellaea sp.*
A. niger	*Phaeococcus sp.*
A. terricola	*Pnoma sp.*
A. versicolor	*Pithomyces sp.*
Cryptococcus sp.	*Rhodotorula glutinis*
Cryptococcus laurentii	*R. rubra*
Epicoccum nigrum	*Trichodermaviride*
Fusarium sp.	*Verticillium sp.*
Geotrichum candidum	*Sterile mycelia*

Apart from the information obtained when doing cultural investigation of water samples and biofilms, the results of electron microscopic research of biofilms from the surface of pipes are very interesting. It is shown that the internal surface of the pipelines was porous and contained crystalline and scaly deposits, bacterial cells and spores of fungi. As is shown in [127], the spores of fungi were main components of the surface of biofilms and incrustations. Analyzing the data of the literature and relying own results we came to set of very important conclusions. First of all, they discovered that water contamination takes place during the operation of distribution systems. Also, fungi may survive in normal processes of water treatment and enter the distribution system through ineffective filters although the information on the impact of chlorine (especially its residual concentration) on the spores of fungi in water is effectively absent.

The species spectrum of fungi described by different authors in a number of cases does not match, but this depends, on the one hand, on the species spectrum of fungi contaminating soil in places where water sources are located, water temperature, the state of distribution systems, and, on the other hand, on non-standardized methods and the methodology of water intake and preparation of the samples for research of quality and variety of nutrient media. The authors of [134] stress that fungi of water origin substantially affect the taste and smell of water, quality of beverages and food from this water, especially when accumulating toxic substances.

The fungi that survive after chlorination colonize the internal surface of the distribution system causing its corrosion, which adversely affects taste qualities of water. On the other hand, the fungi that survive and overcome "stress" under the influence of active chlorine resuscitate after its concentration reduction in water and may realize their pathogenic properties entering the human organism.

In connection with an unfavorable ecological situation in Ukraine [135, 136] and rather poor quality of drinking water, it is necessary to channel the efforts of scientists with regard to the monitoring of the upper part of the rivers Dnieper and Desna, Kiev and Kaniv reservoirs, other sources and reservoirs of drinking water. The results obtained will serve as a basis for the development of drinking water disinfection technologies, taking into account the properties of the mycopathogens detected [122, 137–139].

Confirmation of the possibility of various fungi entering the distribution system may be the result of research presented in paper [140]. This includes the detection and determination of their species composition in the river sand on the banks of the Dnieper River, and in water at different periods of time over 2 years of observation (Chap. 4, Table 4.15).

The analysis of results [140] and their comparison with the data of international researchers leads us to the conclusion that micromycete fungi predominantly from the genera *Aspergillus, Penicillium, Alternaria* and some others are found in water (both natural and from the water-distribution system). This can be explained by the fact that micromycetes, mentioned above, possess a number of advantages in the competitive struggle for the substrate; being polybiotrophes, they are distinguished by a high level of ecological valence, the ability to produce strong biologically active substances and, almost importantly, toxins, which make it possible for them to maintain vitality in extreme conditions and to adapt to functioning on new substrates [141–143].

Having studied the specific composition of the micromycetes isolated both from sand at different depths and from water within 7 months of observation, one should note a particular feature: from October to April one and the same species of micromycetes were isolated to a depth below 10 cm. In May and September, propagules were detected more often at the depth of 5 cm, and in June—August they were predominantly detected on the surface (1 cm) and not deeper than 5 cm. From the data of Fig. 5.37, one can see that in June–August, at the depth of 1–5 cm, there were a great variety of species of fungi detected, whose numbers appreciably increased (up to 18) in the sample studied (per 1 g of sand or 1 cm of water).

In order to provide an objective assessment of the obtained data, we calculated the number of propagules in each sample equal to 1 g of sand or 50 cm^3 of water (see Table 5.24). From the table one can see how the quantitative indices grow in the surface layers of sand (1–5 cm). It is important to stress that the said process

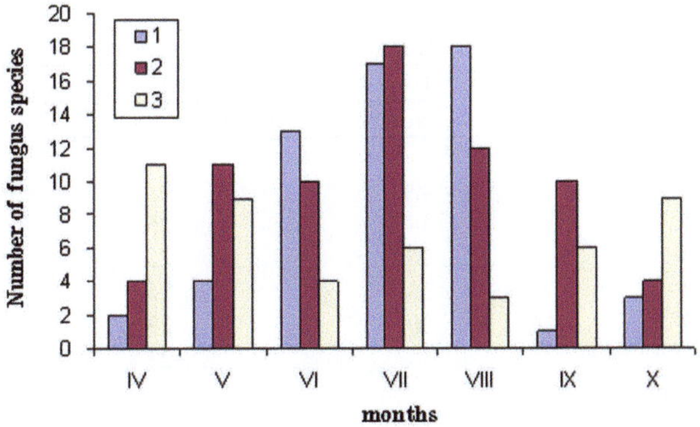

Fig. 5.37 Number of detected micromycetes at different depth of samples taken within 7 months of observation (cm): *1*—1, *2*—5, *3*—10

Table 5.24 Quantitative indexes of sand sporulation at different depth (number of fungus germs in 1 g of sand, thous. units)

Date	Depth, cm		
	1	5	10
	Number of germs, thous. units		
April			
14	2	3	2
28	2	3	2
May			
10	4	3	4
20	5	4	4
June			
22[a]	11	6	4
23[b]	12	8	4
24	12	7	4
25	14	8	4
26	20	10	5
27	26	9	5
28	30	10	5
29[a]	*36*	12	5
30[b]	*38*	16	5
July			
1	38	20	5
2	37	21	6
3	36	20	6
4	38	19	7
5	39	20	8
6[a]	43	20	8
7[b]	44	21	9
13[a]	44	22	12
14[b]	45	23	12
20[a]	44	23	14
21[b]	44	23	14
27[a]	43	24	13
28[b]	44	23	14

Table 5.24 (continued)

Date	Depth, cm		
	1	5	10
	Number of germs, thous. units		
August			
3[a]	43	25	12
4[b]	44	25	12
17[a]	36	25	11
18[b]	38	25	11
24[a]	30	23	8
25[b]	30	23	8
September			
4	12	20	6
5	10	18	6
8	6	12	4
15	4	10	4
22[a]	4	11	3
29[b]	5	11	3
October			
12[a]	3	9	3
13[a]	3	6	2
21	2	5	2

[a] Saturday
[b] Sunday

fluctuates over weekends (Saturdays and Sundays). This once again confirms the significance of the anthropogenic factor of polluting beaches and bank areas of water (5 m from banks).

The data obtained served as the basis for assessing microbiological indices when investigating water samples from water intakes, distribution systems, artesian wells, and open bodies of water located in Kiev and its suburbs.

The safety of drinking water in the epidemiological sense is determined by indices, which, with a sufficiently high degree of probability, characterize the absence in it of bacteria, viruses, and other biological supplements hazardous for human health [144, 145]. However, taking into account of those data, which we reported earlier and the possibility of natural "biological terrorism" on the part of fungi—micromycetes—perhaps, one ought to return to the question of drinking water safety. This issue can be extended within the framework of the state standards and the investigations concerned with detection of pathogenic micromycetes and their toxic substances.

In paper [140] investigated water samples from faucets installed in private sectors, located in various districts of Kiev, samples from artesian wells and water samples whose intake was acquired from open sources. In addition, in paper [143] investigated water samples for domestic-drinking needs taken from faucets in several workshops of one of the enterprises in Kiev. The results of the investigations are shown below.

Mycobiota of water samples in different districts of Kiev and its suburbs (Tap water)	
Isolate fungi	
Mortierella isabelina	*Penicillium expansum*
Penicillium cyclopium	*Penicillium chrysogenum*
Penicillium canescens	*Mycelia sterilia*
Penicillium sp.	*Ulocladium botrytis*
Penicillium notatum	*Aspergillus versicolor*
Aspergillus niger	

Mycobiota of water samples taken from wells in different districts of Kiev (Artesian water)	
Isolate fungi	
Penicillium expansum	*Rhizopus nigricans*
Penicillium implicatum	*Mortierella isabelina*
Aspergillus fumigates	*Penicillium tardum*
Penicillium notatum	*Penicillium funiculosum*
Penicillium multicolor	*Penicillium sp.*
Penicillium versicolor	*Aspergillus glaucus*
Penicillium canescens	*Penicillium cyclopium*
Aspergillus niger	

From these results it follows that in tap water sampled simultaneously (month, day, and hour) in different districts of Kiev and its suburbs, the species spectrum of isolated fungi was not varied. Mainly 5–8 species belonging to different genera were isolated.

The spectrum of fungi species obtained from water samples from the artesian wells was somewhat wider, but they mainly represent two genera—*Aspergillus* and *Penicillium.* Samples of water taken from open sources were distinguished from a greater variety. In addition to molds, they revealed yeast fungi. Such a variety of a species composition and degree of aggression was displayed by fungi isolates from water samples taken from faucets of some workshops at one of the enterprises in Kiev. In 60 % of these cases, the spectrum of isolated fungi species was identical to air samples and surface samples from the hands of employees working in the same shops.

Mycobiota of water samples taken from open water sources
Isolate fungi
Mucor sp.
Rhodotorula sp.
Cladosporium cladosporioides
Penicillium expansum
Mycelia sterilia

Thus, one could admit that the problem of supplying the planet's population with drinking water corresponding to the requirements of epidemiological and hygienic standards in quantities satisfying physiological and household needs of human is of an exceptionally acute nature. In terms of this, research concerning the microbial biocenosis (bacteria, fungi, and viruses) of water reservoirs under the effect of numerous factors (in the first place radiation pollution) is of special importance.

As is known, one of the main conditions for destruction of different contaminants of origin entering reservoirs is the normal functioning of microbial cenosis. However, how the composition of microorganisms-destructors and how the properties of the microbes themselves change under these effects is a question which still needs to be answered.

Ecological illness leads to the enhancement of gene expression responsible for the production of enzymes and toxic substances by microorganisms, which increases the degree of aggression. This is especially so with respect to conventionally pathogenic representatives of the microworld.

5.7.1 Removal from Water of Yeastlike Fungus Candida Albicans by the Method of Coagulation and Flocculation

Yeasts are an important component of natural aqueous ecosystems [146, 147]. Yeastlike fungi of the genus *Candida* is a natural human biota. Thus, for instance, *Candida albicans* were found in the human gastrointestinal tract, in the mouth and pharyngeal region, as well as the skin and mucus membrane of human genitalia; however, with lowered natural protection, mechanisms of the organism's candidas may cause various diseases. The seriousness of the situation is proved by the fact that, for instance, in the USA *Candida* is caused by these kinds of microorganisms among other infectious diseases and ranks fourth place [148].

Recently, many papers have dealt with the detection of *Candida* in drinking water [127, 140, 146, 149, 150] as well as with the issues related to the incidence of various source diseases [148]. However, there are no available papers on water purification of *Candida*. The objective of the present research is to propose the development of the method for removing yeastlike fungi of the genus *Candida* from water. Their ability of surviving generally accepted methods of disinfection [140, 150] and penetrating into the biofilm of the water supply systems [127] is well known.

On the other hand, *Candida*, just as all living organisms, possesses high sorption activity in relation to various surfaces. It was expedient to investigate their removal by means of coagulants and flocculants, which were widely used in the practice of water treatment. A yeastlike fungus, *Candida albicans* was chosen as a test culture. *Candida albicans,* found in water [151], is one of the most frequently separated species of *Candida* in patients [148].

A three-day bouillon culture (Saboro medium) was centrifuged at the speed 5,000 rpm for 10 min. The resultant sediment was washed off thrice in an isotonic solution of sodium chloride and re-suspended in the same solution until the density reached 10^7 of colony-forming units (CFU) in 1 cm^3. The necessary volume of the initial suspension was introduced into sterile tap water prepared in advance. The degree of water contamination constituted 10^4–10^5 CFU/cm^3. The tap water was taken directly from the site of the experiment, and it was settled for 1 day and sterilized in

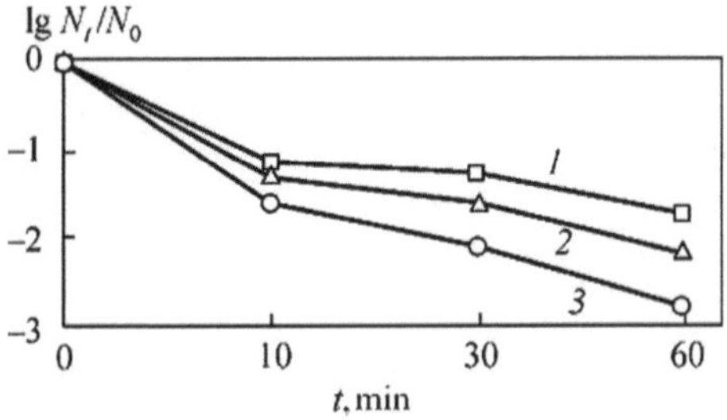

Fig. 5.38 Comparative effect of coagulants AS (*1*), BAC (*2*), ADHS (*3*) when purifying water of *Candida albicans 10231*

an autoclave at 101.325 kPa (T = 121 °C) for 15 min. After that, the water alkalinity and the pH were measured, which constituted respectively 1.6 mg-eq/dm^3 and 8.2. The contaminated water was administered into glasses of 500 cm^3 each. Then, the necessary amount of a coagulant, a flocculant or their composition, was added and mixed on a magnetic stirrer. At the end of the contact time (10; 30; 60 min) the solution was sent through a paper filter, and the first 50 cm^3 were discarded.

The survivability of the species was determined by the presence of CFU during the seeding of selected samples on the Saburo agarized medium and cultivated for 3 days at 27 °C. The results were expressed as the ratio of the logarithm of the yeast (N_t) remaining in the solution from their initial amount (N_0).

Compounds used in tap water practice, such as aluminium sulfate (AS), aluminium dihydrosulfate (ADHS), and basic aluminium chloride (BAC), were used as a coagulant. The concentration of working solutions in terms of Al_2O_3 constituted 1 %.

Polydiallyl dimethyl ammonium chloride—DB-45, which at present is used at the water treatment station in Kiev and Valeus, is representative of polyguanidine bases (PG), which, in the course of water conditioning, still has a controversial nature [152]. Concentrations of these solutions are respectively 0.01 and 0.001 %.

The efficiency of water purification of *Candida albicans* was investigated by means of the coagulants AS, ADHS, and BAC at the concentration 6 mg/dm^3. Contamination of water (hereinafter load in terms of culture) constituted 1.7×10^5 CFU/cm^3. The experiments demonstrated that all three coagulants remove from water upwards of 90 % of the test-microorganisms introduced, but the most effective result was obtained for ADHS. Its purification degree constituted 99.8 %, which corresponds with three orders of the initial value.

This can be explained by the presence of its hydrolysis products which are of a more developed adsorption surface and a higher charge. The degree of purification increases by order of magnitude, with an increase of the contact time of the culture with the reagents (see Fig. 5.38).

The impact of the coagulant concentration on the process of removing *Candida albicans* was studied by means of ADHS. The load on the culture constituted 2×10^5 CFU/cm^3, the coagulant concentration: 6–21 mg/dm^3. The experiments showed that the coagulant concentration affects the degree of water purification of *Candida albicans* insignificantly, and at the maximum concentration of the reagent 21 mg/dm^3 it constituted 99.9 %, at the same time as at the ADHS concentration 6 mg/dm^3–99.8 % (see Table 5.25).

Table 5.25 Impact of the concentration of the ADHS coagulant on the degree of water purification of *Candida albicans 10231*

Contact time, min	ADHS concentration, mg/dm³				
	6	10	14	18	21
	CFU/cm³				
10	4,260	4,230	3,220	3,210	3,206
30	1,688	1,660	1,456	812	549
60	109	98	73	61	33

Table 5.26 Impact of the pH on the degree of water purification of *Candida albicans 10231* by the ADHS coagulant

Contact time, min	pH		
	5.2	7.0	8.2
	CFU/cm³		
10	7,920	3,800	4,000
30	3,640	500	590
60	2,560	200	280

Table 5.26 gives the relationship between the degree of water purification of ADHS with the concentration 6 mg/dm³ and the pH value, which was varied by means of 0.1 M of NaOH and HCl. The most effective result was obtained at pH 7. This is determined by the fact that at pH 5.2 the surface of hydrolysis products of the reagent is less developed and has a smaller charge than at pH 7 and 8.2, while at an increase of the pH, the total adsorbing surface decreases due to an increase of hydration of the particle surface [153]. However, the degree of culture removal for pH 7 and 8.2 differ from each other insignificantly; therefore, the investigations were continued on the sterile tap water whose pH constitutes 8.2.

In the practice of water supplication, flocculants are used for intensification of the coagulation process. In current experiments, cationic flocculants (DB-45, Valeus) were used at the concentration 0.1 mg/dm³ and the load in terms of culture was 1.8×10^5 CFU/cm³. From the data of Fig. 5.39, one can see that Valeus and DB-45 manifest a similar degree of removing *Candida albicans*, which constitutes respectively 90.6 and 92.8 %.

In the investigation of the impact of pH on the degree of water purification of *Candida albicans* by DB-45 and Valeus, it was found that at the concentration 0.1 mg/dm³ the flocculant DB-45 is more active in a weakly alkaline medium of 92.3 % (see Fig. 5.40a), while for Valeus, under the same conditions. The most effective degree of yeast removal was at pH 5.2 and constituted 95.3 % (see Fig. 5.40b). Such variation of the pH is convenient in the case of the selection of reagents for purification of water from different sources of *Candida albicans*.

As for the impact of the concentration of DB-45 on the process of removing yeast cells, as can be seen from Fig. 5.41a, an increase of the concentration of this reagent results in a degree of water purification. Thus, if at the concentration 0.1 mg/dm³ the degree of *Candida albicans* removal from water constitutes 92.3, then at 3 mg/dm³–99.9 %. A similar tendency is observed also for Valeus (under the same conditions) (see Fig. 5.41b).

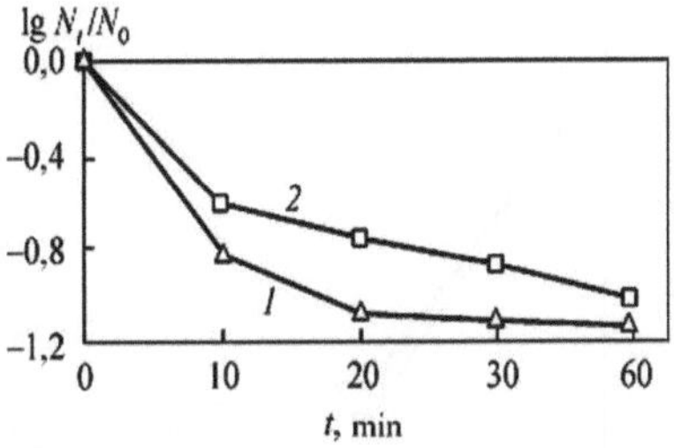

Fig. 5.39 Comparative effect of flocculants DB-45 (*1*), Valeus (*2*) when purifying water of *Candida albicans 10231*

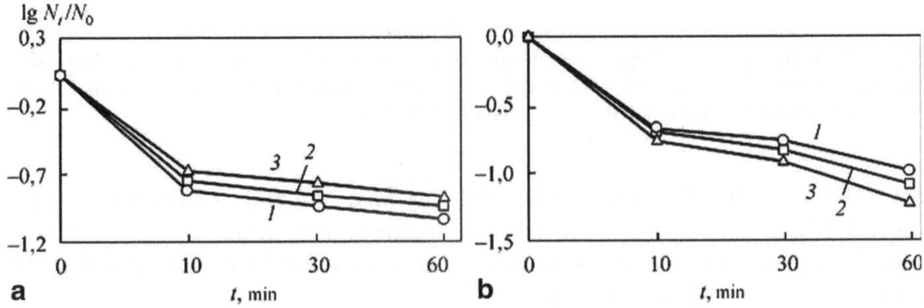

Fig. 5.40 Impact of the pH on the degree of water purification of *Candida albicans 10231* using flocculants DB-45. **a** 8.2 (*1*) 7 (*2*); 5.2 (*3*) and Valeus **b** 5.2 (*1*); 7 (*2*); 8.2 (*3*)

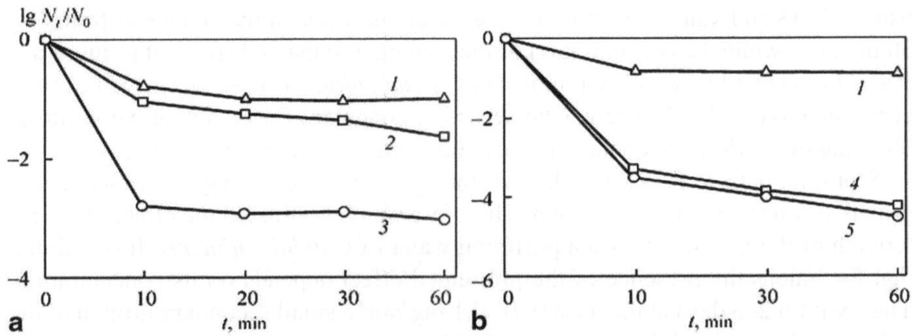

Fig. 5.41 Impact of the concentration of flocculants DB-45. **a** 0.1 (*1*); 0.4 (*2*); 3 mg/dm^3 (*3*) and Valeus **b** 0.1 (*1*); 1.5 (*2*); 2 mg/dm^3 (*3*) on the degree of water purification of *Candida albicans 10231*

In order to increase the degree of water purification and in this case not to increase the dose of the reagents, the joint effect of coagulants and flocculants was studied. In the investigation of the compositions SA + Valeus, ADHS + Valeus, and BAC + Valeus yielded identically effective results. It was established that at the ratio 6 mg/dm^3 of the coagulant to 0.01 mg/dm^3 of Valeus, the removal degree constitutes 99.7 % (see Fig. 5.42a). While in the experiments on using the compositions

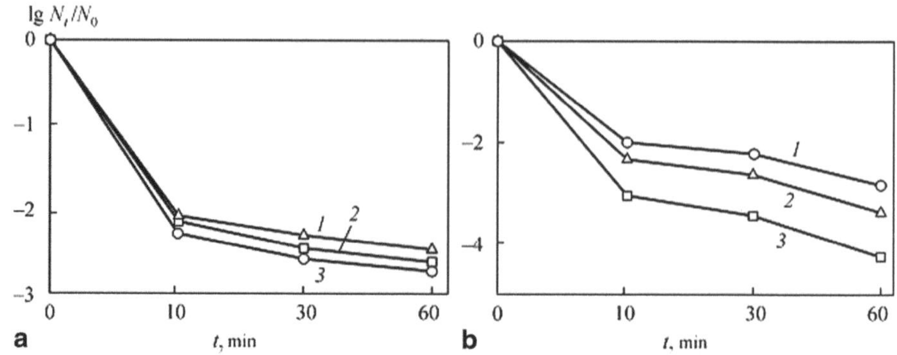

Fig. 5.42 Variation of the degree of water purification of *Candida albicans 10231* when using the composition of reagents: (a)—SA + Valeus (*1*), BAC + Valeus (*2*), ADHS + Valeus (*3*) and (b)—SA + DB-45 (*1*), BAC + DB-45 (*2*), ADHS + DB-45 (*3*)

SA + DB-45, ADHS + DB-45, BAC + DB-45 with a similar dose of coagulants and the DB-45 concentration equal to 0.4 mg/dm^3, the highest degree of water purification (99.99 %) can be achieved in the case of the composition with ADHS at the contact time 60 min (see Fig. 5.42b).

In case of necessity, the conditioning of water of high purity (99.99 %), for instance, for patients in the post-operation period (immunocompromised), concentrations of the above reagents may be raised, but with the subsequent account of residual concentrations of the reagents in water. Thus, an experiment has also been conducted using ADHS and Valeus in which we measured the value of the coagulant from 6 to 30 mg/dm^3, while the concentration of the coagulant remained constant (1 mg/dm^3). It was found that for complete removal of *Candida albicans* from water (at the initial contamination 2×10^5 CFU/cm^3) the necessary coagulant-flocculant ratio constitutes 30: 1 mg/dm^3, while the contact time can be reduced to 30 min (see Fig. 5.43).

From the data of [152], it is known that guanidine bases may cause secondary growth of microorganisms in water. Therefore was investigated the effect of the aftermath of the flocculants when purifying water of *Candida albicans*. It was found that for Valeus the presence of the prolonged effect depends on its concentration. Thus, within 2–3 days at the dose 0.01–0.1 mg/dm^3 a small secondary growth of the culture is detected, while the doses 1.5–2.0 mg/dm^3 ensured the prolonging effect of the flocculant over the same period of time (see Table 5.27).

In the experiments on distilled water (at pH 5.2 and the load of *Candida albicans* 5×10^4 CFU/cm^3), the doses of Valeus 0.1 mg/dm^3 is enough for complete suppression of the 99.99 % growth of the culture studied within the first 10–30 min. This can be explained by the absence in sterile distilled water of an organic matter, which is a known form with PG inactive complexes [152]. The situation with DB-45 is also ambiguous since a substantial growth reduction of *Candida albicans* was recorded for the whole range of concentrations (see Table 5.28). However, when the temperature in the summer period goes up, the resultant data indicate the deficiency of the dose of the flocculant 0.1 mg/dm^3 for suppressing the growth of the culture.

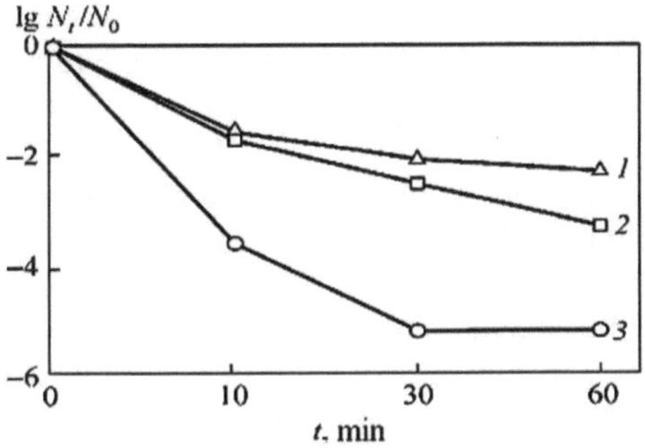

Fig. 5.43 Impact of the concentration of the ADHS + Valeus composition on the degree of water purification of *Candida albicans 10231*, mg/dm³: *1*–6 : 1, *2*–15 : 1, *3*–30 : 1

Table 5.27 Impact of the flocculant concentration on the degree of water purification and the presence of the secondary growth of *Candida albicans 10231*

Contact time, min	Concentration of Valeus, mg/dm³		
	0.1	1.5	2
	CFU/cm³		
10	32,800	110	62
30	28,000	60	30
60	24,000	25	4
1,440	38,720	7	0
2,880	39,630	0	0

Table 5.28 Impact of the concentration of the DB-45 flocculant, mg/dm³

Contact time, min	Concentration of DB-45, mg/dm³		
	0.1	2	3
	CFU/cm³		
10	27,200	12,840	180
30	13,640	6,500	130
60	13,040	3,350	90
1,440	13,000	580	42
2,880	12,720	200	10

It was found that ADHS was the most effective coagulant; in this case the removal of yeastlike fungi *Candida albicans* from water occurs at 99.7 % (see Fig. 5.38). Perhaps, the hydrolysis products of the given coagulant possess a highly developed surface and the highest charge, and as a result the adsorption on its surface affects the purification degree that occurs faster than on other studied coagulants. It is shown that the ADHS concentration affects the purification degree of water insignificantly (see Table 5.25), while the pH range 7–8.2 may be varied (see Table 5.26).

The flocculants DB-45 and Valeus may also be used in the processes of removing yeast from water. Depending on the initial pH value of water both the flocculant and its concentration should be chosen in order to achieve the necessary degree of purification. Thus, even at the concentration 0.1 mg/dm³ of Valeus and DB-45, the degree of *Candida albicans* removal constitutes 90 and 92 %, respectively. It was established that for achieving complete water treatment (99.99 %), the Valeus concentration should be increased to 2, while DB-45 should be increased to 3 mg/dm^3 (see Fig. 5.41). As for the pH in the case of Valeus, the degree of *Candida albicans* removal increases at pH 5.2 and constitutes 95.3 % (see Fig. 5.40b); for DB-45 an optimal pH is 8.2–92.3 % (see Fig. 5.40a). In both cases the concentration of the flocculant constitutes 0.1 mg/dm^3. The optimal conditions of the pH of every flocculant make it possible to choose the maximum effective reagent depending on the initial parameters of the water being purified.

In order to raise the degree of *Candida albicans* removal and not to increase the concentration of the reagents, it is expedient to use the coagulant–flocculant composition. Thus, the ADHS–Valeus composition (6 mg/dm^3 to 0.01 mg/dm^3) and ADHS+DB-45 (6 mg/dm^3 to 0.4 mg/dm^3) makes it possible to remove *Candida albicans* respectively by 99.7 (see Fig. 5.42a) and 99.99 % (see Fig. 5.42b).

Since DB-45 is used at Kiev's water treatment stations it is necessary to know the amount of *Candida* in the water and also to take into consideration the presence of other microorganisms in the pipes of the water supply system. Based on the fact that *Candida albicans* are detected in substantial amounts in drinking water after its purification, and cause serious diseases among humans, this kind of yeastlike fungi ought to be included into the list of microorganisms which should be controlled when assessing drinking water quality.

5.7.2 The Decontamination Effect of UV-Radiation with Respect to Micromycetes

Among the known methods of water decontamination, chlorination and ozonization have been most widely used. In addition, doses of chlorine and ozone for decontamination of the water containing microscopic fungi should be in the order of 10 mg/dm^3 [148, 149, 154, 155], which reduces corrosion resistance of the equipment and increases toxicity of the water. At the same time, UV-radiation brings a number of advantages: reagentless method of water decontamination; power consumption in industrial UV systems 4 to 8 times lower than total power costs in ozonation systems; modern lamp systems which ensure a high degree of reliability and simplicity of operation; the absence of a gaseous harmful ingredient is conducive to system security; UV complexes in terms of compactness are not inferior and are sometimes even superior compared to the chlorination systems and ozonation.

Therefore, the objective of the present research is the study of water decontamination from micromycetes by means of UV radiation [156].

The assessment of the degree of water contamination from fungi was carried out on a Promin'-1 unit on tap water in the follow-through mode. For these tests, tap water was taken directly at the site of the experiments; it was then settled for one 24-h period. Alkalinity, color, and the pH of the water were measured and equalled respectively, 2.5 mg-eq/dm^3, 18 Degrees, and 7.6. The absorption coefficient for

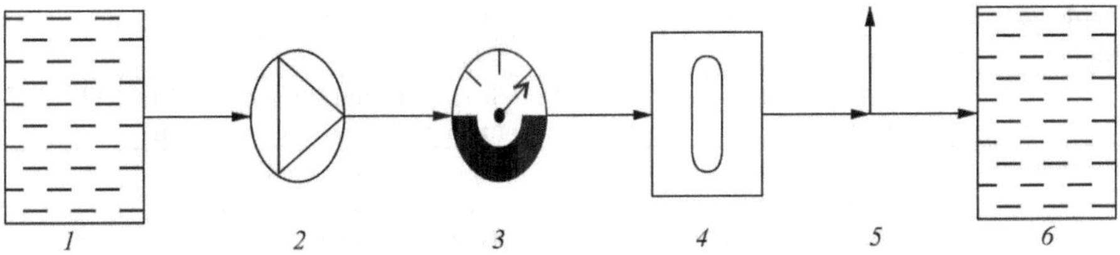

Fig. 5.44 Setup of the unit for decontamination by UV-radiation of the water contaminated with microorganisms

tap water was determined from the absorption spectra in the UV region, taken by a Specord device.

Figure 5.44 shows a schematic flow chart of water treatment with ultraviolet: a water reservoir, capacity 20 dm^3, to which a necessary amount of the suspension of test microorganisms was introduced: a KAMA-5 peristaltic pump by means of which the water velocity was controlled within the range 1.9–27 cm^3/s; the water discharge meter; the unit for water treatment by UV-radiation; the place of water sampling and the storage for the effluent. The experiments were carried out three times. The source of radiation the UV device Promin'-1 employs a mercury low-pressure lamp DRB-8 whose main radiation falls on the wavelength of 254 nm. The radiation intensity in this region is 2.5 W. The liquid layer thickness is 2 cm.

As test cultures for investigating decontamination of water from micromycetes *Candida albicans 10231, Penicillium multicolour, Cladosporium cladosporioides* and *Aspergillus niger* were used.

The suspension *C. albicans* was prepared according to [157]. The required volume of the initial suspension was introduced to the tap water, prepared in advance. The degree of water contamination was 10^4–10^5 CFU/cm^3. The suspension of micromycetes isolated from the water was prepared in the following way: by means of the microbiological loop the culture was transferred to Petri dishes and the Saburo medium was poured into the dishes. The culture was thermostated for 5–7 days at 27 ± 1 °C. The resultant colonies of fungus were twice washed off with sterile tap water and a sterile glass spatula (10 cm^3) and filtered through sterile glass fiber. The suspension density was 1.1×10^7–5.1×10^8 CFU/cm^3. The required volume of the initial suspension was introduced into prepared tap water. The degree of water infection constituted 2.1×10^3–5.5×10^4 CFU/cm^3.

For comparison the *E.coli* 1257 and *Bacillus subtilus* ATSS 6633 cultures were used. The suspension of the *E. coli* 1257 bacteria (10^7 CFU/cm^3) was prepared according to [157]. The suspension of a 48-h bouillon culture *Bac. subtilis* was prepared in a similar way.

The survivability of the microorganisms was determined by the presence of CFU when plating the selected water samples on the agarized Saburo medium for *C. albicans, P. multicolor, C. cladosporioides, A. niger*, the Endo agarized medium for *E. coli,* and nutrient agar for *Bac. subtilis*. The microorganisms were cultivated for 3 or 7 days at 27 °C for yeast-like and microscopic fungi and for 18–24 h for sanitary-indicative test microorganisms. The outcome was expressed as a ratio of

Table 5.29 Dynamics of contamination of tap water with micromycetes

Month	Source 1		Source 2	
	Specific composition of fungi, CFU/cm³	CFU/100 cm³	Specific composition of fungi, CFU/cm³	Total CFU/100 cm³
February	*Aspergillus nidulans, Penicillium expansum*	4	*Aspergillus alternata, Penicillium cyclopium*	3
March	*Aspergillus niger, Penicillium cyclopium*	5	*Aspergillus alternata, Penicillium expansum*	6
April	*Aureobasidium pullulans*	4	*Aspergillus flavus, Penicillium expansum*	5
May	*Aureobasidium pullulans, Penicillium cyclopium*	6	*Aspergillus nidulans, Penicillium cyclopium*	6
June	*Aureobasidium pullulans, Penicillium expansum*	8	*Aspergillus phoenicis, Penicillium expansum*	9
July	*Aureobasidium pullulans*	8	*Aspergillus nidulans, Penicillium fellutanum*	8
August	*Aspergillus niger, Penicillium tardum*	7	*Aspergillus flavus*	10

the logarithm of the concentration of test microorganisms, which remained in the solution after its treatment with ultraviolet (N_t) to the initial amount (N_0).

With the aim of determining the most wide-spread species of aqueous micromycetes, we conducted systematic research of drinking water for detecting in it microscopic fungi. In 2006, for 7 months, water samples were taken from water taps of two residential houses in Kiev. Seasonal dynamics of the content of micromycetes in tap water and specific composition of the fungi were studied. The results given in the Table 5.29 indicate that in terms of quantitative indicator are the genera *Aspergillus* and *Penicillum* prevail.

Similar research of the tap water by the mycological criterion was carried out in 2007–2008 years. Among isolated micromycetes, the most typical representatives were *A. ochraceus, P. griscofulvum,* and *C. cladosporioides.* Quantitative indicators for the content of microscopic fungi in tap water are detected within the interval $1–2 \times 10^2$ CFU/dm³, which, by the opinion of authors of [154], may be a reason for not only obnoxious smell and twang of drinking water, but they might also be a factor in the transference of infectious and allergic diseases [8].

In the experiment the UV dose was determined by the formula

$$F^{\Sigma} = \sum_{\lambda i} \frac{I_{t.sur}^{\lambda i} k_1 \left(1 - 10^{-D_{\lambda i}(R_0 - R_1)}\right)}{2.303 D_{\lambda i} Q} \qquad (5.57)$$

where is light flow with the wavelength λ, falling on the total surface of the protective housing; (R_0–R_1) is the thickness of the layer being radiated, cm; R_0 is reactor radius, cm; R_1 is radius of the protective housing of the UV lamp, cm; $D\lambda_i$ is solu-

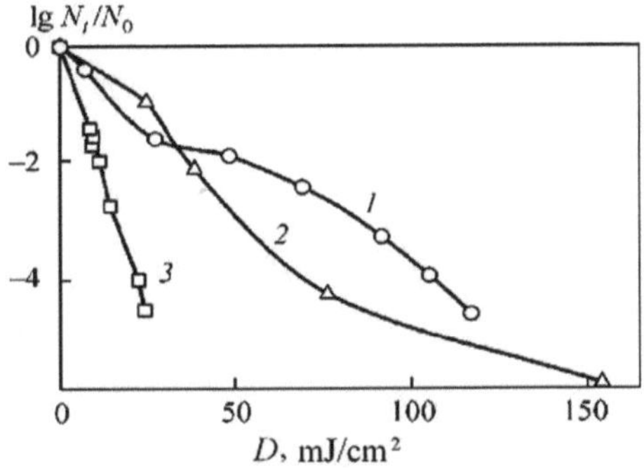

Fig. 5.45 Effects of UV-radiation on the microorganisms: *Bacilus subtilis* (1); *Candida albicans* (2); *Escherihia coli* (3)

tion optical density at the wavelength λ_i, cm^{-1}; Q is water discharge, cm^3/s; k_1 is the transmission factor of the protective housing, which includes both the transmission factor of quartz glass and the coefficient of reflection from the inner surface of the quartz cover.

The expression for the radiation dose with the account of reflection from the walls of the reactor may be written in the form

$$F^{\lambda i}_{t.sur} = F^{\lambda i}_{(R_0-R_1)} + k_{ref}(F^{\lambda i}_{(R_{\max}-R_0)} - F^{\lambda i}_{(R_0-R_1)}) \tag{5.58}$$

Or

$$F^{\lambda i}_{t.sur} = F^{\lambda i}_{(R_0-R_1)}(1-k_{ref}) + k_{ref}F^{\lambda i}_{(R_{\max}-R_0)} \tag{5.59}$$

where k_{ref} is the reflection factor from the reactor walls; R_{max} is the distance of complete light absorption (maximum possible detection).

The reactor is made of stainless steel for which k_{ref}=0.5 [158],

$$F^{\lambda i}_{t.sur} = \frac{F^{\lambda i}_{(R_0-R_1)} + F^{\lambda i}_{(R_{\max}-R_0)}}{2} \tag{5.60}$$

From the data shown in Fig. 5.45, it follows that for reducing water contamination with *C. albicans* by an order of magnitude, a UV dose equals 24 mJ/cm^2 while doses of *E. coli, Bac. subtilis* are equal respectively to 5 and 26 mJ/cm^2. This is close to the data obtained in [159].

For achieving a similar degree of removing microscopic fungi of a dose of UV-radiation there should be an order of magnitude higher (see Fig. 5.46). Thus, if for *P. multicolor* the UV dose equals 40 mJ/cm^2, then for *A. niger* and *C. cladosporioides* they are respectively 180 and 240 mJ/cm^2. The stability of the fungi to UV-radiation may be explained by the presence in their structure of melanin pigment, which protects a fungus from the negative effect of UV.

So, the analyses conducted by the mycological criterion for 2006–2008 years indicate the presence in it of pathogenic microscopic fungi, namely: *A. niger, A. flavus, A. ochraceus, P. cyclopium, P. expansum, C. cladosporioides* and others which

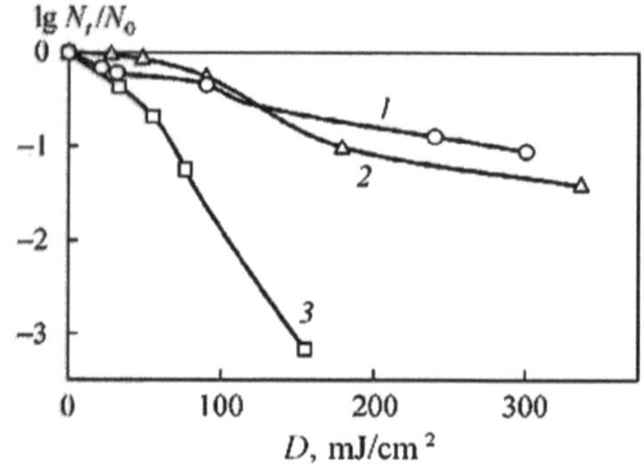

Fig. 5.46 Effects of UV-radiation on micromycetes: *Cladosporium cladosporioides* (*1*); *Aspergillus niger* (*2*); *Penicillium multicolor* (*3*)

present a direct hazard for consumers having weakened immune system. Danger of micromycetes for human is compounded by the fact that, in addition to the micellar form, fungi are capable of existing in the form of spores more resistant to decontamination and, besides form toxins, which remain in water for a long time.

We established that among microscopic fungi in tap water, representatives of the genera *Aspergillus* and *Penicillum* are encountered most often. Quantitative indicators by mycological criterion of tap water vary within the range 1×10^2–2×10^2 CFU/dm^3. The higher level of contamination of tap water with micromycetes in June, July, and August should be noted, which may explain the temperature mode affecting the growth of fungi.

It was found that the decontaminating effect of UV-radiation decreases in the series *E. coli* > *C. albicans* > *Bac. subtilis* > *P. multicolor* > *A. niger* > *C. cladosporioides*. Doses for water decontamination of water from *E. coli, Bac. subtilis* and *C. albicans* are equal to 5; 26 and 24 mJ/cm^2, and for *A. niger* and *C. Cladosporioides,* respectively 180 and 270 mJ/cm^2. From the obtained data it can be seen that *C. albicans* cannot be a sanitary-indicative microorganism when decontaminating water by UV-radiation from microscopic fungi since inactivation of a yeast-like fungus in water does not indicate its cleaning from micromycetes. It is impossible to judge the decontamination of water from micromycetes by the presence of such microorganisms as *E. coli* and *Bac. subtilis*. Resistance of microscopic fungi to the effects of UV radiation can be explained by the presence of a cellular wall of the fungi near the melanin enzyme, which determines the value of the required decontaminating dose of UV. The greater the amount of melanin in a fungus cell, the higher the necessary dose of UV.

Contamination of tap water with micromycetes and resistance to their effects of known disinfectants, are of vital necessity for perfecting existing methods of disinfection and for the introduction of new complex units of water treatment, which will reduce the doses of UV-radiation and secure complete removal of micromycetes from water.

References

1. Grabinska-Loniewska A (2005) Microbial contamination of drinking-water distribution systems: problems and solutions. Eur. Centre of Excellence CEMERA, Warsawa
2. Solodovnikov YuP, Tibekin AT, Cherkasova LV (2004) Spread of salmonellosis in Moscow and the ways of its prophylaxis. J Microbiol Epidemiol Immunobiol 5:36–39
3. Rafiev KhK (1990) Viral hepatitis E: epidemiological characteristics in the republic of Tajikistan. J Microbiol Epidemiol Immunobiol 4:26–29
4. Villanueva CM, Kogevinas M, Grimalt JO (2003) Haloacetic acids and trihalo-methanes in finished drinking water from heterogeneous sources. Water Res 37:953–958
5. Simpon KL, Hayes KP (1998) Drinking water disinfection by-products: an Australian perspective. Water Res 32(5):1522–1528
6. Volodina LA, Burbaev DSh, Lebedev VS et al (1996) Electronic spin resonanse investigation of copper centres in the sites its strong binding with cytoplasmic membrane bacteria *Escherichia coli*. Biophysics 41(5):1075–1081
7. Myrashko EI, Malyshev MK, Savluk OS et al (2006) The mechanism of intensifying the antimicrobial action of activated chloride with cooper. J of Water Chem Technol 28(6):54–60
8. Goncharuk VV, Potapchenko NG, Kosinova VN, Sova AN (2001) Disinfection of water by ozone. 1. Interrelationship between kinetics of the dying-off of bacteria in distilled water and ozonination parameters. J of Water Chem Technol 23(4):57–65
9. Zaidel AN (1986) Eksperimental'nyye otsenki oshibki izmereniy (Experimental errors of measurements). Khimiya, Leningrad
10. American Public Health Association (1995) Standard methods for the examination of water and wastewater. APHA, Washington
11. Mudrak NN, Kolchinskaya LI, Lysko VK (1983) Study of the ion transfer in membrane vesicles by ion exchange gel chromatography and ultrafiltration. Ukr Biochem J 55(2):185–190
12. Kul'skii LA, Savluk OS, Deynega EY (1985) Communication activity of bacteria with their electrokinetic and membrane potential. Rep Soviet Acad Sci 282(1):187–188
13. Savluk OS, Tomashevskaya IP, Bahurskaya LP (1986) Antimicrobial properties of water. J Water Chem Technol 8(6):65–67
14. Savluk OS, Potapchenko NG, Kosinova VN (1998) Disinfection of drinking water. J Water Chem Technol 20(1):99–111
15. Tomashevskaya IP (1988) Intensification of the process of water disinfection with sodium hypochlorite. Abst of Ph.Sc Thesis, AV Dumansky Institute of Colloid and Water Chemistry
16. Lebedev VS, Veselovskii AV, Deynega EYu et al (2002) Role of reactive oxygen species in the copper-induced permeability of the Escherichia coli plasma membrane. Biophysics 47(2):295–299
17. Lebedev VS, Deynega EYu, Savluk OS (1989) Role of transmembrane potential in Cu2+-induced breakdown of Escherichia coli cytoplasmatic membrane. Biolog Membr 6(12):1313–1316
18. Razumovskii SD (1983) Ozon i yego reaktsii s organicheskimi soyedineniyami (Ozone and its reactions with organic compounds). Izd-vo BAN, Sofia
19. Akhmetov NS (1975) Neorganicheskaya khimiya (Inorganic chemistry). Khimiya, Moscow
20. Gillespie RJ (1972) Molecular geometry. Van Nostrand-Reinhold, London
21. Martynenko LI, Spitsyn VI (1986) Izbrannyye glavy neorganicheskoy khimii (Selected chapters of inorganic chemistry). Izd-vo Mosc. un-ta, Moscow
22. Goronoskii IT, Nazarenko YuP, Nekryach YeF (1987) Kratkiy spravochnik po khimii (a concise reference book in chemistry). Nauk. Dumka, Kiev
23. March J (1992) Advanced organic chemistry: reactions, mechanisms, and structure, 4th ed. Wiley, Chichester
24. Klopman G (1974) Reactivity and reaction paths. Wiley-Interscience, New York
25. Schmidt MW, Baldridge KK, Boatz JA et al (1993) The general atomic and molecular electronic structure system. J Comput Chem 14:1347–1363

26. Schmidt MW, Gordon MS (1998) The construction and interpretation of MCSCF wave functions. Ann Rev Phys Chem 49:233–266
27. Davidson ER, Feller D (1986) Basis set selection for molecular calculations. Chem Rev 86:681–696
28. Minkin VI, Simkin BYa, Minyayev RM (1986) Kvantovaya khimiya organicheskikh soyedineniy: mekhanizmy reaktsiy (Quantum chemistry of organic compounds: reaction mechanisms). Khimiya, Moscow
29. Shevchenko SM (1986) Molekula v prostranstve (Molecule in space). Khimiya, Leningrad
30. Roberts G, Casserio M (1977) Basic principles of organic chemistry in 2 vol. Benjamin, New York
31. Gorlov YuI, Goncharuk VV (2006) Electron structure of ozone molecule and reactivity of ozone. J of Water Chem Technol 28(3):1–7
32. Razumovskii SD (1990) Ozone in the process of restoring water quality. J D. I. Mendeleev All-Union Chem Soc 35(1):77–88
33. Dewar MJS, Dougherty RC (1975) The PMO theory of organic chemistry. Plenum Publishing Corporation, New York
34. Mischuk NA, Goncharuk VV, Vakulenko VF (2003) Theoretical analysis of processes occuring at ozonization of water with organic admixtures. J Water Chem Technol 15(1):3–29
35. Mischuk NA, Goncharuk VV, Vakulenko VF et al (2003) Modeling destruction of organic impurities in water by ozonization and ultraviolet radiation. J Water Chem Technol 15(4):1–11
36. Mischuk NA, Goncharuk VV, Potapchenko NG et al (2006) Theoretic analysis of disinfection process of water in ozonation. J Water Chem Technol 28(5):1–13
37. Deryagin BV, Dukhin SS, Rulev NN (1986) Mikrofil'traciya (Microfiltration). Khimiya, Moscow
38. Gyurek LL, Finch GR (1998) Modeling water treatment chemical disinfection kinetics. J Environ Eng ASCE 124(9):783–793
39. Cho M, Chung H, Yoon J (2002) Effect of pH and importance of ozone initiated radical reactions in inactivating bacillus subtilis spore. Ozone Sci Eng 24(2):145–150
40. Collins H, Slleck R (1972) Process kinetics of wastewater chlorination. University of California, Berkeley
41. Montgomery JM (1985) Water treatment and design. Wiley, New York
42. Wei JH, Change SL (1975) Disinfection of water and wastewater. In: Johnson JD (ed) Ann Arbor, Science, MI, pp 34–73
43. Kimball AW (1953) The fitting of multi-hit survival a curves. Biometrics 2:201–211
44. Emanuel NM, Konorre DG (1969) Kurs khimicheskoy kinetiki (The course of chemical kinetics). Vysh Shk, Moscow
45. Kul'skii LA, Strokach PP (1986) Tekhnologiya ochistki prirodnykh vod (Technology of treating natural water). Vyscha Shk, Kiev
46. Sorpiková J, Zajícová V (1990) The effects of ultrasonic disintegrators on cell and bacterial suspensions. Scr Med 63:437–442
47. Jean DS, Chang BV, Liao GS et al (2000) Reduction of microbial density level in sewage sludge through pH adjustment and ultrasonic treatment. Water Sci Technol 42(9):97–102
48. Mornstein V, Davadorzh C, Franc P et al (1992) Effect of low-frequency ultrasound cavitation on concentrated solutions of selected organic compounds. Scr Med 65:411–418
49. Chen JR, Xu XW, Lee AS et al (1990) Feasibility studies of dechlorination of chloroform in water by ultra-sound in the presence of hydrogen peroxide. Environ Technol 11:829–836
50. Lin J-R, Yen TF (1993) An upgrading process through cavitation and surfactant. Energy Fuels 7:111–118
51. Riesz P, Kondo T, Krishna CM (1990) Sonochemistry of volatile and non-volatile solutes in aqueous solutions: e.p.r. and spin trapping studies. Ultrasonics 28(5):295–303
52. Shuslick KS, Hammerton DA (1986) The site of sonochemical reactions. IEEE Trans Ultrasonics Ferroelectrics Freq Contron 33:143–147
53. Goncharuk VV, Malyarenko VV, Yaremenko VA (2004) The mechanism of the effect of ultrasound on aqueous system. J Water Chem Technol 26(3):34–41

54. Dukhin SS, Derjaguin BV (1972) Elektrofores (Electrophoresis). Nauka, Moscow
55. Malyarenko VV (1989) Thermodynamic assessment of the temperature of supercooled water based on two-structure mode. Ukr Chem J 55(8):810–814
56. Skryshevsky AF (1971) Strukturnyy analiz zhidkostey (Structural analysis of liquids and amorphous solids). Vush Shkola, Moscow
57. Rosenberg LD (1968) Moshchnyye ul'trazvukovyye polya (Powerful ultrasonic fields). Nauka, Moscow
58. Fridrihsberg DA (1986) Kurs kolloidnoy khimii (The course of colloid chemistry). Khimiya, Leningrad
59. Kondo T, Kuwabara M, Sato F et al (1986) Influence of dissolved gases of chemical and biological effects of ultrasound. Ultrasound Med Biol 12:151–155
60. Mornstein V, Shponer J (1990) Iodine dosimetry and heat production in low frequency ultrasound. Scr Med 63:409–414
61. Goncharuk VV, Potapchenko NG, Savluk OS et al (2001) Disinfection of water by ozone: effect of inorganic impurities on kinetics of water disinfection. J Water Chem Technol 23(2):55–63
62. Wolfe RY (1990) Ultraviolet disinfection of potable water. Environ Sci Technol 24(6):768–773
63. Goncharuk VV, Chernomorets MP, Potapchenko NG et al (2002) Disinfection of water by UV-radiation generated by various sources. J Water Chem Technol 24(4):12–19
64. Gordon A, Ford R (1972) Chemist's companion. Wiley, New York
65. Zubrilin NG, Potapchenko NG, Savluk OS et al (1991) The combined effect of KrF-laser irradiation and copper ions on the survival of E. coli Cells. J de phusique IV(1):249–251
66. Wekrof A (1991) Treatment of contaminated water, air and soil with UV flashlamps. Environ Progress 10(4):241–247
67. Tikhonov AN, Samarskii AA (1990) Equations of mathematical physics. Courier Dover Publications
68. Linnik PN, Vasilchuk TA, Bolelaya NV (1995) Humic substances in the water of the Dnieper reservoirs. Hydrobiol J 31(2):74–81
69. Camel V, Bermond A (1998) The use of ozone and associated oxidation processes in drinking water treatment. Water Res 32(11):3208–3222
70. Lewis N, Topudurti K, Welshans G et al (1990) A field demonstration of the UV/oxidation technology to treat ground water contaminated with VOCs. J Air Waste Manag Assoc 40(4):540–547
71. Prengle HW (1983) Experimental rate constants and reactor for the destruction of micropollutants and trihalomethane precursors by ozone with UV radiation. Environ Sci Technol 17(12):743–747
72. Glaze WH, Schep R, Chauncey W et al (1990) Evaluating oxidants for the removal of model taste and odor compounds from a municipal water-supply. J Am Water Works Assoc 82(5):79–84
73. Glaze WH, Kang JW, Chapin DH (1987) Chemistry of water treatment processes involving ozone, hydrogen peroxide and UV-radiation. Ozone Sci Eng 9:335–352
74. Glaze WH, Peyton GR, Lin S et al (1982) Destruction of pollutants in water with ozone in combination with ultraviolet radiation. 2. Natural trihalomethane precursors. Environ Sci Technol 16(8):454–458
75. Goncharuk VV, Vakulenko VF, Sova AN et al (2003) Effect of UV-radiation mode on the kinetics and degree of degradation of humic and fulvic acids by ozone. J Water Chem Technol 25(5):1–21
76. Goncharuk VV, Vakulenko VF, Sova AN et al (2004) Removal of organic impurities from natural water by ozonization combined with UV-radiation. J Water Chem Technol 26(1):14–25
77. Westerhoff P, Aiken G, Amy G et al (1999) Relationships between the structure of natural organic matter and its reactivity towards molecular ozone and hydroxyl radicals. Water Res 33(10):2265–2276

78. Topkin YuV (2001) Determination of organic carbon in water by liquid-phase oxidization of impurities and reaction gas chromatography. J Water Chem Technol 23(4):27–32
79. Pedit JA, Iwamasa KJ, Miller CT, Glaze WH (1997) Development and application of a gas-liquid contactor model simulating advanced oxidation processes. Environ Sci Technol 31(10):2791–2796
80. Kamenev SB, Prejs SV, Munter RR et al (1985) The solubility of ozone in aqueous solutions. In Abstract: of the VII all-union symposium on modern prediction problems, water quality control and ozonation, Tallin, 19–21 November 1985
81. Goncharuk VV, Potapchenko NG, Vakulenko VF (1995) Ozonation as a method of drinking water treatment: possible by-products and toxicological analysis. J Water Chem Technol 17(1):3–33
82. Xu S-C, Zhou H-D, Wei X-Y et al (1989) The pH dependence and effects of the oxidative products of some aromatic compounds in ozonation under irradiation. Ozone Sci Eng 11(3):281–296
83. Goncharuk VV, Vakulenko VF, Sova AN et al (2002) The pole of UV-irradiation procedures in the process of O3/UV-treatment of surfactant solution. J Water Chem Technol 24(4):1–11
84. Goncharuk VV, Vakulenko VF, Samsoni-Todorov AO et al (1995) Photooxidation of pesticides by ozone and hydrogen peroxide in drinking water treatment. J Water Chem Technol 17(4):397–410
85. Andreozzi R, Caprio V, Insols A et al (1999) Advanced oxidation processes (AOP) for water purification and recovery. Catal Today 53(1):51–59
86. Prousek J (1996) Advanced oxidation processes for water treatment photochemical processes. Chem List 90(5):307–315
87. Goncharuk VV, Potapchenko NG, Savluk OS et al (2003) Combined action of ozone and ultraviolet radiation for water disinfection. J Water Chem Technol 25(2):53–61
88. Goncharuk VV, Potapchenko NG, Savluk OS et al (2003) A study of the various procedures of water disinfection by O3/UV. J Water Chem Technol 25(5):65–72
89. Ragab-Depre NJ (1982) Water disinfection with the hydrogen peroxide-ascorbic acid-copper (II) system water disinfection with the hydrogen peroxide ascorbic acid-copper (II) system. Appl Environ Microbiol 44(3):555–560
90. Duguet JP (1998) Geosmin and 2-methylisoborneol removal using ozone or ozone/hydrogen peroxide coupling. Paper present at the 9th ozone world congress, New York, 3–9 June, 1989
91. Moss SN, Smith KS (1981) Membrane damage can be a significant factor in the inactivation of Escherichia coli by near ultraviolet radiation. Photochem Photobiol 33(2):203–210
92. Atsushi I, Takashi I (1983) Possible involvement of membrane damage in the inactivation by broad-band near-UV radiation in Saccharomyces cerevisiae cells. Photochem Photobiol 37(4):395–401
93. Wekhof A, Folsom EN, Halpen Y (1992) Treatment of groundwater with UV-flash lamps—the third generation UV-systems. Hazard Mat Contl 5(6):48–54
94. Wekhof A (2000) Disinfection with flash lamps. PDA J Pharm Sci Technol 54(3):264–267
95. Goncharuk VV, Potapchenko NG, Savluk OS et al (2004) Combined antimicrobial effect of ozone and ultraviolet radiation generated by various sources. J Water Chem Technol 26(2):60–67
96. Bader H, Hoigne J (1981) Determination of ozone in water by the indigo method. Water Res 15:449–456
97. Potapchenko NG, Illyashenko VV, Gorchev VF et al (1993) Synergic effect of oxidizers: hydrogen peroxide and ozone with UV-radiation in studies of the survival rate of Escherichia coli 1257 cells. J Water Chem Technol 15(2):146–151
98. Avchinnikov AV, Zhuk EG, Rakhmanin YuA et al (1996) Hygienic assessment of the quality of drinking water conditioned by low-voltage pulse electrical discharges. Gig Sanit 1:14–15
99. Savluk OS, Tomashevskaya IP, Kornieva LP (1988) Change the electrokinetic potential of E. colli 1257 water from the influence of heavy metals. Soviet Surf Eng Appl Electrochem 6:65–67

100. Slipchenko VO, Savluk OS, Belaya VP et al (1975) Nauka i tekhnika v sil's'komu gospodarstvi (Science and engineering in agriculture). Budivelnyk, Kiev
101. Shcherba AA, Podoltsev AD, Kucheryavaya IN (2006) Studying of erosive destruction of materials at eelectrospark processing of the conductive granular media. Tech Electrodynam 1:3–10
102. Asanov UA, Tsoy AD, Shcherba AA, Kazetkin VI (1990) Elektroerozionnaya tekhnologiya khimicheskikh soyedineniy i poroshkov metallov (Electroerosion technology of chemical compounds and metal powders). Ilim, Frunze
103. Rudenko VS, Senko VI (1983) Chizehenko IM Preobrazovatel'naya tekhnika (Conversion equipment) Vyshcha Shkola, Kiev
104. Namitokov KK (1978) Elektroerozionnyye yavleniya (Electroerosion phenomena). Energiya, Moscow
105. Andreev SI, Orlov BI (1966) Development of a spark discharge. J Sov Phys-Techn Phys 10(8):1097–1101
106. Shcherba AA (1996) Preobrazovaniye i stabilizatsiya parametrov elektroenergii (Transformation and stabilization of electric power parameters). EI NAS of Ukraine, Kiev
107. Goncharuk VV, Shcherba AA, Zakharchenko SN et al (1999) Disinfection actions of a space spark discharge in water. J Water Chem Technol 21(3):45–52
108. Kul'skij LA, Savluk OS, Deynega EYu (1980) Vliyaniye elektricheskogo polya na protsessy obezzarazhivaniya vody (Influence of an electric field on the processes of water disinfecting). Nauk Dumka, Kiev
109. Savluk OS, Kul'skij LA, Kosinova VN, Tomashevskaya IP (1982) Intensification of the antimicrobial agents in an electric field. Surf Eng Appl Electrochem 3:73–79
110. Rudenko AV, Koval EZ, Voloshchuk EM (2005) New approaches to the therapy of onychomycosis. Adv Med Mycol 6:242–244
111. Ozerskaya SM, Ivanushkina NE, Kochkina GA (2011) Micromycetes in the connection with problems of biological safety. Probl Med Mycol 13(3):3–15
112. Hoog de GS, Gudrro J, Gene J et al (2000) Atlas of clinically significant fungi, 2nd edn. Universitat Rovira I Virgili Reus, Spain
113. Sutton DA, Fothergill AW, Rinaldi MG (1998) Guide to clinically significant fungi. Williams Wikins, Baltimore
114. Rudenko AV, Koval' EZ, Rvzhko PP et al (2001) Onikhomikozy u zhiteley Ukrainy (Onychomycoses in Ukraine's citizens). OOO "TSK", Kiev
115. Richardson MD, Warnok DW (1997) Fungal infection: diagnosis and management, 2nd edn. Blackwell Science, Oxford
116. Suslov EI, Podgaevskaya TP, Chepil PI et al (1999) Mycotic infection in patients with lung cancer. Ukr Pulm J 4:38–40
117. Sudyachenko AI, Blikhar OI, Voino-Yasenetskaya OV et al (2003) Glubokiye mikozy (Deep mycoses). Hidromaks, Nezhin
118. Klimko NN, Vasil'eve NV, Elinov NP et al (2001) Perechen' osnovnykh metodov i kriteriyev diagnostiki mikozov (The list of main methods and criteria for mycoses diagnosis). OOO "MGK", St. Petersburg
119. Tutelyan VA, Kravchenko LV (1985) Mikeotoksiny: meditsinskiye i sanitarnyye aspekty (Mycotoxins: medical and health-related aspects). Meditsina, Moscow
120. Marfenina OE (2000) Distribution potentially pathogenic micromycetes in environment. Probl Med Mykol 2(2):36–37
121. Rebrova RN (1979) Griby roda Candida pri bakterial'nykh infektsiyakh (Fungi of the genus Candida in bacterial inections). Meditsina, Moscow
122. Abdalov KA, Gedder YuR, Ruvinova LG (2000) Environmental aspects of mycotic defeats of human. Human Ecol 1:13–15
123. Semenova TA, Terenova VA (1992) Seasonal and annual changes in the structure of the microbiota: Micromycetes of Kuibyshev reservoir. Mikol Fitopatol 26(2):112–117
124. Zhuravlev PV, Golovina SV, Alegina VV et al (1997) Barrier role of water-purifying construction of water pipe in relation to opportunistic microorganisms. Hygiene Sanitat 4:15–16

125. Grigor'eva LV (1985) Sanitarnaya mikrobiologiya evtrofnykh vodoyemov (Sanitary microbiology of eutrophic reservoirs). Zdorov'ya, Kiev
126. Rosenzweig WD (1986) Fungi in potable water distribution systems. J Am Water Works Assoc 78(1):53–55
127. Doggett MS (2000) Characterization of fungal biofilms within a municipal water distribution system. Appl Environ Microbiol 66(3):1249–1251
128. Metzer WJ (1976) Sauna-takers disease: hypersensitivity pneumonitis due to contaminated water in a home sauna. J Am Med Assoc 236(19):2209–2212
129. Niemi RM, Knuth S, Lundstrom K (1982) Actinomycetes and fungi in surface water and potable water. Appl Environ Microbiol 43:378–388
130. Nagy L, Olsen BH (1983) The occurrence of filamentous fungi in drinking water distribution system. Can J Microbiol 28:667–671
131. Romanyk FP (1998) Fungal infections in children, caused by opportunistic fungi. Abst of Dr Sci Thesis. St. Petersburg Medical Academy
132. Romanyk FP (2000) Respiratory mycosis and micoallergies in children. Probl Med Micol 2(2):33–34
133. Kurbatova IB (1997) Comparative microbiological characteristics of aspergillosis. In: Abstract of the 7th national congress on respiratory diseases, Moscow, 3–5 July, 1997
134. Bays LR, Burman NP, Leurs WM (1970) Taste and odor in water supplies in Great Britain: a survey of the present position and problems for the future. Water Treatment Exam 19:136–140
135. Chumak AA, Bazyka DA, Tal'ko AA et al (1991) Immunological aspects of the study of contingents of population exposed to ionizing radiation effects as the consequence of the Chernobyl AES accident. Rep Soviet Acad Med Sci 3(8):16–19
136. Feshchenko YuI, Rekalova EM, Kadan LP et al (1999) The Chornobyl accident influence on immune reactions in patients with chronic obstructive bronchitis and healthy Kiev citizen (15 years of observation). Ukr Med J 4(2):116–118
137. Dudka IO, Koval' EZ (1994) Micromycetes of Kremenchug reservoir (Ukraine). Ukr Botanic J 51(4):53–56
138. Roi AA, Koval EZ, Yakushin VM (1995) Microbial contamination in water-treatment system. Microbiol J 4:60–66
139. Avchinnikov AV (2000) Microbial contamination of aquatic environment and their effects on human health. Med Advice 2:43–47
140. Goncharuk VV, Rudenko AV, Koval EZ et al (2004) Problem of water infection with pathogens of mycoses and prospects of resolving this problem. J Water Chem Technol 26(2):1–18
141. Lugauskas AYu, Mikul'skine AI, Shlyauzhene DYu (1987) Katalog mikromitsetov-biodestruktorov (Catalog of micromycetes-biodestructors). Nauka, Moscow
142. Sutton D, Fotergille A, Sodoral L et al (2001) Opredelitel' patogennikh i uslovno patogennikh gribov (The determinator of pathogenic and conventionally pathogenic fungi). Mir, Moscow
143. Cole RJ, Cox RH (1981) Handbook of toxic fungal metabolites. Acad Press, San Diego
144. A.V. Dumansky Institute of Colloid and Water Chemistry NAS of Ukraine (2003) Water: problems of sustainnable development of civilization in the XXIst century. Kiev, Ukraine
145. Goncharuk VV, Potapchenko NG (1998) Water disinfection: the current situation. J Water Chem Technol 20(4):22–42
146. Brinkman NE, Yaugland RA, Wymer LJ (2003) Evaluation of a rapid, quantitative real-time PCR method for enumeration of pathogenic Candida cells in water. Appl and. Environ Microbiol 69(3):1775–1782
147. Hagler AN, Hagler LCM (1978) The yeasts of fresh water and sewage. Ann Microbiol 23:79–103
148. Yelinov NP (2001) Candida species and candidemia: the problem's condition. Probl Med Micol 3(1):4–15
149. Rosenzweig WD, Minnigh H, Pipes WO (1982) Chlorine demand and inactivation of fungal propagules. Appl Environ Microbiol 45(1):182–186

150. Haas CN, Engelbrecht RS (1980) Chlorine dynamics during inactivation of coliforms, acid-fast bacteria and yeasts. Water Res 14(12):1749–1757
151. Arvanitidou M, Kanellou K, Constantinides TC et al (1999) The occurerrence of fungi in hospital and community potable water. Lett Appl Microbiol 22:81–84
152. Goncharuk VV, Potapchenko NG, Kosinova VN et al (2006) Desinfection properties of polyhexamethyleneguanidine chloride and factor influencing its activity. J Water Chem Technol 28(5):49–58
153. Levadnaya TI, Savluk OS, Kosinova VN (2004) The effectiveness of water purification from bacteria using coagulation. Water Water Treat Technol 4(12):36–41
154. Goncalves AB, Paterson RRM, Nelson L (2006) Survey and significance of filamentous fungi from tap water. Int J Hygiene Environ Health 209(3):257–264
155. Matus V, Lyskova T, Sergienko I et al (1982) Fungi: growth and sporulation after a single treatment of spores with ozone. Mikol Fitopatol 16(5):420–423
156. Saprykina MM, Samsoni-Todorov AO, Goncharuk VV (2009) The decontamination effect of UV radiation with respect to micromicetes. J Water Chem Technol 31(5):329–333
157. Saprykina MM, Savluk OS, Goncharuk VV (2009) Removal from water of yeastlike fungus Candida albicans by the method of coagulation and flocculation. J Water Chem Technol 31(1):60–65
158. Hageskal G, Gaustad P, Heier BT et al (2007) Occurrence of moulds in drinking water. Appl Environ Microbiol 102(3):774–780
159. Sokolov VF (1964) Obezzarazhivaniye vody bakteritsidnymi luchami (Water decontamination by bacteria rays). Stroiizdat, Moscow

Chapter 6
Biotest as an Evaluation Method for the Quality of Drinking Water

Abstract The comprehensive critical review of works on drinking water biotests, which is widely applied worldwide as the method to analyse the drinking water quality and risk evaluation for human health. The need to apply complex approach for biotests has been justified. Analysis result of 53 marks of bottled water by biotest methods has been introduced for the first time ever.

Keywords Biomonitoring · Drinking water biotest · Set of test organisms · Ecotoxicological tests · International standards · Complex analysis of bottled water

6.1 An Integrated Approach to Natural Water Biotesting

Assessment of natural water quality by means of biotesting has recently grown in importance because of the fast growth of potentially harmful chemical compounds contaminating water basins. For example, about 70,000 man-made chemical substances had been used by the second half of the eighties. Furthermore, toxic effects of 80% of artificial and considerable amount of natural substances were not yet known [1].

The use of chemical methods for the assessment of natural water quality does not always yield necessary results, due to the lack of information on the biological components of the chemicals analyzed. Biotesting makes it possible to determine water toxicity, and it does not contain any harmful substances according to the chemical analysis. However, according to the biological criteria, the location of the toxicant agent does not indicate water toxicity. Moreover, the chemical examination does not take into consideration the interaction of contaminants and their transformation both in nature and in organisms. Biotesting methods assess objectively and comprehensively the influence of substances on an organism and its vital processes. In many cases biomonitoring is technically simpler and considerably cheaper than other alternatives. It does not require special instrumentation; it is limited in time; and it is more precise and sensitive in comparison with the chemical analysis. On the whole, biotesting will lead both to a reduction in the number of necessary procedures and to the significant simplification of the examination process.

Methods of biological assessment to determine the quality of aqueous environments may be divided into the methods of bioindication and biotesting. Biotesting

V. V. Goncharuk, *Drinking Water*, DOI 10.1007/978-3-319-04334-0_6,

enables a task-oriented use of standard test organisms and methods to determine toxicity level of the waste water, single contaminants and natural water basins [2]. In a broad aspect, biotesting is a methodological approach based on the assessment of environment factors which influence an organism and specific functions of the organism or system of organisms [3]. The biotesting procedure suggests successful completion of an experiment, and this is how it differs from the bioindication according to which water quality is determined by presence or absence of certain organisms.

Certainly biological testing has some disadvantages, especially in the event of separate biotesting. Results obtained have to be complemented with analytical chemistry data. Additionally, an assessment of risks posed by the samples and substances under examination for humans should be supplemented by results of the standard sanitary and hygienic methodologies. However, this does not raise doubts concerning prospectivity and effectiveness of this approach for the rapid testing and monitoring of water of various types and their contaminants.

Several publications [4–7] provide reasonable details as to various problems associated with biotesting.

This leads to the question of what it means to take a complex approach to the biological testing of natural water (both at the organism and cell levels) to achieve an objective evaluation of water quality.

Criteria Used for Selection of Biotesting In order to establish the most effective approach to the assessment of natural water quality by means of biotesting, the International Center for Scientific Research Promotion (Canada) carried out a special "Water Tox" program. The selection criterion which was optimal for certain conditions and, at the same time, simple enough for the application of biotesting sets, are the following:

- Use of test organisms consisting of two or more trophic levels; this is the minimum amount which provides the necessary range of sensitivity regarding the impact of toxic substances;
- Cost of biotesting must be low; the most pertinent factor in this respect is that biotesting utilize cheap equipment, reagents and materials, and performance of certain experiments;
- Test organisms, materials and equipment required for the assessment must be available or easily acquired;
- A biotesting set should possess high sensitivity and a broad range of substances to be analyzed; an optimum test should include various tests for the largest number of potentially harmful substances and register various kinds of toxicity, such as genotoxicity; the sensitivity type-specific characteristics should also complement each other;
- Assessment must not contaminate the environment;
- Biotesting procedures should be simple, their results understandable and well-defined and their performance possible in non-specialized laboratories with minimum equipment;
- Testing methods should be standardized and easy to repeat;

- A biological testing set should be highly effective and multipurpose, i.e., of wide application not limited by one region or ecosystem. Time spent on biotesting and data processing should not exceed 5 days, though this criterion is not considered to be critical.
- It is desirable to receive results quickly because the advantage of fast response is otherwise lost.

"Water Tox" Biotesting Battery Based on these criteria, the following biotesting battery has been proposed for practical use:

- *Allium test.* Ordinary *Állium cépa* onion, a monocotyledonous plant, is used as a test organism. The growth of this onion's rootlets slows down as a result of the toxic impact [8]. During some experiments (usually within a period of 72 h), onions are sprouted using reference or test specimens. Then the average length of the rootlets of various samples is compared.
- *Plant seed development test.* The test object proposed is *Lactuca sativa lettuce* (a representative of bilobed plant). 20 seeds of each sample are cultivated for 5 days in Petri dishes with moist filters. Like the previous test, upon completion of the experiment, the length of various rootlets samples is measured and compared [9].
- *Hydra test.* An invertebrate organism hydra (*Hydra attenuata*) is proposed for the current experiment. This experiment investigates morphological changes (sublethal effects) in the fresh water hydra as well as survivability (lethal effects) upon impact of toxic samples [10].
- *Panagrellus test.* The test organism is *Panagrellus redivivus ascarid.* The test samples are cultivated for 96 h and assessed for the ascarids' ability for healthy reproduction and the survivability of the young posterity at the embryonal stage J2. Additional assessments are based on its ability to develop from stage J2 to stage J4, survivability of the young posterity at stage J4 and the further development of the organism until the grown-up stage [11].
- *Daphnia test.* This test assesses an acute toxicity of test samples by the survivability of daphnia *Daphnia magna* over a period of 48 h [12].
- *Fluctuation (mutation assessment) test.* Test organisms (*Salmonella* bacteria) are incubated for 5 days at 35–37 °C in the growth environment containing diluted test specimens. Components of the environment are selected in such a way that only mutation cells can develop during the incubation period. Their development is registered by the change of the environment's coloring [13].

The approbation test for this battery was carried out for 2 years in the laboratories of eight countries, including the Ukraine.

Toxicity Assessment of Water Samples Applied to a Set of Animal and Vegetation Biotesting Use of the biotesting battery will allow an objective impact assessment of the factor under study on animal and vegetation organisms. For its quantitative interpretation, we proposed a general toxicity index ($I_{g.t}$), which represents a sum of effects calculated per unit of the substance's concentration for all the biotesting, including the battery. In order to generalize any influence of any biotesting, the maximum value of one test function is limited by 100 conventional units. The number of the organisms participating in the experiment automatically determines the highest value $I_{g.t}$ This indicator could not exceed 500 c.u. in our examination with two vegetations and three animal biotesting.

There were 15 substances of organic and non-organic nature assessed by the test organism battery under discussion. The generalized results of this assessment are presented in the Tables 6.1 and 6.2. Note, that these tables only show the initial concentrations of substances assessed (in mg/dm^3); in experiments they could be lower 10 or 100 times lower (due to the high sensitivity of an organism to a given substance). Therefore, there are substantial values in some biotesting when calculating the effect per unit of concentration.

The assessed substances are presented in the Tables 6.1 and 6.2 in accordance with their toxic effect. The tables also provide assessment and general toxicity of each sample on the basis of the total indicator $I_{g.t}$ (see Table 6.2). The highest $I_{g.t}$ values were found in pentachlorophenol with bactericidal and fungicidal activity. It has rather high toxicity values for two vegetation and two animal test organisms. Insecticide aldrin demonstrates high toxicity in only one *Daphnia test*, while on the other four its influence is insignificant. Thus it only occupies the 10th position among the 15 assessed substances. Another insecticide (DDT) having high total toxicity demonstrates specific differences both in animal (maximum effect on daphnia and minimum on hydras) and in vegetation (significant differences between onion and lettuce) biotesting (see Table 6.1). Mercury ions are extremely toxic for hydras and daphnia. At the same time the simplest aromatic amine, aniline, demonstrates the most insignificant effect for all the tests of the battery (see Table 6.1).

On the whole, tested organic and non-organic substances are not clearly differentiated by the biotesting methods as to their toxic properties; they are represented in the groups with maximum and minimum effects. Thus, most toxic substances in the increasing order were p, p′-DDT, 4-Nitrochinolin N-oxide, Cu^{2+}, Hg^{2+}, and pentachlorophenol. The least toxic were—As^{5+}, lindan, metolachlor, and aniline.

A comparative analysis of average values calculated separately as per substances of organic and non-organic origin demonstrated specificity of reactions both in the vegetation and animal biotesting (see Tables 6.1 and 6.2).

For example, in the onion test, the reaction to various groups of substances was more differentiated than in the lettuce. A similar situation was observed in animals: the sensitivity of hydras to organic and non-organic substances was more differentiated than that of daphnia. However, this difference in response of hydras and daphnia was to a great extent caused by excessive effects of copper and mercury ions. Better balanced data were received for the indicator of total toxicity (see Table 6.2).

A correlation analysis was made between the effects obtained during experimental work with all the test organisms. A majority of comparisons do not demonstrate any true correlation between the results registered for different biotesting from the "Water Tox" battery due to the impact of an entire set of organic and non-organic substances. Therefore, in order to produce comprehensive and objective estimates, it is necessary to use an integrated approach to biotesting which includes a range of test organisms.

Table 6.1 Comparison of negative effects registered in animal and vegetation biotesting as a result of toxic substance impact and the ranging of these substances in accordance with their toxic characteristics

Toxicant	The initial concentration mg/dm^3	*Allium test*		*Lactuca saviva*		*Daphnia magna*		*Hydra attenuata*		*Panagrellus redivivus*	
		Effect[a]	Rank	Effect	Rank	Effect	Rank	Effect	Rank	Effect	Rank
2,4-dinitrophenol	1	50.0	3	56.8	1	16.8	11	0	12	1.8	8
Cu^{2+}, sulfate	4 and 8	148.5	4	14.6	7	982.5	3	750.0	2	14.7	2
Cd^{2+}, chloride	5	9.6	8	13.8	8	125.0	8	20.0	7	13.0	4
K_2CrO_7	4	11.9	6	15.4	6	433.4	4	25.0	6	13.6	3
Aniline	100	0.7	15	0.8	14	4.7	15	0.9	11	0.2	15
Zn^{2+}, sulfate	20	2.1	13	2.4	13	141.7	7	0	14	0.2	12
Metolachlor	20	2.4	12	3.1	11	5.0	14	5.0	8	0.4	10
4-Nitrochinolin N-oxide	1.2	14.9	5	34.2	3	73.0	10	1283.0	3	6.8	6
Hg^{2+}, chloride	2	11.5	7	24.0	4	1000.0	1	3200.0	1	21.5	1
Nonylphenol	20	2.0	14	0.1	15	100.0	9	80.0	5	0.2	13
Pentachlorophenol	0.8	97.0	2	43.8	2	178.8	6	125.0	4	4.3	7
Aldrin	8	3.8	10	4.8	10	395.6	5	0	13	0.7	9
As^{5+}	20	6.8	9	4.9	9	12.5	12	1.1	10	0.2	14
p, p′-DDT	0.4	197.1	1	18.0	5	100.0	2	0	15	11.0	5
Lindane	20	3.2	11	2.5	12	10.0	13	5.0	9	0.4	11
The average for organic matters		41.2		18.2		198.2		38.2		2.9	
The average for inorganic substances		10.1		12.5		449.2		666.0		10.5	
Overall average		28.8		15.9		298.6		289.3		5.9	

[a] The effect is calculated per unit of the substance's concentration

Table 6.2 Comparison of toxic substances according to the negative effect for the animal and vegetation biotesting. Ranging of these substances by the total toxicity indicator ($I_{g.t}$)

Toxicant	The initial concentration mg/dm³	Pant test organism (*Allium*+*Lactuca*)		Animal test organism (*Daphnia*+*Hydra*)[a]		$I_{g.t}$	Rank
		Effect	Rank	Effect	Rank		
2,4-dinitrophenol	1	106.8	3	16.8	11	125.4	9
Cu^{2+}, sulfate	4 and 8	33.1	6	1732.5	2	247.8	3
Cd^{2+}, chloride	5	23.4	8	145.0	9	181.4	7
K_2CrO_7	4	27.3	7	458.4	4	165.9	8
Aniline	100	1.5	15	5.6	15	7.3	15
Zn^{2+}, sulfate	20	4.5	13	141.7	10	104.7	11
Metolachlor	20	5.5	12	10.0	14	185.9	14
4-Nitrochinolin N-oxide	1.2	49.1	4	201.0	7	228.9	4
Hg^{2+}, chloride	2	35.5	5	4200.0	1	257.0	2
Nonylphenol	20	2.1	14	180.0	8	182.3	6
Pentachlorophenol	0.8	140.4	2	303.8	6	345.1	1
Aldrin	8	83.6	10	395.6	5	109.3	10
As^{5+}	20	11.7	9	13.6	13	25.5	12
p, p′-DDT	0.4	215.1	1	1000.0	3	226.1	5
Lindane	20	5.7	11	15.0	12	21.1	13
The average for organic matters		59.5		236.4		140.2	
The average for inorganic substances		22.6		1115.2		163.7	
Overall average		44.7		587.9		149.6	

[a] Without effects for *Panagrellus redivivus* as very insignificant by value and most varying in comparison of the summarized results, and for the preservation of the proportionality principle, or in other words, two biotesting each time for the animal and vegetation organisms

Genotoxicity and Cytotoxicity Assessment of Water Samples Using Test Organism Cells In addition to the combination of various biotesting for objective assessment of tested water quality, another very important aspect of this study seems to be the complex development of investigations for mutagenicity, genotoxicity and cytotoxicity of water samples. In order to monitor this, we propose a very simple solution. The Standard Ames test or rather its modification, mutation assessment (fluctuation) method [13], is included in the battery of methods, especially adapted to estimate the mutagenicity of water samples. Additionally, test organisms included in the research battery are studied for the frequency of micronuclei and geminate nuclei (indicators of genotoxicity) and the quantitative characteristics of nucleoles (criteria of cytotoxicity). Monitoring of the proliferative activity of cells is carried out by the mitotic index. In other words, a triplet of methods is used, registering mutations at the genie level, changes in structure and functional activity of the genetic apparatus at the cell level. Selection of these criteria depends on their information content, good reproducibility, and technical simplicity.

The micronuclear test is one of the methods used to detect substances which demonstrate genotoxic properties. Previous publications show that the micronuclear test is sometimes even more informative and quick than chromosome aberration tests. Some other advantages of this test are as follows: fewer artifacts, possibility of higher quality registration, decreased complexity, and greater productivity [14–16].

The quantitative characteristics of nucleoli (number and size) are proposed to be used for the study of the cell genome functioning properties. The nucleoli represent a set of amplificated genes of ribosomal ribonucleic acid (RNA) and their products in the cell. Changes of their morphology are directly connected with the most important molecular genetic processes and objectively reflect the character of cell metabolism [17–20]. It was proved that the most informative characteristics for cells with a small number of nucleoli is the size of a single nucleolus and a share of cells with heteromorphic geminate nucleoli (HGN), while for the cells with several nucleoli it is their quantity [21, 22].

Experiments, meant for the research of genotoxicity and cytotoxicity, were performed on the same substances for which the acute toxicity was studied. Results of this research are provided in Table 6.3, which demonstrate data for various types of toxicity in order to compare the effects. The substances range in decreasing order of toxicity levels, in accordance with $I_{g.t}$ values.

The main conclusion apparent from the analysis of the data provided in Table 6.3 proves that in most cases there is no correspondence between toxic, mutagenic, genotoxic, and cytotoxic effects. For example, the two most toxic substances (according to $I_{g.t}$), organic pentachlorophenol and non-organic mercury, did not significantly change mN and 2N frequencies in plant cells; there was also no significant change in the number of nucleoli and HGN percent in hydras. On the other hand, the least toxic of the analyzed substances, lindan and metolachlor, had a substantial influence on mN and 2N frequencies as well as on the number of nucleolus in the plants. It should be noted that linden has the strongest genotoxic effect.

The nucleole characteristics used in this study were in line with various mechanisms of regulation of nucleole activity in a cell. For example, the number of nucleoli represented the number of active nucleole-forming areas, size of the nucleolus, transcriptional activity of these areas and a share of HGN characterized specific mechanism of the activity regulation of cells with geminate nucleolus. Because of the toxic impact, these characteristics have changed in different ways, in most cases without correlation to each other. Therefore, toxicants examined in the current study changed different processes which control nucleole functioning; it is thus necessary to use a set of characteristics for objective evaluation of citotoxicity by the nucleole biomarker.

Noted differences of toxicity, mutagenicity and genotoxicity indexes imply the necessity of an integrated approach which consists of various methods for the objective and comprehensive evaluation of the assayed substance.

One of the promising trends in terms of practical application of the battery under discussion is biotesting of the quality of water of various origins in the water treatment systems at various stages of the technological process. Biotests included in the battery may be selected for the analysis of contaminants of various classes, or for water of various physical and chemical compositions. In compliance with the

Table 6.3 Comparison of results of toxicity, mutagenicity, and genotoxicity

Toxicant	Toxicity		Mutagenicity	Mitotic index	Genotoxicity		Cytotoxicity					
	Animal biotests	Plant biotests	*Salmonella*	Plant biotests	Plant biotests mN	Plant biotests 2N	Plant biotests nN	Hydra nN	Plant biotests *V1* N	Hydra *V1* N	Plant biotests HGN	Hydra HGN
Pentachlorophenol	+++	+++	+++	++	++	+	++	++	+++	+++	+++	++
Hg^{2+}	++++	++	++++	+++	+	+	+	+	+++	+++	+	++
Cu^{2+}	++++	++	+	++++	++	++	++	+	+++	Undefined	+++	+++
4-Nitrochinolin-N-oxide	++	++	++++	++	+++	++	++	+++	+++	++++	+++	++
p, p′-DDT	++++	+++	+++	+	+++	++	++	+	+++	+	+++	+
Nonylphenol	++	+	+	++	++	++	++	+	++	+	++	+
Cd^{2+}	++	++	++	Undefined	Undefined	Undefined	++	Undefined	++	Undefined	+++	Undefined
K_2CrO_7	+++	++	+++	++	+	Undefined	++	Undefined	++	Undefined	++	Undefined
2,4-dinitrophenol	+	+++	+	Undefined	Undefined	Undefined	++	Undefined	++	Undefined	++	Undefined
Aldrin	+++	+	++	+	+	+	++	+	+	+++	++	+
Zn^{2+}	++	+	+	++	+	++++	+	Undefined	+	Undefined	+++	Undefined
As^{5+}	+	+	+	++	+	+	++	+	++	+	+	++
Lindane	+	+	+	+++	++++	++++	+++	+	++	+	++	+
Metolachlor	+	+	+	++	++	+++	+++	++	+++	Undefined	++	++
Aniline	+	+	+	++	++	Undefined	++	Undefined	+	Undefined	+	Undefined

"+"—practically no effect available, "++", "+++", "++++"—correspond to a single nucleolus
mN micronucleus, *2N* double nucleus, *V1 N* volume of a single nucleolus, HGN heteromorphic geminate nucleoli

tasks to be solved, it is necessary to select sets with minimum necessary, optimal and additional capacity.

Use of Bioassay Methods for Development of Standards of the Natural Water Quality Bioassay results are needed for the development of standards along with the proven and approbated indicators, such as chemical, analytic, chemical, physical, sanitary, hygienic, and the list goes on.

During a biotesting, drinking water should not have any statistically proven deviation from the reference parameters set forth in methodological instructions for the test organisms used. When these instructions are adhered to, water quality is considered to be in line with the biological standards. In the event of proven deviations from the test indicators, water samples under examination are considered toxic according to their biological characteristics. Analyzed assays are classified as low, medium, or highly toxic in terms of biological effect. They are categorized according to the degree of deviation from test values on the one hand and the number of the test organisms which register negative effects on the other hand. The degree of mutagenicity, genotoxicity and citotoxicity of water samples in the process of analyzing both well-developed and approbated as well as most simple and informative subcellular and cellular parameters have also been determined.

In line with the tasks to be solved and technical capability of the laboratory use assay sets with the minimum necessary, optimal or additional capacity.

Taking into consideration the above results, let us review several examples. First of all, it is possible to combine daphnia and ceriodaphnia (as representatives of animals) and onion and lettuce (as representatives of vegetation) in a minimum necessary biotesting set. These test organisms are used to determine the toxicity of water samples by the chemical contaminants. It is possible to use a hydrosulfuric (bacterial) test to carry out a biological contamination test.

An optimal assay set may include: daphnia or ceriodaphnia; hydra (as a representative of invertebrates, which differ by their response to toxins); onion and lettuce (as representatives of monocotyledonous and bilobed plants with differentiated response to chemical substances); and fish (as a representative of vertebrates).

To construct an assay set with additional facilities, the optimum battery is enhanced by the Ames Standard Test (determination of water sample mutagenicity), micronuclear and nucleole tests (assays of structural and functional alterations in genomes of cells of test organisms included in the optimum set: onion, lettuce, hydra, and fish) in order to make an in-depth assay of water samples and detailed characterization of the consequences of their impact on an organism.

For practical use, research institutions and individual scientists generate their own biotesting sets most expediently in line with their technical abilities and accumulated experience (for example, with animals or plants, organism cultures, and depending on the set tasks).

Examples of Some Test Organism Sets Used in Practical Work There are eight methodologies attested for assessment of toxicity of natural and waste water in Ukraine: (A. N. Krainyukova, chief developer, Institute of Ecological Problems, Kharkov). These methodologies include test organisms of various trophic levels and systematic groups, in particular bacterium *Photobacterium phosphoreum* [23], infusorium *Tetrahymene pyriformis* [24], alga *Scenedesmus quadricauda* [25], crustaceans *Daphnia magna* and *Ceriodaphnia affinis* [26–28], fish *Poeccillia reticulate* [29] and fruit-fly *Drosophila melanogaster* for the genotoxicity assay [30].

The U.S. Environmental Protection Agency (EPA) uses a fish survival and growth test as the main set for the water quality assay; for example, *Pimephales promelas*, ceriodaphnia *Ceriodaphnia dubia* at the early stage of growth by the survival and reproduction rate; and alga *Selenastrum capricornutum* by the cell growth.

In Canada for waste water toxicity assessment, the following is recommended: bacterial test estimating changes in the cell glow in *Vibrio fischeri* (Microtox test); microalgae test which registers growth rate of the culture *Selenastrum capricornutum*; reproduction and mortality rate test in a representative of crustaceans *Ceriodaphnia dubia*, and so-called SOS test for genotoxicity by using *E.coli* [32].

Several other sets of test organisms are used in European countries. For example, in France, for acute toxicity monitoring, daphnia is used, as well as the Microtox test and the mutagenicity standard test [33]. In Germany the main test organism is in particular, a fish called orfe (*Leuciscus idus*) [34]; additionally, in Germany daphnia, alga and fluorescent bacterium data are analyzed. The primary screening in England is carried out by using bacteria (Microtox-Test) and determination of *Daphnia* mortality; further assays include assessment of the growth processes in alga *Selenastrum* and registration of mortality of salmon and carp fish.

Use of Biotesting Results for Estimation of Human Health Risk A number of important questions regarding extrapolation of the results produced by the human organism arise in the assessment of drinking water quality by the biotesting method. Is the water toxicity assay data produced by using animals and vegetation test organisms a signal of danger to humans? Which biotests analyzing various toxicity types are the most acceptable and can be extrapolated most efficiently?

The first thing to do is to note that the test organisms used for practical research do not represent the maximum allowable concentrations of chemical substances approved for drinking water by the following organizations: World Health Organization (WHO), Environmental Protection Agency (USEPA) and The European Union (EU). For example, the concentration under discussion for organic compounds metolachlor, kindan, nonylphenol, aldrin, 2,4-dinitrophenol, DDT, and pentachlorophenol is within the range of 0.1–2.0 μg/dm^{3}. The concentration is a little higher for heavy metal ions: mercury—1, cadmium, arsenic and chrome (VI)—50, and copper and zinc—100 μg/dm^{3} [35, 36]. In most cases, the results we received with test organisms were one or two orders higher, or expected concentrations were registered at the level of milligrams per 1 dm^{3} (mg/dm^{3}).

Two conclusions can be drawn from the comparisons provided. First, it is necessary to raise the biotesting sensitivity threshold for the drinking water assays [37]. This requirement can be implemented by combining several test organisms with differentiated sensitivity, by conducting studies both at the organism and cell levels (provided there is compulsory appraisal of the most sensitive functional characteristics) with preliminary concentration of water samples, and so on. Second, water toxicity assessed by the biotesting methods considerably exceeds the maximum allowable for humans, thus prohibiting water consumption before a detailed examination. It should also be noted that biological methods determine the total toxicity of the samples; in other words, the summarized effects of various substances may be synergistic (more toxic than the sum of the individual substance effects), additive (the toxic effects are just summed up) or deductable (some toxic effects compensate or mask others).

The replacement of the long-term tests performed by using animals and assaying carcinogenic and mutagenicity properties of substances with short-term biotesting has some advantages. In particular, the long-term tests are very expensive (about US$ 1 million for one chemical agent), requiring the death of up to 1000 rodents, and producing results which are difficult to interpret [38]; the short-term biotesting, on the other hand, are more accurate in comparison with the long-term assays in assessing human risk [39].

Thus, in order to obtain an objective and quick assessment of the quality of both waste and natural water, including drinking water, it is recommended to use biotesting to implement a more comprehensive approach; that is to say, a battery of test organisms. The inclusion of genetic and cytogenetic methods allows for additional assessment of changes at the cellular and subcellular levels. Due to the comprehensive approach, combining one of the most simple and very importantly cheap methods, it is possible to obtain the most complete data package possible about the toxicity, mutagenicity, genotoxicity, and citotoxicity of water under examination. [40] One of the advantages of this approach is the assessment of water samples' danger to human's health.

6.2 Biotesting Water for Quality at the Cellular Level

Basically, the comprehensive approach, as formulated in our investigations, is as described here. The various types of toxicity are studied, both at the organism level and at the cellular level, with one and the same approach employed at each: these levels have been designed to provide a comprehensive assessment of the toxic impact. In particular, reactions representative of the various system groups and trophic levels (general and acute toxicity) are analyzed at the organism-level, while structural and functional changes of the genome (geno- and cytotoxicity) are studied at the cellular level [41–45].

As a result of experiments carried out on substances of an organic and inorganic nature with a variety of test organisms [46–49], the following biotests were selected for analysis of toxic effects on the organism and its cells: plants—lettuce *(Lactuca sativa)* or bulb onion *(Allium cepa)*; invertebrates—hydra *(Hydra attenuata)*; and vertebrates—fishes, e.g., carp *(Cyprinus carpio)*, crucian carp *(Carassius auratus gibelio)* or other species. In the case of fish, we studied cells of the fin hem to minimize injuries to the fish and disruption to their physiological functions, and also to be able to periodically analyze the same animals [50].

The complex of cellular criteria includes the proportion of cells with micronuclei (registering structural disorders of the hereditary apparatus) and quantitative characteristics of the nucleoli (reflecting functional changes). In addition, taking advantage of the technical possibilities offered by the methods employed, determined the mitotic index (changes in the proportion of dividing cells as an indicator of cytotoxicity) and the number of cells with double nuclei and nuclear disorders (indices of genotoxicity).

On the whole, the more practical choice of ecotoxicological tests with respect to test organisms, test functions, and testing conditions is practically unlimited. For example, more than 120 biotests have been described in the literature for toxicological analyses of fresh water alone [51].

Although some such procedures have been recommended for practical testing, studies at the cellular and subcellular level have so far been largely confined to laboratory research and are not yet prescribed by any international or national standards.

Criteria for Assessment of Geno- and Cytotoxicity of Water Sample The following cytogenetic methods have been used most often for characterizing structural and quantitative variations in the principal components of the cell nucleus (chromosomes and genes), operating as carriers of genetic information: analysis of chromosomal aberrations, sister chromatid exchanges, damage to DNA (adducts, ruptures of DNA strands, COMET-test), fluorescence produced by in situ hybridization (FISH method), frequency of micronuclei and also changes in the quantitative parameters of nucleoli and nucleolus-forming chromosomal organizers.

In biological monitoring, a distinction is made between biomarkers of exposure, effect and susceptibility [52, 53]. Differences between them are not always clear-cut, especially between exposure and effect markers. Susceptibility biomarkers (differences in metabolism, restoration of DNA, changes in oncogenes, etc.) are mostly used to assess variations in individual reactions to the effect of genotoxic carcinogens [4]. Exposure biomarkers (concentration of a substance and its metabolites in biological fluids, mutagenicity of substances released, metabolic effect on hemoglobin and DNA, etc.) assess adsorption of a substance and, if possible, measure its internal and biological effective doses. Effect biomarkers (including most of the above-mentioned cytogenetic disorders, such as chromosomal aberrations, sister chromatid exchanges, micronuclei, DNA ruptures, and adducts, etc.) measure damage to the major genetic structures (genes and chromosomes). An association between the parameters measured and oncological diseases is most probable in the case of effect biomarkers, since genotoxicity is an indicator of the risk of cancer for humans [3]. This is also why we have paid particular attention to these biomarkers in this study.

Structural chromosomal aberrations arise as a result of direct DNA ruptures, replication of hereditary material on a damaged DNA template, and also several other processes. Quantitative chromosomal aberrations (aneuploidy and polyploidy) are caused by various disorders of the cell division and are registered by changes in the number of chromosomes.

The classical cytogenetic methods used to determine structural and quantitative disorders in dividing cells are supplemented by the fluorescence in situ hybridization (FISH) method, which enhances the efficiency and specificity of the identification of certain types of chromosomal aberrations [54–56].

DNA adducts are chemical structures bonded covalently with the DNA strand [57]. They are used, first of all, as biomarkers of the exposure to genotoxic substances. Most substances that are carcinogenic to humans are known to be genotoxic, but not all genotoxic substances can be carcinogenic. This is why DNA adducts determining many genotoxic substances are not interpreted unambiguously as precursors of oncological diseases. Thus, while there is epidemiological evidence indicating that DNA adducts may be indicators of the risk of cancer in humans [58, 59], other data show that DNA adduct levels in certain tissues are associated with exposure to negative factors but are not related to malignant growth in these tissues [60, 61].

Another method used to determine damage in DNA molecules is gel filtration of separate cells (COMET test) [62, 63]. In the process of electrophoresis, under neutral or alkaline conditions, cells applied to the stage of a microscope migrate in different ways, depending on the extent of damage of their DNA. The result of this is that the image seen under a fluorescent or nonfluorescent microscope looks like a comet the length of whose tail reflects the degree of DNA damage in the cell. Owing to the possibility of determining DNA disorders in cells prepared both in vivo and in vitro, this method has been used on an increasingly wide scale for biomonitoring in the interests of man [64, 65].

Micronuclei are small round-shaped extranuclear bodies that form in the process of condensation of acentric chromosome fragments or whole chromosomes excluded from the cell nucleus after completion of cell division. Formation of micronuclei may be caused by disorders of a variety of cell mechanisms. Thus, micronuclei carrying chromosomal fragments arise from direct ruptures of DNA strands, replication on a damaged DNA template or repression of DNA synthesis (clastogenic damage). Micronuclei comprising entire chromosomes form as a result of malfunctions of the spindle, kinetochore, or other parts of the mitotic apparatus (aneugenic damage). Consequently, an increased frequency of cells with micronuclei is a biomarker of genotoxic effects that may arise under the influence of clastogenic or aneugenic agents [3, 66–69].

The question of whether formation of micronuclei plays any significant role in carcinogenesis remains open to this day. In any case, micronuclei indicate a genomic instability [70].

Results of many investigations carried out using various methods proved convincingly that morphological characteristics of the nucleolus reflect the most important molecular-genetic processes and serve as an objective indicator of cell metabolism [18–21]. This relationship is quite understandable, since the nucleolus is the site of synthesis for molecules of ribosomal RNA and formation of preribosomal particles [71–73]. In addition to the synthesis of ribosomes, the nucleolus of a eukaryotic cell participates in a multitude of other processes of gene expression, which was reflected in the so-called multifunctional nucleolus theory [74].

The set of nucleolar parameters proposed for analysis, namely, the number of nucleoli in the cell, the size of an individual nucleolus and the proportion of cells with heteromorphic paired nucleoli, was based on the results of research carried out over many years. These studies were the most objective indices characterizing the biosynthetic activity of cells with low numbers of nucleoli (the most common nucleolar constitution of animal and vegetable cells) and at the same time reflecting various regulatory mechanisms [21].

In particular, formation of an individual nucleolus takes place through convergence of two functionally active nucleolus-forming regions [75]; the merging of nucleoli in an interphase nucleus is in strict correlation with the cell's metabolic activity [76, 77]; and the size of a nucleolus reflects the transcriptional activity of clusters of ribosomal DNA [78–81]. The proportion of cells with heteromorphic paired nucleoli characterizes quantitatively the specific regulatory-mechanism of paired nucleolus-forming regions in interphase nuclei [22]. The number of nucleoli in a cell corresponds to that of active centers of synthesis of ribosomal RNA and serves as a reliable indicator of the proliferative activity of cells [82–84]. Quantitative assessment of the function of nucleolus-forming regions in an interphase nucleus has been described as a new marker of cellular proliferation. This method is technically simple, quick and inexpensive, and readily lends itself to quantitative assessment [85, 86].

The methods mentioned there comply with the current requirements placed on the examination of water samples for quality: they determine their biological characteristics at the (sub)cellular level, register changes in the inheritance apparatus and objectively predict long-range consequences of their effects. However, for practical biotesting purposes (assessment of the various types of toxicity that may be caused by large numbers of known substances and compounds, quality monitoring of waste, natural, drinking water, etc.) most of the methods under discussion have not been taken up because of their high costs (primarily of the equipment) and the labor-consuming procedures involved. Only a few methods combine the requisite characteristics, i.e., relatively low cost and technical simplicity, on the one hand, and informative and comprehensive assessment of the various mechanisms on the other hand. As an optimal set for experiments on the determination of structural and functional changes of the cell genome as a result of exposure to toxicity, we propose the micronuclear test and the nucleolar biomarker.

The European Inventory of Existing Chemical Substances (EINECS) lists more than 100,000 substances. The presence and concentration of only 30–40 of these is regularly monitored in the major ecosystems of European countries. Both short- and long-term toxic effects of 80 % of "man-made" and most natural substances are yet to be studied. A significant proportion of substances cannot be reliably determined in natural and waste water because of lack of corresponding analytical methods or high analysis costs. Toxicological studies on humans, mammals and certain other animals are prohibited. These and other problems can best be solved at the cellular level, thanks to the universal nature of its organization.

Between 1989 and 1996, the Scandinavian Society for Cellular Toxicology sponsored a special program, entitled Multicenter Evaluation of Cytotoxicity in vitro, which was carried out by 31 laboratories throughout the world. Under this program, 50 reference substances (medicines, pesticides, salts, metals, etc.) were analyzed by means of 68 in vitro tests in order to select methods that could be accurately extrapolated to the human organism [37]. On the whole, cytotoxicity for human cells was well predicted by animal tests for cytotoxicity (R^2=0.64 for animal cell lines; R^2=0.62 for primary animal cultures) and slightly worse by ecotoxicological tests (R^2=0.57 for fish cell cultures; R^2=0.43 for small hydrobionts; R^2=0.39 for bacteria; and R^2=0.37 for plants).

The most acceptable methods for extrapolation to the human organism are those that evaluate mutagenicity and geno- and cytotoxicity, i.e., the (sub)cellular effects. This conclusion was corroborated by results of several international programs, such as Gene-Tox, the International Program on Chemical Safety (IPCS), etc., carried out in the 1990s. Even changes in cellular structures of plants, in particular onion *(Allium cepa)*, well predict genotoxic and mutagenic consequences for higher animals, including man [41–45].

Organoids in fish cells are used, e.g., in the submitochondrion test, to evaluate toxicity of water samples for humans [87]. Fish often react to toxic substances much like mammals, including man [88–91]. This is why fish are recommended for use as model organisms in determining potentially dangerous genotoxic substances in drinking water and for screening toxicants capable of inducing deformities and oncological diseases [69].

The micronuclear test on fish cells is one of the biomarkers most frequently used in determining genotoxicity of water samples. This approach has been widely employed both for evaluating biological effects of environmental pollution [92–95] and for testing a variety of chemicals for toxicity after direct or indirect exposure in vivo [96–99].

The importance of studying nucleolar characteristics and their changes under the effect of various factors ensues from the special role played by nucleoli in the vital functions of cells and the entire organism. Plenty of experimental data demonstrate that quantitative parameters of nucleoli and nucleolus-forming regions are useful prognostic and diagnostic indicators of oncological diseases of different types [100–105]. Nucleoli are also viewed as a key factor controlling the mechanisms of aging [106–108].

The approaches discussed in the present chapter are extremely important for this country, as far as their implementation is concerned, since practically no investigations have been undertaken in the Ukraine on geno- and cytotoxicity of natural and drinking water. The two notable exceptions were studies of the consequences of radioactive contamination of water as a result of the Chernobyl nuclear disaster and of genotoxicity of sea water.

At the present stage of social development, biomonitoring of natural and drinking water is of vital importance, as evidenced by ongoing investigations in many countries (Sweden, Spain, Italy, China, the USA, Japan, etc.) [93–95, 109–112]. In the Ukraine, biotesting of natural and drinking water has so far failed to receive proper attention. Here, problems awaiting solution include: algal bloom, i.e., vigorous growth of blue-green algae; genotoxicity of certain physiological products, which requires particularly detailed studies; the search for sensitive and at the same time technically simple biomarkers for evaluating various types of toxicity; biorehabilitation of polluted water, soils, deposits, etc.

Geno- and cytotoxicity and mutagenicity of substances and preparations used at various stages of the water treatment process must be explored further. The orientation to mainly chemical approaches in determining the quality of drinking water [113] is not quite justified from the point of view of the interests of human, its main consumer. Chemical methods can neither detect the whole complex of elements present in an aqueous solution or assess their interaction and transformation in the medium and organism. Biological testing (comprehensive use of optimal sets of both test organisms and cellular parameters) is a tool for objective characterization of the biological component of water quality. Biomonitoring of drinking water at the various stages of its preparation, purification, and disinfection must cover toxicological (acute and chronic toxicity), geno- and cytotoxic and mutagenic parameters. All the above types of toxicity must be controlled. In addition, use of biological criteria can make it possible to determine the optimum (i.e., the most suitable for the functioning of the human organism and its cells) composition of drinking water (salinity, etc.) for the population of a given region. In general, biological criteria for drinking water quality evaluation must conform to the prescribed chemical standards.

6.3 Problems of Quality with Bottled Drinking Water

Consumption of natural mineral drinking water packaged in plastic containers has been soaring in recent years, although doubts persist as to whether this new source of drinking water supply offers any advantages over the traditional ones. A comprehensive analysis by a number of parameters would be a convincing argument for bottled drinking water with an optimized mineral composition, provided it proved that such water really had properties beneficial for human health.

The quality problem of bottled drinking water is closely associated with their chemical composition, radiological safety, and the presence of microorganisms capable of propagation. Since water is packaged in containers for prolonged storage, the processes used for their treatment must not deteriorate their original quality, and they must not degrade in the process of storage.

Peculiarities of Chemical Composition of Bottled Water A major criterion for the quality assessment of bottled water is their chemical composition. In [114], 56 brands of bottled mineral water marketed in Europe were analyzed for 66 chemical elements. The water were found to vary significantly in mineral composition, with only 15 trade brands meeting all the requirements placed by the EU on drinking water. Differences in chemical composition between the individual countries and regions are due to special features of their geological structure, as well as the local taste preferences and recommended quality of mineral water.

Bottled water (mineral, from springs, etc.) retailed in Canada were analyzed for the presence of the most important inorganic substances [115]. Most of the 199 samples met the requirements of the national and international drinking water standards. Content exceeding the respective maximum permissible concentrations was registered for the following substances: boron (in 22 bottled water samples), aluminum (9), chromium (1), manganese (5), nickel (1), arsenic (10), selenium (24) and lead (1). Three samples of distilled water showed unexpectedly high content of copper and nickel ions (88–147 and 16–35 μg/dm^3, respectively), which was probably due to contamination in the process of distillation. Furthermore, in [116] examination of 40 Canadian and other imported water samples by methods of analytical chemistry revealed that some of the water samples failed to conform to the local standards in total content of dissolved solid substances and concentrations of chlorides and lead.

In [117], 182 samples of bottled water consumed in Canada were analyzed for content of volatile organic substances. Benzene (2 μg/kg) was detected only in one sample, and 20 brands were found to contain toluene (average concentration 6.92 μg/kg with a concentration range of 0.5–63.0 μg/kg), 23 brands contained cyclohexane (39.2 and 3–108 μg/kg) and 12 brands contained chloroform (25.8 and 3.7–70.0 μg/kg), and 4 dichloromethane (59 and 22–97 μg/kg). Cyclohexane, which migrates from plastic bottles, was found in 20 samples of spring water. The same samples were also found to contain ethanol, limonene, pentane and tetrachloroethylene.

Analysis of bottled water from Saudi Arabia for a number of metals (Cd, Fe, Hg, Ni, Zn) showed that in some cases their concentrations exceeded the international standards [118]. In [119], the Chilean authors stress the importance of control over the content of copper ions in bottled drinking water. In Spain, concentrations of aluminum ions in drinking water (tap, gassed, and non gassed mineral) varied from 4.2 to 165.3 μg/dm^3, with the average intake of aluminum with water reaching 156 μg per person per day [120].

The average content of fluorides in the gassed mineral water consumed by the inhabitants of Bahrain was 0.28 μg/dm^3 [121]; the authors insist that the chemical composition of bottled water must be checked in view of the high incidence of dental caries in that country. Examination of 65 brands of water bottled in the USA (including samples analyzed for fluoride content, fluoride-free brands and those which were not tested for fluorides) revealed that only 8 samples (12.3 %) had the optimal fluoride level [122]. For 25 commercial brands of bottled water (mineral, spring, and distilled) marketed in northeastern England, the average content of fluorides was 0.08 μg/dm^3 (0.01–0.37 μg/dm^3), whereas the recommended daily consumption of fluorides in England for children aged 4–18 was 0.26 μg/dm^3 [123].

Therefore, chemical analysis of bottled drinking water consumed in the various countries demonstrates the importance of controlling the content a large number of inorganic and organic substances to prevent harmful effects on human health. Deviations in chemical composition from the international norms and guidelines were registered for water bottled in highly developed countries (USA, Canada, Japan, Great Britain, etc.), whose quality control standards are higher than in Ukraine.

Radiation Safety of Bottled Water In 2003, the US Government's Food and Drug Administration (FDA) confirmed permissible limits of alpha- and beta-radioactive contamination of bottled water with Ra-226/228 at the levels prescribed by the US Environmental Protection Agency (EPA) for drinking water [124]. New restrictions were imposed on uranium: this element could be present in bottled water in concentrations not in excess of 30 μg/dm^3.

In paper [125], measurement of the combined alpha- and beta-radioactivity of 21 brands of Mexican and imported bottled water showed that they conformed with the restrictions established for drinking water in beta-activity (1.0 Bq/dm^3), while the alpha-activity exceeded the standard (0.1 Bq/dm^3) in three samples. In addition, it was found that there was a correlation between the water' radioactivity levels and their mineralization. Similar results were obtained in [126] for bottled mineral water from several regions of France, Portugal and Spain, with some of the samples exceeding the standards in alpha- and beta-activity. Elevated radioactivity was mainly related to high mineralization of ground water.

The radionuclide (Ra-226) was detected in nearly every brand of bottled mineral water commercially available in Hungary [127]. Levels exceeding 100 mBq/dm^3 were registered for 6 out of 28 test samples (in one case this value reached 3 Bq/dm^3).

Effect of Container Materials, Coming into Contact with Water and Storage Conditions, on the Quality of Bottled Water Polyethylene terephthalate (PET), because of its physicochemical properties, in particular strength and transparency, is widely used as a packing material for foodstuffs and drinking water. Under the existing regulations, a packing material must not release migratory substances in such quantities as may cause undesirable changes of the product. The controls include both general limits of such migration (60 mg/kg or 10 mg/dm^3), and limits for specific substances whose migration in large quantities may result in damage to human health. For example, the following special limits have been adopted for PET-derivatives (mg/kg): terephthalic acid—7.5, isophthalic acid—5, dimethyl ether of isophthalic acid—0.05, and ethylene glycol—30 [128]. It is known that PET-based materials may undergo a thermal degradation yielding acetaldehyde.

The concentration of acetaldehyde in bottles manufactured with polyethylene terephthalate (hereinafter PET bottles) is 6.3 mg per 1 kg of material, and washout into the water reaches 200 μg/dm^3 [129]. An elevated water temperature (40 °C and higher) significantly enhances the migration of acetaldehyde. The study of a mineral water produced in Poland and bottled in PET bottles showed that formaldehyde, acetaldehyde, and acetone were the most common carbonyl compounds determined in the water; in addition, it was found to contain propanal, nonanal, and glyoxal [130, 131]. The authors note that the concentration of aldehydes may exceed 200 μg/dm^3 and that their migration is influenced by temperature, exposure to solar light, duration and conditions of storage, and the concentration of carbon dioxide. An elevated CO_2 pressure and low pH values make for higher acetaldehyde concentrations in the water.

Migration of volatile products of PET degradation and high-density polyethylene bottles, polypropylene lids and ethyl vinyl acetate liners in ozonized water was measured in [132]. Compounds found in the ozonized water included butanal, pentanal, hexanal, heptanal, octanal, nonanal, 2,2-dimethyl propanal, 3-hexanone, 2-hexanone, and heptanone. The concentration of these substances increased with time of exposure to ozone. On the whole, the content of volatile migrating polymers did not exceed the FDA limits.

Migration of vinyl chloride from polyvinyl chloride bottles into drinking water increased in direct proportion to the storage time at a rate of 1 ng/dm^3 per day [133]. The total dosage of consumption of this monomer with drinking water may reach, in some cases, up to 100 ng per person per day. Under the FDA standards, the maximum daily intake of vinyl chloride for humans is 25 ng per person.

Quality Assessment of Bottled Water by Biological Testing Methods Use of cytotoxicity biotests as the most sensitive analytical methods for determination of most pollutants of the aquatic environment was first proposed as early as 20 years ago [134]. Today, the sphere of application of biotests is very wide: from surface and ground water to bottled mineral water. The authors recommend determining the rate of synthesis of RNA in human cells as a high-sensitivity test for the above-mentioned water.

The French scientists in [135] suggest that an integral assessment of health hazards involved in the consumption of mineral water after packaging and storage can be made on the basis of chemical analysis supplemented with certain cytotoxicity tests. The authors of [136] propose a new sensitive test for cytotoxicity that registers inhibition of RNA synthesis in the cell line of man's hepatoadenoma for assessing the safety of bottled natural water.

Biotesting of Bottled Water Contaminated by Substances Migrating from Plastic Bottles Mineral drinking water is often packed and stored in plastic containers for several months. Studies concerned with possible deterioration of water as a result of prolonged storage in plastic bottles are of special interest since they are important for human health. One line of research consists of analyzing drinking water for the presence of mutagenic, carcinogenic, or genotoxic compounds released from containers.

Bottles manufactured with PET were examined as possible sources of mutagenic contamination. After being filled with mineral water and stored for more than a month, the water tested positive in the Ames test, which is the most common method for detecting mutagenic activity [137]. Exposure to light during storage enhanced the activity.

In paper [138], a number of potentially genotoxic compounds (acetaldehyde, dimethyl terephthalate, terephthalic acid) were identified as substances migrating into water. The results of this migration, i.e., the concentrations of harmful compounds, did not exceed the EU and FDA limits. In this case, the Ames mutagenicity test produced negative results for nonvolatile migratory substances.

Natural mineral water that had been stored for several months in PET bottles were studied using two vegetable biotests for genotoxicity: a micronuclear test on spiderwort *(Tradescantia)* cells, and a test for anaphase chromosome aberrations on onion *(Allium)* cells [139]. The authors of that paper believe that the genotoxic effects (clastogenicity) found in the mineral water after storage was due to the presence of volatile mutagenic substances.

In [140], analysis of samples taken from PET and polyvinyl chloride bottles showed no statistically valid relationship between cytotoxic effects on cells of L-929 fibroblasts and unicellular protozoa *Tetrahymena pyriformis* on the one hand, and the storage time and the type of packing material on the other hand.

Two brands of mineral water stored in both PET and glass bottles under varying ambient conditions (temperature, exposure to light) were examined by using the onion biotest in which macro- and microscopic characteristics of the roots were measured [141]. Water packed in PET bottles induced cytogenetic changes irrespective of the storage conditions. Toxic effects manifested 8 weeks after bottling (which was within the recommended shelf life). The storage conditions of the water were important inasmuch as they altered the genotoxic properties of the samples; for example, the frequency of chromosomal aberrations was the highest after exposure to solar light.

In paper [142], the migration of substances from PET bottles into natural and gassed mineral water and stored for 1–12 months was monitored by using several mutagenicity and genotoxicity tests: a micronuclear test on vegetable cells and a comet analysis for DNA breaks in human leukocytes. The authors registered increases in the numbers of micronuclei in natural mineral water stored for 2 months. Breaks of DNA threads in human cells were found in most natural and gassed water. In addition, di(2-ethyl-hexyl) phthalate, a hepatocarcinogenic plasticizer, was found in mineral water after 9 months of storage in PET bottles.

Biotesting of Bottled Water of Different Mineral Composition A study employing biotesting methods on 14 types of natural mineral water in Italy, varying significantly in mineral composition, showed that the water types in question were not contaminated with microorganisms [143]. They were also nontoxic, as determined by the onion (*Allium cepa*) biotest, with the exception of the sample with the highest mineralization.

The toxicity of fluorides in water has been the subject of much recent debate. Opinions differ on the genotoxicity of these substances and their synergism or antagonism with certain mutagens. For this reason, an experiment was conducted in [144] on lympnocytes of human blood with determination of the frequency of sister chromatid exchanges (SCE). Persons who drank low-fluoride water (0.11–0.23 mg/dm^3) showed a higher frequency of SCE cells than those who consumed water with a higher fluoride content (1.0 and 4.8 mg/dm^3). The results of these experiments indicate that optimal fluoride content in drinking water is a safeguard against genotoxic effects.

Changes in the Mineral Composition of Drinking Water and Impairments of Human Health The data reported in the literature are sufficient to warrant a conclusion on a negative effect of nonoptimal concentrations of a number of substances dissolved in drinking water on the human organism. We will give only a few examples here.

An increase in the content of sodium ions in drinking water from 8 to 107 mg/dm^3 caused statistically reliable blood pressure rises in schoolchildren [145]. Increased concentrations of aluminum ions in drinking water were associated with a higher incidence of Alzheimer's disease [146]. While consumption of aluminum-rich water may be insufficient in itself to directly cause pathological changes, it stimulates processes responsible for the aging of the brain and neurodegradation. Most epidemiological studies reveal a statistically valid relationship between the aluminum content in water and the number of cases of Alzheimer's disease [147].

High iron contents in water produce hemochromatosis. Calcium concentrations significantly higher than those of magnesium cause serious health problems. Finland, where this proportion in the population's diet is extremely high (7:1), has one of the highest rates of heart diseases. An overdose of calcium causes cardiospasms, asthma, arteriosclerosis, headache, hypertension, cataracts, renal calculi, etc. Magnesium is a natural controller of calcium, which is why bottled water enriched with magnesium is used as a medication for kidney calculi.

Biotesting of Bottled Water Produced by Different Technologies It has been demonstrated in a number of studies that not all water treatment methods could remove compounds harmful for human health, e.g., mutagenic and carcinogenic substances [148]. Because of this, the authors recommend controlling the quality of the original and treated water by means of biotests, such as the Ames test. This test is more appropriate because, unlike standard methods of chemical analysis, it furnishes all necessary information.

Dibromoacetic acid is a subproduct of disinfection of drinking water by halogenation and is also found in water after ozonization. Four biotests were used to assess the genotoxicity of this compound in [149]. The results of the experiment showed that dibromoacetic acid increased the proportion of cells with micronuclei and caused breaks of DNA threads; that is, they exhibited genotoxic activity.

A bottled water (one brand out of 6 tested) consumed in Bangkok (Thailand) tested positive in the Ames test [150]. Its mutagenicity was $\sim 26\,\%$ of that of tap water (all the 17 tap water samples after chlorination was found to be mutagenically active).

Cytotoxicity of Bottled Water Contaminated with Microflora Although our purposes in conducting the present investigation did not include analysis of bottled water for bacterial contamination, it is known that bacteria, especially pathogenic, can deteriorate water quality by releasing toxic substances.

Heterotrophic bacteria occur as part of the natural flora in all aquatic environments. These bacteria go through propagation cycles in drinking water, especially in sealed bottles and other containers. The American investigators in [151] found that 95 % of heterotrophic bacteria occurring in drinking water were characterized by a low invasive activity and cytotoxicity for human erythrocytes. Only a small proportion of these bacteria was capable of invasion or could produce cytotoxic effects.

Activity of several strains of the bacteria *Aeromonas* was studied in 61 brands of Italian bottled mineral water samples [152]. Of this number, six were found, by means of a biochemical test, to have cytotoxic properties.

Biotesting of bottled water makes it possible to obtain an integral quality assessment, to summarize and generalize deviations in contents of all organic and inorganic substances and radioactive elements dissolved in an aqueous medium and to take into account the effect of toxic compounds released by pathogenic microorganisms. Even though biological methods still represent a relatively recent line of research, they have been instrumental in revealing the most serious problems associated with the quality of drinking water packaged in containers. This approach enables one to obtain more objective and comprehensive information on the quality of water than, e.g., its analysis for chemical composition. Consequently, biotesting methods, especially at the cell level, must be included among the criteria used for quality assessment of bottled drinking water.

Based on analysis of the literature reports describing the latest scientific achievements in studying the quality of bottled water, it may be concluded that this type of drinking water requires thorough testing for chemical composition and radiation safety as well as possible toxic, cytotoxic, genotoxic, and mutagenic effects; inspection of containers used for water packaging and strict observance of shelf lives and storage conditions are other implications of such studies.

6.4 Assessment of the Quality of Bottled Drinking Water by Biotesting Methods

Biomonitoring (biotesting) of natural surface water used as sources of drinking water supply has been extensively used in many countries as a means of obtaining integral water quality evaluations [31, 34, 51]. Use of ecotoxicological biotests (plant and animal test organisms) and cellular biomarkers is extremely important for objective and comprehensive control over the increasingly numerous xenobiotics that contaminate drinking water. Although most of these xenobiotics are disregarded by the existing standards, they are nevertheless capable of producing a variety of toxic, cytotoxic, genotoxic, or mutagenic effects [152, 153].

Some international water quality standards, in particular ISO 7346, ISO 6341, and ISO 10706, provide for biotests with fish and crustaceans. Based on these standards, the Ukraine has adopted a number of corresponding standards, e.g., DSTU 4074-2001, DSTU 4075-2001, DSTU 4076-2001, DSTU 4173-2003, and DSTU 4174-2003. In Russia, biotesting methods have been employed, together with the physicochemical approaches, in defining water quality requirements. The implementation of these methods form mandatory standard procedures for the organizations concerned [154]. In the Ukraine, several procedures involving use of test organisms of the various trophic levels and systematic groups (bacteria, protozoa, algae, crustaceans, and fish) have been approved for determining acute and chronic toxicity of natural and waste water [23–29].

Recent years have witnessed a boom in the consumption of natural water packaged in containers, including in this country. Deterioration of bottled water quality has been associated with changes in chemical composition and both radioactive and microbiological contamination, as well as storage time, storage conditions and flawed water conditioning processes [155].

International standards for packed drinking water, such as the Codex Alimentarius and European Union directives, provide for hazard analyses, including analyses for contents of toxic substances, by applying the principles of the Hazard Analysis and Critical Control Points (HACCP), a system designed to determine specific hazards and respective preventive controls [156]. Essentially, the HACCP system establishes a direct relationship between the quality and safety of a product.

Based on investigations carried out over many years, the authors propose the optimal procedure of drinking water quality assessment. This includes obligatory experiments (to be conducted using methods of analytical chemistry, microbiology, radiology, and biotesting) for determining the minimum hazard of a water sample for human health, i.e., for verifying a water's safety via different methods. The same procedure is to be used for determining the physiological value of drinking water. Water that, on the one hand, is free from toxic substances and, on the other hand, contains an optimal set of elements needed to maintain the functional activity of the human organism and its cells thereby meets the requirements placed on quality drinking water.

A comprehensive analysis through biotesting methods indicates the absence of any toxic substances harmful for a live organism and its cells in a water type. Biotesting, furthermore, provides convincing evidence of the safety of a bottled water brand as regards its chemical composition.

Our recommended set of cellular parameters includes the proportion of cells with micronuclei (registration of structural irregularities in the cell's hereditary apparatus) and quantitative characteristics of nucleoli (reflecting changes in the cells' metabolic activity). A special feature of our approach is performance of cytological studies on cells of the same test organisms which are used for evaluating general toxicity. In this case, onion (plant), hydra (invertebrate), and the fish (vertebrate) were used.

Cytotoxicity of water samples was determined on cells of the plant and animal test organisms by a nucleolar biomarker in accordance with the procedure presented in [157]. Genotoxicity of the aqueous media was assessed by micronuclear tests on fish cells by the methods described in [69] and on onion cells thereafter [158]. Micronucleus counts were made according to the recommendations given in [159].

Universality of the cell organization opens wide possibilities for toxicological studies involving various groups of animals and plants to be followed by extrapolation of the results to human cells and the human organism. This is why use of cellular biomarkers, whether alone or in conjunction with the traditional organism-level methods, can be extremely useful at the present stage of water quality assessment.

Quality Assessment of Bottled Water by Means of Plant and Animal Test Organisms The 12 brands of packaged drinking water consumed in the Ukraine have been tested on the organism level. Water samples were classified by toxicity as follows: a sample was rated *nontoxic* if the deviation from the standard limits did not exceed 20 %, *slightly toxic* if this was 21–40 %, *medium-toxic*—41–60 %, *very toxic*—61–80 % and *extremely toxic* if the deviation was from 81 to 100 % [160].

This classification is the most suitable for characterizing the toxic properties of natural water, including drinking water, such as those with low contamination levels.

Using biotests with ceriodaphnias, we assessed lethality for these animals after 48 h of exposure in experiments on acute toxicity as well as the mortality of females and the rate of propagation by the increment of young animals after 7 days of experiments on chronic toxicity. In general, out of seven brands of water studied, (58 %) displayed toxic effects for this type of crustacean. Chronic toxicity was registered in 6 water types (50 %) with three brands (25 %) showing slightly toxic effects. The other five water (42 %) were nontoxic by the ceriodaphnia biotest.

Biological testing with the other invertebrate, hydra, was done according to the proportions of lethality (mortality) and sublethality (specific changes in the body morphology) and also in terms of the animals' propagation rate. After incubation of the animals in test solutions during a 96 h period, acute toxicity was not found in any of the bottled water brands. After 3 weeks of exposure, chronic toxicity was registered in 12 brands (100 %) with the following effects: medium-toxic 17 %, very toxic 25 % and extremely toxic 58 %. The two brands designed for baby foods exhibited slight sublethal/lethal effects and produced animal growth rates that were 54.9–61.6 % lower than the control figures. On the whole, the chronic toxicity index for hydras was the most sensitive of all our test functions.

In tests with a vertebrate animal, the aquarium fish guppy, none of the brands of bottled water exhibited toxicity as defined in the analysis criterion, which was the proportion of deaths during 96 h. For fish cultivated in five distinct water brands, the mortality rate did not exceed 10–20 %. On the other hand, all members of the other species, goldfish, used for cytogenetic tests, died in one water sample within 1 h. Thus, 11 water brands (92 %) were found to be nontoxic by the fish biotests, and one brand (8 %) was extremely toxic (the international test ISO 7346 permits use of different fish species in the determination of acute toxicity).

In the plant biotest using onion, the degree of toxicity of water samples was determined by measuring the length and weight of the roots. These two indices both decreased and increased in the experiment in which onions were grown in 12 samples of the bottled water. The more sensitive parameter was the root weight by which slight toxicity was exhibited by three of the brands. This reduced the root weight by 27.0–32.9 %. The other 9 brands (75 %) showed no phytotoxicity and had no negative effect on the development of the plants. It should be emphasized that, of all the plant biotests, this is the test whose results can be extrapolated to the human organism with the greatest accuracy [161].

Assessment of Water Samples for General Toxicity Using a Set of Biotests By using a battery of biotests it is possible to make an objective assessment of the effect of a water sample on animal and plant organisms. A general toxicity index ($I_{g.t}$) which represents the sum of the effects of all biotests, was devised for the quantitative interpretation of this assessment [25]. The maximum value of an individual test function is limited to 100 conventional units, but in a chronic experiment the maximum value is 50 units since its effects indicate a lower level of water toxicity than that produced by acute impact. This value was selected in view of the fact

Table 6.4 Comparison of animal and plant test organisms and their test functions in respect to the sensitivity of drinking water quality

Test organism	Test function (exposure)	Registered effect and extent of its manifestation	Rank
Ceriodaphnia	Female mortality (48 h)	1ET+11NT	8
Ditto	Female mortality (7 days)	1VT+1MT+1ST+7NT	6
Ditto	Increment (7 days)	1ET+1MT+4ST+5NT	4
Ditto	Female mortality and increscent of young animals (7 days)	1ET+2MT+3ST+5NT	5
Hydra	Sublethal and lethal effects (96 h)	12 NT	10
Ditto	Increment (8 days)	1ET+5VT+3MT+2ST+1NT	2
Ditto	Increment (12 days)	1ET+5MT+5ST+1NT	3
Ditto	Sublethal and lethal effects (21 days)	7ET+3VT+2NT[a]	1
Fish (guppy)	Mortality (96 h)	12 NT	10
Onion	Root length change (96 h)	12 NT	10
Ditto	Root weight change (96 h)	3ST+9NT	7

NT nontoxic samples (deviation from control less than 20%), ST slightly toxic (21–40%), MT medium-toxic (41–60%), VT very toxic (61–80%), ET extremely toxic (81–100%)
[a] No increment was registered

that chronic toxicity tests are conducted in case of absence of a 50% toxic effect in an acute experiment (EC/LC_{50}). The number of species involved in an experiment automatically determines the highest $I_{g.t'}$ value. In the present investigation with one plant and three animal biotests, the maximum value of the index could not exceed 400 units.

Sensitivity of the Various Test Organisms to Water with Low Degree of Contamination One of the most important characteristics of a biotest is its sensitivity, i.e., the number of reliable responses to the effect under study and the extent of their manifestation. In the case of our bottled water the sensitivity of the plant and animal test organisms and their test functions (Table 6.4) differed as follows:

- the least sensitive biotests included the plant, vertebrate, and invertebrate test organisms with the test function assessing acute toxicity during 48–96 h of exposure (9–12 nontoxic samples per biotest);
- higher sensitivity was exhibited by ceriodaphnia with the test functions (female mortality rate and increment of young animals) used in chronic toxicity experiments lasting 7 days (5–7 nontoxic samples);
- the most sensitive test organism was the hydra with the set of test functions (increment of young animals, and sublethal and lethal effects) determining the medium toxicity during 8, 12, or 21 days (all samples tested toxic).

Thus, the results of studies indicate that in the chronic toxicity experiments with test functions characterizing the propagation and survival rate of the test animals, the invertebrate organisms were found to be the most sensitive to the quality of drinking water packaged in containers, i.e., water with low contamination levels.

Sensitivity of Test Organisms and MPCs of Chemical Substances In assessing the results of biotesting bottled water, it should be noted that in most cases test organisms used in laboratory studies do not detect chemical substances in the respective MPCs established for drinking water (i.e., their test functions do not show significant responses). For example, for organic compounds metolachlo, lindane, nonyl phenol, aldrin, 2,4-dinitrophenol, DDT and pentachlorophenol, the MFC varies between 0.1 and 2 μg/dm^3; it is somewhat higher for ions of heavy metals: mercury—1, cadmium—5, arsenic and chromium(VI)—50, and copper and zinc—100 μg/dm^3 [35, 36]. The results obtained with animal and plant test organisms in model solutions exceeded the above values in most cases by one or two orders of magnitude, which indicated the concentrations in question at the mg/dm^3 level [48].

These data lead us to several conclusions. First, for an objective quality analysis of a drinking water, the sensitivity threshold of biotests must be raised. The desired effect can be attained in a number of ways; namely, it can be achieved by combining several test organisms with different responses to toxic substances in a standard analytical battery, by conducting tests at both the organism and the cell levels (with a mandatory analysis of the functional parameters, which are the most sensitive indicators) or by using preconcentration of natural water to increase the contents of dissolved substances per test sample volume unit.

Secondly, the toxicity of an aqueous solution, as determined by means of biotests may exceed, sometimes appreciably, the MPCs for the individual substances. As a precaution, water found to be toxic by biotesting should not be consumed pending a more detailed analysis. It must be emphasized that, while biological methods determine the integral toxicity index of a water sample, the combined effect of different substances in solution can be synergic (more toxic than the sum of effects of the individual substances), additive (the toxic effects are simply added), or deductive (some toxic effects offset or mask the others).

Let us now make a comparative assessment of toxic properties of the 12 brands of the bottled water samples, as determined, on the one hand, by chemical analysis and, on the other hand, by biotesting, to answer the following questions: did the registered deviations in chemical composition produce matching changes in the biological test functions? Or were the frequency and degree of the chemical deviations inconsistent with their biological manifestations? The answer to the latter question would be positive if chemical analysis failed to show the presence of toxic substances that could have produced the biotest responses.

The standard ammonium toxicity values (LC_{50}, 96 h of exposure) are 71.1–128.2 mg/dm^3 for guppy [162] and 2.7 mg/dm^3 (LC_{50}, 48 h) for ceriodaphnia [163], which is considerably higher than the maximum figure of 0.68 mg/dm^3 for the 12 test water samples. For newborn ceriodaphnias, the lethal concentration of lithium is 4.0 mg/dm^3 [164], whereas the highest concentration registered in the analyzed water samples was just 0.072 mg/dm^3. The lethal concentration of sodium ions (by sodium carbonate) for the goldfish is 500 mg/dm^3 [165] and for the ceriodaphnia (LC_{50}, 48 h), 1020 mg/dm^3 [166], which is much higher than our maximum figure of 246 mg/dm^3, registered in one brand. The concentration of silver ions (by silver chloride) that produces a 50% death rate in fish of the carp family [167] and in crustaceans, such as daphnias and ceriodaphnias [168], it is measured in mg/dm^3, which is many times higher than our highest silver content, found in one water brand.

Thus, it is unlikely that the deviations found in the 12 water samples among the 40 physicochemical indices analyzed were the only cause of the toxic effects.

Water Mineralization and Results of Biotesting In view of different mineral content of the bottled water, which in these studies varied between 98 and 775 mg/dm^3, as well as other variations in salt content, we should discuss the possible effect of this factor on results of the biological testing. It is known that desalted and, in particular distilled, water adversely affects the vital functions of an organism and its cells. Biotests revealed decreases in the propagation rate of the crustaceans, a reduction of the vegetable root length, and growth of the sizes of nucleolar structures [169]. The biological inferiority of distilled water can be significantly remedied by adding a minimal set of salts. Such media are often used for cultivating test organisms, such as ceriodaphnias and hydras, under laboratory conditions.

At the same time, increases of the mineralization and hardness of an aqueous solution may reduce the toxicity of xenobiotics dissolved in it through bonding, complexation and other processes. Therefore, biotesting results must be corrected for this in cases where the salt content differs substantially from the optimum composition.

However, the mineral content variation range of the water in question is not large enough to suggest any significant effect of the salt composition on the results. This assumption is also corroborated by findings on the basis of which the various brands of water were rated safe or unsafe though they differed in mineral content and hardness. Biotests revealed toxic effects over the entire salt content range of the bottled water. Thus, the toxic, cyto-, and genotoxic properties of the drinking water in this investigation were determined by causes other than their mineral content.

Assessment of Cytotoxicity of the Bottled Water by Using Plant and Animal Cells The quantitative characteristics of nucleoli selected for test functions exhibited different sensitivity to external factors in cells of the plant and animal test organisms. For example, cultivation of hydras in different samples of the bottled drinking water changed the sizes of nucleoli by an average of 15.0–38.0 % (the most sensitive index), the proportion of cells with heteromorphous paired nucleoli by 0.6–15.3 % and the average number of nucleoli in a cell by 1.1–6.5 % (the least sensitive parameter). In fish cells, these figures varied between 5.0 and 52.5 %, 2.4 and 10.9 %, and 0.7 and 5.8 %, and in plant cells between 3.0 and 20.3 %, 0.7 and 20.8 %, and 1.1 and 6.2 %, respectively.

The results of investigations show that exposure to the bottled water produced insignificant increases in nucleolar anomalies in hydra cells.

In cells of vertebrate test animals, the fish, registered (for all studied brands of water) less significant deviations from the quantitative characteristics of nucleoli than the effects observed in the invertebrates' cells. In responses of the nucleolar biomarker to the quality of the bottled water cells of vertebrates (more highly organized animals) were found to be less sensitive or more resistant than those of invertebrates (animals of a lower level of organization).

The plant cell division is assessed by the mitotic index. As in animal tests, the bottled water under study produced only slight increases in the proportion of plant cells with damaged nucleolar structure. Quantitatively, deviations from the control were less significant in the plant cells (−0.8–13.0 % or an average of 5.5 %) than in the animal ones (−0.7–18.2 %; 8.4 %), although the natural level of morphological anomalies of nucleoli in onion cells is much higher than in hydra cells.

Thus, those bottled water samples which exhibited the most pronounced toxic properties at the organism level in most cases also markedly affected the structure and functional parameters of nucleoli at the cell level. Considering the fact that changes in nucleolar activity were registered during the first hours of exposure and toxic effects after four or more days, it would be safe to infer that changes in the vital functions of cells produced by exposure to the aqueous medium were the underlying cause of the subsequent negative impact on the whole organisms.

Assessing Genotoxicity of the Bottled Water by Means of Plant and Animal Cells Study of the genotoxocity of drinking water is very important for an objective assessment of the effect the water in question, or, more specifically, xenobiotics dissolved in it, may have on the health of the population and future generations.

While it is known that genotoxicity does not necessarily cause cancer, it has always been regarded as one of the major carcinogenic factors.

The chromosome aberration index and the micronucleus frequency, as well as one other parameter, the frequency of cells with double nuclei reflecting anomalies of the formation of cell walls in the process of mitotic division, did not increase after cultivation of onion in all brands of the bottled drinking water. Thus, the test water produced no genotoxic effects on the plant cells.

After 4 days of fish cultivation in samples of the bottled water, the percentage of cells with nuclear anomalies in fin tissues increased in several cases. A statistically reliable difference from the control figures was obtained only for one water type, in which the proportion of cells with double nuclei rose. On the whole, the genetic apparatus of the animal cells was found to be more sensitive to impurities contained in the test water than the plant cells. One reason for this may be a structural peculiarity of the plant cell, namely the presence of a cellulose shell that affords additional protection against penetration of genotoxic substances into the cell.

Generalized Characteristics of the Biotests Included in the Test Battery The principles of selecting biotests for determining the various types of toxicity of aqueous media were set forth previously [152]. With the results of the present investigation, we evaluated the efficiency of each individual biotest in examining the drinking water quality.

Biotest with Ceriodaphnias As shown in the data of previous studies, the ceriodaphnia is the most sensitive test organism of this test battery. In experiments with bottled drinking water, the ceriodaphnia test revealed acute toxicity in only one of the brands. At the same time, in the chronic toxicity indices (mortality rate of females after a week of cultivation, numbers of newborn animals), obtained differentiated quality characteristics of the test water that included distinct, moderate, slight, and nontoxic effects. Thanks to its simplicity and high sensitivity, the test with the ceriodaphnia (as also with another crustacean, the daphnia) is one of the world's most widely used biotests. It has been included in both international and national standards, and in many countries it makes part of the standard set of tests prescribed for testing water samples for toxicity, particularly for drinking water quality assessment for acute and chronic toxicity.

Biotest with Hydras Hydra, the other invertebrate in this test battery, did not reveal any acute toxicity in tests with the bottled water samples. At the same time, it was found to be the most sensitive test organism in chronic toxicity experiments in which the test brands produced varying decreases of hydras' propagation rates and increases in sublethality and lethality. Despite the fact that the hydra biotest is not as commonly used as those with daphnias and ceriodaphnias, it still offers certain advantages, such as differentiated responses to toxicity through a variety of morphological changes. In comprehensive biotesting, hydras can determine acute and chronic toxicity, and their cells can be used for a cytological analysis. It is because of the latter convenience that the hydra biotest is recommended for inclusion in the standard test battery. This is because cytogenetic experiments with daphnias or ceriodaphnias are technically complex. We developed a chronic toxicity hydra test that promises to make the hydra a useful tool in drinking water quality assessment.

Biotests with Fish (Guppies, Goldfish, etc.) The fish represented vertebrates in this set of biotests. Even though the fish were found to be the least sensitive of test organisms, they proved useful in acute toxicity tests. Like the (cerio)daphnia biotest, the fish test has been incorporated into the international and the national standards and has been used in many countries for biotesting water samples as part of the standard procedures. In this test battery the fish represented the principal source of animal cell material for cyto- and genotoxicological studies.

Onion Biotest As the only plant used in this set of biotests, onions offered additional possibilities for evaluating the variety of effects and the specificity of substances contained as impurities in aqueous media. The onion test was worked out in detail [41] and has been in considerable use in both organism- and cell-level studies in many countries. A number of reports substantiated the validity of the extrapolation of results obtained with onion for the macroscopic (root length and weight) and the microscopic (mitotic index, chromosome aberrations, occurrence of cells with micronuclei) indices to the organisms and cells of mammals, including humans [161, 170]. In current experiments with the drinking water the plant cells were less sensitive in determining genotoxic effects than the animal (fish) cells.

Nucleolar Biomarker Numerous studies conducted using a variety of methods have provided convincing evidence that morphological characteristics of the nucleolus reflect extremely important molecular–genetic processes and are an objective indicator of cellular metabolism. This relationship can be easily explained by the fact that the nucleolus is the place of synthesis of molecules of ribosomal RNA and formation of preribosomal particles [72, 171]. In addition to the synthesis of ribosomes, the nucleolus of a eukaryotic cell takes part in a multitude of other processes of genetic expression, which was reflected in the conception of the multifunction nucleolus [172]. The quantitative characteristics of nucleoli and nucleolus-forming areas can serve as useful prognostic and diagnostic indicators of various cancers [100, 101, 103]. Also, nucleoli are viewed as a key factor controlling the mechanisms of aging [107, 108].

We selected the set of nucleolar parameters proposed for analyses on the basis of studies completed over many years, and believe it to be the most objective indicator characterizing the biosynthetic activity of cells. We also postulate that, at the same time, it reflects the various mechanisms of its regulation [19]. And also view it as an important indicator of changes in the functional activity of cells.

Micronuclear Test This is one of the most commonly used methods of determining substances that exhibit genotoxic properties. The literature data [14, 15] indicate that the micronuclear test is not inferior to the chromosome aberration tests and may sometimes surpass them in richness of information and analytical rapidity. Its other advantages include fewer artifacts, a more accurate registration of data, a lower labor-intensiveness and a higher level of efficiency.

The responses of fish to the micronuclear test are similar to those of mammals, including human. This is why it is recommended that fish be used to screen substances which are potentially dangerous for humans and can cause deformities and cancer. Therefore, fish can be used as watchdogs looking out for genotoxic impurities that find their way into drinking water [69].

Thus, the proposed procedure for a thorough quality assessment of drinking water by using a set of plant and animal test organisms and registering the structural and functional parameters of their cells provides for an objective and comprehensive characterization of the various types of toxicity (acute, chronic, cyto-, and genotoxicity) that may be caused by pollution of water supply sources, unsatisfactory water treatment technologies, use of xenobiotics, the effect of the containers, storage conditions, etc.

The comparison between the bottled water produced by different technologies by the Ukrainian and foreign companies and the water designed for preparing baby foods did not reveal, by the results of biotesting, any substantial differences. Some of the individual brands were included in groups of both safe and less safe water. Consequently, the toxicological properties of test water do not depend on their production conditions, whether in this country or abroad, or on the more stringent requirements placed on infant foodstuffs.

The results of our experiments indicate that bottled drinking water that has been produced in conformity with the approved standards and is toxicologically safe according to the data of chemical analyses may still produce a harmful, negative effect on the cells of animal and plant organisms. Consequently, use of the methods of analytical chemistry, microbiology, and radiology prescribed by the national standards and regulations may be insufficient for determination of toxicity of aqueous solutions. For an objective water quality assessment, it is necessary to use, in addition to the standard procedures, the biotesting methods.

Quality assessment of drinking water by means of live organisms and cells implies determination of their possible toxicity and, consequently, potential hazards for the human organism. We do not recommend drawing conclusions regarding the physiological adequacy of water solely on the basis of data indicating the absence of the various types of toxicity (acute, chronic, cyto-, and genotoxicity) and the fact that it can be rated safe by the results of biotesting.

Our studies have demonstrated that biotesting with use of both test organisms and cellular biomarkers represent an effective approach to the integral quality assessment of drinking water. Biotests reveal the toxic properties of an aqueous medium and differentiate the various brands of bottled water by the degree of hazard to human health.

Generalized Characteristics of the Test Brands of Bottled Drinking Water Water intended for drinking must not show any reliably determined deviations from the control parameters of biotesting test functions. In other words, an aqueous medium must not produce any toxic, cytotoxic, genotoxic, or mutagenic effects, or seriously impair the vital functions of an organism and its cells. This requirement stems from the most important characteristic of drinking water: "safety in chemical composition," meaning that the concentration of harmful substances in the water must not exert any direct or indirect influence on a person's health during his or her lifetime or on the health of future generations. In the absence of detrimental effects, with the results of comprehensive biotesting, the water quality shall be pronounced to conform to the relevant biological standards. In case of reliably determined deviations from the control figures, the water in question shall be rated toxic by its biological properties.

Based on the results of comprehensive biotesting using a battery of plant, invertebrate and vertebrate test organisms, we have proposed the following quality classification of drinking water by degree of toxicity, as defined by the general toxicity index. If the $I_{g.t}$ value does not exceed 50 units, i.e., if none of the organisms detects acute or chronic (with extremely toxic effects) toxicity, then the drinking water shall be included in the category of *safe* water.

If values of the general toxicity index vary between 50 and 100 units, i.e., if none of the test organisms shows a distinct response to acute toxicity and if chronic toxicity with a maximal effect is not indicated by any two out of the four biotests making part of the test battery, the water in question shall be placed in the category of *relatively safe* water. Therefore, of the 12 test brands of bottled drinking water these groups of relatively safe water the largest one (58 %).

The $I_{g.t}$, value ranges from 100–200 units if the biotesting reveals acute toxicity in at least one biotest or if chronic toxicity has been found in several test organisms. In this case, the water that is analyzed shall be rated as *unsafe*. In our investigation, three brands matched this definition.

If $I_{g.t}$, exceeds 200 units, i.e., if most of the biotests reveal acute and chronic toxicity, the drinking water shall be placed in the category of *harmful* water posing serious hazards to human health. In this study none of the water exhibited such toxic properties.

Water brands that are the most dangerous for consumption are those combining well-pronounced toxic properties with genotoxic effects. One of test brands had a relatively high toxicity index and produced genotoxic effects on animal cells.

We have conducted a comprehensive evaluation of toxicity degree of 53 marks of bottled drinking degassed water and divided them into four categories: safe, unsafe (reversible toxicity), dangerous, and very dangerous toxic water. Acute toxicity appeared at the level of organisms, whose death did not allow us to assess cytogenetic deviation (see Table 6.5).

6.5 Human Health and Chemical Composition of Drinking Water

Opinions differ as to the possible effect of desalted water on human health. For example, popular scientific sources contain contradictory data on both the benefits and hazards of desalted water purified by the various methods, like distillation, ion exchange, or reverse osmosis. On the one hand, Paul Bragg [173] and his followers strongly recommend drinking distilled water to improve health. On the other hand, some popular scientific and internet publications argue that distilled water (or water purified of salts and microelements) has a negative effect on human health. In particular, consumption of such water results in an unbalanced functioning of the organism, increased fragility of bones, susceptibility of teeth to caries, heart diseases, disruption of many physiological functions (from pulse conduction of nerve cells to cell reproduction and transmission of hereditary information) and reduced general resistance of the organism.

Drinking water characterized by both a deficit and an excess of biogenic elements produces negative effects on human health. In particular, calcium deficiency has been found to cause more difficult clinical cases of cardiovascular diseases with a higher rate of lethal outcomes, development of rickets in children and disruption of blood coagulation processes. A deficit of magnesium has been related to exacerbated cardiovascular diseases and sudden deaths of infants; shortage of iodine increases the incidence of diseases of the endocrine system; selenium deficiency is associated with reduced immune stability of the organism, cardiovascular pathologies, oncological diseases, etc. [174].

According to the Nutrition Institute of the Russian Academy of Medical Sciences, the deficits of supply of biogenic elements with food are (%): calcium—30–40, iodine—80 and selenium—80–100. This makes it necessary to compensate for these shortages by consuming these biogenetic elements with drinking water. The effect of chemical composition of drinking water on the population's health and the incidence of diseases have been corroborated in many investigations. As a result, Russia's chief sanitary inspector issued an order "On the Adjustment of Drinking Water Quality in Regard to Contents of Biogenic Elements" [174]. This order recommends that measures be taken to supply the population with natural drinking water and optimal content of biogenic elements.

An indirect negative effect on human health may be produced by drinking water desalting equipment. For example, it has been found that a variety of bacterial colonies, e.g., *Pseudomonas, Acinetobacter, Chromobacterium*, etc., are developed in domestic water filters [175]. Consumption of such water increases the incidence of gastric disorders [176].

It has been suggested that the negative effects of the consumption of distilled and similar water may be remedied by enriching such water with a complex of salts and microelements. But even in this case, artificial water is not a full-value substitute for the natural one. From the viewpoint of human health, it is believed that the optimum solution of the problem under discussion here is natural water with the necessary complex of vitally important elements.

6.5.1 Use of Cytological Biomarkers in Fish to Assay Antropogenic Contamination of Sea and Fresh Water

Due to the increasing anthropogenic impact on sea and fresh water, the detection of its contamination is one of the most urgent tasks of today. Fish are the most convenient object for examination of anthropogenic contaminant impact on water because they usually respond to toxicants similarly to the way the higher vertebrates do and can thus be used to identify substances which can potentially produce toxic or carcinogenic effects in humans. Besides the total toxic impact on biota, fish can be used as a model for determination of potential genotoxic impact of harmful substances contained in water. Evaluation of anthropogenic contamination of sea and fresh water was performed by using standardized toxicity assay methods and a micronuclear test.

As a rule, the increase in the number of cells with micronuclei and division pathology is accompanied by suppression of mitotic activity. This can be preconditioned by depressed viability of cells with micronuclei [153, 177]. Thus, it is revealed that high concentrations of toxic compounds ruin the linear structure of dependence of the number of micronuclei on the dose of mutagen. Toxic doses can inhibit cell division and their death along with the decrease of cells with micronuclei [178].

The purpose of this work [179] was the investigation of the cytological biomarkers of genotoxicity in fish as a response to the anthropological contamination of sea and fresh water, so these tests can later be used in the monitoring of certain water sites. The cytological indicators were assessed on the fish *Mugil cephalus* and *Carassius auratus*, obtained at representative stations in the Saronic Gulf near Athens (Greece) with high and low level of anthropogenic contaminate of the sea and from the rivers Desna and Dnieper. Samples of the blood and gills were taken. The blood and gills were used to produce cytological preparations which were applied to determine the frequency of origination of cells with micronuclei and geminate nuclei in order to determine the level of genotoxic influence of substances contained in the water.

The working material included 32 specimens of *Carassius auratus gibelio* and 30 specimens of *Mugil cephalus*. Fish *Carassius auratus gibelio* were received from the research fishing company in Kiev region. The fish's adaptation to the new living conditions lasted for 30 days. The sea fish *Mugil cephalus* were caught during an experiment in the Saronic Gulf with low and high levels of anthropogenic contamination of the sea. The fish *Carassius auratus gibelio*, weighing 15–20 g, length of the body 80–100 mm were put into containers with the river and reference water. Water was continuously saturated with air to avoid hypoxia. Tissue samples for cytological assay were taken after 5 days of incubation. In order to appraise the impact of natural water on the number of cells with altered genome, samples were taken from gills and blood of each specimen. Blood samples were taken from the tail vein and prepared samples were preserved in 96% ethanol for 30 min and stained by the method of Romanovskii-Gimzsa. Gills' fragments were fixed with a mixture of glycerin, acetic acid, and water (1:1:10). The tissue replicas obtained were stained by the Romanovskii-Gimza method. The presence of cells with micronuclei and geminate nuclei were then determined. The preparations were assayed under a light microscope at x1000 magnification. At least 6000 cells were examined in each preparation. Statistical data treatment was performed by using Student's t-criterion and $p<0.05$ was assumed as statistically significant.

Blood cells with altered genetic apparatus were found during study of anthropogenic contamination of the river water. The number of such cells compared with the reference number was one order higher (1.75 versus 0.17 micronuclei per 1000 cells respectively in the blood of fish who were incubated in the Dnieper water samples and in the laboratory water, $p<0.01$, $n=14$). The same observations were registered for the number of geminate nuclei cells (0.33 versus 363 per 1000 cells respectively for the fish incubated in the laboratory and the Dnieper water, $p<0.01$, $n=14$). After incubation in the river water, the number of micronuclei in the gill cells of crustacian carp increased significantly (0.33 versus 4.5 per 1000 cells respectively for the reference group and the group of fish incubated in Dnieper water, $p<0.01$, $n=14$). The number of gill cells with geminate nuclei also increased in the fish incubated for 4 days in Dnieper water (respectively 1.63 versus 0.167, $p<0.05$, $n=14$). However, the statistical error of this indicator's measuring in the gill tissue was insignificant.

The cytological assay of the gills and blood of the sea fish *Mugil cephalus* aimed at determining the number of cells with micronuclei and geminate nuclei demonstrated the influence of sea water anthropogenic contamination on the level of genome's alteration in tissues of the fish under examination (Fig. 6.1). The comparison of the number of cells with micronuclei in the fish caught at the Perama and Anavyssos stations demonstrates a probable increase of the number of such cells (2 versus 0.75 micronuclei per 1000 cells respectively for the fish caught in the area of Perama and Anavyssos, $p<0.05$, $n=15$). Still more manifest was this difference for the frequency of cells with geminate nuclei (1.67 versus 0.38 per 1000 cells respectively for the Perama and Anavyssos areas, $p<0.01$, $n=14$). The same tendency was observed with regard to the gill tissue; however, the difference discovered between the fish of the two groups was statistically untrue. For example, the number of cells with micronuclei in the gill samples was 3.13 versus 2 per 1000 cells respectively for the Perama and Anavyssos areas ($p>0.05$, $n=16$), the number of cells with geminate nuclei—1.5 versus 0.75 per 1000 cells ($p>0.05$, $n=16$).

Thus, the presence of anthropogenic contamination factors in the river and sea water causes activation of xenobiotic transformation processes on the one hand and

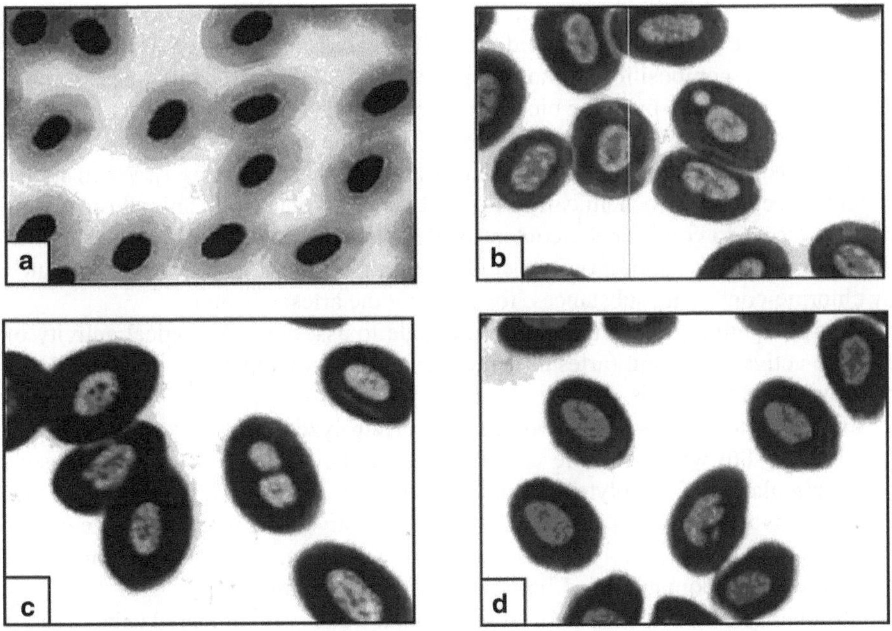

Fig. 6.1 The micronuclear test in fish cells. **a** Control. **b** Cell with micronucleus. **c** Cell with a double nucleus. **d** Cell with the destroyed nucleus

an increase of genetic alterations in the blood and gill cells of the fresh water fish on the other hand. The results of the study of changes in the genetic apparatus of aquatic organisms under the influence of fresh water pollutants can be extrapolated to issues concerning human health, taking into consideration that the river is one of the main sources of drinking water for the population of the Ukraine and other European countries.

6.6 Biotesting of Electrochemically Activated Water

Electrochemically activated water has applications in medicine, agriculture, and construction. According to its properties, it is subdivided into "live" (catholyte) and "dead" (anolyte) water. Anolyte contains active chlorine which makes it a strong disinfectant. The effect of "live" water is more varied. Alkaline cathodic water stimulates processes of cell regeneration and development and, thus, favors wound healing. Catholyte was found to induce appreciable increases of plant biomass and weight increments in animals. The biocatalytic properties of catholyte are due to medium basicity and the presence of free radicals generated on the cathode. The effectiveness of the electrochemical activation of water is usually evaluated by the pH and reduction oxidation potential values (ROP) [180, 181]. However, in some cases this evaluation may be incomplete since the above parameters do not correlate rigidly with the properties of treated water [182] and do not account for the variety of its chemical composition. A more objective assessment of the biological activity of water can be made by using biotesting methods [183].

Biotests based on Water Tox batteries in which the indicators are plants (lettuce) and animals (daphnia) are used in toxicity studies of river water [184]. When studying the effect of electrochemically activated water on plants and animal organisms it is necessary to consider the content of residual chlorine in the source water. In this regard, it seems appropriate to study the toxicity of water that was not treated by chlorine-containing substances, for example the artesian water.

In investigation [185] an attempt was made to assess the biological activity of cathode-activated natural (artesian) water, using a set of vegetable and animal test organisms and their cells.

In order to activate water, a two-chamber electrolyzer with a ceramic separating membrane [186] was used. The unit operated on the principle of separate closed-circuit circulation of catholyte and anolyte 1 and 5 dm^3 in volume, respectively. The electric activation of water was carried out in the galvanostatic mode with currents of 50–200 mA and a catholyte circulation rate of 200 dm^3/min. Water from Kiev's artesian wells Cenomanian and Jurassic aquifers was examined.

The quality of the electrochemically activated water was compared using a set of vegetable and animal test organisms and their cells. At the organism level, the effect of the electrochemical treatment on the water quality was determined by changes in the length of onion roots (*Allium test* [8]), sublethal and lethal effects in hydra (*Hydra attenuata* [10]) and mortality and propagation indices in cenodaphnias (*Ceriodaphnia affinis* [27, 28]). At the cell level studied changes in the functional activity of the cell genome by nucleolar biomarkers on animal (hydra) and vegetable (onion) cells using the method proposed in [49], and also structural anomalies of the genetic apparatus by the micronuclear test on vegetable cells (onion). Cytotoxic effects were additionally analyzed by the proliferative activity of the vegetable cells determining the mitotic index.

The experimental conditions and electrochemical characteristics of water are presented in Table 6.5, which shows that both the artesian wells are characterized by relatively stable pH values and the less stable value ROP. This can be explained by the higher sensitivity of the ROP to impurities present in the water and, in particular, oxygen.

As the quantity of electricity passing through the water increases, the catholyte pH rises monotonously, whereas ROP initially falls sharply, changing, during the first few minutes, from positive to negative values, after which the rate of decrease of ROP slows down (see Fig. 6.2). This is due to the fact that in addition to the decomposition of water on the cathode, there is a reduction of potential-determining impurities present in the water, in particular molecular oxygen and ions of polyvalent metals, which shifts the ROP to negative values. Further changes of the catholyte pH and ROP come only as a result of the reaction of water decomposition on the cathode, with the rate of change of ROP becoming equal to that of pH.

A significant influence on the characteristics of electrically activated water is exerted by the polarization conditions. Tins, for example, the ROP of water electrically

Table 6.5 Experimental conditions and electrochemical characteristics of water

Date (2004)	Well	Test No.	Quantity of electricity, C/dm³	pH	ROP, mV
July	Cenomanian aquifers	Initial	0	7.4	210
July	Ditto	1	150	9.6	−650
July	Ditto	2	21	8.7	−170
July	Ditto	3	15	8.2	−50
August	Ditto	3a	30	9.4	−180
September	Ditto	Initial	0	7.5	380
September	Ditto	3b	30	9.0	−180
October	Ditto	Initial	0	7.5	320
October	Ditto	3c	30	9.4	−180
November	Ditto	Initial	0	7.5	340
November	Ditto	3d	30	9.4	−180
July	Jurassic aquifer	Initial	0	7.9	315
July	Ditto	4	27	9.1	−265
July	Ditto	5	24	9.0	−140
August	Ditto	Initial	0	7.6	460
August	Ditto	5a	18	9.2	−130
September	Ditto	Initial	0	7.6	420
September	Ditto	5b	21	9.0	−120
October	Ditto	Initial	0	7.8	500
October	Ditto	5c	30	9.4	−140
November	Ditto	Initial	0	8.2	340
November	Ditto	5d	36	9.7	−190

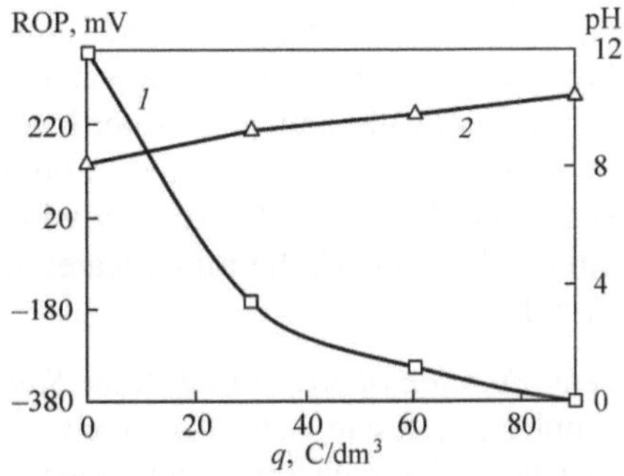

Fig. 6.2 Relationship of ROP (*1*) and pH (*2*) of artesian water (Jurassic aquifer) to quantity of electricity passed through water with a polarizing current of 100 mA

treated with higher densities is less time-variant than that of water treated with low currents (see Figs. 6.3 and 6.4), which is indicative of profound and persistent changes of the oxidative-reducing properties of the aqueous medium under study. The pH values of electrically activated solutions remain practically unchanged over time while ROP relaxes rather slowly but does not reach the initial values remaining below them (see Fig. 6.4). An ROP relaxation time relationship analogous to that shown in Fig. 6.4 was also obtained for water from the well in the Cenomanian aquifer.

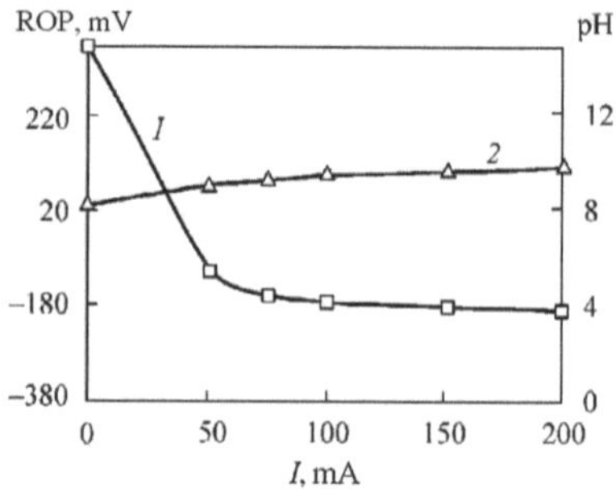

Fig. 6.3 Relationship of ROP (*1*) and pH (*2*) in electrolysis of artesian water (Jurassic aquifer) to polarizing current after passage of 30 C/dm³ of electricity

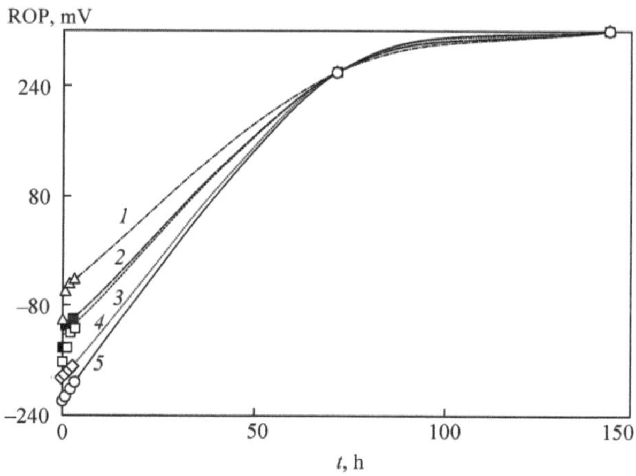

Fig. 6.4 Relationship of ROP of artesian water samples (Jurassic aquifer) polarized by currents (mA): 50 (*1*), 75 (*2*), 100 (*3*), 150 (*4*), and 200 (*5*) after passage of 30 C/dm³ of electricity to time of relaxation at room temperature

For biotesting analyses used samples of electrically activated water Nos. 2 through 5, as well as the initial water from the two wells (see Table 6.5). Since the electrochemically activated water loses its catalytic and biocatalytic activity with the passage of time, we biotested both settled and freshly activated water. Water Nos. 3 and 5 were replaced daily with fresh water designated in Table 6.5 by the letter *a, b, c,* and *d*; the initial water and samples Nos. 2 and 4 were not changed for 5 days.

Determination of Toxicity of Water Samples by the Vegetable Biotest (Onion) Two samples of artesian water from the well in the Jurassic aquifers (Nos. 4 and 5) had, after their electrochemical activation, a stimulating effect on both the length and the weight of onion roots, which increments, respectively, of 8.4 and 10% (No. 4) and 9.9 and 10.7% (No. 5). For the two samples taken on Cenomanian aquifer (Nos. 2 and 3), these figures were as follows: in sample No. 2 the gains in the onion root length and weight were 8.8 and 13.4%, respectively, while in sample No. 3 registered a 6.7% increase of the root length and a 9.1% decrease of the weight. Consequently, the electrochemical treatment of the water slightly stimulated the growth of the vegetable roots increasing their morphological characteristics by an average of 10%.

Table 6.6 Nucleolar activity of hydra cells in electrochemically activated water

Proportion of cells with HPN, %			Average number of nucleoli per cell			Size of individual nucleolus, μ^2		
30 min	90 min	Average	30 min	90 min	Average	30 min	90 min	Average
Negative control (form the Cenomanian aquifer)								
62.1	61.7	61.9	1.167	1.150	1.158	42.4	41.5	42.0
Sample No. 2								
73.2	62.2	67.7	1.083	1.135	1.109	25.6	32.6	29.1
Sample No. 3								
76.0	59.8	67.9	1.044	1.106	1.075	38.9	51.9	45.4
Negative control (from the Jurassic aquifer)								
72.6	75.4	74.0	1.099	1.069	1.084	41.8	43.2	42.5
Sample No. 4								
73.9	53.8	63.8	1.119	1.111	1.115	44.1	30.9	37.5
Sample No. 5								
69.4	58.9	64.2	1.116	1.146	1.131	48.7	33.5	41.1

Determination of Toxicity of Water Samples by the Biotest with Animals (Hydra) Acute toxicity (sublethal and lethal effects) was not found in any of the samples studied (Nos. 2 through 5). In addition, we analyzed the various samples for the hydrae's proliferation rates. Control samples differed in their effect on this indicator: the increment of young hydrae was two units in sample from Cenomanian aquifer and five in sample from Jurassic aquifer. Cultivation of hydrae in electrochemically treated water produced no substantial changes in these figures: the increase in young animals in samples Nos. 2 and 3 were two–three units, and in samples Nos. 4 and 5 four–six. Consequently, the electrochemical treatment produced no appreciable changes in the vital functions of hydrae.

Determination of Toxicity of Water Samples by the Animal Biotest with Ceriodaphnias Acute toxicity (deaths of animals after 48 h of incubation) was not registered in any of the samples analyzed (Nos. 2 through 5). Chronic toxicity, as determined over a 9 day period by the rates of propagation of female ceriodaphnias and their mortality, was not found in three samples (Nos. 2, 4, and 5). In Sample No. 3 a 60 % death rate and a decrease in the proliferation rate (from 56 units in the control sample to 24) were observed, which indicated that this particular sample was toxic.

Determination of Cytotoxicity of Water Samples by the Nucleolar Biomarker on Animal Cells (hydra) Two control samples of artesian water differed in quality, as indicated by the nucleolar biomarker (Table 6.6). These differences were minor in the number of nucleoli per cell and significant in the proportion of cells with heteromorphic paired nucleoli (HPN); the latter index reflects the peculiar mechanism regulating nucleolar activity [22].

It can be seen from the data in Table 6.6 that sample No. 2 had a largely negative effect on the transcriptional activity of ribosomal genes because it slightly increased the proportion of cells with HPN (by 9.4 %), and it decreased the number of nucleoli somewhat, mainly after 90 min of exposure. It also significantly (by

30.7 %) reduced the size of the nucleoli (Table 6.6). With two nucleolar parameters, the proportion of cells with HPN and the number of nucleoli per cell sample No. 3 exhibited the same regularities as sample No. 2, both in the dynamics of changes and in their quantitative assessment. However, the samples differed in the size of an individual nucleolus. On the whole, the effect of sample No. 3 on all the three parameters of nucleolar activity was insignificant. Sample No. 4 caused a decrease of the percentage of HPN cells (by 13.8 %) had practically no effect on the number of nucleoli in cells and reduced the nucleolus size by 11.8 %. Thus, this sample slightly influenced the nucleolar activity of cells in two parameters out of the three studied. Sample No. 5 showed the same regularities (quantitatively and qualitatively) as sample No. 4 in two nucleolar parameters, the proportion of cells with HPN and the number of nucleoli per cell (Table 6.6). The nucleolus size decreased insignificantly. This sample also reduced the nucleolar activity of hydra cells mainly by the HPN index of paired nucleoli.

Thus, samples Nos. 2 and 3 (source: artesian well in Cenomanian aquifer), just as samples Nos. 4 and 5 (source: artesian well in Jurassic aquifer), were similar within the respective pairs in effects produced on the quantitative characteristics of nucleoli in hydra cells. In addition, in each pair one of the samples was more toxic (Nos. 2 and 4) than the other (Nos. 3 and 5).

Determination of Cytotoxicity of Water Samples by the Nucleolar Biomarker and Mitotic Index on Vegetable Cells (Onion) In this experiment only one sample (No. 5) was analyzed. We registered an appreciable inhibition of the nucleolar activity of vegetable cells in all the three parameters under study: proportion of HPN cells (by 13.1 %), number (by 12.5 %) and nucleolus size (by 32.3 %). The greatest effect in the first two indices made itself evident after 180 min of exposure while in the third—the nucleolus size—a sharp decrease was registered after 90 min.

Consequently, as in the case of animal cells (hydra), exposure of onion cells to the electrochemically treated water resulted in decreases of the nucleolar activity parameters, but in the vegetable cells these effects were more pronounced.

All Jurassic aquifer samples studied were found to stimulate the proliferative activity of vegetable cells, as compared to the rates of growth of roots in the control media: for sample No. 4 the mitotic index increased by 35.1 % and for sample No. 4 by 42.3 %. For the samples taken on the well from Cenomanian aquifer obtained the following results: in sample No. 2 the mitotic index value corresponded to the control figure while in sample No. 3 it was lower by 16.5 %. Note that sample No. 3 was also characterized by negative effects at the vegetable organism level.

Determination of Genotoxicity of Water Samples by the Micronuclear Test on Vegetable Cells (Onion) No genotoxic effect (i.e., increase of the proportion of cells with micronuclei and ambiguous nuclei) on vegetable cells was found in any of the samples studied (Nos. 2–5).

Determination of Nuclear/Nucleolar Anomalies Using Animal Cells (Hydra) Control samples (water from the Cenomanian and Jurassic aquifers wells) were identical in morphological anomalies produced by them in the nuclear–nucleolar apparatus: no irregularities in the nucleus structure were registered during the entire period

Table 6.7 Changes in nuclear and nucleolar morphology in hydra cells cultivated in electrochemically activated water

Type of nuclear/nucleolar anomalies	Negative control				Sample No. 2		Sample No. 3		Sample No. 4		Sample No. 5	
	Cenomanian aquifer		Jurassic aquifer									
	Min											
	0	180	0	180	90	180	90	180	90	180	90	180
Completely destroyed nucleus	0	0	0	0	0.6	0.6	1.3	0	1.2	0.6	0.1	1.3
Partially destroyed nucleus	0	0	0	0	2.5	0.8	2.0	0.5	4.5	2.6	2.0	3.3
"Satellites"	1.0	3.0	0.8	2.4	12.1	3.9	3.2	5.8	2.8	5.2	14.2	2.1
"Beads"	0.7	2.9	1.9	3.9	4.0	10.5	11.1	9.8	4.4	4.9	5.1	12.1
"Amorphous" nucleolars	1.8	1.1	0.9	3.0	2.3	6.1	5.5	2.4	11.4	12.7	13.5	9.4
"Vacuole"	2.3	3.3	0.8	2.1	2.7	1.6	4.7	4.5	10.3	0.3	5.3	3.2
Total anomalies of morphology (% of number of cells studied)												
Nuclear	0	0	0	0	3.1	1.4	3.3	0.5	5.7	3.2	2.1	4.6
Nucleolar	5.8	10.3	4.4	11.4	21.1	22.1	24.5	22.5	28.9	23.1	38.1	26.8

of observation, and in the morphology of nucleoli was found a small percentage (10.3–11.4 %) of anomalies, which was higher than the proportion registered in the hydra cultivation medium—4.4–5.8 %.

In sample No. 2, throughout 180 min, a small percentage of destroyed nuclei and a twofold increase of anomalies in the nucleolar morphology were registered (Table 6.7). The new anomalies that accounted for this increase included "satellites," "beads" and amorphous nucleoli.

Cultivation of hydra in sample No. 3 also produced a very small proportion of cells with destroyed nuclei and double number of morphological anomalies in nucleoli (Table 6.7), mainly of cells with nucleoli of the bead type, as well as "satellites" and amorphous nuclei.

Sample No. 4 induced a slight growth of the percentage of cells with partially destroyed nuclei and a 11.7–17.5 % increase of the proportion of cells with morphologically altered nucleoli, mostly by causing anomalies of the amorphous and vacuole types. The latter type of anomalies practically disappeared toward the end of the exposure.

Hydra cells incubated in sample No. 5 showed a small percentage of partially destroyed nuclei and a two–threefold increase in nucleoli with atypical morphology, with "satellites" and amorphous anomalies predominating after 90 min of exposure and "beads" and amorphous ones after 180 min.

Thus, samples of electrochemically treated water induced increases in anomalies in the morphology of nucleoli samples Nos. 2 and 3, like Nos. 4 and 5, were very similar to each other in the observable effects.

Determination of Nuclear/Nucleolar Anomalies Using Vegetable Cells (Onion) In analyzing sample No. 5, which was studied for cytotoxicity by the nucleolar biomarker, we chose a model a bulb containing a high percentage of cells with morphological anomalies of nucleoli, or "amorphous" (43 % in control). Cultivation of the onion roots in Sample No. 5 reduced the negative effects to 29.7 % as a result of a decreased proportion of cells with "amorphous" nucleoli. Consequently, in vegetable cells with a high percentage of nucleolar anomalies electrochemical activation of water reduced cell irregularities.

Thus, analysis of experimental data has shown that, after electrochemical activation in an electrolyzer cathode chamber the artesian water drawn from the Jurassic and Cenomanian aquifers slightly stimulated the growth of vegetable roots without altering, in any substantial way, the vital functions of hydras and ceriodaphnias (for the latter species this was true of three samples out of four studied because one sample exhibited toxic properties). Toxicologically, the samples analyzed were essentially safe for vegetable and animal test organisms. In animal cells, the treated water induced cytotoxic effects that for the various samples were more or less pronounced, as determined by the nucleolar biomarker. All of the studied samples showed no genotoxic effects (i.e., no growth of the numbers of cells with micronuclei and ambiguous nuclei) on vegetable cells. Thus, biotesting of electrochemically activated water did not reveal any negative effects on the vital functions of organisms or the heredity structures of cells.

References

1. Bagrinskii LP (1981) Nauchnyye osnovy kontrolya kachestva vod po gidrobiologicheskim pokazatelyam (Scientific principles of water quality control by hydrobiological characteristics). Gydrometeoizdat, Leningrad
2. Filenko OF (1985) Relationship of bioassaying with normalization and toxicological control of reservoir pollution. Water Resour 3:130–134
3. Postel S (1987) State of the world. WW Norton, New York
4. Butorin NV (ed) (1979) Vliyaniye zagryaznyayushchikh veshchestv na gidrobiontov i ekosistemy vodoyemov (Effects of pollutants upon aquatic ecosystems). Nauka, Leningrad
5. Lukyanenko VI (ed) (1983) Teoreticheskiye voprosy biotestirovaniya (Theoretical questions of biotesting). In-t biologii vnutr vod, Volgograd
6. Moore JV, Ramamoorthy S (1984) Heavy metals in natural water. Applied monitoring and impact assessment. Springer-Verlag, New York
7. Bryzgalo VA, Khorunzhaya TA (eds) (1989) Metody bioindikatsii i biotestirovaniya prirodnykh vod (Methods of bioindication and biotesting of natural water). Gidrometeoizdat, Leningrad
8. Fiskesjo G (1993) A 2–3 days plant test for toxicity assessment of various chemicals by measuring the mean root growth of a series of onions. Environ Toxic Water 8:461–470
9. Dutka B (1989) Short-term root elongation toxicity bioassay. Methods for toxicological analysis of water, wastewater and sediments. National Water Research Institute, Environ Canada
10. Trottier S, Blaise C, Kusui T et al (1997) Acute toxicity assessment of aqueous samples using a microplate-based Hydra attenuata assay. Environ Toxic Water 12:256–271

11. McInnis R (1989) The nematode toxicity assay using *Panagrellus redivivus*. Methods for toxicological analysis of water, wastewater and sediments. National Water Research Institute, Environ Canada
12. McInnis R (1989) *Daphnia magna* 48 h static bioassay method for acute toxicity in environmental samples. Methods for toxicological analysis of water, wastewater and sediments. National Water Research Institute, Environ Canada
13. EBPI (1997) The mutal-chromate kit, Version 2.1. Biodetection Products Inc Brampton, Ontario
14. Heddle JA, Cimino MC, Hayashi M et al (1991) Micronuclei as an index of cytogenetic damage: past, present, and future. Environ Mol Mutagen 18:277–291
15. Ilyinskikh NN, Novitskii VV, Vargunova NI et al (1991) Mikroyadernyy analiz i tsitogeneticheskaya nestabil'nost' (Micronuclear analysis and cytogenetic instability). Izd-vo Tomskogo un-ta, Tomsk
16. Hurna E, Sikelenka P, Hurna S (1997) Effect of selenium on cadmium genotoxicity investigated by micronucleus assay. Vet Med 42:339–342
17. Kiknadze II (1972) Funktsional'naya organizatsiya khromosom (Functional organization of chromosomes). Nauka, Leningrad
18. Chentsov YuS (1978) Obshchaya tsitologiya (General cytology). Iz-vo MGU, Moscow
19. Chelidze PV (1985) Ul'trastruktura i funktsii yadryshka interfaznoy kletki (Ultrastructure and functions of the nucleolus of the interphase cell). Metsniereba, Tbilisi
20. Hadjiolov AA (1985) The nucleolus and ribosome biogenesis. Cell Biol Monographs, vol 12. Springer, New York
21. Arkhipchuk VV (1995) Structural and functional changes of genome in fish phylogenesis. Abst of Dr Sci Thesis, AV Dumansky Institute of Colloid and Water Chemistry
22. Arkhipchuk VV, Palamarchuk JV (1997) Heteromorphic-paired nucleoli in fish cells. Cytobios 88:229–238
23. KND 211.1.4.060-97 (1997) Metodika viznachennya toksichností vodi na bakteríyakh *Photobacierium phosphoreum* (Conn) Ford (Technique for determination of water toxicity by means of bacteria *Photobacierium phosphoreum* (Conn) Ford). Approved by the order of Ukraine Ministry of Nature, N68
24. KND 211.1.4.059-97 (1997) Metodika viznachennya toksichnosti vodi na infuzoriyakh *Tetrahymena pyriformis* (Ehrenberg) Schewiakoff (Technique for determination of water toxicity by means of infusoria *Tetrahymena pyriformis* (Ehrenberg) Schewiakoff). Approved by the order of Ukraine Ministry of Nature, N 68
25. KND 211.1.4.058-97 (1997) Metodika viznachennya toksichnosti vodi na vodoroslyakh *Scenedesmus quandricauda* (Turp) Breb (Technique for determination of acute toxicity of water by means of algae *Scenedesmus quandricauda* (Turp) Breb). Approved by the order of Ukraine Ministry of Nature, N 68
26. KND 211.1.4.054-97 (1997) Metodika viznachennya toksichnosti vodi na rakopodibnikh *Daphnia magna* Straus (Technique for determination of acute toxicity of water by means of crustaceans *Daphnia magna* Straus). Approved by the order of Ukraine Ministry of Nature, N 68
27. KND 211.1.4.055-97 (1997) Metodika viznachennya gostroi letal'noi toksichnosti vodi na rakopodibnikh *Ceriodaphnia affinis* Lilljeborg (Technique for determination of acute lethal toxicity of water by means of crustaceans *Ceriodaphnia affinis* Lilljeborg). Approved by the order of Ukraine Ministry of Nature, N 68
28. KND 211.1.4.056-97 (1997) Metodika viznachennya khronichnoi toksichnosti vodi na rakopodibnikh *Ceriodaphnia affinis* Lilljeborg (Technique for determination of chronic toxicity of water by means of crustaceans *Ceriodaphnia affinis* Lilljeborg). Approved by the Order of Ukraine Ministry of Nature, N 68
29. KND 211.1.4.057-97 (1997) Metodika viznachennya gostroi letal'noíi toksichnosti vodi na ribakh *Poecillia reticulate* Peters (Technique of determining acute toxicity of water on fish *Poecillia reticulate* Peters). Approved by the order of Ukraine Ministry of Nature, N 68

30. Technique of determination of gene toxicity of environmental objects by the incidence of emergence of dominant lethal mutations of the *Drosophila melanogaster Mg* (2001). http://ecology.ks.ua/index.php?module=page&id=161. Accessed 24 Dec 2001
31. Lewis PA, Klemm DJ, Norberg-King TJ et al (eds) (1994) USEPA short-term methods for estimating the chronic toxicity of effluents and receiving water to freshwater organisms, 3rd edn. Environmental Protection Agency, US
32. Costan G, Bermingham N, Blaise C et al (1993) Potential ecotoxic effects probe (PEEP): a novel index to assess and compare the toxic potential of industrial effluents. Environ Toxic Water 8(1):115–140
33. Vassuer P, Ferard JF, Babut M (1991) The biological aspect of the regulatory control of industrial effluents in France. Chemosphere 22(5):625–633
34. Steinhauser KG, Hansen PD (eds) (1992) Biologische testverfahren. Gustav-Fasher Verlag, Stuttgart
35. EC Directive 80/778/EEC (1992) Relating to the quality of water intended for human consumption. Office for official publication of the European communities
36. World Health Organization (1993) Guidelines for drinking-water quality. WHO, Geneva
37. Clemedson C, McFarlane-Abdulla E, Anderson M et al (1996) MEIC evaluation of acute systemic toxicity. 1. Methodology of 68 *in vitro* toxicity assays used to test the first 30 reference chemicals. ATLA-Altern Lab Anim 24(I):251–272
38. Clayson DB (1987) The need for biological risk assessment in reaching decisions about carcinogens. Mutation Res 185:243–269
39. Ennever FK, Rosenkranz HS (1988) Methodologies for interpretation of short-term test result which may allow reduction in the use of animals in careigenicity testing. Toxicol Industr Health 4(2):137–149
40. Knobloch LM, Blondin GA, Harkin JM (1990) Use of submitochondrial particles for prediction of chemical toxicity in man. B Environ Contam Tox 44(5):661–668
41. Fiskesjo G (1985) The Allium as a standard in environmental monitoring. Hereditas 102:99–102
42. Ennever FK, Andreano G, Rozenkranz HS (1988) The ability of plant genotoxicity assays to predict carcinogenicity. Mutation Res 205:99–105
43. Sandhu SS, Serres de FJ, Gopalan HNB et al (1994) Results and recommendations. Mutation Res 310:257–263
44. Grant WF (1982) Chromosome aberration assays in Allium. A report of the U.S. environmental protection agency gene-tox program. Mutation Res 99:273–291
45. Grant WF (1994) The present status of higher plant bioassays for the detection of environmental mutagens. Mutation Res 310:175–185
46. Arkhipchuk VV (1995) The use of nucleolar characteristics in biotesting. Cytol Genet 29:6–12
47. Arkhipchuk VV (1995) Changes in the nucleolar characteristics in fish embrio cells under the action of low doses of radiation. Russ J Radiat Biol Radioecol 29:512–517
48. Arkhipchuk VV, Romanenko VD, Malinovskaya MV et al (2000) Toxicity assessment of water samples with a set of animal and plant bioassays: experience of the Ukrainian participation in the Water Tox program. Environ Toxicol 15(4):277–286
49. Arkhipchuk VV, Malinovskaya MV, Garanko NN (2000) Cytogenetic study of organic and inorganic toxic substances on *Allium cepa*, *Lactuca sativa*, and *Hydra Attenuata* cells. Environ Toxicol 15(4):338–344
50. Arkhipchuk VV (1999) Nucleolar variations during the onto-genesis of diploid and tetraploid cyprinid species. J Fish Biol 54:513–524
51. Keddy CJ, Greene JC, Bonnell MA (1995) Review of whole organism bioassays: soil, freshwater sediment, and freshwater assessment in Canada. Ecotoxicol Environ Saf 30:221–251
52. National Research Council USA (1989) Biologic markers in reproductive toxicology. National Academies Press, Washington.

53. International Programme on Chemical Safety (IPCS) (1993) Biomarkers and risk assessment: concepts and principles, environmental health criteria. World Health Organization, Geneva
54. Pinkel D, Straume T, Gray JW (1986) Cytogenetic analysis using quantitative, high-sensitivity, fluorescence hybridization. Proc Natl Acad Sci USA 83:2934–2938
55. Eastmond DA, Schuler M, Rupa DS (1995) Advantages and limitations of using fluorescence in situ hybridization for the detection of euploidy in interphase human cells. Mutat Res 348(4):153–162
56. Smitt MT, Zhang I, Wang Y et al (1998) Increased translocations and aneusomy in chromosome 8 and 21 among workers exposed to benzene. Cancer Res 58(10):2176–2181
57. Hemminki K et al (ed) (1994) DNA adducts, identification and biological significance. International Academy for Research on Cancer, Lyons
58. Quian G-S, Ross RK, Yu MC et al (1994) A follow-up study of urinary markers of aflatoxin exposure and liver cancer risk in Shanghai, People's Republic of China. Cancer Epidem Biomar 3(1):3–10
59. Kensler TW, Groopman JD, Sutter TR et al (1999) Development of cancer chemopreventive agents: oltipraz as a paradigm. Chem Res Toxicol 12(2):113–126
60. Culp SJ, Gaylor DW, Sheldon WG et al (1996) DNA adduct measurements in relation to tumor incidence during the chronic feeding of coal tar or benzo(a)pyrene to mice. Polycycl Aromat Hydrocarbons 11:161–168
61. Goldstein LS, Wyand EH, Safe S et al (1998) Tumors and DNA adducts in mice exposed to benzo(α) pyrene and coal tars: implications for risk assessment. Environ Health Persp 106:1325–1330
62. Ostling O, Jonson KJ (1994) Microelectrophoretic study of radiation-induced DNA damages in individual mammalian cells. Biochem Bioph Res Co 123(1):291–298
63. Sing NP, McCoy MT, Tice RR et al (1988) A simple technique for the quantitation of low levels of DNA damage in individual cells. Exp Cell Res 175(1):184–191
64. Firbairn DW, Olive PL, O'Neilli KL (1995) The comet assay: a comprehensive review. Mutat Res 339(1):37–59
65. Collins A, Dusinska M, Franklin M et al (1997) Comet assay in human biomonitoring studies: reliability, validation and applications. Environ Mol Mutagen 30(2):139–146
66. Anderson D, Yu T-W, McGregor DB (1998) Comet assay responses as indicators of carcinogen exposure. Mutagenesis 13(6):539–555
67. Heddle JA, Hite M, Kirkahart B et al (1983) The induction of micronuclei as a measure of genotoxicity. A report of the U.S. environmental protection agency Gene-Tox program. Mutat Res 123(1):61–118
68. Heddle JA, Cimino MC, Hayashi M et al (1991) Micronuclei as an index of cytogenetic damage: past, present, and future. Environ Mol Mutagen 18(4):277–291
69. Al-Sabti K, Metcalfe CD (1995) Fish micronuclei for assessing genotoxicity in water. Mutat Res 343(2/3):121–135
70. Inoue A, Yokomori K, Tanabe H et al (1997) Extensive genetic heterogeneity in the neuroblastoma cell line NB(TU)1. Int J Cancer 72(6):1070–1077
71. Schwarzacher HG, Wahtler F (1991) The functional significance of nucleolar structures. Ann Genet 34(3/4):151–160
72. Scheer U, Weisenberger D (1994) The nucleolus . Curr Opin Cell Biol 6(3):354–359
73. Shaw PJ, Jordan EG (1995) The nucleolus. Ann Rev Cell Dev Biol 11:93–121
74. Pederson T (1998) The plurifunctional nucleolus. Nucleic Acids Res 26(17):3871–3876
75. Arkhipchuk VV (1998) Peculiar changes of nucleolar quantitative characteristics in fish cells. Cytobios 94(376):81–96
76. Popp W, Wachtler F (1983) Changes in nucleolar structure, number and size in cellular activation and inactivation. Observations in human phytohaemagglutinin-treated lymphocytes. Cell Tissue Res 234(2):377–388
77. Wachtler F, Schwarzacher HG, Smetana K (1984) On the fusion of nucleoli in interphase. Eur Cell Biol 34(1):190–192

78. Kwiatkowska M, Maszewski J (1985) Functional and structural heterogeneity of nucleoli: the dependence of the activity of transport of newly synthesized RNA on nucleolar size and phase of the cycle. Floria Histochem Cytobiol 23(3):135–144
79. Lafarga M, Villegas J, Crespo D (1985) Changes in nucleolar morphology and volume of the supraoptic nucleus neurons during postnatal development of the rat. Brain Res 354(2):310–313
80. Dzidziguri DV, Chelidze PV, Zarandia MA et al (1994) Transcriptional activity and ultrastructure of morphologically different types of nucleoli isolated from hepatocytes of normal and hepatectomized rats. Epithelial Cell Biol 3(2):54–60
81. Mellink CH, Bosma AA, De Haan NA (1994) Variation in size of Ag-NORs and fluorescent rDNA in situ hybridization signals in six breeds of domestic pig. Hereditas 120(2):141–149
82. Leek RD, Alison MR, Sarraf CE (1991) Variations in the occurrence of silver-staining nucleolar organizer regions (AgNORs) in non-proliferating and proliferating tissues. J Pathol 165(1):43–51
83. Mamaev NN, Mamaeva SE (1992) The structure and function of the chromosomal nucleolus organizer regions: the molecular, cytological and clinical aspects. Cytology 34(10):3–25
84. Ruschoff J, Elsasser HP, Adler G (1995) Correlation of nucleolar organizer regions with secretory and regenerative process in experimental cerulein-induced pancreatitis in the rat. Pancreas 11(2):154–159
85. Tanaka T, Takeuchi T, Nishikawa A et al (1989) Nucleolar organizer regions in hepatocarcinogenesis induced by N-2-fluorenylacetamide in rats: comparison with bromodeoxyuridine immunohistochemistry. J Cancer Res 80(11):1047–1051
86. Trere D (1993) Critical analysis of the methods commonly employed in the assessment of cell proliferation: advantages of the NOR silver-staining technique in routine cyto-histopathology. Anal Cell Pathol 5(4):191–201
87. Knobloch LM, Blondin GA, Harkin JM (1990) Use of submitochondrial particles for prediction of chemical toxicity in man. Bull Environ Contam Toxicol 44(5):661–668
88. Powers DA (1989) Fish as model systems. Science 246(4928):352–358
89. Washburn PC, Di Giulio RT (1989) The stimulation of superoxide production by nitrofuran-tion *p*-nitrobenzoic acid and *m*-dinitrobenzene in hepatic microsomes of three species of freshwater fish. Environ Toxicol Chem 8(2):171–180
90. Yang JH, Kostecki PT, Calabroco EJ et al (1990) Induction of peroxisome proliferation in rainbow trout exposed to ciprofibrate. Toxicol Appl Pharm 104(3):476–482
91. Stegeman JJ, Lech JJ (1991) Cytochrome P450 monooxygenase systems in aquatic species: carcinogen metabolism and biomarkers for carcinogen and pollutant exposure. Environ Health Persp 90:101–109
92. De Flora SL, Vigano F, D'Agostini A et al (1993) Multiple genotoxicity biomarkers in fish exposed in situ to polluted river water. Mutat Res 319(3):167–177
93. Minissi S, Ciccotti E, Rizzoni M (1996) Micronucleus test in erythrocytes of *Barbus plebejus* (Teleostei, Pisces) from two natural environments: a bioassay for the in situ detection of mutagens in freshwater. Mutat Res 367(4):245–251
94. Sanchez-Galan S, Linde AR, Izquierdo JI et al (1998) Micronuclei and fluctuating asymmetry in brown trout (Salmo trutta): complementary methods to biomonitor freshwater ecosystems. Mutat Res 412(3):219–225
95. Ayllon F, Suciu R, Gephard S et al (2000) Conventional armament wastes induce micronuclei in wild brown trout Salmo trutta. Mutat Res 470(2):169–176
96. Bahari IB, Noor FM, Daud NM (1994) Micronucleated erythrocytes as an assay to assess actions by physical and chemical genotoxic agents in *Clarias gariepinus*. Mutat Res 313(1):1–5
97. Nepomuceno JC, Ferrari I, Spano MA et al (1997) Detection of micronuclei in peripheral erythrocytes of *Cyprinus carpio* exposed to metallic mercury. Environ Mol Mutagen 30(3):293–297
98. Campana MA, Panzeri AM, Moreno VJ et al (1999) Genotoxic evaluation of the pyrethroid lambda-cyhalothrin using the micronucleus test in erythrocytes of the fish *Cheirodon interruptus*. Mutat Res 438(2):155–161

99. Ayllon F, Garcia-Vazquez E (2000) Induction of micronuclei and other nuclear abnormalities in European minnow Phoxinus phoxinus and mollie Poecilia latipinna: an assessment of the fish micronucleus test. Mutat Res 467(8):177–186
100. Helpap B (1988) Observations on the number, size and localization of nucleoli in hyperplastic and neoplastic disease. Histopathology 13(2):203–211
101. Yang P, Huang GS, Zhu XS (1999) Role of nucleolar organiser region in differentiating malignant from benign tumors of colon. J Clin Pathol 43(3):235–238
102. Gambini C, Casazza S, Borgiani L et al (1992) Counting the nucleolar organizer region-associated proteins is a prognostic clue of malignant melanoma. Arch Dermatol 128(4):487–490
103. Pich A, Chiusa L, Boccadoro M et al (1994) AgNORs and myeloma prognosis. Leukemia Lymphoma 12(5/6):383–394
104. Iwata H, Otoshi T, Takada N et al (1995) Validation of silver-stained nucleolar organizer regions for evaluation of invasive character of urinary bladder carcinoma in rats and mice. Urol Res 23(1):27–32
105. Nakae S, Nakamura T, Ikegawa R et al (1998) Evaluation of argyrophilic nucleolar organizer region and proliferating cell nuclear antigen in colorectal cancer. J Surg Oncol 69(1):28–35
106. Guarente L (1997) Link between aging and the nucleolus. Genes Dev 19:2449–2455
107. Johnson FB, Marciniak RA, Guarente L (1998) Telomeres, the nucleolus and aging. Curr Opin Cell Biol 10(3):332–338
108. Pedrazzini E, Mamaev N, Slavutsky I (1998) Age related decrease of NOR activity in bone marrow metaphase chromosomes from healthy individuals. Mol Pathol 51(1):39–42
109. Fiskesjo G (1997) Allium test for screening chemicals, evaluation of cytological parameters. Plants Environ Stud 11:307–333
110. Yudong L, Ping W (2008) Biomonitoring: an appealing tool for assessment of metal pollution in the aquatic ecosystem. Anal Chim Acta 606(2):135–150
111. Poongothai K, Shayin S, Usharani MV (1996) Induction of micronuclei in fish by polluted water and heavy metals. Cytobios 86(344):17–22
112. Kusui T, Blaise C (1999) Impact assessment of hazardous aquatic contaminants; concepts and approaches. Ann Arbor Press 7:161–181
113. Fomin GS (2000) Voda. Kontrol′ khimicheskoy, bakterial′noy i radiatsionnoy bezopasnosti po mezhdunarodnym standartam (Water. Control for chemical, bacterial, and radiation safety by international standards). Protector, Moscow
114. Misund A, Frengstad B, Siewers U et al (1999) Variation of 66 elements in European bottled mineral water. Sci Total Environ 243/244:21–41
115. Dabeka RW, Conacher HB, Lawrence JF et al (2002) Survey of bottled drinking water sold in Canada for chlorate, bromide, bromate, lead, cadmium, and other trace elements. Food Addit Contam 19(8):721–732
116. Pip E (2000) Survey of bottled drinking water available in Manitoba, Canada. Environ Health Persp 108(9):863–866
117. Page BD, Conacher HB, Salminen J et al (1993) Survey of bottled drinking water sold in Canada. Part 2: selected volatile organic compounds. J AOAC Int 76(1):26–31
118. Al-Saleh I, Al-Doush I (1998) Survey of trace elements in household and bottled drinking water samples collected in Riyadh, Saudi Arabia. Sci Total Environ 216(3):181–192
119. Araya M, Chen B, Klevay LM et al (2003) Confirmation of an acute no-observed-adverse-effect and low-observed-adverse-effect level for copper in bottled drinking water in a multi-site international study. Regul Toxicol Pharm 38(3):389–399
120. Lopez FF, Cabrera C, Lorenzo ML et al (2002) Aluminium content of drinking water, fruit juices and soft drinks: contribution to dietary intake. Sci Total Environ 292(3):205–213
121. Musaiger AO, Khunji ZA (1990) Chemical quality of drinking water in Bahrain. J R Soc Health 110(3):104–105
122. Johnson SA, DeBiase C (2003) Concentration levels of fluoride in bottled drinking water. J Dent Hyg 77(3):161–167

123. Zohouri FV, Maguire A, Moynihan PJ (2003) Fluoride content of still bottled water available in the North-East of England, UK. Brit Dent J 195(9):515–518
124. Food and drug administration, HHS (2003) Beverages: bottled water. Direct final rule. Fed Regist 68(41):9873–9882
125. Davila Rangel JI, Lopez del RH, Mireles GF et al (2002) Radioactivity in bottled water sold in Mexico. Appl Radiat Isotopes 56(6):931–936
126. Martin Sanchez A, Rubio Montero MP, Gomez Escobar V et al (1999) Radioactivity in bottled mineral water. Appl Radiat Isotopes 50(6):1049–1055
127. Somlai J, Horvath G, Kanyar B et al (2002) Concentration of 226Ra in Hungarian bottled mineral water. J Environ Radioactiv 62(3):235–240
128. Cwiek-Ludwicka K (2003) Polyethelene terephtalate (PET)—health aspects and food packaging application. Rocz Panstw Zakl Hig 54(2):175–182
129. Eberhartinger S, Steiner I, Washuttl J et al (1990) Study of the migration of acetaldehyde from PET bottles into soft drinks containing carvonic acid. Z Lebensm Unters For 191(4/5):286–289
130. Nawrocki J, Dabrowska A, Borcz A (2002) Investigation of carbonyl compounds in bottled water from Poland. Water Res 36(19):4893–4901
131. Song YS, Al-Taher F, Sadler G (2003) Migration of volatile degradation products into ozonated water from plastic packaging materials. Food Addit Contam 20(10):985–994
132. Benfenati E, Natangelo M, Davoli E et al (1991) Migration of vinyl-chloride into PVC-bottled drinking-water assessed by gags chromatography-mass spectrometry. Food Chem Toxicol 29(2):131–134
133. Fauris C, Danglot C, Vilagines R (1985) Rapidity of RNA synthesis in human cells. Water Res 19(6):677–684
134. Sauvant MP, Pepin D, Bohatier J (1995) Chemical and *in vitro* toxicological evaluations of water packaged in polyvinyl chloride (PVC) and polyethylene terephthalate. Food Addit Contam 12(4):567–584
135. Valentin I, Philippe M, Lhuguenot J et al (2001) Uridine uptake inhibition as a cytotoxicity test for a human hepatoma cell line (HepG2 cells): comparison with the neutral red assay. Toxicology 58(3):127–139
136. De Fusco RM, Biscardi D et al (1990) Leaching of mutagens into mineral water from polyethyleneterephthalate bottles. Sci Total Environ 90:241–248
137. Monarca S, De Fusco R, Biscardi D et al (1994) Studies of migration of potentially genotoxic compounds into water stored in PET bottles. Food Chem Toxicol 32(9):783–788
138. Monarca S, Zanardini A, Biscardi D et al (1997) Genotoxicity in mineral water stored in polyethylene terephthalate bottles evaluated using plant vioassays. Mutat Res 379(1):100–101
139. Sauvant MP, Pepin D, Bohatier J et al (1994) Comparative study of two in vitro models (L-929 fibroblasts and tetrahymena pyriformis GL) for the cytotoxicological evaluation of packaged water. Sci Total Environ 156(2):159–167
140. Evandri MG, Tucci P, Bolle P (2000) Toxicological evaluation of commercial mineral water bottled in polyethylene terephthalate: a cytogenetic approach with *Allium cepa*. Food Addit Contam 17(12):1037–1045
141. Biscardi D, Monarca S, De Fusco R et al (2003) Evaluation of the migration of mutagens/carcinogens from PET bottles into mineral water by tradescantia/micronuclei test, comet assay on Leukocytes and GC/MS. Sci Total Environ 302(1/3):101–108
142. Evandri MG, Bolle P (2001) Pharmaco-toxicological screening of commercially available Italian natural mineral water. Farmaco 56(5/7):475–482
143. Li Y, Liang CK, Katz BP et al (1995) Long-term exposure to fluoride in drinking water and sister chromatid exchange frequency in human blood lymphocytes. J Dent Res 74(8):1468–1474
144. Calabrese EJ, Tuthill RW (1981) The influence of elevated levels of Na in drinking water on elementary and highschool students in Massachusetts. Sci Total Environ 18:117–133

145. Campbell A, Becaria A, Lahiri DK et al (2004) Chronic exposure to aluminum in drinking water increases inflammatory parameters selectively in the brain. J Neurosci Res 75(4):565–572
146. Flaten TP (2001) Aluminium as a risk factor in Alzheimer's disease with emphasis on drinking water. Brain Res Bull 55(2):187–196
147. Bilyk A, Kolwzan B, Traczewska TM (1996) Evaluation of mutagenic activity in water disinfected with chlorine. Rocz Panstw Zakl Hig 47(1):77–85
148. Fang C, Wang YP, Jiang S et al (2001) Study on the genotoxicity of dibromoacetic acid in drinking water. Wei Sheng Yan Jiu 30(5):266–269
149. Kusamran WR, Tanthasri N, Meesiripan N et al (2003) Mutagenicity of the drinking water supply in Bangkok. Asian Pac J Cancer P 4(1):31–38
150. Edberg SC, Kops S, Kontnick C et al (1997) Analysis of cytotoxicity and invasiveness of heterotrophic plate count bacteria (HPC) isolated from drinking water on blood media. J Appl Microbiol 82(4):455–461
151. Biscardi D, Castaldo A, Gualillo O et al (2002) The occurrence of cytotoxic aeromonas hydrophila strains in Italian mineral and thermal water. Sci Total Environ 292(3):255–263
152. Arkhipchuk VV, Malinovskaya MV (2000) An integrated approach to the natural water biotesting. J Water Chem Technol 22(4):36–50
153. Arkhipchuk VV, Goncharuk VV (2001) Biotesting water for quality at the cellular level. J Water Chem Technol 23(5):48–54
154. Kuzmich VN, Sokolova SA, Krainyukova AN (2002) Rukovodstvo po opredeleniyu metodom biotestirovaniya toksichnosti vod, donnykh otlozheniy, zagryaznyayushchikh veshchestv i burovykh rastvorov (Guidance for biotesting method of toxicity of water, benthal deposits, pollutants and drilling mud fluid). REFIA, NIA-Nature, Moscow
155. Arkhipchuk VV, Goncharuk VV (2004) Problems of quality of bottled drinking water. J Water Chem Technol 26(4):48–53
156. Zuyev YeT, Fomin GS (2003) Pit'yevaya i mineral'naya voda. Trebovaniya mirovykh i yevropeyskikh standartov k kachestvu i bezopasnosti (Drinking and mineral water. Quality and safety requirements of world and European standard). Protektor, Moscow
157. Arkhipchuk VV, Garanko NN (2002) A novel nucleolar biomarker in plant and animal cells for assessment of substance cytotoxicity. Environ Toxicol 17(3):187–194
158. Reddy NM, Panda KK, Subhadra AV et al (1995) The Allium micronucleus (MNC) assay may be used to distinguish clastogens from aneugens. Biol Zbl 114(4):358–368
159. Fenech M, Chang WP, Kirsch-Volders M et al (2003) HUMN project: detailed description of the scoring criteria for the cytokinesis-block micronucleus assay using isolated lymphocyte cultures. Mutat Res 534(1/2):65–75
160. Wang W (1990) Toxicity assessment of pretreated industrial wastewater using higher plants. Res J Water Pollut C 62(7):853–860
161. Kristen U (1997) Use of higher plants as screens for toxicity assessment. Toxicol in Vitro 11(1/2):181–191
162. Rubin AJ, Elmaraghy MA (1976) Studies on the toxicity of ammonia, nitrate and their mixtures to the common guppy. Water Resources Center, Engineering Experiment Station, The Ohio State University, Columbus
163. Mount DI, Norberg TJ (1984) A seven-day life-cycle cladoceran toxicity test. Environ Toxicol Chem 3:425–434
164. Kszos LA, Crow KR (1997) Identification and treatment of lithium as the primary toxicant in a groundwater treatment facility effluent. Paper presented at the 69th annual conference and exposition, Dallas Convention Center, 5–9 October, Dallas
165. Sanborn NH (1945) The lethal effect of certain chemicals on fresh water fish. Canning Trade 67(49):10–12
166. Mount DR, Gulley DD, Hockett JR et al (1997) Statistical models to predict the toxicity of major ions to *C. dubia, Daphnia magna* and *Pimephales promelas*. Environ Toxicol Chem 16(10):2009–2019

167. EG & G Bionomics Rep N BW-79-12-576 (1979) The Acute toxicity of various silver compounds to the fathead minnow (*Pimephales promelas*). Harrison, New York
168. Rodgers JHJ, Deaver E, Suedel BC et al (1997) Comparative aqueous toxicity of silver compounds: laboratory studies with freshwater species. Bull Environ Contam Toxicol 58(6):851–858
169. Arkhipchuk VV, Goncharuk VV (2003) Effect of desalted water on physiological activity of animal and vegetable organisms and the functioning of their cells. J Water Chem Technol 25(2):62–67
170. Clemedson C, McFarlane-Abdulla E, Andersson M et al (1996) MEIC evaluation of acute systemic toxicity. Part 2: in vitro results from 68 toxicity assays used to test the first 30 reference chemicals and a comparative cytotoxicity analysis. ATLA-Altern Lab Anim 24(1):273–311
171. Schwarzacher HG, Mosgoeller W (2000) Ribosome biogenesis in man: current views on nucleolar structures and function. Cytogenet Cell Genet 91(1/4):243–252
172. Smetana K (2002) Structural features of nucleoli in blood, leukemic, lymphoma and myeloma cells. Eur J Histochem 46(2):125–132
173. Bragg P (2000) Chudo golodaniya. Ochishcheniye i omolozheniye organizma s pomoshch'yu lechebnogo golodaniya (The starvation miracle: purgation and rejuvenation of the organism by nestitherapy). FAIR, Moscow
174. Order N5 of the chef State Sanitary Inspector of the Russian Federation (2000) On the adjustment of drinking water quality in regards to contents of biogenic elements. Moscow, RF
175. Payment P (1989) Bacterial colonization of domestic reverse-osmosis water filtration units. Can J Microbiol 35(11):1065–1067
176. Payment P, Franco E, Richardson L et al (1991) Gastrointestinal health effect associated with the consumption of drinking water produced by point-of-use domestic reverse-osmosis filtration units. Appl Environ Microb 57(4):945–948
177. Cavas T, Ergene-Gozukara S (2005) Micronucleus test in fish cells: a bioassay for in situ monitoring of genotoxic pollution in the marine environment. Environ Mol Mutagen 46(1):64–70
178. Lowe DM, Moore MN, Evans BM (1995) Lysosomal membrane responses in mussels to experimental contaminants exposure. Aquat Toxicol 33:105–112
179. Hellenic Centre for Marine Research (2005) Use of cytological biomarkers in fish to assay anthropogenic contamination of sea and fresh water, The Joint Ukrainian-Greek Research Project
180. Rogov VM, Filipchuk VL (1989) Elektrokhimicheskaya tekhnologiya izmeneniya svoystv vody (Electrochemical technology for altering properties of water). Vyshcha Shkola, Lviv
181. Yakovlev SV, Krasnoborodko IG, Rogov VM (1987) Tekhnologiya elektrokhimicheskoy ochistki vody (Technology of electrochemical water treatment) Stroyizdat, Leningrad
182. Bakhir VM (1992) Elektrokhimicheskaya aktivatsiya (Electrochemical activation). VNIIIMT, Moscow
183. Arkhipchuk VV, Goncharuk VV (2005) Way of assessment the aquatic toxicity. Patent 55842A, 15 June 2003, Ukraine
184. Dutka DJ (1996) Water and sediment ecotoxicity studies in Temuco and Rapel river basin. Environ Toxic Water 11:237–247
185. Goncharuk VV, Bagriy VA, Arkhipchuk VV et al (2005) Biotesting of electrochemically activated water. J Water Chem Technol 27(4):55–62
186. Goncharuk VV, Bashtan SYu, Chebotareva RD (2004) Device for electrochemical water treatment. Ukraine Patent 49535, 15 July 2004

Chapter 7
SOS: Drinking Water. New State Standard "Drinking Water. Requirements and Methods of Quality Control"

Abstract The analysis of modern critical situation with a quality of drinking water in the world has been delivered. The inefficiency of existing approaches to the assessment of drinking water in the world has been justified. Core statements of a new standard for drinking water have been revealed. The key advantage of a new standard is the inclusion to it the integral control methods for the water quality, which is universal for all kinds and types of toxic compounds, irrespective of their origination and pharmacology.

Keywords Drinking water · Toxic substances · Synergism · Integral control methods for the water quality · A new standard for drinking water

7.1 SOS: Drinking Water

Drinking water is the water whose organoleptic, physicochemical, and biological properties meet human biological needs. It has neither color nor smell, and its taste is determined by the presence of physiologically necessary salts (calcium, magnesium, sodium, and potassium) in corresponding concentrations. Without these salts, human metabolism is impossible.

The first attempt to set a standard for drinking water, which included only nine controllable components, was in 1853 in Brussels at the international Congress of Hygienists. Among the components—one should pay special attention to the serendipity and wisdom of scientists as early as 150 years ago—were vitally important substances such as magnesium, calcium, total mineralization (determined mainly by the presence of sodium and potassium salts) the content of oxygen in the water, sulfates, chlorides, nitrates and ammonium.

After almost a century since adoption of the first recommendation for drinking water quality in Brussels, a World Health Organization (WHO) document was published which proposes to bring 95 indexes under regulation. Of these, 26 are toxic substances, whose presence in drinking water is undesirable. The US standard regulates 102 indexes in drinking water, and complete absence is required for as many as 35 toxic indexes.

In all developed countries there is typically an annual correction of national standards in compliance with a varying ecological situation.

V. V. Goncharuk, *Drinking Water*, DOI 10.1007/978-3-319-04334-0_7,

However, increased quality of the matter being controlled in national standards of various countries does not solve the problem of obtaining safe drinking water at centralized water treatment stations. These factors require increasingly expensive equipment and complex technological processes. Special attention should be paid to the fact that over the last two or three decades a sharp increase of chemical compounds in the environment took place. The total amount of anthropogenic chemical compounds increased to 30–40 million due to the rapid development of the pharmaceutical sector, especially the production of new classes of antibiotics, and of the agroindustrial sector, whose development required the synthesis and production of new types of chemicals for plant protection, agricultural chemicals, pesticides (herbicides, fungicides, and insecticides), etc. According to official statistics, more than 1.5 million new compounds are synthesized annually. Among these compounds, antibacterial, antimicrobial drugs as well as household cleaning chemicals are especially dangerous. All these substances are found in the environment and, significantly, in the water basin—a source of the drinking water supply, changing the quality and the amount of conventional and natural microflora, which performs the primary role of self-purification of water bodies in biological processes.

The presence of several dozen substances in drinking water is, to say the least, a surprise. Even increasing the number of controllable components in drinking water by several times, one cannot guarantee that drinking water is safe for human health. Even the use of underground water at present does not guarantee the quality of the drinking water obtained. Especially if one takes into account the fact that the majority of organic matter in the environment, under the effect of various physicochemical and ecological factors, undergoes biotransformation into unknown and, most often, unpredictable compounds. According to the above-mentioned synergism the effects of two or more toxic compounds taken individually increase if this substance initially is not toxic. This demonstrates the futility of using classic approaches to drinking water assessment. The very notion of "normalizing maximum allowable concentrations in drinking water of different toxicants" purportedly safe for human health is, as a matter of fact, immoral.

Therefore, fundamentally new approaches to drinking water quality assessment are needed at the biological level, more quickly than research into living organisms (mice, rats, rabbits, etc.) that requires much time. The proposed new standards include fundamentally new, very effective, and integrated methods of monitoring water quality. They are designed for revealing acute toxicity at the organism level and finding chronic toxicity at the cellular level using cytogenetic methods on biological organisms. These methods apply to all types and kinds of toxic compounds, irrespective of their origin and the nature of their effect.

The new standard for Drinking Water [1] includes three different regulatory documents:

- GOST (State All-Union Standard) for tap water (conditionally drinking)
- GOST for drinking water of increased quality (water absolutely safe for human health)
- GOST for bottled water

During development of the standard for centralized water supply regulation requirements, drinking water quality requirements accepted in the European Union, the WHO, the Codex Alimentarius, etc., have been taken into account.

The primary tenet of the new standard is a new conceptual approach to the water supply for the Ukraine's population with quality drinking water, providing for economic water supply by differentiating the consumption regulation with drinking water on physiological, sanitary–hygienic, and human economic needs over one 24-h period in a specific populated area. The necessity of such an approach is determined by the following objective reasons: unjustified use of quality drinking water for sanitary–hygienic needs; the unsatisfactory state of the transportation and distribution networks for the centralized drinking water supply system; the discrepancy between the technology for drinking water treatment in centralized drinking water supply systems to the water quality in a water source; unjustifiably high economic costs for reconstruction and modernization of centralized drinking water supply systems for the preparation of high-quality drinking water in volume, substantially exceeding physiological and sanitary–hygienically human needs.

The second tenet of the new standard is a new approach to drinking water quality assessment that uses integral indexes of water toxicity, which are set by biotesting methods according to standardized techniques. This makes it possible to obtain objective information on the quality of drinking water before conducting a detailed analysis of all indexes. Such an analysis is useful for detecting toxicity and necessary to determine the cause of its emergence.

The third tenet of the new standard is the requirement for complete absence of all toxic chemical and biological pollutants in drinking water designed for human consumption.

It is a well-known fact that the mini-plants for additional treatment of drinking water from various water supply sources were used on a wide scale for the centralized drinking water supply. In this case the quality of drinking water and its monitoring is not regulated by any state regulatory document. At the same time, an indispensable condition for achieving all regulatory indexes of drinking water safety is the setting of certain requirements for drinking water of both centralized and noncentralized water supply systems. The current regulatory documents regarding centralized drinking water supply cannot be used further without threatening the population's health in all countries of the world.

7.2 Main Aspects of a New State Standard "Drinking Water. Requirements and Methods of Quality Control"

The current regulatory framework in Ukraine, which sets the hygiene and quality control of drinking water in centralized water systems, consists of public health rules and regulations, "Drinking water. Hygiene requirements for the water quality" [2].

Comparative analysis of drinking water quality has shown that the Ukraine is normalizing 30 % fewer parameters than those provided in the European Union.

The new Ukraine State Standard "Drinking water. Requirements and methods of quality control," is aimed at improving the legislative framework to ensure the population of quality and safe drinking water. During development of the new standard, the terms and normative regulations of "State health rules and regulations," [2] and a new concept for providing quality drinking water (see Sect. 4.4) were considered. The new standard is based on the requirements of the current Ukraine legislation [3–5] and the regulatory requirements for drinking water quality adopted in the European Union [6], WHO [7], the Codex Alimentarius [8], the Agency for U.S. Environmental Protection Agency [9], etc. [10, 11]. This standard does not apply to mineral medical, medical table, the natural water table, or to bottled water, which contains specific therapeutic agents (iodine, fluorine, selenium, etc.) for special dietary consumption by children, patients and athletes. The terminology used in the new standard is discussed below.

Water from Uncentralized Drinking Water Supply, Unbottled Water designed for human consumption. It is supplied to individual consumers from underground sources of drinking water or from the centralized drinking water supply systems after add-on purification by drinking water treatment at distribution points (in particular, mobile) to consumers' containers.

Water from Uncentralized Drinking Water Supply, Add-on Purification, Bottled Water from underground sources of drinking water supply or from the centralized drinking water supply systems which have been additionally treated to improve its quality, bottled into water-tight consumer containers and designed to meet physiological, sanitary–hygienic, and economic-household needs.

Drinking Water Designed for Human Consumption Water, which according to organoleptic properties, chemical, microbiological, parasitological, and radiological indexes complies with sanitary legislation (water from centralized water supply systems, bottled, from mine wells, capping sources, well rooms, and distribution points).

Water for Human Economic and Household Needs The water from centralized drinking water supply systems designed for washing linen, cleaning the house, and other similar purposes.

Water for Human Sanitary–Hygienic Needs The water from centralized drinking water supply systems designed for personal hygiene: washing the face, taking showers and baths, removal of contaminants from food staff, washing utensils, and toilet flushing.

Drinking Water for Human Physiological Needs The water which, in its natural state or after treatment, is used to satisfy human drinking and cooking needs. It is supplied from a well room without being fed to the distribution system, after add-on purification in a distribution point for bottles or containers.

Water from the Centralized Drinking Water Supply System The water which meets consumer needs of means of a complex of facilities and distribution networks, connected by a single technological production and transportation process, for drinking water.

Acronyms and Abbreviations

BCG	Group of colibacillus bacteria
Bq	Becquerel
PFU	Plaque-forming units
GMN	(General microbial number)—number of saprophyte microorganisms
CFU	Colony-forming units
NUT	Nephelometric units of turbidity
ASAS	Anionoactive surface-active substances

Requirements for Indexes of Drinking Water Quality Designed for Human Consumption Requirements and norms for the state and composition of drinking water to determine its suitability for satisfying physiological, sanitary–hygienic and economic-household human needs must cover: water safety in an epidemiological respect, harmless chemical composition, favorable organoleptic properties, and toxicological and radiological safety.

A list of indexes and drinking water quality standards is based on the principle of no excess drinking water quality regulation and the values of physical, organoleptic, chemical, microbiological, toxicological, and radiation indicators for drinking water set in this standard and in [2, 12].

A list of drinking water quality indexes designed for human consumption shall be determined in this standard by the source of water supply, secondary contamination as a result of using reagents in the course of water treatment, and evaluation of the toxicity level of add-on purified water from the centralized drinking water supply. For add-on purified bottle water from a noncentralized drinking water supply, the same list of indicators is used as for the add-on purified unbottled water from a noncentralized drinking water supply.

Requirements for the water quality of centralized and noncentralized drinking water supply include 81 indexes and are represented by individual groups: *1*st group—8 microbiological indexes; *2*nd group—virological index; *3*rd group—2 parasitological indexes; *4*th group—mycological index; *5*th group—5 indexes of the toxicity level; *6*th group—2 indexes of radiation safety; *7*th group—4 organoleptic indexes; *8*th group—17 chemical quality indexes, which affect organoleptic properties of drinking water; *9*th group—28 toxicological indexes of the harmless chemical composition (among the 20 inorganic, 6 organic components and 2 integral indexes); *10*th group—13 substances, which are formed and enter drinking water during preparation.

For microbiological, virological, and parasitological indexes, drinking water should conform to the requirements given in Tables 7.1, 7.2 and 7.3.

For mycological indexes (micromycets), drinking water should be in line with the norm given in Table 7.4.

Toxicity of drinking water from a noncentralized water supply is an integral quality index for drinking water in the case of suspicion of water source contamination or a distribution network with toxic compounds. The list of indexes, test objects, and norms for determination of toxicity obtained from the results of biotesting are given in Table 7.5.

Table 7.1 Microbiological indexes of drinking water quality

Indicator	Measurement units	Norm, not more than	
		Water of systems of centralized drinking water supply	Water of uncentralized drinking water supply additionally purified (unbottled, bottled)
Number of bacteria in 1 cm^3 of water under study total microbial number (TMN) at 37°C	CFU/cm^3	100[a]	20[a]
Number of bacteria in 1 cm^3 of water under study (TMN) at 22°C	CFU/cm^3	100[a]	20[a]
Number of bacteria of the colibacillus group (coliform microorganisms) in 1 dm^3 of water under study (BCG index)	CFU/dm^3	3[b]	Absence[b]
Number of thermostable colibacillus (fecal coliform—FC index) on 100 cm^3 of water under study	CFU/100 cm^3	Absence[c]	Absence[c]
Number of pathogenic microorganisms in 1 dm^3 of water under study	CFU/dm^3	Ditto[c]	Ditto[c]
Number of coliphages in 1 dm^3 of water under study	PFU/dm^3	Ditto	Ditto
Spores of sulfate-reducing clostridia	Presence (amount)/20 cm^3	Absence[d]	Absence[d]
Blue pus bacillus (*Pseudomonas aeruginosa*)	CFU/dm^3	Not determined	Absence

[a] An excess of the norm shall not be allowed for 95 % of water sample in the water supply network that is under investigation during the year

[b] An excess of the norm shall not be allowed for 98 % of water samples in the water supply network that is under investigation during the year. In the event the BCG index is exceeded at the stage of colony identification investigations shall be carried out for the presence of fecal coliforms

[c] In the event of the presence of coliform bacteria (or) coliphages in a water sample their amount shall be immediately determined in repeatedly taken water samples. If in these samples general coliform bacteria are found in the amount of $>2/100^3$ and (or) thermostable coliform bacteria and (or) coliphages, pathogenic bacterial count of the colibacillus group and (or) enteroviruses shall be determined. The investigation of drinking water for the presence of pathogenic bacteria of the colibacillus group and enteroviruses shall also be carried out following the decision of the organs of sanitary supervision in the event of the emergence of the epidemic situation

[d] The control shall be exercised at the outlet from the drinking water treatment station in the case of using surface sources of water supply or underground one, which have the hydraulic connection with the surface water body; during the transition period annually the index of the spores of sulfate-reducing clostridia with the norm "absence/20 cm^3 " shall be controlled

Radiation safety of drinking water shall be determined by the acceptable levels given in Table 7.6.

For organoleptic and chemical quality indexes, which affect organoleptic properties, drinking water should correspond to norms given in Tables 7.7, and 7.8, and the EC Directives [12].

According to toxicological indexes for harmlessness of the chemical composition, drinking water should be in line with norms given in Table 7.9.

For the water supply system, which uses reagent water treatment methods prior to its delivery to the distribution network, during pouring-out, transportation, storage for a set time of suitability, in conducting research the indexes given in Table 7.10 are additionally taken into account. The content of substances that are formed and enter the drinking water during its treatment should not exceed norms given in Table 7.10.

The quality of add-on purified unbottled and bottled water from a uncentralized drinking water supply should be in compliance with the requirements of this standard during bottling, transportation, and storage and during the entire life time.

Reagents, introduced as preservatives for drinking water, shall be in accordance with Table 7.11. Other substances and methods for the preservation of add-on purified bottled water from a uncentralized drinking water supply avoid the introduction of carbon dioxide and silver ions.

Thus, the approval and introduction of the new Standard "Drinking water. Requirements and methods of water quality" will allow us to:

- Substantially reduce capital and operational costs for obtaining high-quality drinking water in amounts sufficient for meeting the physiological needs of the population
- Reduce costs for controlling the quality of drinking water intended for human consumption by using integrated methods for assessing water quality without needing to reequip chemical analysis laboratories with expensive equipment
- Improve the state legislative database for providing the population with good-quality drinking water, safe for human health

Thus, in modern civilization the need to create fundamentally new regulations determining the quality of drinking water is dictated by life itself. Development of a new standard for drinking water and control methods is based on novel scientific achievements. The most distinguished experts in the fields of medicine, biology, chemistry, toxicology physiology, etc. have elaborated on it. The new standard excludes the presence of any toxic substances in drinking water. It incorporates new methods for monitoring water quality at the cellular level using cytogenetic techniques on biological objects. The new technologies of water treatment and purification from any significant contamination have been created at the A.V. Dumanski Institute of Colloid Chemistry and Water Chemistry, National Academy of Sciences of Ukraine. These new technologies meet the requirements set by the new standard. In addition, a fundamentally new approach is proposed to ensure that the population has high quality drinking water. It is based on establishing a system of local facilities for water treatment. Such complicated performance measures (0.1–10 m^3/h)

Table 7.2 Virological indexes of drinking water quality

Indicator	Measurement units	Norm	
		Water of systems of centralized drinking water supply	Water of uncentralized drinking water supply additionally purified (unbottled, bottled)
Enteroviruses, adenoviruses, rotaviruses, reoviruses and antigen-virus of hepatitis A	PFU/dm^3	Absence	Absence

Table 7.3 Parasitological indexes of drinking water quality

Indicator	Measurement units	Norm	
		Water of systems of centralized drinking water supply	Water of uncentralized drinking water supply additionally purified (unbottled, bottled)
Number of pathogenic coliform protozoans in 50 dm^3 of water under study	Cells (cysts)/50 dm^3	Absence[a]	Absence[a]
Number of coliform helminthes in 50 dm^3 of water under study	(Cells, eggs, larvae)/50 dm^3	Ditto	Ditto

[a] Determined once a year during full water analysis and according to epidemiological indexes

Table 7.4 Mycological index of drinking water quality

Indicator	Measurement units	Norm	
		Water of systems of centralized drinking water supply	Water of uncentralized drinking water supply additionally purified (unbottled, bottled)
Micromycets	CFU/100 cm^3	Absence	Absence[a]

[a] *Aspergillus fumigatus, Aspergillus niger,* and *Penicillum expangum* must not be present at all

will provide high-quality water for drinking and cooking. This water will not flow into the distribution network. It will be distributed in the same ways as artesian well water. In order to produce such high quality drinking water, the Institute developed stand-alone water treatment systems for collective and individual use for analytical

Table 7.5 Toxicity level of drinking water

Indicator	Measurement units	Norm	
		Water of systems of centralized drinking water supply	Water of noncentralized drinking water supply additionally purified (unbottled, bottled)
Chronic toxicity on *Ceriodaphnia affinis*	Number of died individuals and (or) a decrease of the number of newly born individuals in an experiment compared with the reference during 7 ± 1 days	Not determined	Absence of chronic toxicity
Genotoxicity on *Drosophila melanogaster Mg*	Rate of occurrence of dominant lethal mutations in an experiment compared with the reference during 72 h	Ditto	Genotoxicity absent
Toxicity on *Tetrahymena pyriformis*	A reduction of the increment coefficient of the number infusorians in an experiment compared with the reference during the time set—24 h. (a short-term biotesting) or 96 % (a long-term biotesting)	Ditto	Toxicity absent
Toxicity on *Vibrio fischer*	Reduction of the level of luminescence of bacteria in an experiment compared with the reference over 30 min	Ditto	Toxicity absent
Genotoxicity on *Salmonella thyphimurium*	A deviation of the total mutagenic activity of umuC-gene of bacteria *Salmonella tuphimurium* in an experiment compared with a reference over 4 h	Ditto	Genotoxicity absent

control. Technology and equipment that are offered at cost and the complex problem to be solved are unique and have no analogs in the world.

Table 7.6 Indexes of radiation safety of drinking water

Indicator	Measurement units	Norm, not more than	
		Water of systems of centralized drinking water supply	Water of uncentralized drinking water supply additionally purified (unbottled, bottled)
Total volumetric activity of α-radiation (Σα-activity)[a]	Bq/dm^3	0.1	0.1
Total volumetric activity of β-radiation (Σββ-activity)[a,b]	Bq/dm^3	1.0	1.0

[a] With an excess of the levels of Σα- and Σβ-activities the radionuclide composition of water should be controlled with respect of its compliance noted in [12] to the norms of radiation safety

[b] With an excess of the level of the Σβ-activity it is necessary also to take into account the content of potassium in water since the Σα-activity determined in addition by natural radionuclide, in particular ^{40} K (1 Bq of Σβ-activity corresponds to 3.4 mg K/dm^3 in water). Therefore, with the activity ≥ Bq in the first place it is necessary to determine the content of potassium by the method of atomic-emission spectroscopy and then to carry out detailed radiological research (a relationship between the specific β-activity of the samples under investigation and the concentration in them of potassium is rectilinear, the natural blend of isotopes of radioactive potassium ^{40} K contains 0.0119)

Table 7.7 Organoleptic indexes of drinking water quality

Indicator	Measurement units	Norm, not more than	
		Water of systems of centralized drinking water supply	Water of uncentralized drinking water supply additionally purified (unbottled, bottled)
Organoleptic indicators of quality			
Smell at 20°C	Point	2	0
Smell during heating to 60°C	Point	2	1
Smell and taste	Point	2	0
Color	Deg	20	5
Turbidity	NUT	2.5	0.5

Table 7.8 Chemical indexes of quality, which affect organoleptic properties of drinking water

Indicator	Measurement units	Norm, not more than	
		Water of systems of centralized drinking water supply	Water of uncentralized drinking water supply additionally purified (unbottled, bottled)
Inorganic components			
Hydrogen index (pH)	pH unit	6.5–8.5	6.5–8.5
Fry residue (total mineralization) optimal content	mg/dm^3	1000 (1500)[a]	1000 (100–400)
Total hardness, optimal value	mmol/dm^3	7 (10)[a]	7 (1.5–4.0)
Total alkalinity, optimal value	mmol/dm^3	6.5	6.5 (0.5–5.0)
Sulfates	mg/dm^3	250 (500)[a]	150
Chlorides	mg/dm^3	250 (350)[a]	150
Total iron (Fe)	mg/dm^3	0.2	Absence
Manganese (Mn)	mg/dm^3	0.05	Ditto
Copper (Cu)	mg/dm^3	1.0	Ditto
Zinc (Zn)	mg/dm^3	5.0	Ditto
Calcium (Ca), optimal value	mg/dm^3	100	100 (20–60)
Magnesium (Mg), optimal value	mg/dm^3	30	30 (6.0–15)
Sodium (Na), optimal value	mg/dm^3	200	200 (20)
Potassium (K), optimal value	mg/dm^3	20	20 (2.0–20)
Organic components			
Methyl-*tert*-butyl ether	mg/dm^3	0.015	Absence
Oil products	mg/dm^3	0.05	Ditto
Chlorophenols	mg/dm^3	0.0003	Ditto

[a] The value in the brackets may be established following the ruling by the Chief State Sanitary Physician in the corresponding area for the specific system of drinking water supply based on the evaluation of the sanitary–epidemiological state in a town and the technology of drinking water treatment used in a case, when other sources of drinking water supply are inaccessible

Table 7.9 Toxicological indexes of harmlessness of drinking water chemical composition

Indicator	Measurement units	Norm, not more than	
		Water of systems of centralized drinking water supply	Water of uncentralized drinking water, additionally purified, (unbottled, bottled)
Inorganic components			
Aluminum (Al)	mg/dm^3	0.2 (0.5)[a]	Absence
Ammonia (by NH_4^+)	mg/dm^3	0.5 (1.5)[a]	Ditto
Barium (Ba)	mg/dm^3	0.1	0.1
Beryllium (Be)	mg/dm^3	0.0002	Absence
Boron (B)	mg/dm^3	0.5	Ditto
Cadmium (Cd)	mg/dm^3	0.001	Ditto
Arsenic (As)	mg/dm^3	0.01	Ditto
Nickel (Ni)	mg/dm^3	0.1	Ditto
Nitrates (by NO_3^-)	mg/dm^3	45	5
Nitrites (by NO_2^-)	mg/dm^3	0.1	0.02
Perchlorates (ClO_4^-)	mg/dm^3	0.01	Absence
Mercury (Hg)	mg/dm^3	0.0005	Ditto
Lead (Pb)	mg/dm^3	0.01	Ditto
Selenium (Se)	mg/dm^3	0.01	Ditto
Strontium (Sr)	mg/dm^3	7	2
Stibium (Sb)	mg/dm^3	0.005	Absence
Thallium (Tl)	mg/dm^3	0.0001	Ditto
Fluorides (F^-) for climatic region:[b]	mg/dm^3		
II		1.5	1.5
III		1.2	1.2
IV		0.7	0.7
Total chromium (Cr)	mg/dm^3	0.05	Absence
Cyanides (CN^-), in particular cyanogen chloride	mg/dm^3	0.05	Ditto
Organic components			
Benz(a)pyrene	mg/dm^3	0.00001	Absence
Benzol	mg/dm^3	0.001	Ditto
Pesticides(sum)[c]	mg/dm^3	0.0005	Ditto
Synthetic anionoactive surface-active substances (ASAS)	mg/dm^3	0.1	Ditto
Trichloroethylene and tetrachloroethylene (sum)	mg/dm^3	0.01	Ditto
Tetrachloride carbon	mg/dm^3	0.002	Ditto

Table 7.9 (continued)

Indicator	Measurement units	Norm, not more than	
		Water of systems of centralized drinking water supply	Water of uncentralized drinking water, additionally purified, (unbottled, bottled)
Integral indicators			
Permanganate oxidizability	mg O/dm^3	2	0.75
Total organic carbon	mg C/dm^3	4	1.5

[a] The value in brackets may be set only by the ruling of the Chief State Sanitary Physician in the corresponding area for the specific system of drinking water supply based on the evaluation of the sanitary–epidemiological state in a town and the technology of drinking water treatment used with the account of the specific situation

[b] Normalization of fluorides in drinking water is given in accordance with [13]

[c] Pesticides (sum)–organic insecticides, herbicides, fungicides, nematicides, acaricides, algicides, bactericides, vinicides, rodenticides, slimicides, bound products (in particular growth regulators) and then metabolites and degradation products. The program of control involves only these pesticides, which, most likely, may be contained in this water

Table 7.10 Substances formed and getting into drinking water during water conditioning

Indicator[a]	Measurement units	Norm, not more than	
		Water of systems of centralized drinking water supply	Water of uncentralized drinking water supply additionally purified (unbottled, bottled)
Acrylamide[b]	mg/dm^3	0.001	Absence
Bromates	mg/dm^3	0.1	Ditto
Residual chlorine dioxide	mg/dm^3	0.2–0.8	Ditto
Residual ozone within	mg/dm^3	0.1–0.3	Ditto
Residual polyphosphates (PO_4^{3-})	mg/dm^3	3.5	Ditto
Trihalogenmethanes: chloroform, bromoform, dibromochloromethane, bromodichloromethane (total)	mg/dm^3	0.1	Ditto
Formaldehyde	mg/dm^3	0.05	Ditto
Residual free chlorine, within	mg/dm^3	0.3–0.5	Ditto
Residual bound chlorine, within	mg/dm^3	0.8–1.2	Ditto
Chlorate ion	mg/dm^3	0.7	Ditto
Chlorite ion	mg/dm^3	0.7	Ditto
Chloroform	mg/dm^3	0.06	Ditto
Dibromochloromethane	mg/dm^3	0.03	Ditto

[a] The program of control involves specific indexes with the account of the technology of water treatment from the cited list

[b] The program of control involves specific indexes of using respective reagents and calculate them based on the analysis of the content of the monomer in the commodity flocculant. The reagent doses and the level of the monomer should correspond to the following requirements: at the polyacrylamide dose 1 mg/dm^3 the content of acrylamide in it should not exceed 0.05 %

Table 7.11 Reagents permitted as preservatives for drinking water

Preservatives	Measurement units	Permitted mass fraction of a preservative in water, not more than
Silver (Ag)	mg/dm^3	0.025 (0.05)[a]
Carbon dioxide (CO_2)	%	0.2–0.6[b]

[a] For water containing chloride ions > 50 mg/dm^3
[b] According to DSTU 878, DSanPiN 2.2.4 [14], HPN 4.4.2 [15]

References

1. Goncharuk VV (2010) SOS: drinking water. J Water Chem Technol 32(5):255–283
2. State Sanitary Rules and Standards-383 (1996) Voda pytna. Hihiyenichni vymohy do yakosti vody tsentralizovanoho hospodars'ko-pytnoho vodopostachannya (Drinking water. hygienic requirements for water quality of centralized household and drinking water supply). Ministry of Health, Ukraine
3. The Law of Ukraine N 4004-12 (2008) On provision sanitary and epidemiological well-being of the population. J Verkh Rad Ukraine 27:218
4. The Law of Ukraine N 2918-III (2002) Drinking water and drinking water supply. Off J Verkh Rad Ukraine 16:112
5. The Law of Ukraine N 3933-VI (2005) On General State Program "Drinking water of Ukraine for years 2005–2020". Off J Verkh Rad Ukraine 15:243
6. The Directive of the European Union Council 98/82/EU (1998) Concerning water quality designed for human consumption. Off J Eur Commune 41:2–5
7. World Health Organization (2004) Guidelines on drinking water quality: recommendation, vol 1. WHO, Geneva
8. Food and agriculture organization of the United Nations (2001) CODEX STAN 227-2991. General standard for drinking water bottled/packaged (other than natural mineral water)
9. U.S. Environmental Protection Agency (1998) National primary drinking water regulations: disinfectants and disinfection by-products. Federal Register 63(241):69389–69476
10. SanPin 2.1.4.1116-02 (2002) Pit'yevaya voda. Gigiyenicheskiye trebovaniya k kachestvu vody, rasfasovannoy v yemkosti. Kontrol' kachestva (Drinking water. Hygienic requirements for water quality, packed in containers. Quality control). Ministry of Health and Social Development of the Russian Federation, Moscow
11. SanPin 2.1.4.1074-01 (2001) Gigiyenicheskiye trebovaniya i normativy kachestva pit'yevoy vody (Drinking water. hygienic regulations of the water quality of centralized drinking water supply system. Quality control). Ministry of Health and Social Development of the Russian Federation, Moscow
12. NRBU-97/D-2000 (2000) Normy radiatsiinoi bezpeky Ukrainy: dopovnennya. Radiatsiyjnyyj zakhyst vid dzherel potentsijnoho oprominennya (Norms of radiation safety of Ukraine: supplements. Radiation protection from potential exposure). Ukraine Ministry of Health, Kiev
13. DSanPiN 2.2.4-005-98 (1998) Ftoruvannya vody na vodoprovodakh tsentralizovanogo gospodars'ko-pitnogo vodopostachannya (Water fluorination in systems of centralized household-drinking water supply). Ukraine Ministry of Health, Kiev
14. DSanPin 2.2.4.04 (2004) Voda pytna fasovana. Gigienichni vimogi ta kontrol' za yakistyu (Bottled drinking water. Hygienic requirements and control quality). Ukraine Ministry of Health, Kiev
15. GRN 4.4.2 (2007) Voda pytna, germetichno upakovana/butil'ovana v emnosti. Sanitarni zakhodi bezpechnosti ta yakosti. Gigienichni reglamenti ta normi (Drinking water. Sanitary measures of safety and qyality. Hygienic regulations and norms). Ukraine Ministry of Health, Kiev (to be considered)

Conclusion

The first source and essence of everything—water
Phales Milesian

The importance of water becomes clear for a man when he is deprived of it. Everything is saturated with water. Most the greatest scientists stressed its great importance in life of our planet. Among them American physicist J. Day and chemist K. Davis which called water "the mirror of science".

It is stated in the report of UNO demographic service that the planet population will increase from today's 6.4 billion to 44 billion people in 2100. In 2150 this figure will reach 244 billion, and in 2300–1.34 trillion people. And if mankind experiences the crisis in consumption of fresh water, then, taking into account the above figures, one cannot imagine the scales of the probable future crisis connected with water. The author's call to revise radically our relation to the environment, water, to give a sensible and considerate glance to the surrounding has been inspired by his special concern of water.

Systematic fundamental and applied researches in physics, chemistry and biology of water have been carried out for many years at the Institute of Colloid and Water Chemistry of NAS of Ukraine. Hence the book subject—water as the most urgent need of mankind.

The author has presented information in such a way, that permits one to "see" all the problem complicacies, and above all he has outlined the paths of its urgent solution.

It is evident that the book has not embraced all the sides of the subject, and such task has not been set. But in the context of primary importance, inspired by human life itself, water is presented in the aspect of the most significant, from the author's viewpoint, problems.

More versatile and deep representation of the theme of water requires a multi-volume monograph which will appear without fail, since that is the requirement of life itself. Allowing for the words of the well-known scientist V.V. Shuleikin, a founder of a new science—physics of the sea: "Water is in many respects, most difficult for studying among all substances investigated by physicists and physico-chemists", one can assert that scientists will guess more than one riddle of water, having overcome all difficulties.

V. V. Goncharuk, *Drinking Water*, DOI 10.1007/978-3-319-04334-0,

Index

V. V. Goncharuk, *Drinking Water,* DOI 10.1007/978-3-319-04334-0,

FSC www.fsc.org
MIX
Papier aus verantwortungsvollen Quellen
Paper from responsible sources
FSC® C105338

If you have any concerns about our products,
you can contact us on
ProductSafety@springernature.com

In case Publisher is established outside the EU,
the EU authorized representative is:
Springer Nature Customer Service Center GmbH
Europaplatz 3, 69115 Heidelberg, Germany

Printed by Libri Plureos GmbH
in Hamburg, Germany